Modelling Radiotherapy Side Effects

Practical Applications for Planning Optimisation

Medical Physics and Biomedical Engineering

John G. Webster, E. Russell Ritenour,
Slavik Tabakov, and Kwan-Hoong Ng

Targeted Muscle Reinnervation
A Neural Interface for Artificial Limbs
Todd A. Kuiken, Aimee E. Schultz Feuser, Ann K. Barlow (Eds)

Advanced MR Neuroimaging
From Theory to Clinical Practice
Ioannis Tsougos

Handbook of X-ray Imaging
Physics and Technology
Paolo Russo (Ed)

Quantitative MRI of the Brain
Principles of Physical Measurement, Second Edition
Mara Cercignani, Nicholas G. Dowell, Paul S. Tofts (Eds)

Graphics Processing Unit-Based High Performance Computing in Radiation Therapy
Xun Jia, Steve B. Jiang (Eds)

A Guide to Outcome Modeling In Radiotherapy and Oncology
Listening to the Data
Issam El Naqa (Ed)

Radiotherapy and Clinical Radiobiology of Head and Neck Cancer
Loredana G. Marcu, Iuliana Toma-Dasu, Alexandru Dasu, Claes Mercke

Advances in Particle Therapy
A Multidisciplinary Approach
Manjit Dosanjh, Jacques Bernier (Eds)

Problems and Solutions in Medical Physics
Diagnostic Imaging Physics
Kwan-Hoong Ng, Jeannie Hsiu Ding Wong, Geoffrey D. Clarke

Advances and Emerging Technologies in Radiation Oncology Physics
Siyong Kim, John Wong

Clinical Radiotherapy Physics with MATLAB
A Problem-Solving Approach
Pavel Dvorak

Proton Therapy Physics, Second Edition
Harald Paganetti (Eds)

Modelling Radiotherapy Side Effects

Practical Applications for Planning Optimisation

Edited by
Tiziana Rancati and Claudio Fiorino

CRC Press
Taylor & Francis Group
Boca Raton London New York

CRC Press is an imprint of the
Taylor & Francis Group, an **informa** business

CRC Press
Taylor & Francis Group
6000 Broken Sound Parkway NW, Suite 300
Boca Raton, FL 33487-2742

First issued in paperback 2021

ISBN-13: 978-0-367-77971-9 (pbk)
ISBN-13: 978-1-138-19809-8 (hbk)

Library of Congress Cataloging-in-Publication Data

Names: Rancati, Tiziana, author. | Fiorino, Claudio, author.
Title: Modelling radiotherapy side effects : practical applications for planning optimisation / Tiziana Rancati, Claudio Fiorino.
Other titles: Series in medical physics and biomedical engineering.
Description: Boca Raton, FL : CRC Press, Taylor & Francis Group, [2019] |
Series: Series in medical physics and biomedical engineering | Includes bibliographical references and index.
Identifiers: LCCN 2018048998| ISBN 9781138198098 (hbk ; alk. paper) | ISBN 1138198099 (hbk ; alk. paper) | ISBN 9781315270814 (eBook) | ISBN 1315270811 (eBook)
Subjects: LCSH: Radiotherapy. | Cancer--Radiotherapy.
Classification: LCC RM847 .R27 2019 | DDC 615.8/42--dc23
LC record available at https://lccn.loc.gov/2018048998

Visit the Taylor & Francis Web site at
http://www.taylorandfrancis.com

and the CRC Press Web site at
http://www.crcpress.com

Contents

About the Series, vii

The International Organization for Medical Physics, ix

Preface, xi

Contributors, xiii

Introduction, xvii

CHAPTER 1 ▪ The Importance of the Quality of Data   1

WILMA HEEMSBERGEN AND MARNIX WITTE

CHAPTER 2 ▪ Building a Predictive Model of Toxicity: Methods   23

SUNAN CUI, RANDALL K. TEN HAKEN, AND ISSAM EL NAQA

CHAPTER 3 ▪ Potentials and Limits of Phenomenological Models   53

ARJEN VAN DER SCHAAF

CHAPTER 4 ▪ Pelvis: Rectal and Bowel Toxicity   75

SARAH L. GULLIFORD, JULIA R. MURRAY, AND MARTIN A. EBERT

CHAPTER 5 ▪ Pelvis: Urinary Toxicity   113

TIZIANA RANCATI, CESARE COZZARINI, RICCARDO VALDAGNI,
AND CLAUDIO FIORINO

CHAPTER 6 ▪ Stomach, Duodenum, Liver, and Central Hepatobiliary Tract   137

GIOVANNI MAURO CATTANEO AND LIVIA MARRAZZO

CHAPTER 7 ▪ Central Nervous System (Brain, Brainstem, Spinal Cord),
Ears, Ocular Toxicity   171

FEDERICA PALORINI, ANNA CAVALLO, LETIZIA FERELLA, AND ESTER ORLANDI

CHAPTER 8 ▪ Head and Neck: Parotids   207

MARIA THOR AND JOSEPH O. DEASY

Chapter 9 ■ Head and Neck: Larynx and Structures Involved in Swallowing/Nutritional Problems and Dysphonia   215

Giuseppe Sanguineti

Chapter 10 ■ Thorax: Lungs and Esophagus   243

Daniel Schanne, Jan Unkelbach, and Matthias Guckenberger

Chapter 11 ■ Heart and Vascular Problems   269

Laura Cella and Giovanna Gagliardi

Chapter 12 ■ Adverse Effects to the Skin and Subcutaneous Tissue   289

Michele Avanzo, Joseph Stancanello, and Rajesh Jena

Chapter 13 ■ Bone Marrow and Hematological Toxicity   309

Elena S. Heide and Loren K. Mell

Chapter 14 ■ Predicting Toxicity in External Radiotherapy: A Critical Summary   337

Tiziana Rancati and Claudio Fiorino

Chapter 15 ■ Data Sharing and Toxicity Modelling: A Vision of the Near Future   365

Zhenwei Shi, Rianne Fijten, Zhen Zhou, Andre Dekker, and Leonard Wee

Chapter 16 ■ Quantitative Imaging for Assessing and Predicting Toxicity   401

Maria Thor and Joseph O. Deasy

Chapter 17 ■ Beyond DVH: 2D/3D-Based Dose Comparison to Assess Predictors of Toxicity   415

Oscar Acosta and Renaud de Crevoisier

Chapter 18 ■ Radiobiological Models in (Automated) Treatment Planning   441

Ben Heijmen and Marco Schwarz

Chapter 19 ■ Including Genetic Variables in NTCP Models: Where Are We? Where Are We Going?   455

Sarah L. Kerns, Suhong Yu, and Catharine M. L. West

INDEX, 469

About the Series

THE *SERIES IN MEDICAL PHYSICS AND BIOMEDICAL ENGINEERING* describes the applications of physical sciences, engineering, and mathematics in medicine and clinical research.

The series seeks (but is not restricted to) publications in the following topics:

- Artificial organs

- Assistive technology

- Bioinformatics

- Bioinstrumentation

- Biomaterials

- Biomechanics

- Biomedical engineering

- Clinical engineering

- Imaging

- Implants

- Medical computing and mathematics

- Medical/surgical devices

- Patient monitoring

- Physiological measurement

- Prosthetics

- Radiation protection, health physics, and dosimetry

- Regulatory issues

- Rehabilitation engineering

- Sports medicine
- Systems physiology
- Telemedicine
- Tissue engineering
- Treatment

The International Organization for Medical Physics

THE INTERNATIONAL ORGANIZATION FOR MEDICAL PHYSICS (IOMP) represents over 18,000 medical physicists worldwide and has a membership of 80 national and six regional organizations, together with a number of corporate members. Individual medical physicists of all national member organizations are also automatically members.

The mission of IOMP is to advance medical physics practice worldwide by disseminating scientific and technical information, fostering the educational and professional development of medical physics, and promoting the highest quality medical physics services for patients.

A World Congress on Medical Physics and Biomedical Engineering is held every three years in cooperation with the International Federation for Medical and Biological Engineering (IFMBE) and International Union for Physics and Engineering Sciences in Medicine (IUPESM). A regionally-based international conference, the International Congress of Medical Physics (ICMP) is held between world congresses. IOMP also sponsors international conferences, workshops, and courses.

The IOMP has several programmes to assist medical physicists in developing countries. The joint IOMP Library Programme supports 75 active libraries in 43 developing countries, and the Used Equipment Programme coordinates equipment donations. The Travel Assistance Programme provides a limited number of grants to enable physicists to attend the world congresses.

IOMP co-sponsors the *Journal of Applied Clinical Medical Physics*. The IOMP publishes, twice a year, an electronic bulletin, *Medical Physics World*. IOMP also publishes e-Zine, an electronic news letter about six times a year. IOMP has an agreement with Taylor & Francis for the publication of the *Medical Physics and Biomedical Engineering* series of textbooks. IOMP members receive a discount.

IOMP collaborates with international organizations, such as the World Health Organisations (WHO), the International Atomic Energy Agency (IAEA), and other international professional bodies such as the International Radiation Protection Association (IRPA) and the International Commission on Radiological Protection (ICRP), to promote the development of medical physics and the safe use of radiation and medical devices.

Guidance on education, training, and professional development of medical physicists is issued by IOMP, which is collaborating with other professional organizations in development of a professional certification system for medical physicists that can be implemented on a global basis.

The IOMP website (www.iomp.org) contains information on all the activities of the IOMP, policy statements 1 and 2 and the 'IOMP: Review and Way Forward' which outlines all the activities of IOMP, and plans for the future.

Preface

IN MY PRACTICE AS A RADIATION ONCOLOGIST, I deal with the issues of radiation-induced toxicities every day.

So many patients, so many disease profiles, so many variables to take into consideration in the treatment planning and monitoring. I must admit that so many times I wished to have a book reporting the state of the art on radiation-induced toxicities, a text where I could find all I need to know about the side effects my patients may experience.

It's for all these reasons that I gladly accepted the task of writing the preface to this comprehensive book which fills the gap since the QUANTEC special issues in the *Red Journal* (*International Journal of Radiation Oncology, Biology and Physics*) in 2010. Eight years have passed: eight crucial, long years during which we have been both spectators and actors of multiple significant advances in the concepts of radiobiology, in radiation technology and techniques, in the treatment schemes, and planning systems. In this complex scenario, predictive modelling of toxicity is stepping into adulthood and becoming a full-fledged tool in the hands of radiation oncologists, physicists, and dosimetrists.

Citing the text, "The quite complex phase called *planning optimisation* largely depends on the tissues' response to radiation; in order to better exploit the possibilities of this process, the availability of more accurate, quantitative knowledge of the peculiar responses of the different tissues is of paramount importance."

Besides all the important insights into toxicity profiles, analysis of normal tissue effects and predictive modelling, the book needs a special mention for the new approach to the patient. The specialistic know-how or, better said, the hyperspecialistic know-how does not focus exclusively on the radiobiological and technical issues of radiotherapy and medical physics. As a matter of fact, all starts from (and ends with) the patient as a person to be cured and cared for, which is much more than a tumor to treat. Considering the patient in terms of biological, psychological, and social components, it means to have one's quality of life well in mind, to tailor treatments in order to spare side effects, to care for survivors' quality of life, and to understand its impact eventually on controlling the disease.

Would you care for a tutorial of the professionals involved in the optimisation of the tolerance of organs at risk to radiation treatments and acquire important insights in an applicable, comprehensive summary of the main knowledge on the response of normal tissues to radiation? Not only is this text intended to be a reference book for those already experienced in the management of radiation-induced toxicities, it is also an educational tool for the younger generations stepping into this rapidly evolving, exciting field.

R. Valdagni
Fondazione IRCCS Istituto Nazionale Tumori,
Milan, Italy

Contributors

Oscar Acosta
LTSI - Laboratoire Traitement du
    Signal et de l'Image, Département
    de Radiothérapie
Univ Rennes, CHU Rennes, CLCC
    Eugène Marquis, Inserm
Rennes, France

Michele Avanzo
Division of Medical Physics
Centro di Riferimento Oncologico
Aviano, Italy

Giovanni Mauro Cattaneo
Department of Medical Physics,
San Raffaele Scientific Institute
Milano, Italy

Anna Cavallo
Department of Medical Physics
Fondazione IRCCS Istituto Nazionale
    dei Tumori
Milan, Italy

Laura Cella
Consiglio Nazionale delle Ricerche,
Napoli, Italy

Cesare Cozzarini
Department of Radiotherapy,
San Raffaele Scientific Hospital
Milano, Italy

Sunan Cui
Department of Radiation Oncology
University of Michigan
Ann Arbor, Michigan

Renaud de Crevoisier
LTSI - Laboratoire Traitement du
    Signal et de l'Image, Département de
    Radiothérapie
Univ Rennes, CHU Rennes, CLCC Eugène
    Marquis, Inserm
Rennes, France

Joseph O. Deasy
Department of Medical Physics
Memorial Sloan Kettering Cancer Center
New York, New York

Andre Dekker
MAASTRO Clinic
Maastricht, The Netherlands

and

School of Oncology and Development
    Biology (GROW)
Maastricht University
Maastricht, The Netherlands

Martin A. Ebert
Department of Radiation Oncology
Sir Charles Gairdner Hospital and
    Department of Physics
University of Western Australia
Perth, Australia

Issam El Naqa
Department of Radiation Oncology
University of Michigan
Ann Arbor, Michigan

Letizia Ferella
Department of Radiation Oncology
Fondazione IRCCS Istituto Nazionale dei
    Tumori
Milan, Italy

Rianne Fijten
MAASTRO Clinic
Maastricht, The Netherlands

and

School of Oncology and Development
    Biology (GROW)
Maastricht University
Maastricht, The Netherlands

Claudio Fiorino
Department of Medical Physics
San Raffaele Scientific Institute
Milan, Italy

Giovanna Gagliardi
Department of Medical Physics
Karolinska University Hospital
Stockholm, Sweden

Matthias Guckenberger
Department of Radiation Oncology
Universitäts Spital Zurich
Zurich, Switzerland

Sarah L. Gulliford
Joint Department of Physics
Institute of Cancer Research and Royal
    Marsden National Health Service
    Foundation Trust
Sutton, United Kingdom

Wilma Heemsbergen
Erasmus University Medical Center
Rotterdam, The Netherlands

Ben Heijmen
Erasmus University Medical Center
Rotterdam, The Netherlands

Rajesh Jena
Department of Oncology
University of Cambridge
Cambridge, United Kingdom

Sarah L. Kerns
Department of Radiation Oncology
University of Rochester Medical Center
Rochester, New York

Livia Marrazzo
Medical Physics Unit
Careggi University Hospital
Firenze, Italy

Loren K. Mell
Department of Radiation Oncology
University of California-San Diego
San Diego, California

Julia R. Murray
Royal Marsden National Health Service
    Foundation Trust
Sutton, United Kingdom

Ester Orlandi
Department of Radiation Oncology 2 and
    Department of Radiation Oncology
Fondazione IRCCS Istituto Nazionale dei
    Tumori
Milan, Italy

Federica Palorini
Prostate Cancer Program
Fondazione IRCCS Istituto Nazionale
    Tumori
Milan, Italy

Tiziana Rancati
Prostate Cancer Program
Fondazione IRCCS Istituto Nazionale
    Tumori
Milan, Italy

Giuseppe Sanguineti
Department of Radiation Oncology
Regina Elena National Cancer Institute
Rome, Italy

Daniel Schanne
Department of Radiation Oncology
Massachusetts General Hospital
Boston, Massachusetts

Marco Schwarz
Protontherapy Department,
Azienda Provinciale per i Servizi Sanitari
Trento, Italy

Zhenwei Shi
MAASTRO Clinic
Maastricht, The Netherlands

and

School of Oncology and Development
    Biology (GROW)
Maastricht University
Maastricht, The Netherlands

Elena S. Heide
Department of Radiation Oncology
University of California-San Diego
San Diego, California

Joseph Stancanello
The DLab: Decision Support for Precision
    Medicine
GROW – School for Oncology and
    Developmental Biology
Maastricht University Medical Centre
Maastricht, The Netherlands

Randall K. Ten Haken
Department of Radiation Oncology
University of Michigan
Ann Arbor, Michigan

Maria Thor
Department of Medical Physics
Memorial Sloan Kettering Cancer Center
New York, New York

Jan Unkelbach
Department of Radiation Oncology
Universitäts Spital Zurich
Zurich, Switzerland

Arjen van der Schaaf
University of Groningen,
University Medical Center Groningen
Groningen, The Netherlands

Riccardo Valdagni
Fondazione IRCCS Istituto Nazionale
    Tumori
Milan, Italy

Leonard Wee
MAASTRO Clinic
Maastricht, The Netherlands

and

School of Oncology and Development
    Biology (GROW)
Maastricht University
Maastricht, The Netherlands

Catharine M. L. West
University of Manchester
The Christie NHS Foundation Trust
Manchester, United Kingdom

Marnix Witte
Department of Radiation Oncology
The Netherlands Cancer Institute
Amsterdan, The Netherlands

Suhong Yu
Department of Radiation Oncology
Boston University Medical Center
Boston, Massachusetts

Zhen Zhou
MAASTRO Clinic
Maastricht, The Netherlands

and

School of Oncology and Development
  Biology (GROW)
Maastricht University
Maastricht, The Netherlands

Introduction

Claudio Fiorino and Tiziana Rancati

A LMOST THREE DECADES AGO, in the early period of modern 3D conformal radiotherapy (3DCRT) the milestone publication by Emami et al. (1991) first tried to accomplish a systematic summary of the knowledge regarding the dose-volume effects of organs at risk in fractionated radiotherapy.

The huge merits of this work and of other "companion papers" introducing the concept of "normal tissue complication probability" (NTCP) (Burman et al. 1991; Kutcher and Burman 1989) largely overbalanced the limits: We may say that, also thanks to this work, 3DCRT became quickly the standard, preparing the advent of the next step of intensity-modulated and image-guided radiotherapy.

Of great relevance, these early papers introduced the "idea" that the risk of an adverse effect could be quantified starting from the 3D dose-volume information of the organ.

Nowadays the situation has radically changed but a lot remains to explore. The availability of 3D individual dose-volume information gave the possibility to model the dose-volume effects for many organs, based on clinical, real life data collected from large cohorts of patients. The growing amount of publications concerning dose-volume relationships reflects the need of continuously updating our knowledge in the field of quantitative modelling of normal tissue effects. This is more and more manifest in the current era, where plan optimisation is driven by numbers (i.e., the constraints) that directly reflect our knowledge (and often our ignorance) in this field.

In this context, in 2010 the QUANTEC group (Marks et al. 2010) tried to summarize this huge and growing amount of information in a relatively short and usable guide.

From the point of view of the recommendations, differently from the Emami work, QUANTEC tried to report what we knew at that moment following a quite cautious approach. The document gave clear and exhaustive recommendations in the (few) situations where consistent results were available. In the case of controversial results or still more of lack of results, the document critically discussed the controversial points, often suggesting urgent lines of research and giving clear warnings around the uncertainty of the suggested recommendations.

During the "post-QUANTEC" years, the progress of the field has been relevant, confirming its vitality, with many research groups continuously contributing ideas and new data. In addition, new challenges entered into the arena, substantially modifying the traditional aspects dealing with clinical dose-volume effects studies.

Among them, probably the most important is the shift from NTCP dose-based modelling to the wider field of more "comprehensive" predictive models. In the speculative case that two patients receive exactly the "same dose distribution," the risk of toxicity is always modulated by the single individual profile.

The fact that "dose is not enough" was clear from the early days of radiobiology and it is receiving a constantly growing attention in the current "omics" era (Bentzen 2006): the availability of individual information characterizing the patients and potentially influencing their reaction to radiation is more and more essential, especially in the era of image-guided IMRT, in which organs are efficiently spared in most patients.

This implies the need to have access to data including individually assessed clinical, biological and genetic information and to face the issue of modelling response of normal tissue to radiation in a more and more "phenomenological" approach (van der Schaaf 2015), requiring robust methods for the selection of the most predictive variables (both dosimetric and non-dosimetric) and the adoption of advanced data mining/machine learning methods to manage large databases, including large numbers of patients and lots of variables.

The availability of platforms able to quickly and safely connect networks of institutes is becoming a reality. This path also entails a "cultural" process, with a migration from a "doctor/institutional data" culture to a "pooling data" culture (Deasy et al. 2010). It is likely that this process will be accompanied by a radical automation of the process of data collection, allowing the sharing of data with much less effort compared to the past (Deasy et al. 2010; Skripcac 2014; McNutt et al. 2018), consistently with the rapid evolution toward personalized radiation oncology.

On the other hand, the outcome of the process in terms of robustness and reliability of the models does not only depend on the "numbers," but also (and maybe more importantly) on the "quality" of data. Differently from the "easy" score of the success of a therapy (the patient is dead or alive, is under control or not), toxicity is a much more complex and demanding problem that deserves attention and the prospective and careful collection of patient-reported and physician-reported information for years.

Based on these short considerations, we felt the need for a book like this, with a double intent: from one side, the ambition to give to the readers an updated, rigorous and usable summary of recommendations for guiding planning optimisation, aiming to be a practical tool for dosimetrists, radiation oncologists and medical physicists who are committed to obtaining high-quality plans every day, with or without the help of automatic or semi-automatic optimisation tools. This first goal is accomplished in the central part of the book.

On the other hand, the book also aims to be an introductory lecture to the peculiar methods and the present challenges for students, professionals and researchers in the field; for this reason, the first part of the book is devoted to basic aspects, while the third part is focused on advanced and audacious perspectives, looking more to the coming future.

Thanks to these features, we believe that this book could be used as a textbook in advanced courses of radiotherapy, and medical physics applied to radiotherapy, for radiation oncologists, medical physicists, dosimetrists and radiation therapists.

We are grateful to all the contributors; without their enthusiastic commitment, this book would simply not exist! We are sure that they share with us the sincere hope that this work could be appreciated by the community and, above all, used in practice in many hospitals, all around the world.

CHAPTER 1

The Importance of the Quality of Data

Wilma Heemsbergen and Marnix Witte

CONTENTS

Introduction 2
Dimensions of Data Quality 3
 Study Design 3
  Cohort Study 3
  Causality 3
  Prospective and Retrospective Study 4
  Studying Long-Term Effects 4
  Case-Control Study 5
 Defining a Complication of Interest 5
 Toxicity Scales 6
  RTOG/EORTC and LENT/SOMA Scales 6
  CTCAE Scale 6
  Subjective Components in Toxicity Scoring 6
 Patient-Reported Outcome Measures (PROMs) 7
  Quality of Life 7
  Electronic Questionnaires 8
 Measuring a Complication 8
  Quantitative Versus Qualitative Measurements 8
  Unstable Uncertainties in Measurements 9
  Timing of Measurements 10
  Disease (Progression) and Complication Scoring 10
  Cross-Sectional versus Longitudinal Measurements 10
 Factors Associated with Complication Rates 11
  Risk Factors versus Prognostic Factors 11
 Recording and Collection of Data 11
 Quality and Validity 12
The Quality of Dose and Volume Data 13
 Sources of Uncertainties 13

Collection of Dose and Volume Data 14
Organs at Risk (OAR) 14
    Organ Delineation 14
    Hollow OAR 15
    Dose Reconstruction in the OAR.. 15
Patient Setup and Anatomical Variations.. 16
    Changes in Volume and/or Location During Treatment... 17
Fractionation Effects 17
Practical Implications 18
General Aspects 18
Complication Registration Procedures.. 18
Compilation of Dose and Volume Data 19
Data Warehouse 19
Potential Pitfalls... 20
    Big Data: Real-Time Monitoring?... 20
Conclusions.. 20
References... 21

INTRODUCTION

Models quantifying the relationship between dose-volume parameters and normal tissue complications were developed on a large scale once three-dimensional radiotherapy became available in the 1990s. The QUANTEC project (quantitative analysis of normal tissue effects in the clinic) summarized the available clinical data and models on acute and late radiation-induced complications with the goal of improving patient care by providing useful tools. However, this project also revealed the shortcomings of the available models and underlying concepts and data (Bentzen et al., 2010; Deasy et al., 2010; Marks et al., 2010; Jackson et al., 2010). The quality of the data is paramount in Normal Tissue Complication Probability (NTCP) modeling. It is related to several methodological aspects like the nature of the applied study design, the definition of the clinical endpoints, consistency in toxicity scoring, data collection procedures, and inclusion of all relevant variables. Bad data quality inevitably leads to inaccurate analyses and may lead to biased results, bad models, and false conclusions.

The systematic, data-driven collection and evaluation of complications is essential for knowledge-based treatment optimisation in radiotherapy. For the development of adequate NTCP models, we need preferably large datasets of high-quality clinical and dose-volume data that represent all relevant information on the dose distribution in the organ(s) at risk of interest, and the radiation-induced complications of interest (also referred to as side effect, adverse event, or toxicity endpoint). The data collection and the underlying concept therefore have to fulfill a number of essential criteria:

a. The data collection is part of an appropriately designed study.

b. The data recording and collection procedures are well defined and executed.

c. The data are collected in relevant patient populations with sufficient numbers.

d. The complication of interest is parametrized in an optimal way.

e. The complication is measured with a suitable instrument.

f. The complication of interest is measured at relevant time points.

g. Other factors that might affect the complication probability are registered.

h. The origin of the complication (organ/tissue at risk) is known.

i. Relevant dose and volume information can be extracted from the treatment plans.

j. The available information allows a sufficiently accurate estimate of the actual absorbed dose in tissues.

These aspects of data quality concerning clinical, dose, and volume parameters will be further discussed in the next sections.

DIMENSIONS OF DATA QUALITY

Study Design

Cohort Study

Collecting dose and complication data for a defined group of patients treated with radiotherapy can be considered as an observational cohort study. Which patients have to be included in the cohort of interest is specified with in- and exclusion criteria. Inclusion criteria could, for instance, be based on the diagnosis of the patient and the applied treatment protocol. Exclusion criteria could refer to specific conditions and circumstances, like re-irradiation, multiple tumors, or not speaking the native language. Defining the patient group, one should keep in mind to which patient group the results will be applied, in order to obtain a representative study population.

In a cohort study, typically the association between exposure and the development of health-related events is studied, including the assessment of (potential) risk factors and effect modifiers. Translated into the radiotherapy setting, the exposure is the radiotherapy course and the event is the complication. The key factor for demonstrating causal relationships in an epidemiological observational cohort is a valid comparison group. A cohort lacking a control group is referred to as a descriptive cohort or a case-series, which has a lower level of evidence.

With data from descriptive cohorts, patterns over time and among cases and subgroups can be described without providing evidence for a causal relationship between observed events and factors. Establishing the outcomes of interest in descriptive patient cohorts defined by its disease and treatment is a broadly accepted and applied study design in medicine, since the treating physician is mainly interested in knowledge about the prognosis of treated patients rather than a comparison with an untreated control group.

Causality

Consider the following situation: We observe 35 events of urinary obstruction during follow-up in a prostate cancer patient group treated with radiotherapy, and we observe

12 events in a similar patient group without radiotherapy (a valid control group). Now (a) we can estimate the true treatment effect, and (b) we have level II evidence that a causal relationship exists between radiotherapy for prostate cancer and urinary obstruction. As mentioned earlier, in radiotherapy we usually evaluate descriptive cohorts (all exposed) for NTCP modeling. However, if we are able to establish a dose–effect relationship in a descriptive cohort, this is also regarded as a piece of evidence for causal relationships, as described in Austin Bradford's famous paper (Bradford, 1965), which describes eight criteria to assess causality for an observed association between a cause and an effect: temporality, strength, dose-response, reversibility, consistency, biological plausibility, specificity, and analogy. In studies concerning NTCP modeling, the identified correlation is usually similar to as reported in other studies (analogy); the underlying mechanism of radiation-induced damage may be known (biologic plausibility); for some endpoints, it has been demonstrated that the effect can be reduced by reducing exposure, i.e., lower dose levels (reversibility); and obviously the exposure preceded the observed effect (temporality). We should however keep in mind that without a control group we have no estimate of the number of non-radiation-induced events among the observed events.

Prospective and Retrospective Study

There are two conceptual types of observational cohort studies: prospective and retrospective. A cohort study may also have retrospective and prospective phases. The main essence of a prospective cohort is that a subject is recruited for the cohort prior to developing the complication of interest, according to a predefined set of in-/exclusion criteria. Retrospective studies are considered as having a relatively low level of evidence, but this depends on the specific design of the retrospective study. In a hospital setting, retrospective data collection from medical patient records is often of poor-quality. In case of frequent follow-up visits, limited dropout of patients, and the availability of extensive information in the patient files, a dataset of reasonable quality may be achieved as long as it concerns clinically relevant information that is reported in the patient records in a reasonably consistent and systematic way by the treating physicians. However, obtaining consistent and systematic data on relevant baseline information (especially potential prognostic factors and effect modifiers) remains an issue in such studies.

Studying Long-Term Effects

Radiotherapy can cause long-term effects with a long latency time that are impossible to capture during a standard follow-up period at the outpatient clinics. Well-known examples of such long-term effects are heart failure and secondary tumors. Such effects are mainly studied in retrospectively established cohorts. Such cohorts can be established in existing registrations with a prospective nature, such as national cancer registries, to avoid a selection bias. The endpoint of (secondary) cancer or diagnosis of disease is of such a nature that it can be extracted from medical patient records and/or existing prospective registries in a reliable way. However, obtaining consistent and systematic data on relevant baseline information (i.e., potential prognostic factors and effect modifiers) remains an issue in such studies.

Case-Control Study

For rare complications or complications with a long latency, a case-control study can be an attractive, fast, and effective method to study relevant prognostic factors and dose-effect relationships. The level of evidence is considered lower compared to cohort studies because causality cannot be assessed, and the validity of the study can be seriously affected by selection bias, information bias, confounding, and selective choices for cases and controls. However, carefully designed nested case-control studies within established radiotherapy cohorts can be a very useful tool to generate hypotheses and estimate dose-effect relationships for rare complications. Examples of such nested case-control studies are the quantification of the effect of radiation dose to the heart for developing coronary heart disease (van Nimwegen et al., 2016), and assessing the effects of the delivered radiation dose to the lung and the risk of second primary lung cancer (Grantzau et al., 2014).

Defining a Complication of Interest

Irradiation of healthy tissue may eventually lead to (temporary or chronic) clinical symptoms that can be experienced by the patient, diagnosed by the physician, and/or observed through a function test or blood test. The level of the health damage can be observed as an event (qualitative endpoint: present, not present), as a ranked ordinal outcome (none, mild, moderate, severe), or as a continuous, quantitative outcome (e.g., loss of organ function). Furthermore, an event (present, not present) can be scored as "yes" or "no" regardless the time, or it can be studied in relation to the timing (time-to-event), censoring patients with limited follow-up. A quantitative endpoint can be defined at a fixed time point (e.g., function loss after one year) or it can be followed longitudinally, establishing organ function loss as a function of time. Therefore, the required data depends not only on the definition of the endpoint but also on the relevant timing of the measurement and the concept of the NTCP model.

Complications of interest in NTCP models published up to now often model the presence or absence of a certain clinically observed event or symptom. Such information can be roughly divided into physician-reported data and patient-reported data. Physician-reported information typically includes not only events or symptoms but also initiated treatments and interventions (e.g., medication prescription, physiotherapy). This information can be graded according to an internationally established nomenclature (toxicity scale).

Apart from physician-reported or patient-reported complications, other endpoints for NTCP modeling are possible as well. A complication can be defined based on blood tests, imaging, or function tests (Figure 1.2). Examples of such endpoints are: measuring inflammation of the esophagus on fluorodeoxyglucose positron emission tomography (FDG-PET) (Nijkamp et al., 2013), measuring organ expansion to establish the degree of inflammation on a computed tomography (CT) (Niedzielski et al., 2016), measuring the kidney function in blood samples (Trip et al., 2014), measuring the saliva flow rate to establish xerostomia (Miah et al., 2013), and measuring the mouth opening to establish the presence of trismus (Gebre-Medhin et al., 2016).

Toxicity Scales

Standardized toxicity scales are broadly applied in reporting normal tissue complications in radiotherapy studies and NTCP modeling. Standardizing the reporting of complications enables comparisons between studies, treatments, and hospitals. Moreover, it allows data pooling and meta-analyses of published data. For collaborations between institutes to do joined studies, such common criteria to score toxicities are essential. In general, all systems assign grade 0 to no toxicity, grade 1 to mild, grade 2 to moderate, grade 3 to severe toxicity, and grade 4 to adverse events that are disabling, life threatening, need immediate surgery, and/or have a major impact on activities of daily living (ADL). Grade 5 is death directly related to the adverse event. In radiotherapy, different toxicity scales are used; thus, calculated toxicity rates as well as quantified dose-volume effects may depend on the applied grading system. In a recent study it was demonstrated that adverse event reporting in oncology trials is very heterogeneous and suboptimal, clearly showing the need for more broadly accepted standardized nomenclature (Sivendran et al., 2014).

RTOG/EORTC and LENT/SOMA Scales
The Radiation Therapy Oncology Group (RTOG) and European Organization for Research and Treatment of Cancer (EORTC) developed toxicity scales for uniformly scoring of acute and late effects of normal tissue irradiation, for different types of tissues; the ROG/EORTC scales as published by Cox et al. are still in use (Cox et al., 1995). In the same year the Late Effects Normal Tissue Task Force subjective, objective, management, and analytic (LENT/SOMA) system was introduced as an improved successor to the RTOG/EORTC system (Pavy et al., 1995). Both systems have been used since for late toxicity reporting in radiotherapy in general and randomized trials in particular, often using slightly adapted versions of the published scales.

CTCAE Scale
More recently, the Common Terminology Criteria for Adverse Events (CTCAE) grading system of adverse events of cancer treatment is becoming a new standard in reporting radiotherapy complications. Early versions of this system mainly focused on acute adverse effects of treatment, therefore regarded as unfit for adverse event scoring in radiotherapy. In the third version originally published in 2003, one of the missions was to achieve a uniform document that covers all modalities including radiotherapy, merging acute and late effects into one list of criteria (Trotti et al., 2003). This system was adopted worldwide for comprehensive adverse event recording in cooperative group trials at a consensus meeting (Chen et al., 2006).

Subjective Components in Toxicity Scoring
Despite the standardization of toxicity scales, there can still be a large subjective component in the assignment of toxicity scores. Consider the following situation: a patient complains to the physician about painful abdominal cramps and blood in their stool during a follow-up visit. There are now several options for additional systematic adverse event scoring according to CTCAE version 4.0.3: With regard to bleeding: "anal hemorrhage," "rectal hemorrhage,"

"colonic hemorrhage," or "lower gastrointestinal hemorrhage" are legitimate options. With regard to the abdominal cramps: "abdominal pain," "rectal pain," "gastrointestinal pain," or "pain," are options (abdominal cramps itself is not a listed item). With regard to the whole picture of symptoms: "proctitis," "gastrointestinal mucositis," or "enterocolitis" are options. There are now several ways to proceed: the physician may or may not ask about the impact of the complaints on ADL, and he may or may not prescribe medication for the symptoms. All these decisions are relevant for the final grading according CTCAE. In addition, some physicians may have their own opinion about the severity of a certain complaint and indicate a grade 1 even though the patient judges it as moderate or severe. Another issue is that the physician might decide to refer the patient for a sigmoidoscopy and this may lead to the diagnosis of hemorrhoids in the rectum. Now the physician might score hemorrhoids according CTCAE as well, 3 months later, but only for referred (selected) patients.

Patient-Reported Outcome Measures (PROMs)

Side effects as experienced and reported by the patient, and their impact on Quality of Life (QoL) aspects, have recently been recognized as important outcomes that deserve more attention (Bottomley et al., 2016). Moreover, studies have demonstrated that the collection of this information directly from patients improves the precision and reliability of the detection of adverse events. Recently, the CTCAE-PRO list has been developed with the main goal to include patient-reported adverse events in future oncology trials (Dueck et al., 2015).

Patient-reported outcomes on standardized questionnaires enable data-pooling and comparisons between studies. A disadvantage of patient-reported complications is that it usually does not include information about medication prescription, medical intervention, referral for therapy, and the status of the patient with regard to new diseases, tumor progression, and new treatments. Since interventions might influence the level of symptoms such information should ideally be taken into account as well. An example of NTCP models based on patient-reported outcomes is the study of Schaake et al. (2016) and DeFraene et al. (2012) where data on fecal incontinence and increased stool frequency were derived from questionnaires.

Quality of Life

Clinical symptoms may affect daily functioning of the patient in terms of physical, social, and mental wellbeing of the patient, deteriorating the quality of life of the patient. A QoL instrument (QLQ-C30) for use in international clinical trials in oncology was published in 1993 (Aaronson et al., 1993). In addition to this general oncological questionnaire, tumor site specific modules have been developed by the European Organisation for Research and Treatment of Cancer (EORTC) Quality of Life Group, again designed for use in clinical trials. One of the first published modules was the QLQ-PR25 for prostate cancer (Borghede et al., 1996). These modules typically contain a mix of questions related to anticipated (site specific) adverse events, patient bother, and QoL. Other broadly applied instruments related to QoL (not specific for oncology) are: the Eastern Cooperative Oncology Group (ECOG) performance scale, pain scales, and the EQ-5D–5 L (Herdman et al., 2011).

QoL aspects are usually not considered as an endpoint in NTCP models although in theory this is a possibility. In the context of NTCP modeling, the complications that bother patients the most, i.e., affecting QoL domains, could be regarded as a top priority during treatment optimisation. In order to do so, we need more datasets in which both adverse events and quality of life items are collected.

Electronic Questionnaires
Currently there are many exciting developments in the field of electronic data collection by means of web-based patient questionnaires, scoring health-related symptoms and QoL aspects (Campbell et al., 2015). Well organized electronic data collection is very attractive since it requires no time-consuming and expensive data-entry afterwards, and it enables data collection at chosen time points, not depending on scheduled RT visits. Also, the new touch screen technology is interesting, since it may create a more easy and patient-friendly way of scoring. However, there are also down sides. For instance, patients may receive many questionnaires from different departments and different hospitals, especially during the first year post-treatment, which may seriously affect the compliance of the patient and therefore the completeness of the database. It will probably also affect the quality of the submitted answers if the patient experiences an overload of questions. Another practical problem is that especially older patients may not be able to handle electronic question-naires and need assistance. Furthermore, little is known about all kind of aspects of elec-tronic questionnaires because it is such a relatively new phenomenon. For instance, does it make a difference whether the patient assesses quality of life at home or in the hospital.

Measuring a Complication
The applied complication measurements in radiotherapy are often *indirect* measurements of the radiation damage, as represented in Figure 1.1. There are many types of measure-ments involved in complication scoring. Each type of measurement (questionnaire, medical record, image, blood test) has its own unique properties in terms of uncertainties, errors, types of potential bias, reliability, inter- and intra-observer, and test retest variation. One of the key issues in measuring a complication in a clinical setting is that one usually has to rely on the information available from the medical practice whereas other types of mea-surements might be much more relevant and valid to detect the complication of interest. Or, for instance, in case of imaging, the quality and settings of the available images might be suboptimal for this purpose. Another important issue is that relevant measurements are often not systematically performed for all patients in a clinical cohort. For example, during an endoscopy severe radiation-induced damage of the rectal mucosa is observed, but only a few patients of the total cohort underwent this procedure. So from a scientific point of view, we cannot evaluate this specific endpoint of hemorrhoids as a complication of inter-est because it has not been systematically evaluated.

Quantitative Versus Qualitative Measurements
In radiotherapy research we are often focused on quantitative measurements of complica-tions, i.e., observations that can be expressed in numbers, since we want suitable data to run statistical analyses on. However, paying serious attention to qualitative measurements as well during data collection, could in the end lead to improvement of the quantitative

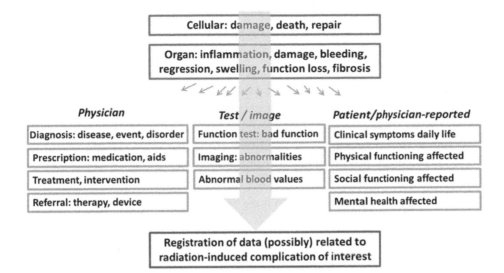

FIGURE 1.1 The radiation damage (top two rows) can be indirectly measured or observed in clinical practice (middle three rows). In the database (bottom row) we have to capture all relevant data related to the complications of interest by defining relevant parameters.

measures. For instance, when we would systematically record patient remarks about the most bothersome complications outside the predefined boxes, we might reveal interesting and clinically very useful information that could be translated into quantitative measures for future patient cohorts.

Unstable Uncertainties in Measurements

An important question to consider during and after data collection is: are the underlying uncertainties in the gathered and measured data the same for all patients? If not, this can seriously affect the internal validity of the study. This is a very realistic scenario if one collects data from hospital records during a period of years. Examples are listed below:

- The quality and settings of relevant images has changed over time.

- Guidelines to prescribe medications and interventions have changed.

- A new method has been introduced in the laboratory to assess certain blood values.

- The frequency and timing of patient visits has changed.

- The procedure to perform a functional test has changed.

- A new version of a patient questionnaire was implemented.

One should therefore always check whether or not the information of interest was collected according to the same procedures for each patient, and if not, record the information for the different calendar periods.

Timing of Measurements

In study protocols, strict time schedules for evaluation are often included (e.g., every three months in the first year, every six months in the second and third year, once a year thereafter). In daily practice, these requirements are often not met and the patient is evaluated according to the schedule of the hospital for follow-up and the preference of the treating physician. Ideally, timing of measurements should be driven by the research question and the hypothesis. Designing data collection procedures, we should therefore be aware of this conflict and come up with practical solutions that guarantee acceptable timing of the measurements involved. Furthermore, as indicated in Figure 1.2, we would like to obtain information in the period before and after the time window in which the regular RT visits take place; for instance, we would like to have baseline information collected prior to the start of any cancer treatment.

Disease (Progression) and Complication Scoring

As indicated in Figure 1.2, a patient may experience recurrence of the disease, or develop a new disease or comorbidity, and receive treatment for this. In such situations, complication scoring can become very complex, since the patient may experience many symptoms related to the disease and/or the treatment, and obviously it does not make sense to register all these events as possible radiation-induced complications. One should address how to handle this situation in the study protocol. One could for instance decide to censor at some point in time when new treatment is initiated, ignoring all complications thereafter. One can also decide to register all new treatments along with the complications and include this information in the later analysis.

Cross-Sectional versus Longitudinal Measurements

In a cohort study, exposures and events are typically studied in chronological order and each subject in the cohort is potentially at risk to develop the event of interest. In this setting, complication incidences and rates can be established as a function of time. An example of such a longitudinal observation is counting the events (and time to event) of

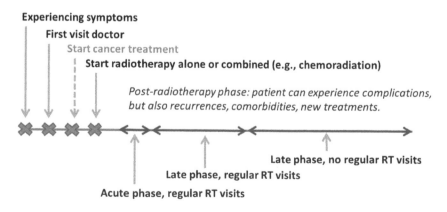

FIGURE 1.2   Schematic representation of the time line for a patient receiving radiotherapy; data collection during regular RT visits may cover only partly the time points of interest of the study.

having a diagnosis of heart disease during a follow-up period. However, in radiotherapy cohorts, the scoring of complications may be obtained in a more cross-sectional setting as well, for instance, if we distribute a questionnaire at two years post-treatment in which we ask the patient to score certain symptoms.

Factors Associated with Complication Rates

Baseline factors that are considered and collected in databases for NTCP modeling include patient characteristics (e.g., age, comorbidity), treatment characteristics (e.g., prescribed dose, concomitant and/or adjuvant treatment), tumor characteristics (e.g., tumor volume), and baseline symptoms (e.g., scores from a baseline questionnaires). For each complication of interest, at least known factors from literature have to be established and included in data collection procedures. In clinical practice the following situations can occur that may affect the validity of the data collection of baseline information:

- When the patient is evaluated for the first time, treatment for the disease already started (e.g., surgery, chemotherapy). In such a setting, scoring of baseline aspects can become quite challenging, and it is too late to ask the patient to fill out a "baseline" questionnaire. This can be tackled in advance by scheduling the baseline assessment and questionnaire prior to the first treatment in collaboration with other departments.

- The patient presents at baseline with symptoms that are likely to be caused by the tumor (e.g., shortness of breath), and these symptoms resemble radiation-induced symptoms that may occur later. If this is the case, additional data collection on the course of the presented symptoms is needed, and you may consider the post-treatment situation (after acute reactions have resolved) as the baseline situation.

Risk Factors versus Prognostic Factors

Variables predictive for the onset of disease in a (healthy) population are referred to as risk factors, and are usually associated with (very) low absolute event rates (e.g., 1 per 10,000). Prognostic factors are related to counting the consequences of (the treatment of) a disease, i.e., relatively frequent observed events. This includes events like disease recurrence and death, but also events like complications, disabilities, and suffering. Moreover, the same factor can have a different effect being a risk factor and a prognostic factor. For instance, a high age can be a risk factor associated with a higher probability to develop cancer, and at the same time a high age can be a prognostic factor associated with a lower probability to develop metastatic disease after diagnosis and treatment.

Recording and Collection of Data

In order to create a high-quality database, data recording and collection procedures are critical. A prerequisite for a prospective study design is that the data collection procedures are designed carefully, in order to have accurate and complete baseline information on all relevant aspects of patient, disease, and treatment characteristics, including relevant comorbidities and baseline symptoms. If data collection of baseline variables (i.e., relevant cofactors) is heavily depending on the (retrospective) availability in the patient hospital records, poor consistency and missing data will inevitably affect the validity of the study. For this purpose, the research questions of interest should be determined in advance, in order to list all essential baseline information that will be needed later during the analysis. In case the data collection is embedded in the clinical workflow, procedures to collect the relevant information in a systematic way should be straightforward, simple, and take little extra time, and the involved physicians and assistants should be trained and get feedback to record information in a similar fashion.

In general, it is preferred to record relevant information as accurate as possible, since coding into relevant categories (binary, ordinal, categorical) can always be done afterwards, whereas changing the definition of predefined categories cannot easily be done afterwards if the original numbers were not stored in the research database.

A finalized database ready for further analysis should fulfill the following demands:

- Completeness: missing data are acceptable below a certain threshold according to predefined requirements.

- Consistency: the information of different variables does not contain contraindications.

- Constant quality: the quality of the data is not depending on certain factors like calendar periods or treating physician.

- Integrity: the data trail can be traced, connected, and checked.

- Conformity: the definition, label, and format of the data within one variable field is the same for all data rows and over time.

- Uniqueness: no duplicates of data.

- Accuracy: information is included and updated as much as possible, and the number or errors is limited (through data check procedures).

Furthermore, one should be aware of the legislations and laws concerning the privacy of the patient (e.g., anonymization of data), the ownership of routine clinical data that are used for research purposes, the demands for informed consent to use routine patient data for research, and the demands for secure data storing, and authorized data access.

Quality and Validity

The dimensions of data quality may have consequences for both the internal and external validity of obtained study results. Internal validity refers to how well a study was designed, performed, and analyzed, with respect to delivering the correct answer to the research question with certainty. External validity concerns the generalizability of the results to other similar patient populations outside the database. Poor study design, poor data recording and collection procedures, suboptimal measures, bad timing of follow-up: it will all have an impact on the validity of subsequent modeling. Figure 1.3 shows the different

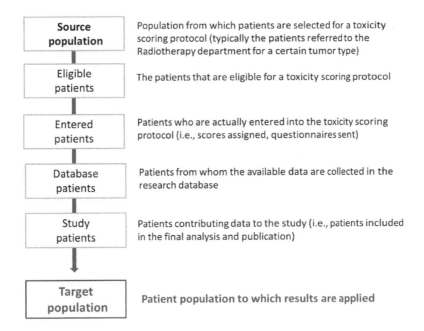

FIGURE 1.3   The different levels of patient group selections in a complication scoring protocol, going from a source population to the final selection of the study patients that are included in the analysis and publication.

steps of patient group selection, going from a source population to the final selection of the study patients included in the analysis, and all steps in between. In case of well-organized prospective procedures, patients entered in the protocol are a representative sample of the source population, and the study patients ending up in the database for analysis are the same patients as the patients originally entering the protocol. In case of poor procedures to identify (retrospectively) the patient data of interest or in case of many missing data, the patient group analyzed at the end can be quite different from the patient group entered earlier (risk of selection bias, missing data bias), which can potentially threaten the internal and external validity. Furthermore, the external validity may also be affected in case the characteristics of the source population and its treatment are different for future patient groups in the hospital, and for patient groups outside the hospital. For instance, in case of other equipment, other techniques, other chemo-radiation schedules, or when pronounced differences in the characteristics of the population are present (e.g., older, more advanced tumors, genetic differences).

THE QUALITY OF DOSE AND VOLUME DATA
Sources of Uncertainties

To derive dose-effect models with high predictive power, it is necessary to use an accurate estimate of dose that was actually absorbed in tissues during the delivery of the radiation treatment. NTCP modeling is generally based on the pre-treatment imaging information collected for the purpose of treatment planning. Typically, the information available comprises a planning CT scan on which organs relevant for treatment optimisation have been (hand-) delineated, and a 3D dose distribution extracted from the treatment planning system (TPS). While at the pre-treatment planning stage this information may provide the best possible prediction of the expected dose deposition during treatment, several sources of uncertainty may cause a discrepancy between the true absorbed dose and the estimation thus acquired. Additional information collected during the treatment fractions could potentially improve this.

Collection of Dose and Volume Data

Each radiotherapy department has a designated TPS platform to prepare the radiation treatment for its patients, and clinical protocols for its use to ensure safe and adequate treatment. While the internal representations of geometries may differ between the software platforms of different institutes, the collection of dose and volume data should be performed in a standardized way. Nowadays, all treatment planning software should be able to export the relevant information in the standardized DICOM (Digital Imaging and COmmunications in Medicine) format, and the database of volumetric data will consist of a large number of DICOM image slices for every patient. Depending on the tumor site, 3-dimensional (3D) dose distributions should be exported on a sufficiently high resolution (in general between 4×4×4 mm and 2×2×2 mm). The size of these image databases is typically several gigabytes, and efficient computational methods are needed to analyze them.

Organs at Risk (OAR)

A particular type of toxicity is (usually thought to be) related to the radiation damage of a specific organ, which is no longer able to function properly. Depending on its function, an organ may suffer more from the maximum absorbed dose levels (serial organs, e.g., spinal cord) or more from the amount of dose absorbed on average in the whole organ (parallel organs, e.g., lung). Depending on this characteristic, the impact of uncertainties in the estimations of dose and of volume on the quality of the resulting NTCP model may vary. Organs at risk are delineated on the planning CT and/or MRI (Magnetic Resonance Imaging) images by hand or using automated procedures, based on the gray values, potentially aided by models or an atlas. Many organs can be clearly distinguished on these images; therefore, identification of such organs should be much less prone to observer variations as is the case for tumor and (positive) lymph node delineations. On the other hand, some structures with a specific physiological function may not be easily distinguished from their surroundings (e.g., hippocampus), leading to larger uncertainties. In some cases (e.g., anus, rectum, sigmoid) the distinction between (sub-) organs depends on protocol rather than on visible tissue interfaces, and discrepancies may arise when protocols or their interpretation differ between institutes.

Organ Delineation

In radiotherapy a delineation is usually constructed as a number of closed 2-dimensional (2D) contours on the planning CT slices, which are then exported as DICOM structures. Evaluation of the encompassed volume requires an interpretation step and is not unique. One strategy is to convert each contour into a mask (slab) on the underlying CT's resolution, constructing the total volume as the union of such slabs. A drawback may be that CT pixels which are intersected by the contour are implicitly considered to lie entirely within the volume, leading to an overestimation of the organ's true volume. With a CT resolution of typically 1 mm in a slice and 3 mm between slices, this effect may be relevant for small or thin structures (optic nerve, chiasm, blood vessels, membranes). Without correction, missing contours (skipped CT slices) will lead to an erroneous volume estimate. An alternative approach which may prevent bias in the measured volume is to convert the contours into a surface triangulation (mesh), followed by the creation of a cloud of (random) points within this surface representing the organ's volume. It is a non-trivial task to determine which dots to connect between contours, and for complex surface shapes the algorithms which solve this problem may fail. On the other hand, skipped slices may be bridged by the triangulation and still lead to an accurate volume, and partial volume effects of the CT voxels may be prevented.

Hollow OAR

In some cases, the OAR consists of a hollow organ (e.g., rectum, bladder). Typically, during treatment planning the entire volume (including filling) is used for the purpose of dose optimisation, and the clinical protocol dictates that only the outer organ contour shall be delineated. This is reasonable, as additional delineation of the inner wall increases workload, while dose optimisation based on the wall only could lead to clinically undesirable hotspots inside such an organ (as organ filling varies these may yet end up inside critical tissues during plan delivery). However, for the purpose of NTCP modeling it may not be optimal (in terms of the predictive power of the resulting NTCP model) to consider the dose deposition to the solid organ including filling. Several approaches may be followed to better approximate the dose absorbed in the wall tissues, including manual re-delineation or automatic generation of the inner wall contour. Care should be taken that the slab- or mesh-based volume reconstruction methods are able to handle such nested contours. A caveat for the mesh-based method is that a common approach to create a closed surface from a set of contours is to add 'caps' cranial to the most superiorly and caudal of the most inferiorly delineated contour, by connecting all dots from these contours with an extra dot placed at, e.g., half a CT slice distance centrally above or below. This works well for somewhat spherical volumes but may lead to erroneous volumes including caps for a partially delineated hollow organ such as the rectum wall. (If both the inner and outer contour are closed with coplanar triangles, i.e., a cap distance of 0, this issue can be prevented.) Alternatively, the use of dose-surface parameters rather than dose-volume parameters could be considered; as the dose gradients over the wall thickness are likely small these may provide a sufficiently accurate approximation of the true dose-volume parameters of the wall tissue.

Dose Reconstruction in the OAR

Within a treatment planning system, the dose engine is the piece of software that computes the estimated dose deposition to tissues, based on physical models of the delivery beams and the interactions of radiation with matter. For photon radiotherapy the electron density values which are required for these computations can be accurately derived from the planning CT set. Contrarily, for particle therapies the stopping powers inside the patient's tissues are not as easily derived from the CT Hounsfield units. More advanced techniques such as dual energy CT and proton CT are under investigation to improve on these, yet uncertainties in particle beam ranges typically remain.

As dose needs to be recomputed many times within the dose optimisation loop, dose engines are usually very highly optimized for computational speed. It is a common approach in treatment planning software to commence optimisation with a fast pencil beam algorithm, switching to more accurate but slower methods (e.g., collapsed cone convolution) as the optimizer converges towards the final dose distribution. Still, these latter algorithms have to strike a balance between accuracy and speed. The most realistic dose distribution may be acquired using Monte Carlo methods. In general, these are computationally prohibitive and not available in clinical practice; however, developments in computer graphics hardware are lowering this threshold and practical solutions are becoming viable. For the purpose of NTCP modeling a final Monte Carlo dose re-computation of a (historical) plan generated using a lower quality dose algorithm could in principle improve the quality of the model; however, the infrastructure and treatment machine specific model to perform such computations will rarely be available.

Patient Setup and Anatomical Variations

Accurate daily repositioning of the patient to match the planned geometry as well as possible is crucial for a high-quality radiation treatment, and has been the focus of the thriving field of image guided RT over the last decade or so. In the most ambitious online image guidance, (volumetric) imaging inside the treatment room is performed usually just before commencing treatment, and corrections to the patient setup (such as an automated table shift) are made, or in case geometric discrepancies are too large the patient is repositioned or treatment is aborted and a new treatment plan created. Alternatively, in an offline setting patient scans are acquired on treatment days but only accommodated for during subsequent treatments, thus correcting for potential systematic deviations of the planning image set but not for daily variations.

Regardless of the strategy even after corrections, the patient geometry on any given treatment day will still differ to some extent from that during acquisition of the planning CT, and from the geometries on other treatment days. Image guidance protocols usually aim to reposition the tumor as accurately as possible, but in doing so the surrounding organs may become displaced if relative motions between tumor and/or OARs occurred. Such relative motions cannot be corrected for using table displacements and remain as residual geometric uncertainties on the locations of the OARs. An example of this is found in prostate radiation treatment using implanted gold markers, and daily (e.g., cone beam CT) imaging. In such a case the prostate's center of gravity can be realigned to high precision even when rectum distention caused a large prostate displacement (relative to bony structures), but at the same time the surrounding bladder and rectum wall structures could receive a dose much different than in the planned geometry. Similarly, for head-and-neck patients whose neck flex during treatment is different than during planning (these patients regularly lose weight during treatment weeks such that the facial fixation mask becomes less effective), a table shift cannot realign all relevant organs simultaneously.

Changes in Volume and/or Location During Treatment
Besides the unavoidable discrepancies due to patient mobility, errors may arise when the radiation (and/or chemo) treatment affects the volumes of the tumor or OAR themselves. A shrinking tumor loses its mass effect on surrounding structures, which may then be pulled into the high-dose region as the tumor recedes. Similarly, some OAR (salivary glands in particular) may react to radiation dose by shrinking, complicating the estimation of overall absorbed dose.

Using the treatment room volumetric imaging acquired for the purpose of image guidance, a more accurate estimation of the absorbed doses might be established. Evaluation of a deformation vector field between planned and daily patient geometries would allow one to track the daily motions of OARs and to correctly accumulate dose over treatment fractions. Depending on the quality of the image guidance scans, the deformable image registration (DIR) algorithms used, and the OARs of interest, such methods may be more or less accurate. Sliding tissues (e.g., in the abdomen) are in general difficult to accommodate in DIR methods, for such sites finite element modeling is sometimes explored. Besides the changing shapes and positions of the OARs, also the shape of the dose distribution may react to a changing patient geometry. The errors introduced by this effect are generally assumed to be small for photon RT of deep-seated tumors but may become prominent in regions with heterogeneous tissues and air-tissue interfaces, and for particle therapies due to the steep dose fall-off behind the Bragg peak. A proper inclusion of these effects would require re-evaluation of the 3D dose distributions for the daily patient geometries. However, image guidance scans are usually not immediately suitable for dose re-computation due to a limited field of view and/or the use of cone beam CT, which cannot directly produce correct Hounsfield units. An intermediate step is then required, typically deforming the planning CT to match the daily geometries. In general, the additional daily imaging data and the advanced methods required to process them are institute specific, and not collected in multi-institute trial databases.

Fractionation Effects
The rationale of fractionated radiotherapy is the difference in radiobiological sensitivity of tumors versus healthy tissues, where tumors often have impaired radiation damage repair mechanisms, while healthy tissues can more easily (although not in all cases fully) recover from radiation dose exposure. A realistic NTCP model should acknowledge the effect of dose fractionation on the complication rate and should then be predictive for toxicity across treatments using different fraction sizes. In the data of a clinical trial in which all patients were treated with equal daily tumor dose prescription, this fractionation effect may still be relevant for the OARs, as these – in contrast to the tumor – receive a highly non-uniform dose, which may strongly vary between patients. Therefore, some part of an OAR may be subject to the same fraction dose as the tumor, while another part of the same OAR may receive only half the dose per fraction, and effectively be in a much lower dose

fractionation regime. The non-linear way in which the dose should be corrected for fraction size before accumulation is different from the non-linear relation between this biologically corrected total dose and the toxicity rate. In an NTCP model constructed using only the data of such a trial, it may be difficult (if not impossible) to isolate these two non-linear contributions, and in general an effective fractionation effect (α/β ratio) is assumed and applied or disregarded altogether; in the latter case, this fractionation effect of dose heterogeneity is absorbed in the parameters of the non-linear NTCP model. In case daily dose variations as described above are important, the effect of fractionation should be carefully evaluated, since a single fraction with much increased dose could cause a large contribution to the total biologically effective dose.

PRACTICAL IMPLICATIONS

General Aspects

There are different strategies to compile a database for the purpose of NTCP modeling. Such a database can be the result of a dedicated clinical study in which the complications of interest are part of the study objectives described in a study protocol, including carefully planned measurements and follow-up. Another option is to re-use data from clinical trials for the purpose of NTCP modeling. Another approach is to (retrospectively) extract data from the medical notes in the hospital records, allowing collection of data for relatively large number of patients. This is however time-consuming and may generate data of questionable quality. More recently, a different approach for the collection of complication data has been adapted by several research groups: retrieving data from routine clinical practice, through the implementation of procedures enhancing a prospective and systematic registration of complications.

Clinical radiotherapy protocols rapidly evolve over time, as a result of developments in oncology in general, and technical developments in the field of radiation oncology in particular. At the same time, personalized medicine and shared decision making are becoming the new standard of care. In order to facilitate knowledge-based medicine, and to inform patients accurately about their prognosis in terms of adverse events risks, we need good-quality data from daily practice on adverse events on a large scale, since the number of clinical trials providing such data is limited.

Complication Registration Procedures

Procedures for registration of complications following the radiotherapy treatment (including the distribution of questionnaires) are nowadays more and more embedded in the clinical workflow at radiotherapy departments. In case of complication scored by the physician according to a toxicity scoring system, consistent and systematic scoring can be enhanced by using predefined toxicity item lists. One should be aware that collection of complication data as fixed grades, according to a certain standardized toxicity scale directly, provides little flexibility when the data have to be translated into another toxicity scoring system or updated to a new version of the same scoring system. Furthermore, it does not give insight into underlying inter-observer differences in assigning scores. A more robust alternative would be to record the pieces of relevant information itself in a systematic way and translate the information into grading scales later.

The prospective nature of such a complication scoring procedure does not guarantee a dataset of good quality. As argued in the previous chapters, it demands a systematic, unbiased approach. As a starting point, prospective registration of information relevant for defining and identifying membership of a specific cohort is critical (e.g., diagnosis, applied treatment protocol), and the recording of patient, treatment, and tumor characteristics should be well organized. During ongoing data collection and scoring regular checks and feedback are needed to identify and solve problems at an early stage.

Compilation of Dose and Volume Data

Depending on the source of the volumetric treatment planning data, various intermediate steps may be required to bring a database containing the collected scans of a cohort of patients into a state that allows large scale analysis. To answer a specific research question, e.g., concerning a particular dose-volume histogram (DVH) point of a given OAR, automated methods are needed to compute the relevant values for each patient. Typically, some kind of scripting environment is used to perform such computations. For such scripts to execute correctly, the data should be brought into a uniform format, e.g., with respect to the names of delineated organs. For some analyses an OAR may be considered which was not (for all patients) delineated during treatment planning, and additional delineations need to be created. Further optimisation of these analyses can be reached by caching intermediate computational results, e.g., saving a cloud of random points that was created from a delineation, so it can be re-used for subsequent DVH analyses. Also, a cumulative 3D dose distribution combining the different stages of a sequential boost protocol may be computed and stored; the use of an alternative image format (e.g., Neuroimaging Informatics Technology Initiative, NIfTI) could reduce the overhead associated with the somewhat verbose multi-file DICOM format. However, all such intermediate steps increase the distance to the original data, and strict quality checks should be implemented to ensure errors are not introduced without going noticed.

Data Warehouse

In this information age, the big data surrounding us have enormous potential. In radiotherapy departments, large volumes of patient data are generated in patient registration systems, treatment planning systems, at the linear accelerator, and in electronic hospital records. The inspirational vision to process these big data into assembled radiotherapy outcome databases using existing frameworks has been described by several authors (Lustberg, 2017; Aapro et al., 2017). Such data warehouses would serve the needs of clinicians, managers, and researchers. Strategies could be developed and implemented for automatic identification of specific patient cohorts, dose exposure levels, and other relevant key information, enabling data mining opportunities on a large scale. Obviously, data quality assurance processes have to be implemented to obtain acceptable levels of reliable data for research purposes. There is however no such thing as a 'master plan' that will create one big radiotherapy healthcare database which will provide all the relevant data for every potential research question. Every set of research questions has to be operationalized into carefully defined endpoints and measurements which will rarely match completely with routinely obtained data. However, one can adapt the research questions to clinical practice, and adapt routine clinical procedures to increase the availability of data with sufficient quality for research purposes.

Potential Pitfalls

The competence of the doctor to tailor medical procedures to the prognosis, health risks, and needs of the individual patient, will create optimal care for the patient. It is however potentially conflicting with the key condition of scientific research, demanding uniform and systematic evaluations for each study patient in a cohort. This makes complication scoring for scientific purposes in a clinical setting a real challenge. In addition, the simple fact that patients will often be seen during follow-up by different doctors makes it even more complicated and subject to inconsistencies. Developing complication scoring protocols, one should therefore realize that (a) the time and effort a doctor and patient can spend on systematic complication scoring during patient visits is limited, and (b) the scoring procedure should be straightforward with no room for different interpretations and misunderstandings.

Another potential threat to the quality of the data is that patients are often seen by other clinicians outside radiotherapy (e.g., dietician, oncologist, gynaecologist, urologist, speech therapist). They might identify potential radiation-induced complications as part of regular checks and visits, but this may not be included in the complication registration procedures of the radiotherapy department. Comprehensive complication scoring should be the starting point of any toxicity scoring protocol and capturing useful clinical data outside the radiotherapy department is therefore essential.

Big Data: Real-Time Monitoring?

Patient cohorts in daily clinical practice will reflect subgroups of patients treated with partly different protocols and guidelines. The radiation-oncologist and clinical physicist might be especially interested in the effects of such protocol changes on treatment outcomes and NTCP models and request for monitoring the effect of such changes on outcome and modeling, since the promise of "big data" is real-time knowing. Apart from the issues concerning the quality of the data, there are other issues here as well: small effects will be difficult to measure with statistical certainty and small effects may be overshadowed by bigger effects resulting from alterations in the logistics and procedures around complication scoring. Moreover, a constant flow of changes will make it difficult to judge causality between one particular change and measured outcomes.

CONCLUSIONS

Collecting clinical and dosimetric data from patients treated with radiotherapy for the purpose of NTCP modeling demands a thorough investigational plan addressing how to obtain a valid database that contains relevant, complete, consistent, and accurate data for a defined cohort of patients, preferably in concordance with applicable (inter)national standards and definitions. Current developments in radiotherapy departments are very promising: assembled healthcare databases can become very powerful and valuable sources of data for the development and validation of NTCP models.

REFERENCES

Aapro M, Astier A, Audisio R, Banks I, Bedossa P, Brain E, Cameron D, Casali P, Chiti A, De Mattos-Arruda L, et al. Identifying critical steps towards improved access to innovation in cancer care: a European CanCer Organisation position paper. *Eur J Cancer.* 2017 Sep;82:193–202.

Aaronson NK, Ahmedzai S, Bergman B, Bullinger M, Cull A, Duez NJ, Filiberti A, Flechtner H, Fleishman SB, de Haes JC, et al. The European organization for research and treatment of cancer QLQ-C30: a quality-of-life instrument for use in international clinical trials in oncology. *J Natl Cancer Inst.* 1993 Mar 3;85(5):365–76.

Bentzen SM, Constine LS, Deasy JO, Eisbruch A, Jackson A, Marks LB, Ten Haken RK, Yorke ED. Quantitative Analyses of Normal Tissue Effects in the Clinic (QUANTEC): an introduction to the scientific issues. *Int J Radiat Oncol Biol Phys.* 2010 Mar 1;76(3 Suppl):S3–9.

Borghede G, Sullivan M. Measurement of quality of life in localized prostatic cancer patients treated with radiotherapy. Development of a prostate cancer-specific module supplementing the EORTC QLQC30. *Qual Life Res.* 1996;5:212–22.

Bottomley A, Pe M, Sloan J, Basch E, Bonnetain F, Calvert M, Campbell A, Cleeland C, Cocks K, Collette L. et al. Analysing data from patient-reported outcome and quality of life endpoints for cancer clinical trials: a start in setting international standards. *Lancet Oncol.* 2016 Nov;17(11):e510–4.Bradford Hill, A. 1965. The Environment and Disease: Association or Causation? *Proceedings of the Royal Society of Medicine.* 58 (5): 295–300.

Campbell N, Ali F, Finlay AY, Salek SS. Equivalence of electronic and paper-based patient-reported outcome measures. *Qual Life Res.* 2015;24(8):1949–61.

Chen Y, Trotti A, Coleman CN, Machtay M, Mirimanoff RO, Hay J, O'brien PC, El-Gueddari B, Salvajoli JV, Jeremic B. Adverse event reporting and developments in radiation biology after normal tissue injury: international atomic energy agency consultation. *Int J Radiat Oncol Biol Phys.* 2006;64(5):1442–51.

Cox J, Stetz J, Pajak T. Toxicity criteria of the radiation therapy oncology group (RTOG) and the European organization for research and treatment of cancer (EORTC). 1995.

Deasy JO, Bentzen SM, Jackson A, Ten Haken RK, Yorke ED, Constine LS, Sharma A, Marks LB. Improving normal tissue complication probability models: the need to adopt a "data-pooling" culture. *Int J Radiat Oncol Biol Phys.* 2010 Mar 1;76(3 Suppl):S151–4.

Defraene G, Van den Bergh L, Al-Mamgani A, Haustermans K, Heemsbergen W, Van den Heuvel F, Lebesque JV. The benefits of including clinical factors in rectal normal tissue complication probability modeling after radiotherapy for prostate cancer. *Int J Radiat Oncol Biol Phys.* 2012;82(3):1233–42.

Dueck AC, Mendoza TR, Mitchell SA, Reeve BB, Castro KM, Rogak LJ, Atkinson TM, Bennett AV, Denicoff AM, O'Mara AM. et al. Validity and reliability of the US National Cancer Institute's Patient-Reported Outcomes Version of the Common Terminology Criteria for Adverse Events (PRO-CTCAE). *JAMA Oncol.* 2015;1(8):1051–9.

Gebre-Medhin M, Haghanegi M, Robért L, Kjellén E, Nilsson P. Dose-volume analysis of radiation-induced trismus in head and neck cancer patients. *Acta Oncol.* 2016;55(11):1313–7.

Grantzau T, Thomsen MS, Væth M, Overgaard J. Risk of second primary lung cancer in women after radiotherapy for breast cancer. *Radiother Oncol.* 2014;111(3):366–73.

Herdman M, Gudex C, Lloyd A, Janssen M, Kind P, Parkin D, Bonsel G, Badia X. Development and preliminary testing of the new five-level version of EQ-5D (EQ-5D-5L). *Qual Life Res.* 2011;20(10):1727–36.

Hill AB. The environment and disease: association or causation. *Proc R Soc Med.* 1965;58:293–300.

Jackson A, Marks LB, Bentzen SM, Eisbruch A, Yorke ED, Ten Haken RK, Constine LS, Deasy JO. The lessons of QUANTEC: recommendations for reporting and gathering data on dose-volume dependencies of treatment outcome. *Int J Radiat Oncol Biol Phys.* 2010;76(3 Suppl):S155–60.

Lustberg T, van Soest J, Jochems A, Deist T, van Wijk Y, Walsh S, Lambin P, Dekker A. Big Data in radiation therapy: challenges and opportunities. *Br J Radiol.* 2017;90(1069):20160689.

Marks LB, Yorke ED, Jackson A, Ten Haken RK, Constine LS, Eisbruch A, Bentzen SM, Nam J, Deasy JO. Use of normal tissue complication probability models in the clinic. *Int J Radiat Oncol Biol Phys.* 2010;76(3 Suppl):S10–9.

Miah AB, Gulliford SL, Clark CH, Bhide SA, Zaidi SH, Newbold KL, Harrington KJ, Nutting CM. Dose-response analysis of parotid gland function: what is the best measure of xerostomia? *Radiother Oncol.* 2013 Mar;106(3):341–5.

Niedzielski JS, Yang J, Stingo F, Martel MK, Mohan R, Gomez DR, Briere TM, Liao Z, Court LE. Objectively quantifying radiation esophagitis with novel computed tomography-based metrics. *Int J Radiat Oncol Biol Phys.* 2016 Feb 1;94(2):385–93.

Nijkamp J, Rossi M, Lebesque J, Belderbos J, van den Heuvel M, Kwint M, Uyterlinde W, Vogel W, Sonke JJ. Relating acute esophagitis to radiotherapy dose using FDG-PET in concurrent chemo-radiotherapy for locally advanced non-small cell lung cancer. *Radiother Oncol.* 2013;106(1):118–23.

Pavy JJ, Denekamp J, Letschert J, Littbrand B, Mornex F, Bernier J, Gonzales-Gonzales D, Horiot JC, Bolla M, Bartelink H. EORTC Late Effects Working Group. Late Effects toxicity scoring: the SOMA scale. *Int J Radiat Oncol Biol Phys.* 1995 Mar 30;31(5):1043–7.

Schaake W, van der Schaaf A, van Dijk LV, Bongaerts AH, van den Bergh AC, Langendijk JA. Normal tissue complication probability (NTCP) models for late rectal bleeding, stool frequency and fecal incontinence after radiotherapy in prostate cancer patients. *Radiother Oncol.* 2016;119(3):381–7.

Sivendran S, Latif A, McBride RB, Stensland KD, Wisnivesky J, Haines L, Oh WK, Galsky MD. Adverse event reporting in cancer clinical trial publications. *Clin Oncol.* 2014;32(2):83–9.

Trip AK, Nijkamp J, van Tinteren H, Cats A, Boot H, Jansen EP, Verheij M. IMRT limits nephrotoxicity after chemoradiotherapy for gastric cancer. *Radiother Oncol.* 2014;112(2):289–94.

Trotti A, Colevas AD, Setser A, Rusch V, Jaques D, Budach V, Langer C, Murphy B, Cumberlin R, Coleman CN, et al. CTCAE v3.0: development of a comprehensive grading system for the adverse effects of cancer treatment. *Semin Radiat Oncol.* 2003;13(3):176–81. Review.

van Nimwegen FA, Schaapveld M, Cutter DJ, Janus CP, Krol AD, Hauptmann M, Kooijman Kl, Roesink J, van der Maazen R, Darby SC, et al. Radiation dose-response relationship for risk of coronary heart disease in survivors of hodgkin lymphoma. *J Clin Oncol.* 2016;34(3):235–43.

CHAPTER 2

Building a Predictive Model of Toxicity

Methods

Sunan Cui, Randall K. Ten Haken, and Issam El Naqa

CONTENTS

Introduction 24
Definition of Radiation-Induced Toxicities 25
Analytical Models 25
    Mechanistic Models 26
    Phenomenological Models 27
Data-Driven Models 29
    Regression Models 30
        Logistic Regression 30
        Ridge and Lasso Regression 31
    Machine Learning Models 31
        Support Vector Machine (SVM) 31
        Multi-Layer Neural Network 32
        Bayesian Networks 34
Variable Selection in Data-Driven Approaches 35
    Why Feature Selection? 35
        Filtering Methods 36
        Wrapper Methods 37
        Embedded Method 37
Model Evaluation 38
    Bias Variance and Model Complexity 38
    Metrics of Prediction Performance 39
        Confusion Matrix-Based Metrics 39
        Metrics Based on Receiver Operating Characteristic Curve (ROC) 40
    Trade-Off Between Goodness of Fit and Complexity 41
    Prediction Error Estimation by Resampling 41

NTCP Modeling Examples 44
    NTCP Modeling in Prostate Cancer Using Traditional Methods 44
    NTCP Modeling in Lung Cancer Using Machine Learning 44
Discussion and Conclusions 47
References 49

INTRODUCTION

The goal of radiotherapy is to deliver high doses of ionizing radiation to eradicate tumour cells, while at the same time minimizing the risks of damaging surrounding normal tissue (Halperin et al. 2008). Radiotherapy outcomes are usually characterized by two indices: *tumour control probability* (TCP) (Zaider and Hanin 2011), which is the probability of the extinction of clonogenic tumour cells after radiotherapy, and *normal tissue complication probability* (NTCP), which is the probability of healthy normal tissue injury (Steel 2002). One of the key components of modern radiation oncology research is to predict treatment outcomes (e.g., TCP and NTCP) during treatment planning or during a fractionated course of therapy to personalize prescription and optimize response. Outcome models can also inform clinicians when weighing different treatment options with their patients or guiding/adapting radiotherapy fractionation subject to patient-specific variables.

Radiation-induced toxicities remain the major limiting factor for promising dose escalation/intensification studies such as intracranial stereotactic radiosurgery or extracranial stereotactic body radiation therapy. Hence, prediction of radiation-induced toxicities is a crucial step for successful treatment and improved patient outcomes. Although, it has been widely recognized that toxicities are determined by complex interactions among patient-specific anatomic, biological and treatment variables, traditionally, NTCPs are modeled with information derived only from dose distributions (Bentzen et al. 2010), e.g., dose-volume parameters (Blanco et al. 2005; Bradley et al. 2004; Jackson et al; Marks et al. 2010). More recently, biological variables have been suggested to play an important role in the prediction of radiation-induced toxicities. For instance, a number of studies have reported associations between certain genetic variants (genomics) and NTCP after radiotherapy (Alsner 2008; West et al. 2007). Also, features extracted from imaging (radiomics) (Deasy and El Naqa 2008), which can provide patient-specific anatomic and metabolic information (El Naqa et al. 2009) have also been suggested as prognostic factors to aid in the prediction of outcomes. With the availability of clinical, dosimetric and biological factors, data-driven models utilizing advanced statistical learning and informatics techniques have recently taken centre stage in NTCP modeling.

In this chapter, an overview of clinical toxicity endpoints is first provided. Then, NTCP models in radiotherapy, including analytical methods and more recent data-driven approaches, are presented. This is followed by a section on feature and model selection in NTCP modeling. A discussion of current issues and highlighting of future perspectives in NTCP modeling completes the chapter.

DEFINITION OF RADIATION-INDUCED TOXICITIES

Radiation-induced toxicity can be categorized according to its onset time into early and late effects. *Early effects* can occur during or a few days to weeks after irradiations, typically in the rapidly proliferating tissues. These effects include skin erythema, mucositis, esophagitis, diarrhoea and immunosuppression. *Late effects* typically occur months to years after treatment, usually in slowly or non-proliferating tissues. Common late effects are lung fibrosis, kidney damage, heart disease, liver disease, spinal cord injury and proctitis (Halperin et al. 2008).

Radiation-induced toxicities are usually categorized using clinical standards such as RTOG (the Radiation Therapy Oncology Group), LENT-SOMA (late effects of normal tissue-SOMA) or the National Cancer Institute CTCAE (common terminology criteria for adverse events) (Jackson et al. 2010). Some common side effects metrics like patient symptoms (e.g., shortness of breath), formal clinical/functional assessments (e.g., quality of life tools) and laboratory tests (e.g., pulmonary function tests (PFTs)) are usually considered in these standards for evaluating toxicities.

RTOG and CTCAE are both comprehensive dictionaries for the grading of toxicities for all major cancer therapies. They combine multiple signs or symptoms into a single score. RTOG grades acute effects and late effects separately, while CTCAE does not. LENT-SOMA has separated components for each endpoint. Four domains including subjective findings (e.g., symptoms like shortness of breath), objective findings (e.g., changes in respiratory rate), management interventions and analytic data are independently quantified, and then incorporated into a global toxicity score.

More recently patient-reported outcomes (PROs) (Sloan et al. 2016) are being proposed for grading toxicities, to serve as more pertinent indicators of a patient's quality of life. PRO domains are generally based on validated questionnaires and can include functional status, symptoms (intensity, frequency), satisfaction (with medication), multiple domains of well-being and global satisfaction with life. Incorporating PROs into clinical practice may be anticipated to lead to better decision guidance.

ANALYTICAL MODELS

Toxicity of tissue may relate to dose and fractionation, tissue structure and architecture and DNA repair ability. In radiobiology, organs may be thought of as being arranged in serial, parallel or some combination of functional subunits (FSUs). In this context, they might be classified as parallel or series organs based on their arrangements. Parallel organs would exhibit a functional reserve and can still survive with damage to some fraction of its subunits; thus small volumes may tolerate very high doses. However, serial organs may fail due to damage of a single subunit, thus maximum doses may determine their radiosensitivity. DNA repair mechanisms (e.g., homologous recombination or non-homologous endjoining) are tightly related to the radiosensitivity of the cells and thus can affect the toxicity of normal tissue and the organ-at-risk as well (Hall and Giaccia 2012).

Dose-response analytical models for NTCP are roughly categorized as mechanistic or phenomenological. The former approach mathematically formulates toxicity based on simplified biophysical understanding of radiation effects on cells primarily from *in vitro* cell culture experiments. The latter one attempts to fit the available dosimetric data to an empirical and parametric model.

Mechanistic Models

Mechanistic models are based on the understanding of underlying biological processes. The most prevalent example is the linear quadratic (LQ) model attributing cell killing to DNA damage in the nucleus (Figure 2.1).

In LQ models (Hall and Giaccia 2012), the logarithm of survival fraction (SF) is a combination of a linear term and a quadratic term of physical dose. Parameters α and β are related to radiosensitivity and their values can emphasize the difference between early- and late-responding tissues. Moreover, the model can be practicably extended to applications in fractionated radiotherapy (Joiner et al. 1993).

For a single acute dose D, the biologic effect E is given by:

$$E = \alpha D + \beta D^2 \tag{2.1}$$

For n fractions (well separated in time) of dose d, we have:

$$E = n\left(\alpha d + \beta d^2\right) \tag{2.2}$$

A quantity called the *biological effective dose* (BED) is defined for very large number of fractions with very small doses and is used to simplify the conversion between different radiation fractionation regimens:

$$\mathrm{BED} = \frac{E}{\alpha} = nd \times \left(1 + \frac{d}{\alpha/\beta}\right) \tag{2.3}$$

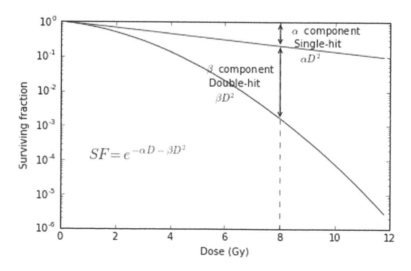

FIGURE 2.1   Logarithm of survival fraction in Linear Quadratic (LQ) models.

To convert BED back to a physical quantity, an equivalent dose at some standard fractionation is used (e.g., EQD2 for 2 Gy fraction):

$$EQD2 = \frac{BED}{1 + \dfrac{2}{\alpha/\beta}} \qquad (2.4)$$

Phenomenological Models

From *in-vitro* cell survival experiments exhibiting exponential dependence of survival with dose, a sigmoid-shaped relationship between dose and NTCP can be directly derived (Willers and Held 2006). The Lyman model (Lyman 1985) is the most commonly used NTCP model, where the cumulative distribution function of a Gaussian distribution (a Probit function) is chosen to represent the empirical sigmoid dependence of NTCP on dose. In this model, two parameters D_{50} and m can be adjusted to change the position and shape of NTCP curve respectively.

$$\text{NTCP}(D, D_{50}, m) = \frac{1}{\sqrt{2\pi}} \int_{-\infty}^{t} \exp\left(-\frac{u^2}{2}\right) du \qquad (2.5)$$

Where $t = \dfrac{D - D_{50}}{m D_{50}}$,

D_{50} is the dose related to 50% toxicity probability, if the whole organ is irradiated uniformly and m is a parameter to control the slope of the curve. D_{50} can also be expressed as function of partial organ uniform irradiation, using a third parameter n which describes the magnitude of volume effect:

$$D_{50}(V) = D_{50}(1)/V^n, D_{50}(1) \text{ is } D_{50} \text{ for the whole volume,} \qquad (2.6)$$

Small values of n correspond to serial volume effects, while large ones correspond to parallel volume effects. When the organ is irradiated with inhomogeneous dose distribution described by a dose-volume histogram (DVH), the Kutcher–Burman DVH reduction to an effective partial volume (V_{EFF}) irradiated at a reference dose (often the maximum or isocentre dose) is commonly used (Kutcher and Burman 1989), which is known as the Lyman-Kutcher-Burman (LKB) model:

$$V_{EFF} = \sum_i v_i \left(\frac{D_i}{D_{max}}\right)^{\frac{1}{n}} \qquad (2.7)$$

where:

$\{(v_i, D_i), i \text{ from 0 Gy to maximum dose}\}$ are the points of the direct DVH

D_{max} is maximum dose to the organ

n is the previously cited volume effect parameter

A variation of LKB could be presented by replacing physical dose metric with the equivalent uniform dose (EUD), discussed below. Of note EUD and V_{EFF} are related by

$$EUD = V_{EFF}^{n} \cdot D_{max} \tag{2.8}$$

The parameters of LKB model are typically estimated by using maximum likelihood estimation (MLE) methods.

The Critical Volume model (Niemierko and Goitein 1993) is another common NTCP model. It assumes that organs are composed of functional subunits which are arranged in parallel or serial architectures.

$$NTCP(\bar{\mu}_d, \mu_{cr}, m) = \frac{1}{\sqrt{2\pi}} \int_{-\infty}^{t} \exp\left(-\frac{u^2}{2}\right) du \tag{2.9}$$

$$\text{where } t = \frac{-\ln(-\ln\bar{\mu}_d) - \ln(-\ln\mu_{cr})}{\sigma}$$

$\bar{\mu}_d$ is the mean damage volume, which is the product of relative irradiated volume and the probability of failure of FSUs. μ_{cr} is the relative critical volume and σ accounts for variability among patients.

An equivalent uniform dose (EUD) (Niemierko 1997) (in Eq. 8) can be applied to each toxicity model (Seppenwoolde and Lebesque 2001). It is basically a single uniform dose for whole organ irradiation that might be expected to have the same outcome as the non-uniform dose distribution actually observed. In a LKB-type formalism, a generalized EUD (gEUD) for different tissues can be computed given tissue-specific parameters.

$$gEUD = \left(\sum_i v_i D_i^a\right)^{1/a} \tag{2.10}$$

where:
  v_i is the fractional volume receiving dose D_i
  a is the tissue-specific parameter that describes the volume effect (=1/n in LKB)

Large negative 'a' makes gEUD correlate with low-dose region of DVH, which is appropriated for response of tumour. Positive 'a' can be thought of as the volume effect parameter for normal tissues. Large positive 'a' correlates gEUD with the high-dose region of DVH, which is appropriate for tissue which can be damaged by even a small volume of high dose (i.e., serial organ). When 'a' approaches 1, gEUD is close to the mean dose value (i.e., parallel organ).

Dose modifying factors (DMF) (Coates et al. 2015; Tucker et al. 2013; Defraene et al. 2012) incorporate variables other than dose into LKB models.

$$\text{NTCP}(D, \text{DMFs}, D_{50}, m) = \frac{1}{\sqrt{2\pi}} \int_{-\infty}^{t} \exp\left(-\frac{u^2}{2}\right) du \qquad (2.11)$$

$$t = \frac{D_{\text{eff}} \text{DMF}_1 \ldots \text{DMF}_k - D_{50}}{m D_{50}}, \qquad (2.12)$$

where the DMFs reflect the impact of covariates other than dose (e.g., single-nucleotide polymorphism [SNPs] genotype, copy number variations [CNVs], smoking status, etc.). Alternatively, one can write an equivalent form where the DMFs are reciprocals of those above:

$$t = \frac{D_{\text{eff}} - D_{50} \text{DMF}_1 \ldots \text{DMF}_k}{m D_{50} \text{DMF}_1 \ldots \text{DMF}_k} \qquad (2.13)$$

in which DMF is defined as follows,

$$\text{DMF}_{\text{SNP}} = e^{\delta_{\text{SNP}} R_{\text{SNP}}} \qquad (2.14)$$

$$\text{DMF}_{\text{CNV}} = e^{\delta_{\text{CVN}} R_{\text{CNV}}} \qquad (2.15)$$

$$\text{DMF}_{\text{smoke}} = e^{\delta_{\text{smoke}} R_{\text{smoke}}} \qquad (2.16)$$

R_z could be binary variables representing, for example, allele status of SNPs, dichotomous CNV or smoking status. δ_z are risk-factor weights to be identified using a generalized version of the MLE approach.

In principle, analytics models can appear to be scientifically sound. However, considering the complexity involved in radiotherapy response, mathematically formulating such a process may not be as accurate or complete as intended.

DATA-DRIVEN MODELS

As radiotherapy outcomes are determined by complex interactions among patient-specific anatomic and biological and treatment conditions, analytical models which consider limited numbers of variables and sometimes depend on tuning parameters by hand, may not be able to provide a complete or accurate prediction of NTCP. Recently, data-driven models using statistical and machine learning methods, which allow the incorporation of more information and 'cross-talk' among the variables into the model building process have gained popularity in NTCP modeling.

In the context of machine learning, the prediction of NTCP can be viewed as a supervised learning problem. The goal of supervised learning is to infer a function that maps inputs to outputs from a labeled training dataset. In supervised learning, each example is a pair consisting of input object (feature) and output object (label). The learned function can be applied to predict the label of some unseen (out-of-sample) data in the future. In the case of NTCP modeling, inputs are variables related to toxicity; the output is NTCP.

Mathematically, this can be formulated as follows: $f(x,w):x->Y$, where $x \in d^n$, is an input variable vector of N-dimension, composed of the inputs metrics such as dosimetric, clinical and biological variables. Label Y is a scalar representing NTCP and w denotes the parameters to be optimized in the model. The optimal parameters w^* of model $f(\cdot)$ are obtained by optimizing a certain objective function given observed training data. For instance, one can estimate optimal w^* by minimizing the least-squared differences between the model predictions and observed outcomes (least-square approach) or maximizing likelihood function that gave rise to the observed (labeled) data (MLE approach).

Although, NTCP modeling is largely a supervised learning problem, some unsupervised learning approaches such as principal component analysis (PCA) can be applied before supervised learning to improve performance and increase robustness. Unsupervised learning is a type of learning that infers a function to describe hidden structure from "unlabelled data." It can be adapted to reduce dimensionality and extract composite features from the original variables.

Regression Models
Logistic Regression
Logistic regression is a specific type of generalized linear models (GLMs) (McCullagh and Nelder 2000), with binomial random component and logit link function.

Compared to other GLMs, logistic regression is a more appropriate model for NTCP as radiation outcomes have been observed to follow an S-shaped (sigmoidal) curve. In logistic regression:

$$f(x_i) = \text{sigmoid}(g(x_i)) = \frac{e^{g(x_i)}}{1+e^{g(x_i)}} \qquad (2.17)$$

where:

n is the number of samples (patients)

x_i is the input vector for ith patient

$g(\cdot)$ is a weighted sum of entries in the input vector, which can be written as:

$$g(x_i) = \beta_0 + \sum_{j=1}^{s} \beta_j x_{ij}, \qquad i = 1,\ldots,n, \; j = 1,\ldots,s \qquad (2.18)$$

where s is the dimension of the input variables vector. $\beta = (\beta_0, \beta_1, \ldots \beta_s)$ are parameters to be optimized by minimizing a cross-entropy loss function, which is defined by taking the negative logarithm of the likelihood:

$$E(w) = -\ln p(f \mid \beta) = -\ln \prod_{i=1}^{n} f_i^{y_i} (1-f_i)^{1-y_i}$$

$$= -\sum_{i=1}^{n} \left[y_i \ln f_i + (1-y_i)\ln(1-f_i) \right], \qquad (2.19)$$

where y_i is the true label (binary) for patient i. This MLE optimisation problem is usually solved by an iterative reweighted least squares (IRLS) algorithm.

Sometimes, to account for the interaction effects between variables (cross-talks), extra terms can be added to expression of $g(\cdot)$:

$$g(x_i) = \beta_0 + \sum_{j=1}^{s} \beta_j x_{ij} + \sum_{m=1}^{s}\sum_{n=1}^{s} \gamma_{mn} x_{im} x_{in} \qquad (2.20)$$

Ridge and Lasso Regression

Ridge and Lasso (Tibshirani 1996) regressions are two methods that can introduce regularization into the regression analysis. They enhance prediction accuracy and interpretability of a model by introducing an extra penalty term of complexity in the objective function, which allows for 'shrinking' of redundant variables in the model and hence reduces variable selection bias.

Ridge regression adds a L_2 penalty:

$$E(w) = -\ln p(f \mid \beta) + \lambda \sum_{j=0}^{s} \beta_j^2 \qquad (2.21)$$

Lasso regression adds a L_1 penalty:

$$E(w) = -\ln p(f \mid \beta) + \lambda \sum_{j=0}^{s} |\beta_j| \qquad (2.22)$$

where λ is a fixed regularization parameter during optimisation of β and can be determined by grid search with cross-validation. Large values of λ help avoid overfitting, but may lead to over-simplified models; small λ means less regularization, but may lead to large prediction variance (overfitting). These two approaches can be combined into one formalism as in the Elastic net model (Zou and Hastie 2005) frequently used in the genetic analysis literature.

Machine Learning Models

Support Vector Machine (SVM)

A SVM (Steinwart and Christmann 2008) is a classifier formally defined by a separating hyperplane, which can categorize labeled data. In practice, as it is usually not feasible to completely separate samples from different classes, some tolerance errors ξ are allowed.

Optimisation of SVM can be formulated by minimizing a loss function as follows,

$$L(w, \xi) = \frac{1}{2} w^T w + C \sum_{i=1}^{n} \xi_i \qquad (2.23)$$

with constraints,

$$y_i\left(w^T\phi(x_i)+b\right)\geq 1-\xi_i$$

$$\zeta_i \geq 0, \forall i \qquad (2.24)$$

This would result in a classifier of the form:

$$f(x)=w^T\phi(x)+b, \qquad (2.25)$$

to be trained by an optimisation problem with a hinge loss,

$$\min_{w\in\mathbb{R}^d} \|w\|^2 + C\sum_i^N \max\left(0, 1-y_i f(x_i)\right) \qquad (2.26)$$

In the loss function, the first term is correlated with the size of margins between two classes, the second term is an error-tolerance term. Parameter C is for regularization purposes and is responsible for trade-offs between these two terms. $\phi(\cdot)$ is a non-liner mapping function, which maps variables from an original space to a higher dimension space for a better separation.

Weights, w, for a SVM are usually optimized by solving an equivalent dual optimisation problem. In the dual problem, the SVM classifier could be represented by:

$$f(x)=\sum_i^s \alpha_i y_i K(x, x_i)+b, \qquad (2.27)$$

which can be learned by solving a convex optimisation problem over $\alpha=(\alpha_1,...,a_s)$. The number s is for so-called support vectors, which is a subset of training data that is on the border of or outside the error margin as being circled in Figure 2.2. K is the kernel function, which corresponds to an inner product in a feature (Hilbert) space based on the mapping function $\phi(\cdot)$. Some common kernel functions are: linear kernel $K(x,x')=x^T x'$, polynomial kernel $K(x,x')=\left(x^T x'+c\right)^q$ and radial basis function kernel $K(x,x')=\exp\left(-\frac{1}{2\sigma^2}\|x-x'\|^2\right)$.

In polynomial SVMs, there are three parameters (c, C, q) to be adjusted, while in RBF-SVMs, there are two parameters (C, σ) to be adjusted. Again, optimal parameter settings are typically determined by an exhaustive grid search and cross-validation as in the case of λ for logistic regression.

Multi-Layer Neural Network

Multi-layer neural networks or multi-layer perceptrons (MLPs) are a type of artificial neural networks which are feed-forward and fully-connected. These methods have witnessed

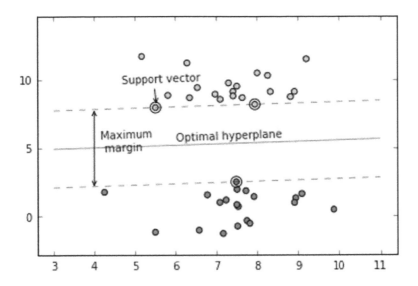

FIGURE 2.2   Support vector machine finds optimal hyperplane separating two classes.

revived interest in recent years with the advent of deep learning methods and their popularity particularly in computer vision applications. An MLP consists of several layers and neurons, where every neuron in the following layer is connected to all the neurons in its former layer. The connection is unidirectional and no circles exist in the network architecture (Figure 2.3).

The value of a neuron in the hidden layer and output layer is calculated by taking a weighted sum of all the neurons in its former layer followed by a non-linear activation function as shown in Figure 2.4 and is given by:

$$a = g\left(\sum_i x_i w_i + b \right) \tag{2.28}$$

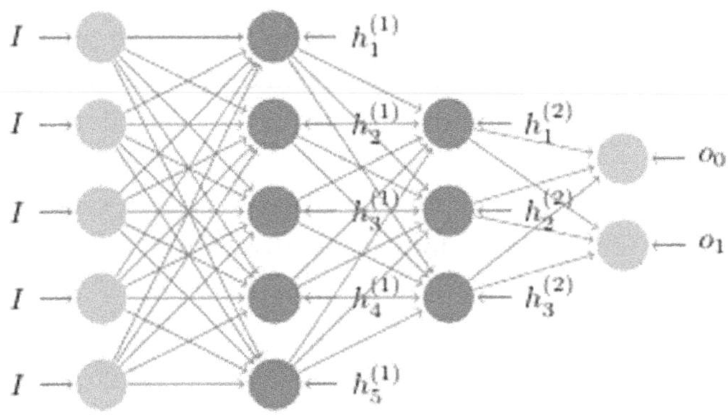

FIGURE 2.3   Diagram of multi-layer neural network.

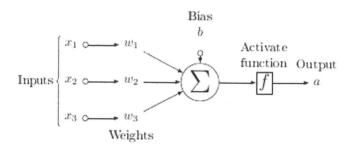

FIGURE 2.4 Activation function in multilayer perceptron (MLP).

Some common activation functions are, sigmoid $g(t) = \dfrac{1}{1+e^{-t}}$, ReLU $g(t) = \max(0, t)$ and softmax $g_i(t) = \dfrac{e^{t_i}}{\displaystyle\sum_j e^{t_j}}$. The softmax activation function scales the value of a neuron by a sum of all the neurons in the same layer. It guarantees the sum to be 1, which is appropriate to be implemented in the output layer for solving classification problems.

A MLP can also be seen as a function $f(\cdot)$ mapping inputs to outputs. Its objective loss function is usually cross-entropy for classification problems. Below is an example of the cross-entropy in a 2-class problem.

$$E(w) = -\sum_{i=1}^{n}\left[y_i \ln f_i + (1 - y_i)\ln(1 - f_i)\right] \tag{2.29}$$

There are two phases in the training of MLP. One is forward-propagation, which, starting from inputs, obtains the values of all the neurons in hidden layers and in the output layer. The other one is backward propagation which propagates the error signal, computed at the output, all the way back to the inputs, updating weights value.

The basic training algorithm of MLP is stochastic gradients descent (SGD). However, as there are numerous local minima, advanced SGD algorithms such as Adagrad (Duchi et al. 2011), RMSprop (Dauphin et al. 2015), and Adam (Kingma and Ba 2014), which consider momentum and adaptive steps to avoid such pitfalls, yield better prediction performance.

Sometimes, to avoid overfitting, the dropout technique (Srivastava et al. 2014) is applied in MLP. Dropout is a technique that randomly turns off a certain ratio of neurons during forward propagation such that they do not contribute to the activation of downstream neurons and so their weights get temporarily disconnected in subsequent updates. This process has been shown to make MLP more generalizable and mitigates overfitting issues.

Bayesian Networks

A Bayesian network (BN) (Neapolitan 2003) is a directed acyclic graph (DAG) whose nodes represent a set of random variables and edges represent conditional dependencies between nodes. It can be applied to explore interaction between NTCP and patients-specific variables in a transparent and interpretable manner compared to other machine learning methods (Figure 2.5).

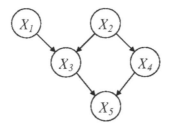

FIGURE 2.5 Diagram of a Bayesian network.

In a BN, if there is an edge from node x_i to node x_j, a value taken by x_j is affected by a value taken by x_i. x_i is then referred to as a parent of x_j, likewise x_j is referred to as child of x_i. Concepts of "parent" and "child" can be extended to "descendants" and "ancestors," which only require the presence of a direct path between two nodes. In BN, each variable is independent of its non-descendants given its parents.

Mathematically, a BN can be defined as an annotated acyclic graph that represents the joint probability distribution over a set of random variables $X = (x_1,...,x_s)$, denoted as $B = \langle G, \Theta \rangle$, where graph G encodes random variables and dependence assumptions and Θ denotes the set of parameters describing how the nodes depend on their parents. The joint probability is given by:

$$P_B(x_1,...,x_s) = \prod_{i=1}^{s} Q(x_i \mid \pi_i, \theta_i)$$

(2.30)

where:
   π_i is the set of parents of node x_i
   θ_i is a vector of parameters in the conditional probability Q

Given a BN that is specified in its full probability form, in principle, one can infer the conditional distribution of any variable (e.g., NTCP in our case) or combination of variables by joint distribution marginalization. However, due to the complexity of such distributions, exact inference is generally not feasible. Therefore, Markov chain Monte Carlo (MCMC) methods are usually adapted for approximate inference.

BN "learning" consists of two components, structure learning, which learns the graph topology, and parameter learning, which estimates the parameters Θ of the conditional probabilities of the BN. Generally speaking, structure learning is a harder problem than parameter learning. Recovery algorithms and optimisation-based algorithms can be applied to learn the BN structure. For parameter learning, approximate MLE methods such as the expectation-maximization algorithm are the most commonly used approach.

VARIABLE SELECTION IN DATA-DRIVEN APPROACHES

Why Feature Selection?

In the current domain of NTCP modeling, hundreds of variables are available to be explored, but with a limited sample size, which can impose a big challenge for prediction modeling (under-powered analysis). Under these circumstances, feature selection, which constructs and selects subsets of features, can be crucial for improving predictive performance.

A complex model can easily lead to overfitting when the sample size is limited. Therefore, it is better to do dimensionality reduction and resort to a simpler model. Feature selection is one of several dimension reduction techniques. There are basically three categories of feature selection methods: filter methods, wrapper methods and embedded methods. It should be noted, one should adopt feature selection within appropriate model evaluation schemes (as discussed in the following section) to avoid selection bias.

Filtering Methods

Filtering methods (Sánchez-Maroño et al. 2007) usually select features according to univariate measures among the variables irrespective of the learning algorithm.

Distance or separation methods are applied to classification problems. They measure class separability to decide the importance of features. For example, in the Fisher linear discriminant (FLD) method, the score of feature i is given by:

$$S_i = \frac{\sum n_j \left(\mu_{ij} - \mu_i\right)^2}{\sum n_j \rho_{ij}^2} \tag{2.31}$$

where:

μ_{ij} and ρ_{ij} are the mean and the variance of the ith feature in the jth class
μ_i is the mean of the ith feature over the whole sample
n_j is the sample size of the jth class

That is, each score is given by the ratio of the variance between the classes to variance within the classes. Higher quality features have high scores due to their high variances between the classes.

Correlation-based methods compare feature candidates to response to select features that have relatively large correlation coefficients. One of the simplest methods is Pearson correlation, which measures the linear correlation between an individual feature and a response variable. There are also non-linear correlation coefficient methods such as Spearman rank correlation assessing monotonic relationship and Kendall's tau coefficient measuring ordinal associations.

Information gain-based methods are entropy-based feature evaluation methods. A common measure is mutual information (MI) given by:

$$MI(X,Y) = \sum_{y \in Y} \sum_{x \in X} p(x,y) \log \frac{p(x,y)}{p(x)p(y)} \tag{2.32}$$

MI can assess mutual dependence between features (**x**) and response (**y**). However, MI is not a metric and is not normalized. For continuous variables, its value is sensitive to bin selection. Maximal information coefficient (MIC) (Reshef et al. 2011) is a technique which can overcome the above shortcomings and is robust in application of feature selection.

Generally speaking, filter methods are univariate, which consider each feature independently. They are usually easy to implement, robust to overfitting and efficient in computation, but they don't consider redundancy among features.

Wrapper Methods

Wrapper methods (Ron Kohavi 1997) select the candidate features using a trained model (not necessarily the model of interest for modeling outcomes), which takes available features as inputs. Based on the inferences drawn from a previous model, one can decide to remove or add a feature from a current subset.

Forward selection and backward elimination both add or remove one feature at each iteration respectively. Forward selection starts with no features in the model, adds one feature at a time that best improves the model until a new feature does not improve the model. Similarly, backward elimination starts with all the features, excludes one feature at a time until no improvement is observed.

Sequential floating selection takes into account the net-effect, which is ignored in naïve forward and backward selection. In sequential floating forward selection, after each forward step, backward steps are performed to evaluate the features already selected, deleting any selected feature when necessary. This can prevent selection of less powerful combinations of features.

In wrapper methods, there is almost no limitation on what learning model is used; common methods like random forest, logistic regression and SVM can all be applied. Wrapper methods generally achieve a better classification rate than filter methods since they are tuned to detect specific interactions between the classifier and the dataset. However, they may not be computationally efficient, and may lack generality as they are tied to the bias of the classifier.

Embedded Method

Embedded methods (Lal et al. 2006) insert feature selection steps into the training of a model.

Lasso and ridge methods are embedded feature selection methods. As we mentioned in the logistic regression section, one can add a regularization term to an objective function. This regularization can be either L_1 in lasso regression or L_2 in ridge regression. For example, in L_1 regularization, the penalty term is $L_1 = \lambda \sum_{i=1}^{s} |\omega_i|$, and since the penalty term grows with the magnitude of the weight parameter, one can induce sparsity in weights by this term, which can be considered as an intrinsic way of selecting a feature. The same trick can be also played in methods other than logistic regression, such as SVM.

Decision trees (Breiman 2001) which choose a variable at each step that best splits the data can also be seen as an embedded method. Some common "best" measurements are Gini impurity and information gain.

Since a decision tree selects a feature at each step, and usually terminates as it transverses the dataset without including all the features in the dataset, it can be intrinsically considered an embedded feature selection approach.

Embedded methods perform feature selection and classification simultaneously. They consider the interactions among features but are more computationally efficient than wrapper methods. However, embedded methods are usually specific to a given learning model.

Besides feature selection, feature extraction methods such as Principal Component Analysis (PCA) (Jolliffe 2002) can be applied for dimension reduction as well. The difference between feature extraction and selection is that the former one involves transforming original features and constructing a new set of features, the latter selects a subset of the original feature set.

PCA finds the k largest eigenvalues and their corresponding eigenvectors of the covariance matrix of data. It can project data from an original space of dimension D to a space of lower dimension of $k(D \geq k)$. The basic idea is that these k largest eigenvector can describe most of the variability (variance) in the data.

Model Evaluation

To make a trained model useful in practice, it is important to evaluate its performance and potential generalizability. Model evaluation guides the choice of a learning method, ultimately making the chosen model more generalizable to independent datasets. In this section, some fundamental concepts in model selection, including bias, variance and model complexity, are first discussed. Then, common metrics for performance assessments and model complexity are introduced. In the end, key methods for performance assessment are described.

Bias Variance and Model Complexity

The generalization performance of a learning method is crucial to its prediction capability on an unseen dataset. A complex model may describe data on which it is trained perfectly, but may not perform well on the independent (out of sample) dataset.

For example, assume one has a response variable Y, and a vector of features X, such that $Y = f(x) + \epsilon$, where we have introduced noise ϵ satisfying $E(\epsilon) = 0$. One could decompose the expected prediction error of a regression fit $\hat{f}(x)$ at an input point $X = x_0$ into three terms (Tibshirani and Friedman 2009).

$$\mathrm{Err}(x_0) = E\left[\left(Y - \hat{f}(x_0)\right)^2 \mid X = x_0\right]$$
$$= \sigma_f^2 + \mathrm{var}\left[\hat{f}(x_0)\right] + \mathrm{Bias}^2\left[\hat{f}(x_0)\right] \tag{2.33}$$

The first term is the variance of intrinsic noise, which is not avoidable in practice. The second term $\mathrm{var}\left[\hat{f}(x_0)\right] = E\left[\hat{f}(x_0) - E\hat{f}(x_0)\right]^2$ is called variance, which is the expected squared deviation of the learned model to its mean. The third term $\mathrm{bias}^2\left[\hat{f}(x_0)\right] = \left[E\hat{f}(x_0) - f(x_0)\right]^2$ is the squared bias describing how much the expectation of the learned model differs from the ground truth. Typically as model $\hat{f}$ become more complex, bias will be lower but variance will be higher (Figure 2.6).

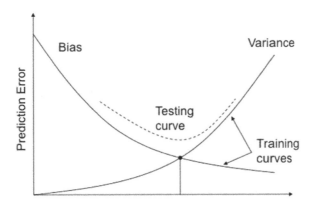

FIGURE 2.6   Trade-off between bias and variance.

Too complex of a model is expected to overfit the data. However, too simple of a model usually underfits the data. Generally speaking, one needs to consider the trade-off between variance and bias to choose the 'right' model.

Metrics of Prediction Performance

In the problem of classification or regression, methods based on a so-called confusion matrix and additional information can be used for performance evaluation.

Confusion Matrix-Based Metrics

A confusion matrix is a table describing the performance of classification. For a binary classification problem, a confusion matrix has four entries.

$$\begin{pmatrix} TN & FP \\ FN & TP \end{pmatrix}$$

True positives (TP): cases are predicted to be yes and are positive in reality
True negatives (TN): cases are predicted to be no and are negative in reality
False positives (FP): cases are predicted to be yes but are negative in reality
False negatives (FN): cases are predicted to be no but positive in reality

Several metrics for performance can be deduced from the confusion matrix. They include, accuracy: $\dfrac{FP+TN}{FP+TN+FP+TP}$, which describes how often the classifier is correct; true positive rate (TPR) $\dfrac{TP}{TP+FN}$, which describes how often the classifier predicts yes when the sample is positive; false positive rate (FPR) $\dfrac{FP}{FP+TN}$, which describes how often the classifier predicts no, when the sample is actually negative. Sensitivity and specificity which contain the same information as TPR and FPR are defined as TPR and 1-FPR respectively.

A pure accuracy criterion is simple, but can cause problems when the dataset is imbalanced, a common situation in NTCP prediction. Training a classifier with an imbalanced dataset leads to a better prediction of the samples in the majority class (so as to get better accuracy overall), but at the expense of less prediction power in the minority class. In this case, the classifier is deemed to be not properly trained and may not be able to give any useful information. That's why one generally takes into account the combination of TPR and FPR or sensitivity and specificity to evaluate a model in the case of an imbalanced dataset.

Metrics Based on Receiver Operating Characteristic Curve (ROC)
Although metrics based on a confusion matrix can reflect the performance of model to some extent, they potentially lose relevant parts of information by requiring a pre-selected fixed threshold for classification determination. Particularly, to calculate the number of FP, TN, FN and TP, one needs to set a fixed discrimination threshold, which is usually 0.5 on a prediction scale 0 to 1, for instance.

A ROC (Metz 2006) is a graphical plot that illustrates the diagnostic ability of a binary classifier with varied threshold. It may also provide some robustness in situations of an imbalanced dataset (Figure 2.7).

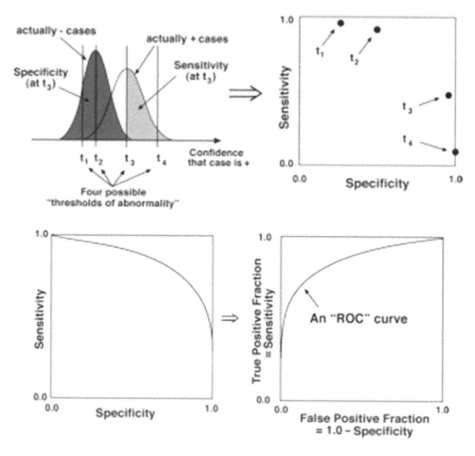

FIGURE 2.7   Plotting of Receiver-operating characteristic (ROC) curve. *Source:* Adapted from E. Metz C, "Receiver Operating Characteristic Analysis: A Tool for the Quantitative Evaluation of Observer Performance and Imaging Systems." *Journal of the American College of Radiology 2006*; 3(6): 413–22.

A ROC curve is created by plotting the TPR against the FPR at various thresholds. Instead of setting a fixed threshold to get a pair of TPR and FPR, as in the case of confusion-matrix-based methods, one sets several thresholds to measure (TPR, FPR) pairs. A scatter plot can then be created based on the values of those pairs. Finally, the points in the plot are joined by a smooth curve, which is referred to as the ROC curve. Note that although many thresholds may yield exactly the same pairs, the duplicates will not affect the ROC curve as they coincide at the same location.

The area under a ROC curve (AUC) is a summary statistic of a ROC. It is equal to the probability that a classifier ranks a randomly chosen positive instance higher than a randomly chosen negative one. AUC has a range of 0.5 (random) to 1 (perfect prediction), the closer AUC is to 1, the better the quality of the classifier is.

Trade-Off Between Goodness of Fit and Complexity

Several methods are available to help determine the optimal model complexity, such as Akaike information criterion (AIC), which is founded on information theory. Suppose that we have a statistical model M of some data x. Let k be the number of estimated parameters. Let $\hat{L}(\hat{\theta}) = P(x|\hat{\theta}, M)$ be the maximum likelihood function of the model, where $\hat{\theta}$ are the optimal parameters for the model. Then

$$\text{AIC} = 2k - 2\ln(\hat{L}) \tag{2.34}$$

AIC rewards the goodness-of-fit of the model, and penalizes increasing the number of parameters in the model. Like AIC, the Bayesian information criterion (BIC) is also applicable in settings where the fitting is carried out by maximization of a log-likelihood. The generic form of BIC is:

$$\text{BIC} = (\ln N)k - 2\ln(\hat{L}) \tag{2.35}$$

It can be seen that the first term in BIC is proportional to that in AIC, with the factor 2 replaced by $\ln N$, where N is the sample size. BIC is founded on Bayesian theory; it is equivalent to choosing the model with the largest (approximate) posterior probability. For a large dataset, BIC tends to choose the right model relative to the sample size while AIC tends to choose models that are too complex. On the other hand, BIC often chooses models that are too simple when sample size is small due to the heavy penalty.

Minimum description length (MDL) and Vapnik–Chervonenkis (VC) dimension are also useful model selection metrics that can guide trade-offs between complexity and goodness-of-fit (Tibshirani and Friedman 2009).

Prediction Error Estimation by Resampling

To make the classifier evaluation meaningful to its application in clinical practice, one shouldn't evaluate the model on the dataset on which it was trained. Instead, resampling is done to evaluate the expected performance of a classifier in unseen dataset. Unless the dataset is large, one can hold out a representative portion of data reserved for testing by randomly sampling or by other criteria that are not susceptible to selection bias.

Cross-validation is the most widely used method for estimating prediction error. In K-fold cross-validation, one splits the data into K roughly equal-sized parts, then for each kth part, one first fits the model with the rest, $K–1$ of the parts, and then evaluates the model on the kth part. Thus, the model is trained and tested for K times and the average error is calculated as:

$$CV\left(\hat{f}\right) = \frac{1}{N}\sum_{i=1}^{N} L\left(y_i, \hat{f}^{-k(i)}\left(x_i\right)\right) \tag{2.36}$$

where $\hat{f}^{-k(i)}$ is the learned classifier without the kth part of the data and L is the designated loss function.

Typically, K is set to be 5 or 10; in the case of $K=N$, the method is known as leave-one-out cross-validation (LOOCV) or Jackknife (Figure 2.8).

Cross-validation can be used for choosing the optimal parameters for a set of models. Given a set of models parameterized by α, the estimation of error can be written as,

$$CV\left(\hat{f},\alpha\right) = \frac{1}{N}\sum_{i=1}^{N} L\left(y_i, \hat{f}^{-k(i)}\left(x_i,\alpha\right)\right) \tag{2.37}$$

The function of $CV\left(\hat{f},\alpha\right)$ provides a test error curve. Thus, a final model can be decided by finding the minimum of the curve.

There is a variant of K-fold cross-validation called stratified (or partitioned) K-fold cross-validation, which takes into account situations of an imbalanced dataset. In the plain K-fold cross-validation, the random division of the data may yield almost no minority data in one subset; and, the performance of the classifier on this subset can be misleading. In stratified K-fold CV, the distribution of classes in each subset is fixed to be the same as that in the whole dataset, which can guarantee a reasonable estimation of error.

Bootstrapping is another way for assessing statistical accuracy. Bootstrapping basically means randomly drawing datasets with replacement from the training data, typically each

FIGURE 2.8 *K*-fold cross-validation.

time sampling the same size as the original dataset. One may train the model on each bootstrap sample and average the errors. Suppose one has B bootstrap samples for a dataset of size N. One can compute the expected error as:

$$\widehat{\text{err}}_{\text{boot}} = \frac{1}{B}\frac{1}{N}\sum_{b=1}^{B}\sum_{i=1}^{N}L\left(y_i, \hat{f}^{*b}(x_i)\right)$$ (2.38)

where $\hat{f}^{*b}$ is the classifier trained with bth bootstrap samples. However, this error is not a good estimation of prediction error, since the overlap between training data and testing data can make the results unrealistically good.

To solve this issue, a better bootstrap is proposed, which is called leave-one-out bootstrap, in this setting, error can be written as:

$$\widehat{\text{Err}}^{(1)} = \frac{1}{N}\sum_{i=1}^{N}\frac{1}{|C^{-i}|}\sum_{b\in C^{-i}}L\left(y_i, \hat{f}^{*b}(x_i)\right)$$ (2.39)

where C^{-i} is the set of indices of the bootstrap samples b that do not contain observation i. The setting prevents using the same observation repeatedly for training and testing. But the error is upward biased due to non-distinct observations in the bootstrap samples resulting from sampling with replacement. To solve this issue, a ".632 estimator" has been proposed, where error is defined as:

$$\text{Err}_{0.632} = 0.368\overline{\text{err}} + 0.632\widehat{\text{Err}}^{(1)},$$ (2.40)

where $\overline{\text{err}} = \frac{1}{N}\sum_{i=1}^{N}L(y_i, \hat{f}(x_i))$,

of note 0.632 is the average ratio of distinct observations in a sample. However, with a highly overfit prediction function (e.g $\overline{\text{err}}=0$), even the .632 estimator will be downward biased. As a result, the .632+ estimator is designed to be a less-bias compromise between the $\overline{\text{err}}$ and $\widehat{\text{Err}}^{(1)}$

$$\text{Err}_{0.632+} = (1-w)\overline{\text{err}} + w\widehat{\text{Err}}^{(1)}$$ (2.41)

Where

$$w = \frac{0.632}{1-0.368R} \text{ and } R = \frac{\widehat{\text{Err}}^{(1)} - \overline{\text{err}}}{\gamma - \overline{\text{err}}}$$

Where γ is no-information error rate, estimated by evaluating the prediction model on all possible combination of x_i and y_j.

$$\gamma = \frac{1}{N^2}\sum_{i=1}^{N}\sum_{j=1}^{N}L\left(y_i, f(x_j)\right)$$ (2.42)

NTCP MODELING EXAMPLES

NTCP Modeling in Prostate Cancer Using Traditional Methods

The study by Coates et al. (2015) demonstrated that adding genetic variables to dosimetric variables can improve predictive performance of NTCP in both analytical and data-driven models. In the study, a dose modifying model integrating genetic risk factors (e.g., single nucleotide polymorphisms [SNPs] and copy number variations [CNVs]) into the LKB model via DMFs and a data-driven logistic regression model were both evaluated for prediction of radiation-induced rectal bleeding (RB) and erectile dysfunction (ED), respectively, in prostate cancer patients.

It was found that cross-validated prediction performance was significantly improved when the CNV of DNA repair gene XRCC1 was included as a DMF in dose modifying models for RB. Similarly, prediction performance was optimized when the SNP of the same gene was included in modeling ED. A comparison of LKB and dose modifying models for RB and ED is shown in Figure 2.9A,B, respectively.

As for a data-driven model, logistic regression was applied with an optimal order estimated by forward selection and resampling by cross-validation. Figure 2.10 shows the best fit RB and ED model alongside octile plots.

As mentioned earlier, ROC plots can be used to visually gauge the trade-off in sensitivity for specificity according to cut-off values. Figure 2.11 shows two examples of ROC plot classification performance in the case of data-driven models for both late RB and ED.

NTCP Modeling in Lung Cancer Using Machine Learning

The study by Lee et al. (2015) developed an ensemble (combination) of Bayesian networks (BNs) for predicting radiation pneumonitis (RP) as shown in Figure 2.12a.

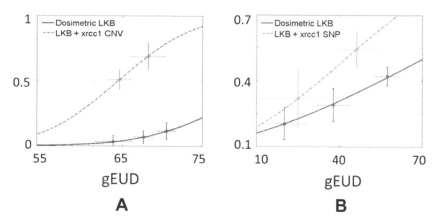

FIGURE 2.9   Lyman-Kutcher-Burman (LKB) models modified to include genetic variables relevant to the outcomes of interest: (A) severe (Grade ≥ 3) late rectal bleeding with two main patient groups stratified by copy number of DNA repair gene XRCC1 and (B) late erectile dysfunction model (Grade ≥ 1) demonstrating increased risk for patients with a polymorphism in the XRCC1 gene. *Source*: adapted from James Coates, "Contrasting analytical and data-driven frameworks for radiogenomic modelling of normal tissue toxicities in prostate cancer," *Radiotherapy and Oncology* 2015; 115(1): 107–13.

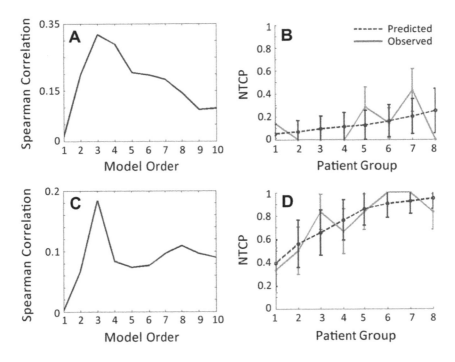

FIGURE 2.10 Model order estimation plots (A,C) and corresponding octile plots (B,D) for late rectal bleeding (RB) model (top row) and late erectile dysfunction (ED) (bottom row). Variables that were included in the forward-search as part of the optimal model order estimation algorithm included dosimetric, clinical as well as genetic variables. *Source*: adapted from James Coates, "Contrasting analytical and data-driven frameworks for radiogenomic modelling of normal tissue toxicities in prostate cancer," *Radiotherapy and Oncology* 2015; 115(1): 107–13.

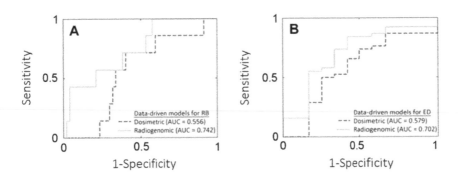

FIGURE 2.11 Receiver-operating characteristic (ROC) plots demonstrating the classification performance of data-driven cross-validated radiogenomic models for (A) rectal bleeding (RB) and (B) late erectile dysfunction (ED) induced by hypofractionated prostate radiotherapy. Note the improvement in all-round classification performance when genetic variables were included in the cross-validated framework. *Source*: adapted from James Coates, "Contrasting analytical and data-driven frameworks for radiogenomic modelling of normal tissue toxicities in prostate cancer," *Radiotherapy and Oncology* 2015; 115(1): 107–13.

54 NSCLC patients who received curative 3D-conformal radiotherapy were used to train a BN. Serum concentration of the four candidate biomarkers were measured at baseline and mid treatment. These four biomarkers were alpha-2-macroglobulin (α2M), angiotensin converting enzyme (ACE), transforming growth factor (TGF-β1) and interleukin-6 (IL-6). Dose-volumetric and clinical parameters were also included as covariates.

Feature selection was performed using a Markov blanket approach based on the Koller-Sahami (KS) filter. The Markov Chain Monte Carlo (MCMC) technique was adapted for estimation of the posterior probability distribution of BN graphs built from the observed selected variables with prior causality constraints. These constraints were based on known knowledge of biophysical interactions among the selected variables as shown in Figure 2.12b. A resampling method based on bootstrapping was applied to training and validation in order to control under- and over-fitting pitfalls. RP prediction power of the BN ensemble approach reached its optimum at a size of 200 BNs. The optimized performance of the BN ensemble model recorded an area under the ROC curve

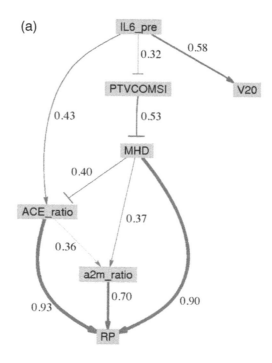

FIGURE 2.12 (a) Variables connected by directed edges with a confidence level higher than random. Edge thickness is proportional to its confidence level. Arrow-headed and bar-headed edges are assigned to positive and negative correlations, respectively. (b) Diagram of allowed causal links between variable categories used for accepting/rejecting graph samples during Markov chain Monte Carlo (MCMC) simulation. (c) Receiver-operating characteristic (ROC) metrics using the Bayesian network model ensemble with varying sizes. Black: prediction with a complete dataset, gray: prediction without intra-treatment biomarker measurements. Error bars: bootstrap-estimated 95% confidence intervals. *Source:* Adapted from "Bayesian network ensemble as a multivariate strategy to predict radiation pneumonitis risk," *Medical Physics* 2015; 42(5): 2421–30.

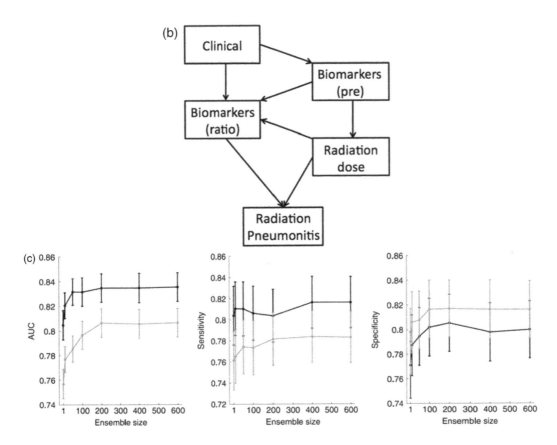

FIGURE 2.12 (CONTINUED)

(AUC) of 0.83 as shown in Figure 2.12c, which was significantly higher than multivariate logistic regression (0.77).

DISCUSSION AND CONCLUSIONS

The definition of clinical endpoints is a crucial step for modeling of radiation-induced toxicities. On one hand, the definition should be objective and quantitative, e.g., based on functional imaging and testing. On the other hand, for the model to be useful, the definition should be also related to clinically relevant outcomes. Many efforts have been made to analyze radiation-induced toxicity endpoints. As mentioned earlier, criteria such as RTOG, LENT-SOMA and CTCAE are commonly used for quantifying toxicities. However, even if the same standard is applied, there still exists some noise in endpoints due to the differences in how physicians apply the toxicity standard in their practice. This noise will cause difficulty in the prediction of accuracy of toxicities.

In the domain of radiation oncology, one usually is faced with limited sample size (e.g., number of patients) compared with other areas (e.g., computer vision) where machine learning is prevalent. Limited sample size restricts utilization of complex predictive models and causes difficulty in incorporating large numbers of available variables into a predictive model. As a result, feature selection, with its caveats, can increase model generalizability, interpretability and computational efficiency and is an essential component of any current successful NTCP modeling exercise.

In an analytical model, the shape of the response curve is pre-defined. It is easy to understand but can lead to poor prediction results. In the data-driven model, there are more degrees of freedom regarding the shape of response, which is adjusted by data and the selected modeling algorithm. The degrees of freedom and the complexity involved in data-driven model enable itself to better describe more complex biophysical interactions among dose and patients-specific variables compared to analytical models.

The complexity varies in different data-driven models. Generally speaking, complex models lead to high predictive accuracy at the cost of sacrificing interpretability. A debate about the preference of interpretability versus predictive accuracy (Breiman 2001) has been ongoing and remains unresolved in the society of statistical learning. In the field of medicine, the importance of interpretability cannot be understated, as it is vital to convincing medical professionals of the action recommended by the predictive model. Also, interpretability of a model may provide insights into the underlying pathophysiological mechanisms of the disease. However, this should not be done at the expense of sacrificing accuracy. In all, model predictive performance and interpretability should be equally emphasized in NTCP modeling to attain its goals.

For evaluation and validation of NTCP models, realization of testing the model on some external (independent) sources can be an issue. In the medical society, data sharing is not easy due to legal, political and administrative barriers. Distributed (rapid) learning (Jochems et al. 2017) can be a solution to this obstacle. Distributed learning is defined as learning the model from multiple hospitals without requiring data leaving these hospitals. In the distributed learning framework, infrastructures are implemented at each hospital to enable learning, a learning portal is responsible for transmitting the model and validating prediction results. Distributed learning can help obtain sufficient patient data for the training of a complex NTCP model, if desired.

In the future, there is the potential for applying deep learning into radiotherapy outcome modeling. Deep learning is a class of machine learning algorithm that learns multiple levels of representations that correspond to different levels of abstraction. It has gained great success in fields like imaging recognition, natural language understanding, and artificial intelligence. Deep learning usually learns from data in their raw form, detecting intricate structures automatically, which avoids cumbersome feature selection procedure. However, deep learning is really data-hungry. Efforts should be made to share patients' data among different institutions. Otherwise, expert domain knowledge (e.g., associations between biomarkers) should be utilized to develop deep architectures with decreased demand for data volume. Another concern is that deep learning models generally lack interpretability. To enable their wider application in clinical systems, meaningful interpretable architectures to bridge the gap between deep learning prediction accuracy and model interpretability are still needed.

REFERENCES

Alsner J, Andreassen CN, Overgaard J. Genetic markers for prediction of normal tissue toxicity after radiotherapy. *Semin Radiat Oncol* 2008; **18**(2): 126–35.

Bentzen SM, Constine LS, Deasy JO, et al. Quantitative Analyses of Normal Tissue Effects in the Clinic (QUANTEC): An introduction to the scientific issues. *International Journal of Radiation Oncology* 2010; **76**(3 Suppl): S3–9.

Blanco AI, Chao KS, El Naqa I, et al. Dose-volume modeling of salivary function in patients with head-and-neck cancer receiving radiotherapy. *International Journal of Radiation Oncology* 2005; **62**(4): 1055–69.

Bradley J, Deasy JO, Bentzen S, El-Naqa I. Dosimetric correlates for acute esophagitis in patients treated with radiotherapy for lung carcinoma. *International Journal of Radiation Oncology* 2004; **58**(4): 1106–13.

Breiman L. Random Forests. *Machine Learning* 2001; **45**(1): 5–32.

Breiman L. Statistical Modeling: The Two Cultures. *Statistical Science* 2001; **16**(3): 199–215.

Coates J, Jeyaseelan AK, Ybarra N, et al. Contrasting analytical and data-driven frameworks for radiogenomic modeling of normal tissue toxicities in prostate cancer. *Radiotherapy and Oncology* 2015; **115**(1): 107–13.

N. Dauphin Y, Vries H, Chung J, Bengio Y. RMSProp and equilibrated adaptive learning rates for non-convex optimisation; 2015.

Deasy JO, El Naqa I. Image-based modeling of normal tissue complication probability for radiation therapy. *Cancer Treat Res* 2008; **139**: 215–56.

Defraene G, Van den Bergh L, Al-Mamgani A, et al. The benefits of including clinical factors in rectal normal tissue complication probability modeling after radiotherapy for prostate cancer. *International Journal of Radiation Oncology* 2012; **82**(3): 1233–42.

Duchi J, Hazan E, Singer Y. Adaptive subgradient methods for online learning and stochastic optimisation. *Journal of Machine Learning Research* 2011; **12**: 2121–59.

El Naqa I, Grigsby P, Apte A, et al. Exploring feature-based approaches in PET images for predicting cancer treatment outcomes. *Pattern Recognit* 2009; **42**(6): 1162–71.

Hall EJ, Giaccia AJ. Radiosensitivity and Cell Age in the Mitotic Cycle. *Radiobiology for the radiologist*: Lippincott Williams & Wilkins; 2012.

Hall EJ, Giaccia AJ. Cell Survival Curves. *Radiobiology for the radiologist*: Lippincott Williams & Wilkins; 2012.

Halperin EC, Perez CA, Brady LW. *Perez and Brady's Principles and Practice of Radiation Oncology*. 5th ed. Philadelphia: Wolters Kluwer Health/Lippincott Williams & Wilkins; 2008.

Jackson A, Ten Haken RK, Robertson JM, Kessler ML, Kutcher GJ, Lawrence TS. Analysis of clinical complication data for radiation hepatitis using a parallel architecture model. *International Journal of Radiation Oncology Biology Physics*; **31**(4): 883–91.

Jackson A, Marks LB, Bentzen SM, et al. The lessons of QUANTEC: recommendations for reporting and gathering data on dose-volume dependencies of treatment outcome. *International Journal of Radiation Oncology* 2010; **76**(3 Suppl): S155–60.

Jochems A, Deist TM, El Naqa I, et al. Developing and validating a survival prediction model for NSCLC patients through distributed learning across 3 countries. *International Journal of Radiation Oncology Biology Physics* 2017; **99**(2): 344–52.

Joiner MC, Marples B, Johns H. The Response of tissues to very low doses per fraction: A reflection of induced repair? In: Hinkelbein W, Bruggmoser G, Frommhold H, Wannenmacher M, eds. *Acute and Long-Term Side-Effects of Radiotherapy*; 1993; Berlin, Heidelberg: Springer Berlin Heidelberg; 1993. pp. 27–40.

Jolliffe IT. *Principal Component Analysis*. New York: Springer-Verlag ; 2002.

Kingma DP, Ba J. Adam: a method for stochastic optimisation. *CoRR* 2014.

Kutcher GJ, Burman C. Calculation of complication probability factors for non-uniform normal tissue irradiation: the effective volume method. *International Journal of Radiation Oncology* 1989; **16**(6): 1623–30.

Lal TN, Chapelle O, Weston J, Elisseeff A. Embedded methods. In: Guyon I, Nikravesh M, Gunn S, Zadeh LA, eds. *Feature Extraction: Foundations and Applications*. Berlin, Heidelberg: Springer Berlin Heidelberg; 2006: 137–65.

Lee S, Ybarra N, Jeyaseelan K, et al. Bayesian network ensemble as a multivariate strategy to predict radiation pneumonitis risk. *Medical Physics* 2015; **42**(5): 2421–30.

Lyman JT. Complication probability as assessed from dose-volume histograms. *Radiation Research Supplement* 1985; **8**: S13–9.

Marks LB, Bentzen SM, Deasy JO, et al. Radiation dose-volume effects in the lung. *International Journal of Radiation Oncology* 2010; **76**(3 Suppl): S70–6.

McCullagh P, Nelder JA. *Generalized Linear Models*. Boca Raton, Fla.: Chapman & Hall/CRC; 2000.

Metz CE. Receiver operating characteristic analysis: a tool for the quantitative evaluation of observer performance and imaging systems. *Journal of the American College of Radiology* 2006; **3**(6): 413–22.

Neapolitan RE, editor. *Learning Bayesian Networks*. Upper Saddle River, NJ, USA: Prentice-Hall, Inc; 2003.

Niemierko A, Goitein M. Modeling of normal tissue response to radiation: the critical volume model. *International Journal of Radiation Oncology* 1993; **25**(1): 135–45.

Niemierko A. Reporting and analyzing dose distributions: a concept of equivalent uniform dose. *Medical Physics* 1997; **24**(1): 103–10.

Reshef DN, Reshef YA, Finucane HK, et al. Detecting novel associations in large data sets. *Science* 2011; **334**(6062): 1518–24.

Ron Kohavi GHJ. Wrappers for feature subset selection. *Artificial Intelligence* 1997; **97**: 273–324.

Sánchez-Maroño N, Alonso-Betanzos A, Tombilla-Sanromán M. Filter methods for feature selection – a comparative study. In: Yin H, Tino P, Corchado E, Byrne W, Yao X, eds. *Intelligent Data Engineering and Automated Learning - IDEAL 2007: 8th International Conference, Birmingham, UK, December 16–19, 2007 Proceedings*. Berlin, Heidelberg: Springer Berlin Heidelberg; 2007: 178–87.

Seppenwoolde Y, Lebesque JV. Partial irradiation of the lung. *Seminars in Radiation Oncology 2001*; **11**(3): 247–58.

Sloan JA, Halyard M, El Naqa I, Mayo C. Lessons From large-scale collection of patient-reported outcomes: implications for big data aggregation and analytics. *International Journal of Radiation Oncology* 2016; **95**(3): 922–9.

Srivastava N, Hinton G, Krizhevsky A, Sutskever I, Salakhutdinov R. Dropout: a simple way to prevent neural networks from overfitting. *Journal of Machine Learning Research* 2014; **15**(1): 1929–58.

Steel GG. *Basic Clinical Radiobiology*. 3rd ed. London New York: Arnold; Oxford University Press; 2002.

Steinwart I, Christmann A. *Support Vector Machines*: Springer Publishing Company, Incorporated; 2008.

Tibshirani R. Regression Shrinkage and Selection via the Lasso. *Journal of the Royal Statistical Society Series B* 1996; **58**(1): 267–88.

Tibshirani R, Friedman J. *The Elements of Statistical Learning: Data Mining, Inference and Prediction*: Springer; 2009. New York city

Tucker SL, Li M, Xu T, et al. incorporating single-nucleotide polymorphisms into the lyman model to improve prediction of radiation pneumonitis. *International Journal of Radiation Oncology Biology Physics* 2013; **85**(1): 251–7.

West CML, Elliott RM, Burnet NG. The genomics revolution and radiotherapy. *Clinical Oncology* 2007; **19**(6): 470–80.

Willers H, Held KD. Introduction to clinical radiation biology. *Hematology/Oncology Clinics of North America* 2006; **20**(1): 1–24.

Zaider M, Hanin L. Tumor control probability in radiation treatment. *Medical Physics* 2011; **38**(2): 574–83.

Zou H, Hastie T. Regularization and variable selection via the elastic net. *Journal of the Royal Statistical Society Series B* 2005; **67**(2): 301–20.

CHAPTER 3

Potentials and Limits of Phenomenological Models

Arjen van der Schaaf

CONTENTS

Introduction 54
    What Determines the Probability of Side Effects in Radiotherapy? 54
    The Phenomenological Aspect of Modeling Side Effects 54
    Typical Datasets 55
        Response Variables 56
        Explanatory Variables 56
        Collinearity 56
        Missing Data 56
        How Much Information Does a Dataset Provide? 57
    Modeling Methods 57
    Fundamental Limits and Potentials 57
Overfitting and Internal Validation 58
    Goodness of Fit and Prediction Performance 58
        Important Measures of Model Performance 59
        Effect of Dataset Size 59
    Techniques to Restrict Model Freedom 60
        Variable Selection 60
        Regularization and Shrinkage 60
        Model Averaging 61
    Internal Validation 61
        Cross Validation 61
        Bootstrapping 62
        Permutation Test 63
Model Instability 63
Generalizability and External Validation 65
    Validation and Updating Methods 66
        Calibration 66
        Refit 67

Further Modifications 67
Combining Datasets 67
Causality and Rapid Learning Healthcare Systems... 68
Models as Conjectures that We Need to Test 68
Rapid Learning Healthcare Systems... 68
The Importance of Data Registration Programs.. 69
Further Potentials and Opportunities.. 69
Discovering New Predictors.. 70
Getting More Out of Response Variables 70
Radiobiological Research... 71
Evidence-Based Medicine 71
Conclusion 72
References... 72

INTRODUCTION

In this chapter we discuss general potentials and fundamental limits of prediction models for side effects in radiotherapy, in particular those models that result from data-driven analysis. The aim is to learn how and when we can trust these models and safely utilize them to optimize the treatment of patients.

What Determines the Probability of Side Effects in Radiotherapy?

Prediction models are commonly probabilistic in nature: they describe the probability for each individual patient to develop a certain complication, given the information that is available at a particular time, generally including characteristics of the patient and specification of the disease and the treatment. Most importantly, we are interested in the relationship with the three-dimensional dose distribution, because this is the variable that we can control in radiotherapy. Unfortunately, the underlying mechanisms by which radiation dose in normal tissues leads to observable side effects are complex and often involve unknown factors. For example, the dose distribution drives various interrelated radiobiological processes, such as cell loss, inflammation, fibrosis, and repair, that lead to different forms of tissue damage and function loss in different organs. Furthermore, multiple organs may be involved, with possible interactions, e.g., between heart and lung (Ghobadi et al., 2012), or sensitive substructures, e.g., a stem cell rich compartment in the parotid glands (van Luijk et al., 2015). Moreover, individual patients differ in various inherent characteristics, including radiation sensitivities, risk factors, and reserve capacities. All these components and their complex relationships are commonly only partially understood. Finally, the probability of side effects may include fundamentally unpredictable components originating from inherently random processes or hidden (unobserved) factors.

The Phenomenological Aspect of Modeling Side Effects

The lack of fundamental knowledge to understand the underlying biological mechanisms in sufficient detail hinders the construction of models from first principles (bottom-up approach). These approaches tend to be oversimplified or underspecified, and still need to fit unknown model parameters to clinical data. Therefore, increasingly often, a top-down approach is followed in which models are built primarily from statistical analysis of clinical observations with only limited biological motivation. These models can be described as phenomenological (van der Schaaf et al., 2015), meaning that they are consistent with existing knowledge and clinical data, i.e., they describe the observed phenomenology, but that they are not fully understood in terms of fundamental radiobiological mechanisms. (See Figure 3.1.)

Typical Datasets

Data is the main source of information for phenomenological models. A dataset typically consists of observations, including objective and subjective measures, from a sample cohort that is a subset of a particular patient population. This means that, besides information, the data contains sampling noise, selection bias, observation errors, and observation bias. These degrading qualities are particularly profound in datasets that are acquired retrospectively. Importantly, a phenomenological model can only be as good as the data. Noise and bias may lead to inaccuracies of the model that do not necessarily vanish asymptotically when more data is added. Moreover, due to the nature of phenomenological modeling to describe observations rather than the underlying reality, it is difficult to detect effects of observation errors in the analysis and it is often impossible to discriminate them from real

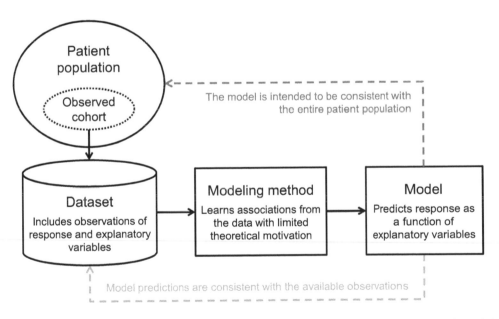

FIGURE 3.1 Characteristic method of phenomenological modeling. Models are developed from data rather than theoretical knowledge. The data originates from observations of a cohort that is a sample of a patient population. The model is consistent with the data and intended to be consistent with the entire patient population. However, the model is not necessarily fully understood, and its general validity is not necessarily evident. (Modified from: van der Schaaf A, Langendijk JA, Fiorino C, Rancati T. Embracing phenomenological approaches to normal tissue complication probability modeling: A question of method. *Int J Radiat Oncol Biol Phys.* 2015 Mar 1;91(3):468–71).

effects. Therefore, the quality of the data collection is essential for the quality of phenomenological models.

A dataset always contains response variables, i.e., the outcomes that we want to predict, and explanatory variables, i.e., the information that is available to base our predictions on.

Response Variables

Often multiple response variables are observed for each patient, such as physician rated and patient rated outcomes (often based on multiple questionnaires with many items) and objective measures at various time points. However, response models usually describe only a single item or a composite score that is constructed from only a few items. Response variables may be measured using various scales, but often these are reduced before analysis to a single dichotomous scale (complication vs. no complication) defined over a specific time range. This greatly simplifies the statistical analysis and the interpretation of the results, however, at the cost of losing information. The resulting models for such dichotomous endpoints are commonly referred to as normal tissue complication probability (NTCP) models.

Explanatory Variables

The explanatory variables that are usually observed are items that are already recorded in the treatment process or variables that are known or expected to be potentially predictive. These include patient characteristics, such as demographic information, anatomy, risk factors, comorbidities, and baseline toxicities; disease characteristics, such as tumor location and stage; and treatment characteristics, such as treatment modality and dose distribution. Medical images of the patient provide additional or supporting information that is increasingly utilized. Before analysis this large quantity of data is usually reduced to a comprehensive set of candidate predictors. For example, the three-dimensional dose distribution is often reduced to a set of dose-volume parameters in selected organs, which are assumed to provide a biologically relevant and consistent dosimetric description.

Collinearity

A particular property of dosimetric data is that usually dose distributions are very similar in comparable patients, resulting in a limited range of observed dose distributions and high correlations between dosimetric variables. This property, generally referred to as 'collinearity', tends to decrease modeling accuracy and disrupt the ability to identify unambiguous associations. Often, highly collinear variables are eliminated or combined prior to modeling.

Missing Data

Often, parts of the data are missing for various reasons, including skipped items, failed recordings, missed appointments, loss to follow-up, or death of patients. If cases with missing data are removed from the analysis (complete case analysis), the resulting model may suffer from loss of information and selection bias if data is missing not complete at random. With substitution of best guesses for the missing data (multiple imputation) the deteriorating effects of missing data can be minimized but not completely eliminated.

How Much Information Does a Dataset Provide?

The size of datasets in radiation oncology research typically ranges from nearly one hundred patients to hundreds of patients. These datasets with many response variables and explanatory variables appear large, usually requiring many megabytes of storage. However, what we want to do is essentially to compare various prediction models and decide, based on the data, which model provides the best prediction of an actual response. Therefore, for dichotomous response variables, the essential information is contained in a binary value that we want to compare with a predicted probability, which provides at most one bit of information per patient, and often even less (e.g., if the average response rate is low). In summary, in radiation oncology the datasets often appear extensive, but actually contain limited information to resolve phenomenological models.

Modeling Methods

Model fitting, which is at the heart of phenomenological modeling, is the act of adjusting model parameters such that the model predictions correspond as good as possible to the actually observed outcomes. The most frequently used methods for fitting phenomenological models are based on common regression techniques, of which logistic regression analysis is the most common for dichotomous response variables. These methods are well known from the field of statistics. However, many alternative methods exist, e.g., from the field of machine learning, that all have distinct qualities. In this chapter we try to keep a broad perspective and focus on general properties that are independent of specific methods.

Fundamental Limits and Potentials

To appreciate one of the fundamental difficulties of phenomenological model building, imagine that all model parameters are combined to form a model space. We can regard a model as a point in this space, and model building as finding the optimal point in this space. For phenomenological models, which are minimally constrained by fundamental knowledge, the number of possible parameters can be huge. Every additional linear candidate variable essentially results in one extra parameter in the model, adding one more dimension to the model space. For categorical and nonlinear variables this may be even more. The high dimensional model space effectively represents a huge number of different possible models, of which all but one must be discarded based on the limited available data. This problem of too many possible models and too little data inevitably leads to model uncertainty, which may translate into overfitting and model instability. On the other hand, the large model freedom allows discovering unanticipated relationships.

Another limit of observational studies in general is that the discovered associations may not be generalized (extrapolated) beyond the range of the observed data. Furthermore, only associations and not causal relationships may be inferred. A benefit of observational studies is that they are usually performed in existing clinical settings, such that the results can be more readily translated into clinical practice than those of most pre-clinical experiments.

The fundamental limits of phenomenological models are explored in more detail in the next paragraphs in conjunction with practical methods to deal with them, while in the last paragraph before the conclusion we elaborate on further potentials.

OVERFITTING AND INTERNAL VALIDATION

Overfitting occurs if a model is fit to a particular dataset in such specific detail that it loses its ability to describe the general relationships of the whole population. The telltale aspect of an overfitted model is that it predicts (or describes) the training data much more accurately than new data that was acquired from the same population but that was not used to train the model. The general cause of overfitting is that the model fitting procedure is allowed too much freedom to describe details of the training data (i.e., the model space is too large) while the training data provides insufficient power to rule out the models that are actually inconsistent with the rest of the population.

Goodness of Fit and Prediction Performance

A measure of how accurate a model describes a dataset is called 'model performance'. If this measure is applied to the training data it is often referred to as 'goodness of fit'. We will use the term 'prediction performance' to refer to the model performance that is measured on new data, which was not part of the training dataset. Generally, if a model fit is allowed more freedom (i.e., it can choose from a larger model space), it can achieve a better (higher) goodness of fit (see Figure 3.2). Similarly, the prediction performance will initially rise as the model is allowed to describe more detail. However, the gain in prediction performance will diminish once the extra model freedom is used to describe details of the training data rather

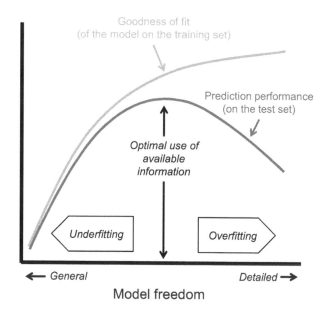

FIGURE 3.2   Goodness of fit (of the model on the training set) and prediction performance (on the test set) as function of model freedom, showing that goodness of fit increases monotonically with increasing model detail, while prediction performance peaks at a specific value. At this point, between underfitting and overfitting, the available information in the data is used optimally. (Modified from: van der Schaaf A, Langendijk JA, Fiorino C, Rancati T. Embracing phenomenological approaches to normal tissue complication probability modeling: A question of method. *Int J Radiat Oncol Biol Phys.* 2015 Mar 1;91(3):468–71).

than general properties of the whole population. Eventually, the prediction performance will decrease with increasing model freedom. In the transition zone, between underfitting and overfitting, the prediction performance reaches a maximum value. Modeling methods generally aim to find this point of maximum prediction performance. The optimum, however, is not evident from looking at the goodness of fit to the training data, but can only be approximated by estimating the prediction performance for the population.

Important Measures of Model Performance

The most commonly used model performance measure is the likelihood function, which is defined as the probability of the data given the model parameters. For a single dichotomous response variables this is simply the predicted probability of the actual outcome. Most model fitting methods are based on maximizing the likelihood. The optimisation process can be visualized as hill climbing in a likelihood landscape that stretches over the many dimensions of model space. The top of the hill represents the estimated model, and the 95% confidence intervals of the model parameters are determined by the most likely part of model space with a total likelihood of 95%.

The likelihood can be transformed very easily from a goodness of fit measure into a simple estimator of the prediction performance by adding a penalty for each degree of freedom of the model. Well-known examples are the Akaike information criterion (AIC) and the Bayesian information criterion (BIC).

The likelihood is a strong measure of model accuracy and particularly useful to compare different models. However, its value depends heavily on details of the dataset, making it hard to judge by its value alone if the model performance is good or bad. For this purpose alternative measures are better suited, particularly the discrimination and calibration.

The discrimination is most often measured with the c-statistic (for dichotomous response variables also known as 'the area under the curve'), which can be interpreted as the probability that two patients with different outcomes are ranked correctly by the model. The c-statistic ranges between 0.5 (discrimination equal to chance) and 1 (perfect discrimination). Often a c-statistic around or above 0.7 is considered as a good performance. Because the discrimination is a rank statistic, the actual predicted probabilities do not need to be accurate, but only in the right order.

Calibration describes if predicted probabilities of patients with specific high or low predicted risk correspond with equally high or low response rates in the data. It usually consists of two numbers: the calibration intercept, which is related to the overall response rate, and the calibration slope, which is related to the strength of the modeled relationships. Furthermore, the total calibration curve can be evaluated with the Hosmer-Lemeshow test.

Effect of Dataset Size

If observations are added to a training dataset, and the model is refitted, then the corresponding goodness of fit (as a function of model freedom) will essentially not change much. The model still describes both the general properties of the population and the details of the data (in slightly altered proportions). On the other hand, because more data is available to estimate the model parameters more accurately, the prediction performance will peak at a higher model freedom and at a relatively higher value. So, with more data we can build models with more model freedom and better prediction performance. However, this gain levels off asymptotically for very large datasets (van der Schaaf et al., 2012). Although data is essential for phenomenological modeling, and adding more data generally improves the prediction performance, large datasets will not automatically result in models with perfect prediction performance.

Techniques to Restrict Model Freedom

To avoid overfitting it is important that the modeling method includes a technique to restrict the model freedom in such a way that it focuses on the main effects in the population and not on irrelevant details. We describe the principles behind three general techniques.

Variable Selection

Model freedom can be restricted by reducing the dimensions of model space, i.e., by eliminating parameters from the model. A common method to achieve this is by variable selection, which enters only the few most relevant candidate variables in the model (fitting their related parameters to the data) and eliminates the rest (effectively fixing the corresponding parameters to zero). The trick is to select the right number and the right combination of variables that maximizes the estimated prediction performance. Usually, if the number of candidate predictors is large, the combinatorial complexity is so huge that not all combinations can be examined. Therefore, stepwise procedures have been developed that add or eliminate variables one by one. These procedures can be quite fast and usually result in good models, but they are not strictly guaranteed to find the best solution. Forward methods start with an empty model and iteratively add a variable that increase the prediction performance as much as possible until no further improvement is possible. The advantages are: relatively fast calculation, and the avoidance of severe overfitting during the procedure. In addition, the selection order of the variables provides information about their relative importance. Backward methods work the other way around, starting with a full model (with all candidate variables in it) and iteratively eliminating variables to maximize the estimated prediction performance. The main advantage of backward elimination is its potential to discover combinations of variables that predict well together but not individually. However, if the number of candidate variables is large, the initial models can be severely overfitted. In conjunction with collinearity this may result in adverse elimination of important variables at an early stage of the process. Sometimes a pre-selection to eliminate seemingly unimportant candidate variables, based on crude associations with the response (univariable analysis), is conducted prior to modeling (multivariable analysis).

Regularization and Shrinkage

Variable selection restricts model space by eliminating entire dimensions. A somewhat different technique to reduce the size of model space is to restrict the amplitude of the model parameters. This effectively confines the search for the best model to a sub-region in model space with the most general or least specific models. This general method is called regularization, and the corresponding model fitting procedures include a regularization factor (or penalty) for the parameter amplitudes in their optimisation criterion. The regularization factor needs to be tuned to obtain the best prediction performance, which is commonly accomplished with cross validation (see below). Familiar methods are 'ridge regression' and 'lasso'. In comparison to variable selection, regularization methods tend to produce models with more variables but with smaller parameters. The reduction of the model parameters by adding regularization is commonly referred to as 'shrinkage'. Shrunken estimates of model parameters are generally biased and have a reduced variance. Therefore, it is difficult to express meaningful confidence intervals for this type of model parameters.

Model Averaging

A final method to restrict model freedom is to combine models into averages. Effectively this means that individual models from model space cannot be independently selected, but only in conjunction with other models, thus restricting the model freedom. The principle idea is that model averaging enhances the similar (general) aspects of individual models, while reducing the dissimilar (detailed) aspects. Familiar methods that use model averaging are 'random forest', 'bagging', and 'Bayesian networks'. A disadvantage is that averaged models are generally more complex to develop, to interpret, and to report.

Internal Validation

When we develop a model using the available training data, we can directly calculate the goodness of fit to the training data, but we cannot directly calculate the prediction performance if we do not have independent test data from the same population. Moreover, no general accurate analytic solutions exist to estimate the prediction performance of a model based on the training data. So, we resort to numerical solutions. The most simple trick is to split the dataset randomly into a training and a test set, then train the model on the training set and calculate the model performance on the test set (see Figure 3.3). This has two important limitations: not all available data was used to train the model, and the prediction performance is based on a single small subset. More advanced schemes use repeated resampling to estimate the prediction performance, while the final (nominal) model is trained on the complete dataset. In these internal validation schemes, it is important to include not only the model fitting procedure, but also all additional techniques that are used to restrict the model freedom, including pre-selection and pre-processing of variables.

Cross Validation

The most easy to understand resampling scheme is cross validation. With this technique the original dataset is divided into a specific number of subsets. In turn, each subset is held out once, while a model is trained on the remaining data. For each subset the responses are predicted by the corresponding model that was trained on the remaining data. Finally all predicted responses are merged into a single complete test set from which the prediction performance can be estimated. Multiple repetitions with different random splits of the observations may be required to obtain reliable estimates. (This is not possible if the hold-out subsets include only one observation, i.e., with leave-one-out cross validation.) A

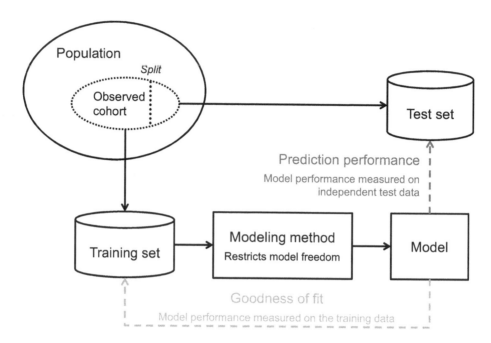

FIGURE 3.3   Principal method of internal validation. The cohort is split in two independent data-sets: a training set and a test set. The goodness of fit is evaluated on the training set and the prediction performance on the test set. The model is trained using proper restriction of model freedom with the aim to maximize the prediction performance and to avoid overfitting. (Modified from: van der Schaaf A, Langendijk JA, Fiorino C, Rancati T. Embracing phenomenological approaches to normal tissue complication probability modeling: A question of method. *Int J Radiat Oncol Biol Phys*. 2015 Mar 1;91(3):468–71).

disadvantage of cross validation is that the dataset size of the training data does not equal that of the original data, especially if the number of subsets is small. Furthermore, each subset requires a full training procedure, which is computationally expensive, especially if the number of subsets is large.

Cross validation is often used in regularization methods to determine the penalty value that maximizes the prediction performance. However, because tuning this parameter is an essential element to restrict model freedom and avoid overfitting, it must be included in the internal validation to estimate the final prediction performance. This requires double cross validation techniques with one inner loop to tune the model and an outer loop to estimate the final prediction performance (Xu et al., 2012).

Bootstrapping

The merit of bootstrapping is to mimic sampling from the population by resampling with substitution from the original data, i.e., observations may be picked more than once. An advantage of bootstrapping is that the training set has the same size as the original dataset, although some observations are missing and others appear multiple times. The complete original dataset is used as the test set, which is not independent of the training set. Therefore, the model performance measured on the test set overestimates the real prediction performance. However, the model performance on the training data similarly overestimates the goodness of fit of the nominal model, because the training set contains predictable duplicated observations. Therefore, the difference between the model performances on the training set and the test set is a consistent estimator of the difference between the goodness of fit and the prediction performance, which is usually referred to as 'optimism'. If we determine the optimism with bootstrapping, we can correct the goodness of fit of the nominal model to obtain a consistent estimate of the prediction performance. For an accurate estimate multiple random bootstrap repetitions are required. However, each repetition results in a full validation at the computational cost of a single training procedure, which is much more efficient than using cross validation.

Deviations in the calibration that are found with bootstrap validation are sometimes corrected by recalibrating (shrinking) the nominal model, such that the expected calibration of the model for new data is perfect. This, however, introduces a bias similar to other regularization techniques, as discussed above. It may also affect other model performance measures, but not the discrimination, which is independent of calibration.

Permutation Test
Even after careful internal validation it is possible to obtain relatively high prediction performance scores by chance, even if no real associations exist between the response and the explanatory variables. The likelihood of such a situation to occur can be estimated by using a permutation test. For this test the outcomes are randomly permutated (scrambled) before performing the modeling and internal validation procedures. This is repeated multiple times, with different random permutations, and the resulting prediction performances are aggregated into a numerical distribution, which represents the distribution of prediction performance scores that are expected if the responses are purely random. The actual performance of the original model (using the original data) can be compared with this distribution to obtain the probability (p-value) that a similar or better result is found by pure chance. These permutation tests are particularly useful if there is doubt about the prediction performance, e.g., because of low performance scores, high variability (model instability, see below), or methodological deficiencies.

MODEL INSTABILITY

Phenomenological modeling may suffer from instability, which means that small changes in the input data result in large changes of the output models. For example, the selected variables may differ between models that are fitted to different cohorts of the same population, resulting in a completely different description of the dose response relationship. Similarly, for a single dataset, different modeling procedures may result in very different models. All these different resulting models may produce very different predictions for individual patients. However, they are all tuned to achieve a prediction performance that is close to maximum, which will therefore be, actually, very similar for all models. (See Figure 3.4.)

The most important apprehension of model instability is that multiple models may exist that are consistent with the available data. This means that we may not infer from finding one model that all other models must be incorrect. If one study finds a certain model and

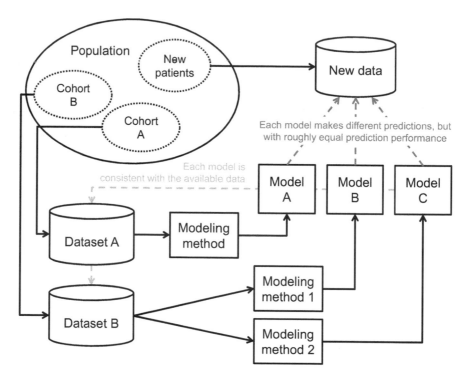

FIGURE 3.4 Schematic figure of model instability. Phenomenological model uncertainty depending on cohorts and learning methods. The resulting models may produce very different predictions while being consistent with the available data and having roughly equal prediction performance. (Modified from: van der Schaaf A, Langendijk JA, Fiorino C, Rancati T. Embracing phenomenological approaches to normal tissue complication probability modeling: A question of method. *Int J Radiat Oncol Biol Phys.* 2015 Mar 1;91(3):468–71).

another study finds a very different model, both may be correct. This does not imply, however, that incorrectness of a model cannot be determined. As we shall see later, appropriate methods exist to test if models are consistent with the available data.

Model instability is enhanced by discreet steps or bifurcations in the model development process, such as variable selection, particularly near critical decisions boundaries where small variations result in very different models. Also, collinearity of the data enhances model instability. Collinearity is known to increase the variance of the model parameters, as described by the 'variance inflation factor'. But collinearity also impedes variable selection, because it is more difficult to recognize the truly best predictor from a set of similar (correlated) variables than from a group of dissimilar (independent) ones. Instability is partly related to sampling of the data (the composition of the patient cohort) and partly by the choice of modeling method. With small datasets and data-driven analysis the model instability is usually dominated by sampling effects, but the choice of modeling method is rarely negligible.

It is useful to explore the range of models that can be produced by changes in the dataset, or by different modeling methods, but that are still consistent with the data. Such analysis can be integrated efficiently with an internal validation procedure, for example by analyzing the selection frequencies of variables (or variable combinations) from resampled datasets.

A method to reduce model instability is to aggregate multiple models that are consistent with the data into a combined averaged model. However, the resulting averaged models will still vary between data samples and modeling methods.

GENERALIZABILITY AND EXTERNAL VALIDATION

So far we have discussed the situation that the training cohort and the subsequent new patients whose outcome we want to predict are sampled from the same population. However, when predicting for patient populations or conditions that differ from the training data, the validity of the model is not self-evident. The quality that a model predicts well in different populations is referred to as 'generalizability'. (See Figure 3.5.)

Phenomenological models rely on mimicking the statistical relationships of the training data, and, therefore, may not be generalized into populations with a different statistical structure. Statistical differences between datasets may be observable from the characteristics of the independent variables, but they may also arise unpredictably from hidden factors. Patient populations or conditions may differ between hospitals, between treatment methods, between patient subgroups, or over time. In fact, generalizability cannot be guaranteed without further assumptions or tests.

To test that a model is valid for a new population we need to assess that the model is consistent with data from that population. More specifically, we test if the predictions of the

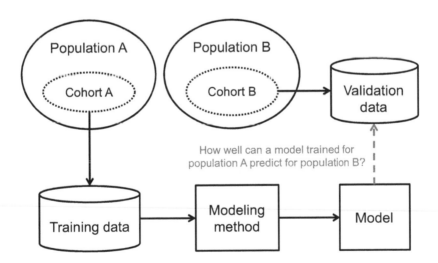

FIGURE 3.5   Illustration of generalizability and external validation. Consider a phenomenological model that is derived from data of population A and is used to predict for patients from a different population B. Without further assumptions, there is no guarantee that the model is generalizable to population B, i.e., that the model predicts well for population B. However, this can be evaluated by comparing the predictions of the model for cohort B with the actual outcomes in an external validation. (Modified from: van der Schaaf A, Langendijk JA, Fiorino C, Rancati T. Embracing phenomenological approaches to normal tissue complication probability modeling: A question of method. *Int J Radiat Oncol Biol Phys.* 2015 Mar 1;91(3):468–71).

model are statistically consistent with the actual outcomes in the data. This is commonly referred to as 'external validation'. To externally validate a model for a new population we need data from a sample cohort of that population. Particularly, if we want to apply models from literature to the patient population in our own clinic, we need to assess the data of our own patients. Standardized prospective data registration programs are paramount to have this data available at an early stage. External validation of published models before applying them in the clinic has been referred to as quality assurance of prediction models (Lambin et al., 2013b).

However, it is debatable if models always need to be externally validated for each individual population. Generalizability may also be inferred from similarities between populations, although these decisions need to be considered carefully. When using phenomenological models for patient care, it is recommended to follow these patients in a prospective data registration program, such that the model validity can be assessed at a later stage.

External validation is a statistical process with common limitations that may possibly produce false positive or false negative conclusions, particularly if the validation cohort is small. If the validation cohort is large, however, even irrelevant details may statistically falsify the model. Therefore, it is important in validation assessments to evaluate the confidence intervals of the results with care. Furthermore, a statistical procedure can only assess general statistical properties, such that a validated model can still deviate from the actual data in specific details. Moreover, validation of a model only implies that the model is generally consistent with the data and not necessarily that the prediction performance is good or optimal.

Validation and Updating Methods

When evaluating data to externally validate a model it is tempting to develop a new model from the data and compare both models. However, because of model instability, we cannot conclude that one of both models is wrong if they are different. Another tempting method is to measure the model performance on the new data and compare it with the prediction performance of the model in the original data. However, the model performance does not only depend on the model, but also on the case-mix of each dataset, i.e., the particular composition of the cohort whose outcomes are predicted. Therefore, it is better to compare the actual model performance with the expected performance for the same cohort, which can be calculated analytically or by simulation.

Calibration

The most basic external validation method is to test if the actual response rate is statistically consistent with the predictions of the model. This is known as 'calibration in-the-large'. If the actual response rate differs from the one predicted, the model can be updated by adjusting (usually) only a single parameter (the intercept).

One step further is to also test the steepness (or slope) of the calibration curve. Using generalized linear models this is equivalent with fitting a model to the data with the linear predictor of the model as the only variable. The calibration slope then equals the model parameter that corresponds to that variable, which can be tested to be within statistical boundaries of the expected value (usually one). The model can be updated (recalibrated) by adjusting the model parameters based on the measured calibration intercept and slope (usually two parameters).

Refit

To add additional detail to the external validation, we may consider to adjust all model parameters independently, i.e., to refit the model to the new data. To evaluate if this is necessary, we can compare the performance of the original model on the new data with that of the refitted model, e.g., by using a likelihood ratio test. Combing the assessments of calibration and refitting into a closed testing procedure (Vergouwe et al., 2016) enables correction for multiple testing.

Further Modifications

A step beyond refitting is to consider the role of each individual variable in the model. Maybe the model can be improved by transforming the variables. For example, if the relationship between a variable and the response appears to be non-linear, or, if the original levels of a categorical variable are not adequate to describe the new data. Maybe the prediction performance of the model can be improved by adding extra variables or by replacing certain variables with better predictors, or by removing variables that appear useless. These actions, however, closely resemble developing a new model rather than performing an external validation.

We must be aware that modification of a model to fit the new data may render it less consistent with the original data. This is not a problem if we intend to build a new model specifically for the new population, e.g., when we know that the new population is really different. However, if we intend to construct a general model, or, if we aim to use the information contained in the original model to aid model development, then we should restrict modification of the model such that the updated model represents a balanced average of the information from the original and the new data.

Combining Datasets

If the data of the original model is available, it is possible to validate and update the model on the combined original and new datasets. This ensures that all available information is used and properly balanced. However, the results of the combined validation cannot be interpreted as a demonstration of generalizability, because the analysis partly included data that was used to train the model.

Before combining the original and new datasets, we can test for heterogeneity, i.e., dissimilarity of their statistical properties. Heterogeneity may be related to generalizability, but they are not equivalent. Better generalizability is expected for more homogenous populations, but on the other hand, generalizability across heterogeneous populations may be perceived as a favourable quality of a model. To test if population differences are a significant factor in the model, a variable that encodes for the cohort can be included in the analysis.

CAUSALITY AND RAPID LEARNING HEALTHCARE SYSTEMS

Phenomenological models typically originate from observational studies and not from repeatable or controlled experiments. In the eighteenth century, David Hume described the induction problem, which states, fundamentally, that from observation we can only infer relationships, but not the causality of those relationships. When applying phenomenological models purposely to modify the treatment of patients, e.g., to reduce the toxicity, we assume that the modeled relationships are causal. However, because of the induction problem, we are fundamentally uncertain that the actual risk reduction will be as predicted by the model. This verdict may appear to ruin the applicability of phenomenological models, just as the induction problem appeared to put an end to science in the eighteenth century. Fortunately, science has found a way around the induction problem.

Models as Conjectures that We Need to Test

In the twentieth century, Karl Popper realized that science does not work by induction at all. Instead, the scientific method consists of formulating simple but daring hypotheses (conjectures) that are subsequently tested by critical experiments. The experimental results are compared with the predictions that follow from the conjecture. If the outcome of the experiments differ from the predictions, the conjecture is falsified and needs to be revised or replaced by a more complex hypothesis. If, however, the outcome agrees with the predictions, we consider this as a tentative proof of the conjecture, until new critical experiments are performed.

Likewise, we must accept that phenomenological models are conjectures that appear to be correct but that are not fully and generally proven. This has two important consequences for the application of phenomenological models to optimize the treatment of patients. First, we must realize that that the model predictions may be incorrect. In order not to harm patients we must be careful with changes towards previously unexplored forms of treatment, especially if these are potentially more toxic. Second, we must collect patient outcome data and compare these to the model predictions to attempt to falsify the conjectured model.

Rapid Learning Healthcare Systems

We can put the above ideas in the context of rapid learning healthcare systems (Lambin et al., 2013a), as outlined in Figure 3.6. We acquire data from a first cohort in a patient population, use it to build a phenomenological model, which we apply to optimize the treatment of new patients. However, because these new patients are treated differently, we are not sure about the generalizability and causality of the model. Therefore, existing safety constraints of the treatment still need to be respected. Furthermore, the subsequent cohort must be considered as patients from a different population, requiring external validation of the model. Therefore, we collect outcome data for this cohort and validate the model for this population of patients that received the optimized treatment. As soon as we have collected enough evidence to falsify the model, it can be updated based on the new data. We can then, again, apply the updated model to optimize the treatment, starting a new cycle.

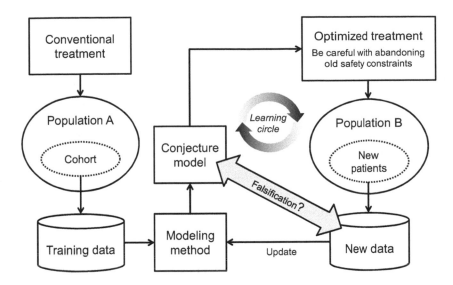

FIGURE 3.6 Schematic of a rapid learning healthcare system. Based on data from conventional treatment an initial model is created. This model is regarded as conjecture and is used to optimize treatment of subsequent patients. Because these patients receive different treatment compared to the previous patients, they must be considered as a separate population. After prospective data registration, the observed outcomes are compared with the model predictions, enabling model validation. If the model is falsified, it is replaced by an updated model that includes the new data. (From: van der Schaaf A, Langendijk JA, Fiorino C, Rancati T. Embracing phenomenological approaches to normal tissue complication probability modeling: A question of method. *Int J Radiat Oncol Biol Phys*. 2015 Mar 1;91(3):468–71).

Each iteration expands the information on which the model is founded and enhances our trust in the generalizability and causality of the model. This repeated process constitutes a learning circle in which experience from old patients is accumulated and used to optimize the treatment of new patients.

The Importance of Data Registration Programs

As we have seen, data registration programs are important tools that help us to rely on the generalizability and the causality of phenomenological models. Without these programs, the introduction of phenomenological models in the clinic is unreliable or even unsafe. Not only do data registration programs enable reliable application, they also serve to improve phenomenological models. However, data registration programs require extensive efforts, and should, therefore, be kept within efficient proportions. At some point, maybe, certain models will be so extensively tested that data registration programs are no longer strictly necessary to accompany their application.

FURTHER POTENTIALS AND OPPORTUNITIES

So far we have discussed mainly the limits of phenomenological models and how to deal with them. In this final part we focus on potentials and connections with other research areas.

Discovering New Predictors

One of the great advantages of phenomenological modeling is its potential to explore large datasets and to discover new, potential, or better predictors. Genomics is a well known and successful related field that uses exploratory statistical methods. Genetic information has great potential to improve phenomenological models in radiotherapy, but is still barely used. Recently, a similar field call 'radiomics' (Lambin et al., 2012) has emerged in which prediction models in radiotherapy are enhanced by adding predictive features derived from medical images. Thus, specific signatures in the anatomy or pathology of patients can be derived and quantified from medical images and used to predict the outcomes of these patients.

Another exciting innovation is the use of voxel-based analysis. In this technique the dose response relationship is not restricted to dosimetric parameters of delineated structures. Instead, the dose response relationship of each voxel of the anatomy can be assessed through fusion of individual dose distributions onto a common anatomical atlas. A challenge in these methods is the multiple comparison problem, which can be managed with, permutation tests and clustering techniques (Monti et al., 2017).

A promising and swiftly emerging new machine learning technique is 'deep learning', which is capable of learning to recognize complicated patterns from examples. Potentially this method may learn to recognize complicated predictive patterns that we have failed to discover so far. However, it is yet unknown if such complicated patterns exist and how much they can improve the prediction performance. Furthermore, these learning techniques require particularly large training sets, while only limited data is yet available.

Getting More Out of Response Variables

Another still largely unexploited potential lies in using more information from the available datasets, particularly those of the response variables. Most datasets include multiple recorded responses, with many different measures of related symptoms at various time points. In current standard practice, however, this wealth of information is reduced to a poor single dichotomous variable before analysis. There are several options to increase the information that we use from the available data by applying more elaborate analysis techniques.

Instead of a dichotomous response we may use an ordinal response that encodes the severity (grade) of the outcome, which can be analyzed, e.g., with ordinal logistic regression. Such analysis uses information from common low-grade events to estimate model parameters, which are simultaneously used to model less common, but more relevant, high grade events. Similarly, multiple response items can be combined into a composite response that can be analyzed as continuous variable using linear regression. Even more interesting is the idea to search in the data for clusters of similar response variables and try to attribute these to various toxicity syndromes, which may be related to more fundamental biological processes than the individual responses (Steineck et al., 2017). Furthermore, because clustered response variables are different measures of the same syndrome, they may serve to filter out observation noise and bias. Additionally, response measurements that are repeated over time can be analyzed using longitudinal methods with optional corrections for competing risks and modeling of temporal patterns. Finally, the grand combination of all response measures, split into syndromes and time patterns, or joined in a multivariate toxicity profile, may be analyzed in relation to quality of life, which is, next to survival, the ultimate outcome measure to optimize for patients.

Radiobiological Research

Phenomenological modeling and radiobiological research appear to be very different approaches to unravel dose effect relationships. Modeling works top down, looking at the phenomenology, trying to understand the relationships, and only guessing at the possible biological mechanisms. Radiobiology, on the other hand, works bottom-up, trying firsts to understand the mechanisms before considering how and when mechanisms translate to symptoms in patients. These approaches are complementary rather than competing. They can inspire one another with new findings and by generating hypotheses that may be validated or falsified by the other. In conjunction they enable the elucidation of biological mechanisms in clinical situations, which is the key to really understand radiobiology in clinical radiotherapy and a prerequisite to truly trusting phenomenological or radiobiological models. An opportunity to bring both research approaches closer together may be in studies that observe physiological and radiobiological processes in patients in detail, for example with minimal-invasive measurements or with functional imaging techniques. When observing in detail the joint dynamics of multiple interrelated physiological processes before, during, and after treatment, even small cohorts of patients may produce large quantities of rich data. In combination with computer simulations, this may be enough to unravel the important underlying processes and mechanisms.

Evidence-Based Medicine

Phenomenological modeling and rapid learning healthcare systems have great potential to aid evidence-based medicine as alternative approaches to randomized controlled trials. The fundamental difference between randomization and modeling is that randomization relies on averaging out differences between patients while modeling relies on describing the effect of differences between patients, enabling to correct outcomes taking differences into account. Modeling approaches allow more flexible and more efficient study designs (KNAW, 2014), but are also restricted by the limitations that we have discussed earlier in this chapter.

Rapid learning healthcare systems may have an important role in the introduction of emerging treatment modalities for which randomized controlled trials are unethical, impractical, difficult, or for any other reason not performed. To efficiently allocate new modalities with limited availability, phenomenological models can be used to select patients that likely benefit most in terms of predicted outcome. Establishing the effectiveness of the new treatment and the validity of the models under the new conditions may be part of a rapid learning healthcare process. An example is the model based approach (Langendijk et al., 2013) that is used for the introduction of proton therapy in the Netherlands. Within the model-based approach the effectiveness of the new technique is assessed by comparing the actual outcomes of patients with the corresponding predictions, but also by comparing outcomes of the new technique with corresponding predictions for the old technique, which are expected to differ if the new technique is really better than the old one (Christianen et al., 2016).

Phenomenological modeling can also be combined with randomized controlled trials, for example to enrich patient cohorts such that predicted differences between randomized trial arms are large enough to be detectable with practical cohort sizes (Widder et al., 2016).

CONCLUSION

Phenomenological models should not be regarded as general abstractions of an absolute truth but as conjectures that originate from accumulated experience. When applied with caution and accompanied by prospective data registration programs, these models are potentially powerful tools to learn from patients in the past and to optimize treatment of patients in the future.

REFERENCES

Christianen, M.E., van der Schaaf, A., van der Laan, H.P., Verdonck-de Leeuw, I.M., Doornaert, P., Chouvalova, O., et al. 2016. Swallowing sparing intensity modulated radiotherapy (SW-IMRT) in head and neck cancer: clinical validation according to the model-based approach, *Radiother Oncol* 118(2):298–303.

Ghobadi, G., van der Veen, S., Bartelds, B., de Boer, R.A., Dickinson, M.G., de Jong J.R., et al. 2012. Physiological interaction of heart and lung in thoracic irradiation, *Int J Radiat Oncol Biol Phys* 84(5):e639–46.

KNAW, 2014. Evaluation of new technology in health care. In need of guidance for relevant evidence. Amsterdam, KNAW. https://www.knaw.nl/en/news/publications/evaluation-of-new-technology-in-health-care-1.

Lambin, P., Rios-Velazquez, E., Leijenaar, R., Carvalho, S., van Stiphout, R.G., Granton, P., et al. 2012. Radiomics: extracting more information from medical images using advanced feature analysis, *Eur J Cancer* 48(4):441–6.

Lambin, P., Roelofs, E., Reymen, B., Velazquez, E.R., Buijsen, J., Zegers, C.M., et al. 2013a. 'Rapid Learning health care in oncology' – an approach towards decision support systems enabling customised radiotherapy, *Radiother Oncol* 109(1):159–64.

Lambin, P., van Stiphout, R.G.P.M., Starmans, M.H.W., Rios-Velazquez, E., Nalbantov, G., Aerts, H.J.W.L., et al. 2013b. Predicting outcomes in radiation oncology — multifactorial decision support systems, *Nat Rev Clin Oncol* 10(1):27–40.

Langendijk, J.A., Lambin, P., de Ruysscher, D., Widder, J., Bos, M., Verheij, M. 2013. Selection of patients for radiotherapy with protons aiming at reduction of side effects: the model-based approach, *Radiother Oncol* 107(3):267–73.

Monti, S., Palma, G., D'Avino, V., Gerardi, M., Marvaso, G., Ciardo, D., et al. 2017. Voxel-based analysis unveils regional dose differences associated with radiation-induced morbidity in head and neck cancer patients, *Sci Rep* 7(1):7220.

Steineck, G., Skokic, V., Sjöberg, F., Bull, C., Alevronta, E., Dunberger, G., et al. 2017. Identifying radiation-induced survivorship syndromes affecting bowel health in a cohort of gynecological cancer survivors, *PLoS One* 12(2):e0171461.

van der Schaaf, A., Xu, C.J., van Luijk, P., van't Veld, A.A., Langendijk, J.A., Schilstra, C. 2012. Multivariate modeling of complications with data driven variable selection: guarding against overfitting and effects of data set size, *Radiother Oncol* 105(1):115–21.

van der Schaaf, A., Langendijk, J.A., Fiorino, C., Rancati, T. 2015. Embracing phenomenological approaches to normal tissue complication probability modeling: a question of method, *Int J Radiat Oncol Biol Phys* 91(3):468–71.

van Luijk, P., Pringle, S., Deasy, J.O., Moiseenko, V.V., Faber, H., Hovan, A, et al. 2015. Sparing the region of the salivary gland containing stem cells preserves saliva production after radiotherapy for head and neck cancer, *Sci Transl Med* 16;7(305):305ra147.

Vergouwe, Y, Nieboer, D., Oostenbrink, R., Debray, T.P., Murray, G.D., Kattan, M.W., et al. 2016. A closed testing procedure to select an appropriate method for updating prediction models, *Stat Med* 28. doi: 10.1002/sim.7179 [epub ahead of print].

Widder, J., van der Schaaf, A., Lambin, P., Marijnen, C.A., Pignol, P., Rasch, C.R., et al. 2016. The quest for evidence for proton therapy: model-based approach and precision medicine, *Int J Radiat Oncol Biol Phys* 95(1):30–6.

Xu, C.J., van der Schaaf, A., van't Veld, A.A., Langendijk, J.A., Schilstra, C. 2012. Statistical validation of normal tissue complication probability models, *Int J Radiat Oncol Biol Phys* 84(1):e123–9.

CHAPTER 4

Pelvis

Rectal and Bowel Toxicity

Sarah L. Gulliford, Julia R. Murray, and Martin A. Ebert

CONTENTS

Introduction 76
  Clinical Significance of the Injury 76
  Relevant Toxicities 76
  Clinician-Reported Outcomes 77
  Patient-Reported Outcomes 77
  Relevant Structures and Dosimetry 78
    Relevant Structure Definitions 79
  Anorectum and Anal Canal 79
  Large and Small Bowel 79
    Dose-Volume Data 80
Recommendations at the Time of QUANTEC 80
  Rectum 80
    Dose-Volume Constraints 80
    NTCP Models 81
  Small Bowel 81
    Acute Radiotherapy-Induced Small Bowel Toxicity 81
    Dose-Volume Constraints 81
    Late Radiotherapy-Induced Small Bowel Toxicity 81
    Stereotactic Body Radiotherapy 82
Review of the Literature after QUANTEC (Conventional Fractionation) 82
  Prostate Cancer 82
    Anorectum 82
    Acute Rectal Toxicity 82
    Late Rectal Toxicity 85
    Rectal Toxicity Following Proton Therapy 86
    Anal Canal 86
    NTCP Models 86
  Bladder Cancer 87

Bowel and Rectal Toxicity..87
IMRT, IGRT, and Adaptive Planning..88
Pelvic Lymph Node Irradiation in Prostate Cancer..89
Gynaecological Malignancies...90
Rectal Cancer 91
Review of the Literature after QUANTEC (Altered Fractionation) 92
Bowel – Accelerated Treatment 92
Bowel – Hypofractionation 92
Rectum – Moderate Hypofractionation (2.5–3.5 Gy/Fraction) 92
Rectum – Extreme Hypofractionation (>3.5 Gy/fraction) 94
Suggested Parameters...99
Rectum...99
Bowel...100
Open Issues and Future Directions of Investigation 101
Non-Conventional Anatomical Association 101
Planned vs. Accumulated Dose..101
Molecular Factors Impacting Toxicity 101
Improved Outcomes Reporting...102
References...102

INTRODUCTION

Approximately one million people are treated worldwide with pelvic radiotherapy annually (Andreyev, 2016). The irradiation field usually encompasses healthy intestinal tissue, which can result in gastrointestinal radiation-induced toxicity and up to half of pelvic-radiation treated patients say that their quality of life is affected by gastrointestinal symptoms (Andreyev et al., 2010). A reduction in toxicity is seen in parallel with radiation technological advances such as image guided intensity modulated radiotherapy.

Clinical Significance of the Injury

The ano-rectum and bowel are long muscular tubes in linear continuity from the gastric pylorus to the anus. The small bowel (intestine) is about 7 m long and gradually diminishes in diameter and the large bowel is about 1.5 m long. The bowel performs specialised digestion and absorption functions, regulates fluid and electrolyte movement, and absorbs vitamins and minerals. The pathophysiology of the consequences of radiation exposure include involvement of the intestinal immune system and microvascular endothelium in acute mucositis and adverse tissue remodelling and vascular insufficiency (intestinal fibrosis). Cell kinetic turnover times of the gastrointestinal system determine the latent period before clinical syndromes and manifestations occur.

Relevant Toxicities

The constellation of gastrointestinal (GI) symptoms experienced by patients who have received radiation treatment for a pelvic malignancy, from transient to long term, mild to very severe, have been comprised in the definition of pelvic radiation disease. These complications include rectal bleeding, increased frequency of bowel movements, bowel urgency, tenesmus, diarrhoea, faecal incontinence, bowel obstruction and fistulae. "Acute" toxicity expresses symptoms during or up to approximately 12 weeks after completion of radiotherapy. Incidence rates of acute toxicities are often, though not always, inversely correlated with the overall time of treatment. Symptoms subsequent to this will be referred to here as the result of "chronic" or "late" toxicities. Many late toxicities are believed to be related to acute toxicity – referred to as a "consequential" injury.

Clinician-Reported Outcomes

The most prevalent clinician-reported instruments used for reporting GI toxicities, providing integer grades from 0 (absence of any radiation-induced toxicity) through to a maximum of 5 (death related to that toxicity), are:

- RTOG/EORTC (Cox et al., 1995): Based on a consensus of the Radiation Therapy Oncology Group (RTOG) and the European Organization for Research and Treatment of Cancer (EORTC), provides a single grade for generic acute and late GI symptoms.

- LENT-SOMA (Pavy et al., 1995; "Lent Soma Scales for All Anatomic Sites," 1995): Late Effects Normal Tissues – Subjective Objective Management Analysis. Based on a consensus of the RTOG and EORTC, designed to incorporate the diversity and degree of symptom and anatomy-specific radiation injury.

- CTCAE (NCI 2010): Common Terminology Criteria for Adverse Events. Developed by the National Cancer Institute, providing anatomy-specific symptom-based grades.

It is common for these instruments to be combined (including with patient-reported outcomes), amended or modified completely for specific institutional or trial use.

Reports of trial outcomes and toxicity models based on clinician scored toxicity commonly group individual symptom scores together, which carries a risk of ambiguity and loss of association as symptoms may have different etiology potentially leading to a loss of casual associations (Yahya et al., 2014). With differences in the time-course of individual symptoms, a different group of associated predictive factors can be found if the duration, prevalence or peak of symptoms are used (Fiorino et al., 2012). Factors unrelated to radiotherapy can be accommodated, such as baseline (pre-treatment) toxicity (e.g., presence of rectal bleeding due to haemorrhoids) by not including patients with baseline grades >0 in the associated analysis.

Patient-Reported Outcomes

Asking patients to report their own symptoms using validated patient outcome questionnaires can achieve added value by improving symptom control, physician-patient communication, and decision making (Sonn et al., 2013). A systematic review confirmed that the incidence, severity, and distress of the symptoms expressed by cancer patients was often underestimated (Xiao et al., 2013). The most frequently used in the literature for evaluating gastrointestinal symptoms in patients receiving pelvic radiotherapy are:

- European Organisation for Research and Treatment of Cancer (EORTC) QLQ-C30: translated and validated in over 100 languages and is used in more than 3000 studies worldwide, supplemented by disease-specific modules.

- Functional Assessment of Cancer Therapy – Prostate (FACT-P): assesses the impacts of cancer therapy in four general domains (physical, social/family, emotional, functional) with the addition of specific symptoms possibly arisen by therapies for prostate cancer. There are additional questions measuring cancer-specific factors that may affect quality of life.

- Expanded Prostate Cancer Index Composite (EPIC): developed to measure health related quality of life and has been validated in men with localized prostate cancer who underwent surgery, external beam radiotherapy, or brachytherapy. EPIC is sensitive to specific health related quality of life effects of these therapies and of cancer progression.

Ideally, clinician-derived scores would correlate with the quality of life (QOL) impact reflected in patient reporting though frequently only a vague association is found with a tendency for clinicians to under-report symptoms of concern to patients (Di Maio et al., 2016).

Relevant Structures and Dosimetry

For any method used to derive or utilise dosimetric features based on delineated anatomy, it is important to ensure consistent definition of the structure for delineation (Onal et al., 2009). For this reason, definitions are provided below (see Figures 4.1 and 4.2) and,

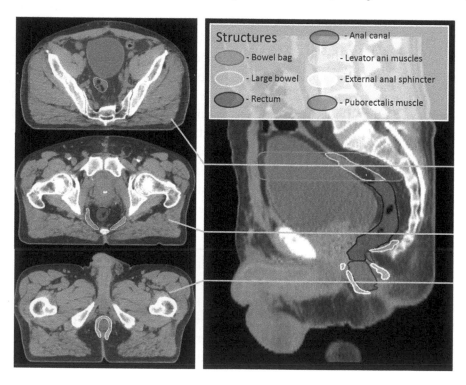

FIGURE 4.1 Male pelvic computed tomography (CT) scan showing sagittal view (right) with locations for three axial views (left), indicating defined relevant anatomy.

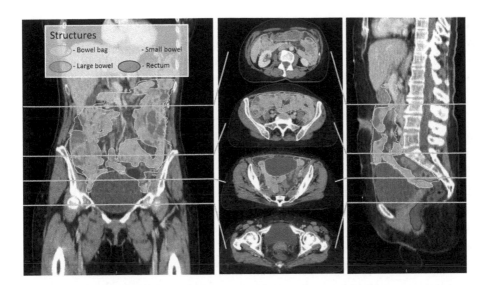

FIGURE 4.2 Female pelvic/abdominal computed tomography (CT) showing coronal (left) and sagittal (right) and multiple axial (centre) views, indicating relevant defined anatomy.

where required, are translated from reported studies. Reporting and use of either absolute or relative quantities (such as volume and dose) (Fiorino et al., 2003) and either physical or biologically effective/equi-effective dose should be consistent.

Relevant Structure Definitions

Anorectum and Anal Canal

The superior extent of the anorectum is defined as the recto sigmoid junction where the rectum turns left and anteriorly. Inferiorly, it is common to use the bony landmark of the ischial tuberosities to define the lower limit. The anal canal is considered to be the 3 cm superior to the ischial tuberosities. It is important to consider the outlining definition for a specific study since some groups choose to only contour the rectum which lies within close proximity to the prostate/Planning Target Volume (PTV). Also of note is the inconsistency between defining the anorectum as a volume (including contents) compared to defining the rectal wall.

Large and Small Bowel

The large bowel is contoured as individual visible bowel loops. Inferiorly, this will typically be from the axial slice immediately above the end of the anorectum or the sigmoid (Gay et al., 2012). The loops of the small bowel will generally be visible on computed tomography (CT) and can be made more distinct with contrast. The small bowel is contoured as individual visible bowel loops. Note that the degree of attention to detail of the structure of the bowel loops will vary between observers.

The complexities of the actual bowel structure and movement throughout treatment can make contouring challenging and ambiguous. For this reason, the contouring of the peritoneal cavity region containing bowel loops, the "bowel bag," can be easier and more robust, with the caveat that detailed dose-volume relationships for actual bowel structures can then only be implied (Xu et al., 2016). The region can be contoured using CT covering the area and incorporates abdominal contents excluding muscle and bones (Gay et al., 2012). The inferior extent will be the axial slice showing the most inferior bowel loop or the slice immediately above the anorectum.

Dose-Volume Data

Dose-volume histograms (DVHs) have been the standard reporting method for dosimetric features and to correlate with toxicity outcomes. As the filling of these structures is typically assumed irrelevant, many investigations have examined the role of dose histograms to the surface, using surface-area approximations (dose-surface histogram, DSH) or the structure "wall" (dose-wall histogram, DWH) (Meijer et al., 1999). Similar or slightly better predictive power for DSH- over DVH-based metrics has been found [e.g., (Casares-Magaz et al., 2017; Buettner et al., 2011)]. Often there is a strong correlation between histogram types and similar associations and predictive models result from the use of each (Fiorino et al., 2003). With the advent of stereotactic techniques for pelvic treatments combined with irradiation that can produce steep dose gradients, the benefit of more precise structure definition may become more apparent, as found for example by Kim et al. (2014) (see Table 4.4). Spatial descriptors of dose distributions, such as the rectal dose-surface map (DSM) can also be used to define dosimetric features. Such spatial descriptors shall be discussed in Chapter 17.

RECOMMENDATIONS AT THE TIME OF QUANTEC

Rectum

The QUANTEC (Quantitative Analyses of Normal Tissue Effects in the Clinic) report relevant to rectal injury (Michalski et al., 2010) drew information solely from prostate cancer cohorts and generally 3D conformal radiotherapy. In the report, issues of inconsistency in delineating/defining the rectum were highlighted, especially in relation to the superior and inferior extent, with some groups only considering the rectum within a centimetre of the PTV. The challenges of inter- and intra-fraction rectal motion/filling were also noted.

Dose-Volume Constraints

The review of dose-volume data indicated that the majority of reports described late grade 2 RTOG toxicity, with dose constraints for doses ≥60 Gy. A plot of published dose-volume constraints was included with constraints converted to equivalent dose in 2 Gy per fraction as required. Overall the recommended constraints were $V_{50Gy} < 50\%$, $V_{60\,Gy} < 35\%$, $V_{65Gy} < 25\%$, $V_{70Gy} < 20\%$ and $V_{75Gy} < 15\%$. The paper emphasised that since different dose levels are highly correlated the statistical significance of some intermediate dose constraints may be as a result of correlations to higher dose levels. It is however acknowledged that there may be a biological explanation with volumes exposed to intermediate doses playing a role in repair.

It was noted that data from a Dutch randomised trial (Peeters et al., 2006) had been used to report rectal bleeding, stool frequency, and fecal incontinence separately and that dosimetric analysis had resulted in inconsistencies in parameter fits. In the case of faecal incontinence, the organ of risk defined was the anal canal. The report highlighted factors that may modify the radiotherapy induced toxicity including diabetes mellitus, haemorrhoids, inflammatory bowel disease, prior abdominal surgery, age, androgen deprivation therapy, and severe acute toxicity.

NTCP Models

A number of studies had published parameter values fitting rectal toxicity data (grade ≥ 2 rectal bleeding defined following RTOG) to the Lyman-Kutcher-Burman (LKB) model (Burman et al., 1991). A meta-analysis of these results derived values (95% Confidence Interval [CI]) of $D_{50} = 76.9$ (73.7–80.1) Gy, $n = 0.09$ (0.04–0.14) and $m = 0.13$ (0.10–0.17) indicating that the highest doses are the most important to predict for toxicity and do not differ significantly from the original values of the Emami publication of $D_{50} = 80$ Gy, $n = 0.12$ and $m = 0.15$ (Emami et al., 1991).

Data from hypofractionated prostate radiotherapy studies was limited at this time and the recommendation was to convert the DVH to conventional fractionation schemes using the linear quadratic (LQ) model with an α/β ratio of 3 Gy.

Small Bowel

Acute Radiotherapy-Induced Small Bowel Toxicity

QUANTEC data for acute toxicity relating to the small bowel (Kavanagh et al., 2010) centered around 6 studies from rectal and gynaecological malignancies with quantitative dose-volume analysis. Toxicities included diarrhoea, obstruction or constriction, and fistula or perforation, although details on grades and scales were sparse. Acute grade 3 toxicity was the most commonly reported endpoint. There was a systematic difference in contouring whereas in most cases, individual bowel loops were delineated however, for some studies the entire peritoneal cavity defined the contour. Most patients received concomitant chemotherapy making it difficult to isolate the contribution of each treatment modality to toxicity. A limited number of studies were included which compared acute grade 3 toxicity incidences between cohorts who were treated with/without chemotherapy. Concomitant chemotherapy increased toxicity.

Dose-Volume Constraints

Baglan et al. (2002) reported data for a cohort of 40 rectal cancer patients where individual loops of bowel were delineated. Volume thresholds to distinguish between patients with/without acute grade 3 toxicity were established for a range of doses. This study was validated with a cohort of 96 patients (Robertson et al., 2008). The main constraint to predict acute grade 3 toxicity was $V_{15Gy} < 127$ cm^3. The results from the other cited publications were generally consistent.

Late Radiotherapy-Induced Small Bowel Toxicity

There is consistent evidence that pre-operative short course rectal radiotherapy (25 Gy in 5 fractions) increases the risk of bowel obstruction. There is also some evidence that extending pelvic fields superiorly above the sacral promontory increases the risk (Mak et al., 1994). Overall, risk of toxicity (late obstruction/perforation) for conventional radiotherapy of 50 Gy where $V_{50Gy} < 5\%$ is reported as 2 to 9%, which is considered consistent with the original Emami partial volume estimate of 50 Gy to one third of the small bowel as associated to 5% probability of bowel toxicity within the first 5 years after radiotherapy treatment (Emami et al. 1991). The only study cited presenting a model to predict late toxicity described a power law where doubling of the volume of a bowel in the radiotherapy field required a reduction of dose of 17% to maintain isotoxicity (Letschert et al., 1990).

Stereotactic Body Radiotherapy

In the QUANTEC paper dedicated to stomach and small bowel dose-volume relationships, Stereotactic Body Radiotherapy (SBRT) was considered under the heading of "special situations." One reference (Schellenberg et al., 2008) described a threshold of 30 cm³ receiving 12.5 Gy in a cohort of pancreatic cancer patients (following 25 Gy single fraction). A subsequent publication (Chang et al., 2009) applied constraints of <5% receiving 22.5 Gy and <50% receiving 12.5 Gy with the 50% isodose line to not reach the opposite luminal wall. Late toxicity for duodenum and stomach was 9%.

REVIEW OF THE LITERATURE AFTER QUANTEC (CONVENTIONAL FRACTIONATION)

Prostate Cancer

The majority of published studies relating radiotherapy to rectal toxicity are from prostate cancer radiotherapy cohorts and the long-term data collection ensures a full scope of the toxicity profile for this cohort. Landoni et al. (2016) provide a summary of publications relating the dosimetry of organs at risk to toxicity following prostate radiotherapy. Commonly, the anorectum is the anatomical structure of interest but the definition varies and the anorectum can be split in to the sub regions of anal canal and rectum. Significant relations between dosimetry and GI toxicity derived since QUANTEC for conventional prostate radiotherapy dose fractionation, are summarized in Table 4.1.

Anorectum

Acute Rectal Toxicity

Although not uncommon in the setting of prostate cancer radiotherapy, being generally low grade and resolving within 6 months of starting treatment, acute rectal toxicity is not usually considered to be dose limiting. A recent study by Mirjolet et al. (2016) assessed both absolute (volume in cm³) and relative (%) descriptors of rectal dosimetry against CTCAE v3 reported acute toxicity in a cohort of prostate-only Image Guided (IG) Intensity Modulated Radiation Therapy (IMRT) patients. 53.3% of patients reported grade ≥1 toxicity in at least one of the endpoints assessed. Unusually the results indicated that absolute volumes receiving doses in the range 35–50 Gy were predictive of toxicity whilst none of the relative dosimetry parameters were statistically significant. A cumulative metric of area under the DVH between 25 and 50 Gy was also tested and a cut off value of 794 cc x Gy (sic) was also found to be predictive. Palumbo et al. (2017) found no relative dosimetric parameters were statistically related to acute CTCAE v3 rectal toxicity in a cohort of 195 localised prostate patients with dose-volume constraints integrated into their treatment planning. 79 patients reported toxicity grade 1–3.

TABLE 4.1 Derived Associations Between Dosimetric Risk Factors and Gastrointestinal Toxicity Following Conventional Prostate Radiotherapy Since the QUANTEC Report

	Number of Patients	Total Prescription Dose (Gy)/Total Fractions/Total Weeks	Dose per Fraction (Gy)	Treatment Technique	Toxicity Scoring	Identified Predictive Dosimetric Risk Factors	Identified Modifying/Predictive Risk Factors
Fellin et al. (2009)	718	≥70/not specified/not specified	1.8–2	3DCRT	In-house questionnaire	Rectal Bleeding $V_{75Gy} < 5\%$ $V_{70Gy} < 15-20\%*$	Faecal Incontinence-acute toxicity *patients treated with previous pelvic/abdominal surgery
Ebert et al. (2015)	754	66–78/33–39/6.5–8	2	3DCRT	LENT/SOMA	Rectal Bleeding: Anorectum $V_{30-65Gy}$ Anal Canal V_{40Gy} Proctitis: Anorectum: $V_{50-62Gy}$ Anal Canal $V_{36-63Gy}$ Stool Frequency: Anal canal V_{8-56} Gy	
Gulliford et al. (2010)	388	64 or 75/32 or 37/6.5 or 7.5	2	3DCRT	RMH/RTOG/ LENT/SOMA/ UCLA-PCI	$V_{30Gy} \le 80\%$, $V_{40Gy} \le 65\%$, $V_{50Gy} \le 55\%$, $V_{60Gy} \le 40\%$, $V_{65Gy} \le 30\%$, $V_{70Gy} \le 15\%$, $V_{75Gy} \le 3\%$.	
Cicchetti et al. (2017)	1336	66–80/not specified/not specified	1.8–2	3DCRT	LENT/SOMA	Stool Frequency: EUD ($n=1$) mean dose Rectal Pain EUD ($n=0.35$)	Stool Frequency: Cardiovascular disease Both: Acute GI toxicity
Fonteyne et al. (2014)	637	74–78/not specified/not specified	Not stated	IMRT	RILIT (combined toxicity score)	Late Rectal Toxicity $V_{40Gy} < 64-35\%$, $V_{50Gy} < 52-22\%$, $V_{60Gy} < 38-14\%$ $V_{65Gy} < 5\%$	
Peterson et al. (2014)	111	72–79/not specified/not specified	1.8 Gy	IG-IMRT	CTCAE vs. 4	$V_{67.5\,Gy} < 1.29$ cm^3 $V_{70Gy} < 0.73$ cm^3 $V_{72.5Gy} < 0.45$ cm^3	

(Continued)

TABLE 4.1 (CONTINUED) Derived Associations Between Dosimetric Risk Factors and Gastrointestinal Toxicity Following Conventional Prostate Radiotherapy Since the QUANTEC Report

	Number of Patients	Total Prescription Dose (Gy)/ Fractions/Total Weeks	Dose per Fraction (Gy)	Treatment Technique	Toxicity Scoring	Identified Predictive Dosimetric Risk Factors	Identified Modifying/Predictive Risk Factors
Chennupati et al. (2014)	327	69–79.2/5 fraction per week	1.8–2 Gy	IMRT	CTCAE vs. 3	$V_{70Gy} < 10\%$, $V_{65Gy} < 20\%$, $V_{40Gy} < 40\%$.	
Mendenhall et al. (2012)	211	78–82 GyE/39–41/not specified	2Gy	Protons	EPIC IPSS IEFF-5	Rectal wall V_{70Gy}	
Colaco et al. (2015)	1285	72–82.3 RBE/not specified/not specified	Not stated	Protons	CTCAE vs3	$V_{75Gy} < 9.4\%$ Rectal wall $V_{75Gy} < 9.2\%$	Aspirin usage Anticoagulant therapy
Smeenk et al. (2012a)	60	67.5 or 70 Gy	2.25 or 2.5 Gy	3DCRTor IMRT	RILIT	Urgency: Anal Wall mean dose < 38 Gy Incontinence: Anal Wall Dmin < 5 Gy $V_{50Gy} < 39\%$	
Alsadius et al. (2012)	414	70 Gy/35/7	2 Gy	3DCRT	In-house questionnaire	Fecal Leakage: Anal Sphincter Mean dose < 40 Gy	

IMRT=Intensity Modulated Radiation Therapy; 3DCRT=3 dimensional Conformal Radiation Therapy; IG=image guided; Dmin=minimum dose; V_{XGy}=percent or absolute organ volume receiving ≥ X Gy; RBE=relative biological effectiveness; RILIT=radiotherapy-induced lower intestinal toxicity; EUD=Equivalent Uniform Dose; EPIC=Expanded Prostate Cancer Index Composite; IPSS=International Prostate Symptom Score; IEFF-5= simplified International Index of Erectile Function (5 questions); CTCAE vs3= Common Terminology Criteria for Adverse Events version 3.0; RMH= Royal Marsden hospital scale: RTOG= Radiation Therapy Oncology Group;
LENT/SOMA= Late Effects of Normal Tissues/ Subjective Objective Management and Analytic; UCLA-PCI= University of California Los Angeles – Prostate Cancer Index

Late Rectal Toxicity

The QUANTEC report for rectal injury (Michalski et al., 2010) indicated consistent results in terms of constraints in the region of 60 Gy to reduce toxicity and it is still widely accepted that the high-dose region of the DVH is a strong predictor for rectal bleeding (Fellin et al., 2009). However, a number of studies have supported the QUANTEC observation that lower doses may also be relevant (Ebert et al., 2015; Gulliford et al., 2010). Despite the comprehensive body of publications relating to rectal bleeding, other toxicities are considered to have a potential for more serious impact on quality of life. Symptoms such as urgency and loose stools are less objective endpoints with potential for co-morbid causes. Since these endpoints are harder to quantify it is also harder to find definitive relationships with dosimetry. Development of multi-symptom syndromes which describe the patient experience have evolved (Capp et al., 2009). However, a number of publications have indicated that rectal urgency and loose stools have a more complex dosimetric relationship than rectal bleeding.

Cicchetti et al. (2017) reported on a pooled analysis of stool frequency and rectal pain between two large multicentre trials (Airopros 0102 and TROG03.04 RADAR). Incidence was markedly different in the two trials but was consistently low (<5% in both cohorts). Dosimetric data was converted to Equivalent Uniform Dose (EUD) (Niemierko, 1997) and the volume parameter n was optimised by maximising log likelihood. For stool frequency (reported as a longitudinal score) n was found to be 1 suggesting that mean dose was the best descriptor to relate dosimetry to toxicity. For rectal pain n was found to be 0.35 suggesting a serial response however still much higher than the values of n associated with rectal bleeding ($n = 0.09$) (Michalski et al., 2010).

Fonteyne et al. (2014) derived dose-volume constraints for prostate patients treated using IMRT, both radical and post prostatectomy patients were included. A composite endpoint of late rectal toxicity was considered which included diarrhoea, rectal blood loss, mucus loss, faecal incontinence, abdominal cramps, urgency stool frequency, and rectal pain mostly reported according to RTOG. 11% of the 637 patients included in the study reported grade 2 toxicity. Dose-volume constraints were derived using Cox proportional hazards regression for an estimated risk of 5% and 10% of developing grade 2 toxicity at 60 months. Helpfully the results are presented alongside other published studies. Of note is the choice of rectal wall rather than rectal volume as the outlined structure. Despite this difference there is concordance between the constraints for a 10% estimates risk of grade 2 late rectal toxicity and other published sets of constraints derived from conformal prostate radiotherapy.

Peterson et al. (2014) considered absolute volume of the anterior rectal wall when deriving constraints for CTCAE late rectal toxicity. This toxicity scale includes a diverse range of endpoints with toxicities including diarrhoea, proctitis, and rectal haemorrhage. 27% of the 111 patients who were treated with IMRT with daily image guidance reported late rectal toxicity. Statistically significant constraints were derived for doses in the region of 67.5–72.5 Gy although these dose levels are likely to be very highly correlated and suggest a serial response. Chennupati et al. (2014) demonstrated that there was a significant difference (p = 0.03) in reporting grade 2 CTCAE toxicity 2 years after prostate radiotherapy between patients whose prostate IMRT treatment plan met a set of dose-volume constraints and those who did not (see Table 4.1).

Rectal Toxicity Following Proton Therapy

Relatively low GI toxicity rates from proton and ion beam radiotherapy have been reported [e.g., (Ishikawa et al., 2017)], with only limited dosimetric analyses undertaken. The RTOG 03–12 study (Coen et al., 2011) tested late toxicity after prostate proton therapy delivered with lateral beams to 82 Gy Cobalt Gray Equivalent (CGE). Of the 84 evaluable patients, 17 reported acute GI toxicity, whilst 24 reported late toxicity (RTOG scoring). A comparison was made of V_{70Gy}, $V_{75Gy,}$ and V_{80Gy} for patients with and without grade 2 toxicity and no statistical differences were found.

Rectal bleeding was reported as the dominant toxicity in a cohort of 1285 prostate patients treated with proton therapy (Colaco et al., 2015). 217, 187, and 11 patients reported CTCAE rectal bleeding of grades 1, 2, and 3 respectively. Multivariate analysis of dosimetric and clinical factors indicated that the volume of rectum or rectal wall receiving $\geq$ 75 Gy was statistically significant with thresholds of 9.4% and 9.2% respectively.

Anal Canal

Buettner et al. (2012) created dose-surface maps of the anal canal and demonstrated that the mean dose and lateral extent of anal canal receiving at least 53 Gy predicted for subjective sphincter control (LENT SOMA).

A study by Smeenk et al. (2012a) (N=60) found that the dose to the anal wall was associated with both urgency and incontinence. Alsadius et al. (2012) studying 414 patients compared the dose distribution to the anal sphincter with the responses to a patient-reported outcome measure of fecal leakage at least once a month and demonstrated that a mean dose constraint of 40 Gy was predictive of toxicity with a prevalence ratio of 3.8 (95% CI 1.6–8.6). A large retrospective analysis of patients from the RADAR study (Ebert et al., 2015) found that dose to the anal canal was identified as being related to bleeding, proctitis, stool frequency and urgency and tenesmus.

NTCP Models

Traditionally NTCP models for rectal toxicity have been represented using the LKB model. For rectal bleeding there are consistent results in the literature which indicate that maximum doses are the dominant predictors (represented by small values of the volume parameter n). However for other endpoints and for composite endpoints data is more varied (Gulliford et al., 2012; Rancati et al., 2011).

An alternative approach is to use a multivariate logistic regression. Schaake et al. (2016) developed separate models for grade $\geq$2 rectal bleeding, stool frequency, and faecal incontinence as defined using CTCAE v3, for a cohort of 262 prostate patients. Rectum, anal canal, and pelvic floor muscles were studied together with a range of clinical factors. The final model for rectal bleeding included the volume of anorectum receiving $\geq$ 70 Gy and anticoagulant use. In contrast the model for stool frequency included the volume of the iliococcygeal muscle receiving $\geq$ 45 Gy and the volume of the levator ani muscle receiving $\geq$ 40 Gy.

Faecal incontinence was best predicted using volume of the external sphincter receiving ≥ 15 Gy and the volume of iliococcygeal muscle receiving ≥ 55 Gy. The use of EUD as a summary metric for dosimetry was also tested, but the results were found to be inferior to using individual dose metrics.

D'Avino et al. (2015) compared LKB and multivariate NTCP models for a cohort of 84 patients. The parameters for the LKB model are consistent with other publications, $D_{50} = 87.3$ (95% CI : 75.9–102.2), $m = 0.37$ (95% CI :0.26–0.64), $n = 0.1$ (95% CI :0.02–0.26), with the multivariate model including the volume of rectum receiving ≥ 65 Gy, the use of hypertensives/anticoagulants and acute toxicity. The area under the Receiver Operating Characteristic curve (AUC) for the two models were 0.6 (LKB) and 0.75 (multivariate). A similar comparison was made by Troeller et al. (2015) for grade ≥ 2 CTCAE rectal toxicity. Models were built separately for cohorts treated with 3D conformal radiotherapy and IMRT. It was observed that parameter values in the final models did not agree, underlining that models should only be used on similar datasets. Ospina et al. extended the comparison to include a third model built using a random forest (Ospina et al., 2014) using data from 261 patients who received 3D conformal prostate radiotherapy to rectal bleeding and overall rectal toxicity. In each case the LKB model parameters were similar to those published in QUANTEC. However, both the random forest and logistic regression models incorporated a range of dosimetric and clinical parameters and were shown to have superior model performance compared to a conventional LKB model. This may be in part due to the ability to include clinical factors. Peeters et al. (2006) introduced the concept of a dose modifying factor to include clinical variables into the LKB model. This concept was extended by Rancati et al. (2011) to combine clinical factors. The dose modifying factor was defined as the ratio of D_{50} with and without the feature in the model and, again, previous pelvic surgery and acute toxicity were identified as improving model predictions along with a past history of bowel pathology which were predictive for incontinence. Additional clinical dose modifying factors were observed by Defraene et al. (2012) including cardiac history and diabetes.

Bladder Cancer

In whole-bladder radiotherapy for invasive bladder cancer, often delivered concurrently with chemotherapy, the small bowel volumes receiving appreciable radiation doses are sufficient to cause injury. Methods to account for organ motion, and to deliver more conformal dose distributions, have been implemented recently. Toxicity assessment can be complicated by pre-morbid conditions and coincidental symptoms.

Bowel and Rectal Toxicity

Few relevant studies have separated rectal toxicity from specific bowel toxicity. Specific reported rectal toxicity rates tend to be low. Efstathiou et al. (2009) reported on a combination of RTOG studies representing 285 patients treated during 1990–2002. In this study, no clinical variable could be associated with pelvic toxicity.

Majewski and Tarnawski (2009) reported on 487 patients treated between 1975 and 1995 with prescribed doses between 59.2 Gy and 72 Gy in 1.2–2.5 Gy per fraction. Whole-pelvis treatment was provided to 303 patients. Five year risk of grade ≥ 2 and grade 3 bowel toxicity was 7% and 3% respectively, with a typical 10–15 months latency to toxicity. On univariate Cox proportional hazards analysis, incidence of late bowel toxicity was influenced by patient age (Hazard Ratio (HR): 0.94, $p = 0.01$), beam energy (^{60}Co vs. 9–20 MV X-rays, HR: = 10.84, $p = 0.02$), whole-pelvis irradiation (HR: 11.7, $p = 0.01$), and incidence of acute bowel toxicity (HR: 1.8, $p = 0.004$). The onset of acute bowel toxicity was accelerated when the dose was delivered quickly.

In one of the few in-depth bowel dose-volume analyses in the context of bladder cancer (McDonald et al., 2015), 55 patients were treated with a 2×2 combination of whole-bladder or focal radiotherapy with or without concurrent chemotherapy. Bowel was outlined as individual bowel loops on axial images within 2 cm from the PTV with late RTOG toxicity data available for dosimetric correlation for 47 patients (15% grade 1, 6% grade 2). Distinct dose-volume relationships were observed above 30 Gy. A table of thresholds for specific accepted rates of late grade ≥1 RTOG toxicity was generated as presented in Table 4.2.

IMRT, IGRT, and Adaptive Planning

IGRT techniques permit adaptation to bladder or tumour position typically via the application of a "plan of the day" according to daily-assessed anatomy against a library of potential treatment plans (Thornqvist et al., 2016), allowing a reduction in the amount of bowel exposed to high doses with an estimated reduction in grade ≥ 2 bowel toxicity (diarrhoea) from 35% to 7% (Zhang et al., 2015). Similarly, the steeper dose gradients and normal tissue avoidance afforded by IMRT allows reduction in toxicity as well summarised in Zhang et al. (2015).

Multiple trials and clinical studies of adaptive IGRT and/or IMRT have demonstrated the reduction of GI toxicity from an adaptive approach or no increase in the presence of dose escalation, with acute grade ≥ 2 GI toxicities typically of the order of 20% and a near-absence of grade ≥ 3 toxicities [e.g., (Hafeez et al., 2017; Lutkenhaus et al., 2016)].

Lutkenhaus et al. (2016) found a significant ($p = 0.05$) decrease in late grade ≥ 2 toxicity with IMRT relative to 3D Conformal Radiation Therapy (3DCRT). Only tumour size was a predictor of acute grade ≥ 1 GI toxicity. As reported by Sondergaard et al. (2014), patients

TABLE 4.2   Constraint Volume (cm³) of Bowel Receiving a Given Dose to Keep Grade ≥ 1 Late Toxicity Rates Below the Level (%) Specified in the Left-Most Column

	Dose Levels							
	V_{30Gy}	V_{35Gy}	$V4_{0Gy}$	V_{45Gy}	V_{50Gy}	V_{55Gy}	V_{60Gy}	V_{65Gy}
Constraints for grade ≥ 1 toxicity risk <20%	149	136	126	116	104	91	73	23
Constraints for grade ≥ 1 toxicity risk <25%	178	163	151	139	137	115	98	40
Constraints for grade ≥ 1 toxicity risk <30%	202	187	174	158	146	136	121	56
Constraints for grade ≥ 1 toxicity risk <40%	246	228	213	193	181	172	169	83

Source: McDonald et al., 2015.

treated consecutively with 3DCRT (N=66) were compared to those treated with IMRT (N = 50). Several factors impacted on CTCAE toxicity – grade ≥ 2 acute diarrhoea was 56% for 3DCRT vs. 30% for IMRT (p = 0.008); inclusion of pelvic nodes increased acute toxicity; bowel (loops) DVH was significantly greater in the group with acute diarrhoea in the range 10–50 Gy; for the bowel, V_{45Gy} of 200 cm³, 400 cm³, 600 cm³, and 800 cm³ corresponded to a risk of acute grade 2 diarrhoea of 27%, 38%, 50%, and 62% respectively; there were no significant differences in late toxicity between 3DCRT and IMRT.

Pelvic Lymph Node Irradiation in Prostate Cancer

The merit of elective pelvic lymph node radiotherapy (PLNRT) compared with treatment of prostate and seminal vesicles alone remains controversial. IMRT makes it possible to increase bowel sparing, which is the dose-limiting normal tissue when treating the pelvis (Fiorino et al., 2009). A phase I/II dose-escalation study for PLNRT in patients at high risk of lymph node metastases has shown acceptably low toxicity using doses up to 60 Gy in 37 fractions and in modest hypofractionation schedules (47 Gy in 20 fractions over 4 or 5 weeks), if the bowel constraints were adhered to recommendations (Reis Ferreira et al., 2017). Mandatory bowel dose constraints in conventionally and (hypofractionated) cohorts were 158 cm³ for the absolute bowel volume receiving ≥ 45 (39) Gy (V_{45Gy} and $V39_{Gy}$ for conventionally and hypofractionated patients, respectively), 110 cm³ for V_{50Gy} (V_{43Gy}), 28 cm³ for V_{55Gy} (V_{47Gy}), 6 cm³ for V_{60Gy} (V_{51Gy}), and 0 cm³ for V_{65Gy} (V_{55Gy}). At 2 years, RTOG grade ≥2 rates for the five reported cohorts were between 8.3 and 16.4%.

In radical series, the majority of retrospective analyses have shown some increase in principally GI toxicity with pelvic radiotherapy compared with prostate or prostate bed radiotherapy. Deville et al. (2010) in a retrospective study on 67 patients found a significant increase in acute bowel toxicity (61% vs. 29% grade ≥ 2 toxicity in whole pelvis irradiation and prostate bed radiotherapy respectively), with no difference in late toxicity (3% vs. 0%). A few retrospective analyses of patients treated with radiotherapy to the prostate or prostate bed and whole pelvis radiotherapy have been undertaken to evaluate the correlation between bowel dose and bowel toxicity:

- Fiorino et al. (2009): 191 patients; conformal or IMRT to pelvic lymph nodes (50.4–54 Gy in 1.8–2 Gy/fraction). 22/191 patients experienced toxicity. Univariate analysis showed significant correlation between V_{20Gy}-V_{50Gy} and toxicity, with a higher discriminative power for V_{40Gy}-V_{50Gy}. Previous prostatectomy (p = 0.066) and abdominal/pelvic surgery (p = 0.12) also correlated with toxicity. Multivariate analysis in this cohort showed that the most predictive parameters were V_{45Gy} (p = 0.002) and abdominal/pelvic surgery (HR: 2.4, p = 0.05).

- Perna et al. (2017): 96 patients; whole pelvis IMRT (50.4 Gy in 1.8 Gy/fraction). On multivariate analysis, acute grade ≥2 RTOG diarrhoea was predicted by $V_{50Gy} \geq 13$ cm³ (for delineated bowel loops, Odds Ratio [OR]: 8.2, p = 0.009) and age (OR: 1.13, p = 0.021).

- Longobardi et al. (2011): 178 patients; PLNRT tomotherapy (median dose: 51.8 Gy in 28 fractions). Acute grade 2 toxicity rate was 8.4%. Main predictors on univariate analysis were nodal Clinical Target Volume (CTVN), treatment duration (<40 days; OR: 6.2, p = 0.006) and acute grade 2 rectal toxicity (OR: 6.5, $p = 0.015$). A multivariate analysis including only pre-treatment variables revealed an independent role of CTVN and age; if including treatment-related factors the best predictors were age, treatment duration, and grade 2 rectal toxicity.

- Mcdonald et al. (2014): 212 patients treated with PLNRT dose 50.4 Gy in 1.8 Gy/fraction. Acute grade ≥ 2 lower GI toxicity occurred in 37% patients receiving PLNRT versus 17% in those who did not (p = 0.001). The Kaplan-Meier estimate of grade ≥ 2 lower GI toxicity at 3 years was 15.3% for patients receiving elective nodal irradiation (ENI) versus 5.3% for those who did not (p = 0.026). Across all patients, V_{70Gy} (in cm^3) of the rectum was the only predictor of late GI toxicity. For patients receiving ENI, $V_{70Gy} > 3$ cm^3 was associated with increased GI events.

- Sini et al. (2017): 206 patients; whole PLNRT to a median dose of 51.8 Gy. Dose-volume parameters for delineated bowel loops were correlated with incidence of acute loose stools toxicity on univariate analysis. A multivariate predictive model was formed by classifying patients according to those meeting "high-risk" DVH criteria ($V_{20Gy} > 470$ cm^3, $V_{30Gy} > 245$ cm^3, $V_{42Gy} > 110$ cm^3). The best cut-off values discriminating patients with or without diarrhoea were $V_{42Gy} > 110$ cm^3 and age ≤66 years.

Gynaecological Malignancies

Chopra et al. (2014) investigated small and large bowel constraints for post-operative cervix cancer radiotherapy. Individual bowel loops were outlined to a level 2 cm superior to the PTV. Late bowel toxicity was scored using CTCAE v3 for a cohort of 71 patients treated with either conformal or IMRT. Both grade ≥ 2 and grade ≥ 3 toxicity were reported with incidence of 30.9% and 12.6% respectively. Constraints were derived for V_{15Gy}, V_{30Gy}, and V_{40Gy} for small bowel and large bowel independently. Reducing the volume of small and large bowel receiving ≥15 Gy (V_{15Gy}) to less than 275 cm^3 and 250 cm^3, respectively, could reduce the incidence of grade 3 toxicity. A subsequent publication (Chopra et al., 2015) used an enlarged dataset (N=103) and found that, for acute toxicity only, volume of peritoneal cavity receiving ≥ 30 Gy ($V_{30Gy} > 900$ cm^3) and concurrent chemotherapy remained statistically significant on univariate analysis whilst no dosimetric constraints remained statistically significant for the multivariate analysis of late toxicity.

Isohashi et al. (2013) reported on dose-volume constraints for small bowel loops derived for a cohort of early stage cervix cancer patients who received postoperative chemoradiotherapy. Both bowel loops and peritoneal cavity were outlined on post-op treatment planning CT for all patients. The median follow-up in the cohort of 97 patients was 43 months (range 4–111 months). Univariate analysis of clinical factors indicated that smoking and low Body Mass Index (BMI) were related to grade ≥ 2 RTOG/EORTC late radiation morbidity. Volumes of small bowel receiving a range of doses from 15–45 Gy were significantly different (p < 0.001) for patients with and without toxicity. Statistically significant results were not observed for dosimetric descriptors of the large bowel and peritoneal cavity. On multivariate analysis, V_{40Gy} was statistically significant with a threshold value of 340 cm^3. It should be noted that the CT scans were limited to 2 cm above PTV and therefore the bowel was not fully outlined. A subsequent publication (Isohashi et al., 2016) expanded the cohort to include IMRT patients and found results consistent with the previous publication with the addition of statistically significant results for the dosimetric descriptors of the bowel bag. However, the authors concluded that the dosimetric descriptors for the small bowel were better predictors of chronic GI toxicity than those derived using the bowel bag.

Rectal Cancer

The relationship between volume of irradiated small bowel and toxicity in rectal cancer radiotherapy is sparsely quantified and complicated by the majority of patients treated with concurrent chemoradiotherapy. A dose-volume relationship has been shown to exist between the volume of irradiated small bowel and acute diarrhoea, with the strongest correlation at lowest doses. Mok et al. (2011) showed retrospectively that IMRT reduced the mean dose to small bowel (25.2 Gy to 18.6 Gy), with a reduction in acute gastrointestinal toxicity (Samuelian et al., 2012). The largest prospective late toxicity data after treatment with IMRT and simultaneous boost (SIB) had a median follow-up time of 54 months with a 9% rate of grade $\geq$ 3GI toxicities (Engels et al., 2014). This retrospective analysis of IG-IMRT and 3DCRT showed a significant reduction in grade $\geq$ 3 acute GI toxicity (IG-IMRT 6.7% vs 3DCRT 15.1%, p = 0.039) with median follow up of 53 months. A further single centre retrospective analysis showed that Volumetric Modulate Arc Therapy (VMAT) treatment substantially reduced high grade acute and late toxicity (Droge et al., 2015).

RTOG 0822 was initiated to determine whether the use of IMRT could decrease the rate of GI toxicity when radiotherapy was combined with neoadjuvant chemoradiation in locally advanced rectal cancer (Hong et al., 2015). Organs at risk included the small bowel, i.e., peritoneal space containing the small bowel, with the following dose-volume constraints: $V_{35Gy} < 180$ cm^3, $V_{40Gy} < 100$ cm^3, $V_{45Gy} < 65$ cm^3, maximum point dose < 50 Gy. The study showed a 51.5% rate of grade $\geq$2 GI toxicity, which substantially exceeded the observed rate of 40% in RTOG 0247. This substantial toxicity is thought to be the consequence of multi-agent chemotherapy (capecitabine and oxaliplatin). It has been postulated that the volume of bowel receiving low-dose radiation may be more important with the confounding elements of multiagent chemotherapy and that the bowel constraints used in RTOG 0822 may be insufficient. Although acute gastrointestinal toxicity is multifactorial, some studies have described a significant relationship between irradiated small-bowel volume and treatment induced diarrhoea during chemoradiotherapy for rectal cancer. Therefore, an optimisation of radiotherapy planning to spare small bowel should be aspired. Reis et al. (2015) performed a study to analyse the differences between the irradiated small-bowel volumes and the occurrence of acute diarrhoea during combined chemoradiotherapy for rectal cancer. There was a statistically significant difference between irradiated small-bowel volumes and the severity of therapy related diarrhoea. Patients with $V_{5Gy} > 291.94$ cm^3 had significantly more often National Cancer Institute Common Toxicity Criteria (NCI CTC) v3.0 grade ≥ 2diarrhoea, than patients with V_{5Gy} below this cut-off value (82% vs. 29%, p < 0.0001). These patients received capecitabine, irinotecan, and cetuximab with the radiotherapy.

A cut-off dose of 15 Gy was found for prediction of grade 3 acute diarrhoea during chemoradiotherapy with 5-fluorouracil (Robertson, Sohn, and Yan, 2010). Studies have shown consistently a volume effect in radiation-induced diarrhoea, however, there is a paucity of data regarding volume effect for small bowel obstruction (Letschert et al., 1994).

REVIEW OF THE LITERATURE AFTER QUANTEC (ALTERED FRACTIONATION)

Interpreting reports of dosimetric correlates can be difficult if the application of 2 Gy Equivalent Dose (EQD2Gy [Bentzen et al., 2012]) in the reporting is ambiguous. Unless otherwise indicated, correlates discussed here are as reported in uncorrected dose per fraction. When converting to EQD2Gy for GI structures (particularly rectum), a reasonable value of α/β to use would be in the range of 3–5.4 Gy (Brenner, 2004).

Bowel – Accelerated Treatment

Accelerated and short-duration radiotherapy for bladder cancer has been found to lead to excessive bowel toxicity (Majewski and Tarnawski, 2009). Lutkenhaus et al. (2016) found these rates greatly reduced (< 2%) when IGRT and IMRT were used for treatment delivery.

Bowel – Hypofractionation

Hypofractionated treatment delivery (>2.5 Gy/fraction) appears to not increase bowel toxicity incidence in the context of bladder radiotherapy provided overall treatment time remains long. Kouloulias et al. (2013) found low rates of acute GI toxicity (5.6% for grade ≥ 1) for 36 Gy in weekly 6 Gy fractions with a 3DCRT technique. Turgeon et al. (2014) reported on mild hypofractionation (2.5 Gy per fractions) delivered daily with IMRT resulting in 4% acute grade 3 GI toxicity. The study by Hafeez et al. (2017) reported outcomes of hypofractionated treatment of 36 Gy in 6 Gy weekly fractions using plan-of-the-day 3DCRT with again 4% acute grade 3 GI toxicity though no grade ≥ 3 late GI toxicity.

Rectum – Moderate Hypofractionation (2.5–3.5 Gy/Fraction)

Considerable evidence on factors influencing rectal toxicity during hypofractionation has emerged in the context of prostate radiotherapy. A comparison of rates observed after hypofractionation with rates after conventional fractionation are somewhat confounded by the changes in technique that can be associated with hypofractionation, such as routine IGRT. A summary of GI toxicity outcomes for earlier non-randomised studies of moderate hypofractionation can be found in several compilations (Landoni et al., 2016; Koontz et al., 2017). A strong consequential relationship has been found for moderate hypofractionation as indicated in Table 4.2.

Di Franco et al. (2017) undertook a meta-analysis of outcomes for studies reporting GI toxicity for moderate hypofractionation and conventional fractionation. However only one of the more recently published large randomized trials comparing moderate hypofractionation and conventional fractionation was included and this study was at odds with their findings which infer a significant reduction in acute GI toxicity for moderate hypofractionation regardless of technique (i.e., 3DCRT or IMRT), a reduction in late GI toxicity for hypofractionated relative to conventionally fractionationated 3DCRT, and a reduction in acute and late GI toxicity for IMRT relative to 3DCRT for hypofractionated treatments.

Here we summarise outcomes comparisons for trials with conventional and hypofractionation arms and otherwise equivalent treatment methods. Treatment schedules are described as total dose (Gy)/number of fractions/dose-per-fraction (Gy)/total treatment time (weeks):

- (Arcangeli et al., 2011; Giorgio Arcangeli et al., 2017): 168 patients, 80/40/2.0/8 vs. 62/20/3.0/4 with 3DCRT. No significant difference in acute or late GI toxicity was found.

- (Norkus et al., 2013): 130 patients, 76/38/2.0/7.6 vs. 63/20/3.15/5 with IGRT (CT)+IMRT including treatment of pelvic lymph nodes. Other than earlier onset for the hypofractionated relative to conventional fractionated schedule there was no significant difference in EPIC-scored acute GI toxicity.

- (Pollack et al., 2013): 303 patients, 76/38/2.0/7.6 vs. 70.2/26/2.7/5.2 with IMRT. No significant difference in acute or late GI toxicity was found.

- (Hoffman et al., 2014): 203 patients, 75.6/42/1.8/8.4 vs. 72/30/2.4/6 with IGRT + IMRT. No significant difference in acute or late GI toxicity was found.

- HYPRO (Aluwini et al., 2016; Incrocci et al., 2016): 820 patients, 78/39/2.0/7.8 vs. 64.6/19/3.4/6.5 3DCRT. No significant difference in late GI toxicity was found. Acute $\geq$ grade 2 GI toxicity was higher in the hypofractionation arm (OR: 1.6, $p = 0.0015$).

- RTOG 0415 (Lee et al., 2016): 1092 patients, 73.8/41/1.8/8.2 vs. 70/28/2.5/5.6 IMRT + 3DCRT. No significant difference in acute GI toxicity was found. Late grade 2 GI toxicity was more likely in the hypofractionation arm (RR = 1.59, $p = 0.005$).

- CHHiP (Dearnaley et al., 2016): 3216 patients, 74/37/2.0/7.4 vs. (60/20/3.0/4 or 57/19/3.0/3.8) with IMRT ± IGRT. The peak GI toxicity rate occurred earlier in the hypofractionated arms. Acute peak grade $\geq$ 2 RTOG bowel toxicity in the conventional fractionation arm (prescribed dose: 74 Gy) was 25%, which was lower than that in the 57 Gy (prescribed dose) hypofractionation arm (38%, $p < 0.0001$) as well as the 60 Gy hypofractionation arm (38%, $p < 0.0001$). Estimated cumulative incidence of late grade $\geq$ 2 bowel toxicity was not significantly different between study arms across multiple scoring scales, except for a difference in LENT-SOMA scores between the 60 Gy relative to 57 Gy hypofractionated schedules (HR: 1.39, p = 0.010).

- PROFIT (Catton et al., 2017): 1206 patients, 78/39/2.0/7.8 vs. 70/28/2.5/5.6 IGRT with IMRT or 3DCRT. No significant difference in acute or late GI toxicity was found.

In these studies, acute and late GI toxicity rates are either equal or higher among the hypofractionated arms (and note that multiple studies demonstrate earlier onset of acute toxicity with hypofractionation). Peach et al. (2015) reported that acute toxicity was not predictive of late toxicity in the hypofractionated arm of (Arcangeli et al., 2011), in contradiction to previous reports from conventional fractionation studies, though a strong consequential relationship was found for moderate hypofractionation by (Arunsingh et al., 2017). Arcangeli et al. have suggested increased acute high-grade GI toxicity seen following hypofractionation represents a transient response which is the result of the reduced overall treatment time rather than an increased dose-response.

Multiple observational studies have been undertaken on moderate hypofractionation for prostate cancer radiotherapy where observed GI toxicities have been related to predictive factors including rectal dosimetry. A summary from multiple related studies yielding statistically-significant associations is provided in Table 4.3.

Rectum – Extreme Hypofractionation (>3.5 Gy/fraction)

A steady move to increased dose per fraction treatment has been supported by the development of treatment planning and delivery techniques that can achieve precise and accurate delivery of highly conformal radiation doses (Kishan and King, 2017). A summary by Koontz et al. (2017) demonstrates a potential correlation of increasing toxicity rates with increasing EQD2Gy in the context of extreme prostate hypofractionation. The much reduced overall treatment time demands consideration of overall treatment time in calculations of EQD2Gy.

Emerging evidence suggests that with contemporary stereotactic delivery methods, genitourinary toxicity may limit clinical utility of extreme hypofractionation for prostate cancer rather than GI toxicity (Kishan and King, 2017). Kishan et al. (2017) have provided a very comprehensive review of single-arm studies of stereotactic prostate radiotherapy where the incidence of grade $\geq$ 3 GI toxicity is observed to be below 2%. A meta-analysis of outcomes for studies collated by Di Franco et al (Di Franco et al., 2017) indicates acute and late GI toxicity for extreme hypofractionation using stereotactic techniques have a lower incidence relative to IMRT moderate hypofractionation (OR: 0.17, p < 0.00005 and OR: 0.39, p < 0.00005, for acute and late toxicity, respectively).

Several analyses have been undertaken of dosimetric and clinical features that predict for GI toxicity following extreme hypofractionated radiotherapy for prostate cancer. Relevant studies are summarised in Table 4.4, including studies comparing alternative fractionation schedules.

TABLE 4.3 Summary of Dosimetric and Other Factors Associated to Acute and Late Gastrointestinal Toxicity Following Prostate Radiotherapy with Moderate Hypofractionation

Study	Number of Patients	Total Prescription Dose (Gy)/no. of fractions/Total weeks	Dose per Fraction (Gy)	Treatment Technique	Toxicity Scoring	Identified Predictive Dosimetric Risk Factors	Identified Modifying/ Predictive Risk Factors
Arcangeli et al. (2009)	102	56/16/4	3.5	IGRT (ultrasound and bony-anatomy) + IMRT	RTOG/EORTC	For acute grade 2 toxicity, rectal $V_{53Gy} > 8\%$ (p=0.0056) in multivariate analysis.	
Fonteyne et al. (2012)[late toxicity following (S. Arcangeli et al., 2009)]	113	56/16/4	3.5	IGRT (ultrasound and bony-anatomy) + IMRT	RTOG/EORTC and LENT-SOMA		Acute grade≥ 1 GI toxicity predictive of late grade ≥2 GI toxicity (p=0.05 on univariate, non-significant on multivariate analysis)
Pervez et al. (2010)	60	68/25/5	2.72	IGRT (MVCT) + IMRT (Tomotherapy)	RTOG	For acute grade ≥1 GI toxicity, rectal V_{60Gy} (p=0.085[a])	
Kong et al., (2014)	70	75/30/2.5	2.5	IGRT (MVCT) + IMRT (Tomotherapy) + rectal balloon fixation	RTOG	For acute grade ≥ 1 GI toxicity, most commonly anal pain: maximum rectal dose>76.5 Gy (p=0.042), by ROC analysis For late toxicity, most commonly rectal bleeding:rectal V_{70Gy}≥2.8% (p=0.032)	Age (p=0.018) and pre-treatment symptoms (p=0.023) predictive of late grade ≥ 1 GI toxicity on univariate analysis
Arunsingh et al. (2017)	101	65 or 60/25 or 20/5 or 4	2.6 or 3.0	IGRT (volumetric, CT) + IMRT	RTOG/EORTC	For acute grade 2 GI toxicity: $V\_EQD2Gy_{60Gy}{}^{b}> 9.7$ cm³ (p=0.003); $V\_EQD2Gy_{50Gy}{}^{b}> 15.9$ cm³ (p=0.005)	Acute grade ≥2 GI toxicity predictive of late grade ≥ 2 GI toxicity (p=0.007)

(Continued)

TABLE 4.3 (CONTINUED) Summary of Dosimetric and Other Factors Associated to Acute and Late Gastrointestinal Toxicity Following Prostate Radiotherapy with Moderate Hypofractionation

Study	Number of Patients	Total Prescription Dose (Gy)/no. of fractions/Total weeks	Dose per Fraction (Gy)	Treatment Technique	Toxicity Scoring	Identified Predictive Dosimetric Risk Factors	Identified Modifying/Predictive Risk Factors
Hoffman et al. (2014)	102	72/30/6	2.4	IGRT + IMRT	A modified version of RTOG	For late grade ≥ 2 GI toxicity (almost all rectal bleeding): $V\_EQD2Gy_{40Gy}^{d} > 50\%$ (p=0.026) $V\_EQD2Gy_{50Gy}^{d} > 30\%$ (p=0.028) $V\_EQD2Gy_{70Gy}^{d} > 20\%$ (p=0.016) $V\_EQD2Gy_{80Gy}^{d} > 10\%$ (p=0.016)	Late grade ≥2 GI toxicity was lower for higher Gleason score (7–8 vs. 6, HR: 0.17, p=0.01) or intermediate/high-risk vs. low-risk disease (HR: 0.59, p=0.015), and increased with prostate volume (HR 1.02/cm², p=0.045)
Delobel et al. (2017), prospective compilation from two trials	281 (hypofractionation, included in combined analysis with conventional fractionation to total of 972)	70/28/7	2.5	3DCRT	RTOG/CTCAE V3.0, LENT-SOMA	For late grade ≥ 2 rectal toxicity: dose-per-fraction (2 Gy vs. 2.5 Gy, RR:3.29, p=0.03)	IMRT and IGRT reduced rectal toxicity, though these options were not included in the hypofractionated treatments

^a with p<0.1 significant for Kendal's Tau-b test

^b using α/β=3.0 Gy.

^c Multiple cut-points investigated, most significant presented here.

^d using α/β=5.4 Gy.

no.=number; IGRT = Image Guided Radiation Therapy; IMRT= Intensity Modulated Radiation Therapy; RTOG/EORTC= Radiation Therapy Oncology Group/ European Organisation for Research and Treatment of Cancer; DVH=Dose-Volume Histogram; LENT/SOMA= Late Effects of Normal Tissues/ Subjective Objective Management and Analytic; MVCT=Mega-Voltage Computed Tomography; GI=gastrointestinal; V_{XGy}= percent or absolute organ volume receiving ≥ X Gy; V_EQD2Gy_{XGy}=percent or absolute organ volume receiving ≥ X Gy with organ doses corrected to 2 Gy equivalent doses using the linear quadratic model; HR=Hazard Ratio; RR=Relative Risk.

TABLE 4.4 Summary of Dosimetric and Other Factors Predictive of Acute and Late Gastrointestinal Toxicity Following Prostate Radiotherapy with Extreme Hypofractionation

Study	No. of pts	Total Prescription Dose (Gy)/Fractions/Total Weeks	Dose/Fraction (Gy)	Treatment Technique	Toxicity Scoring	Identified Predictive Dosimetric Risk Factors	Identified Modifying/Predictive Risk Factors
Gomez et al. (2015)	75	40/5/unspecified	8.0	IGRT (fiducials + planar and CT imaging) + VMAT	EPIC	For short-term (median 12 months) reduction in bowel QOL $V_{90\%} > 4.2$ cm^3 (p < 0.01) $V_{100\%} > 1.5$ cm^3 (p = 0.01)	
Zilli et al. (2017) reporting on results from (Zilli et al., 2011)	96	56/14/7 or 60/15/7.5	4.0	IMRT	CTCAE v3	No significant difference in GI toxicity between dose schedules. For late grade ≥ 2 GI toxicity: $V_{50Gy} > 19\%$ (p = 0.031),	
Elias et al. (2014)	84	35/5/4/1	7.0	IGRT (fiducials + orthogonal imaging) + IMRT	RTOG, EPIC	For late grade ≥ 2 GI toxicity: rectal $V_{31.8\ Gy} > 10\%$ (p = 0.0189); $D_{1cc} > 35$ Gy (p = 0.0324, OR:3.53)	For late grade ≥ 2 GI toxicity acutegrade ≥ 2 GI toxicity (OR:5.9, p = 0.0017)
Kim et al. (2014)	91	45 or 47.5 or 50/5/ unspecified	9.0, 9.5 or 10.0	SBRT (details unspecified)	CTCAE V3	For late grade ≥ 3 GI toxicity: rectal wall $V_{50Gy} > 3$ cm^3; rectal wall V_{35Gy} and V_{50Gy} (OR: 1.7 - 2.7 for increase of 1 cm^3, p < 0.013)[b]; >35% circumference of rectal wall receiving ≥ 39 Gy (p = 0.003). For late grade ≥ 2 GI toxicity: >50% of circumference of rectal wall receiving ≥ 24 Gy (p = 0.01)	

(Continued)

TABLE 4.4 (CONTINUED) Summary of Dosimetric and Other Factors Predictive of Acute and Late Gastrointestinal Toxicity Following Prostate Radiotherapy with Extreme Hypofractionation

Study	No. of pts	Total Prescription Dose (Gy)/Fractions/ Total Weeks	Dose/ Fraction (Gy)	Treatment Technique	Toxicity Scoring	Identified Predictive Dosimetric Risk Factors	Identified Modifying/ Predictive Risk Factors
Quon et al. (2015) (PATRIOT trial)	152	40/5/1.6 (every-other-day) or 4.1 (once per week)	8.0	SBRT (unspecified)	RTOG	For acute bowel toxicity: treatment duration (90% vs 68.1%, short vs long treatment arm, p<0.01). No predictor for late bowel toxicity	
Katz et al. (2013)	304	35 or 36.25/5/1	7.0 or 7.25	IGRT (fiducials and same-day CT) + Cyberknife	EPIC, RTOG	No predictor for late grade≥ 2 rectal toxicity	
Widmark et al. (2017) (HYPO-RT-PC trial)	866	78/39/8 vs. 42.7/7/2.5	2.0 vs. 6.1	IGRT (fiducials) + 3DCRT or VMAT	RTOG, PCSS	For acute grade ≥ 2 bowel toxicity: hypofractionation (9.4% vs. 5.3%, p=0.023). No predictor for late grade ≥ 2 bowel toxicity Similar pattern for patient-reported outcomes	

[a] authors Zilli et al. 2017) suggest that this is equivalent to the QUANTEC Michalski et al. (2010) recommendation of rectal $V_{70Gy} < 20\%$, if the doses to the rectum are corrected to 2 Gy equivalent doses using the linear quadratic model and an $\alpha/\beta = 3$ Gy.
[b] Similar results reported for anterior rectal wall.

No.=number; pts=patients; IGRT = Image Guided Radiation Therapy; IMRT= Intensity Modulated Radiation Therapy; VMAT= Volumetric Modulated Arc Therapy; SBRT=Stereotactic Body Radiation Therapy; CT=computed tomography; QOL=Quality of Life; VXGy= percent or absolute organ volume receiving ≥ X Gy; VY%= percent or absolute organ volume receiving ≥ Y% of the prescribed dose; DNcc= maximum dose to N cm3 of organ; GI=gastrointestinal; EPIC= Expanded Prostate Cancer Index Composite; CTCAE v3= Common Terminology Criteria for Adverse Events version 3.0; RTOG= Radiation Therapy Oncology Group; PCSS= Prostate Cancer Symptom Scale; OR=Odds Ratio.

SUGGESTED PARAMETERS

Rectum

There is strong evidence that applying dose-volume constraints over a range of mid to high doses reduces the incidence of toxicity. It is clear that different regions of the DVH are related to specific manifestations of toxicity. Rectal bleeding and proctitis are normally associated with high doses to the anorectum whilst endpoints such as loose stools and urgency are more commonly related to lower doses often ~20 to 30 Gy. Correspondingly LKB NTCP models of rectal bleeding have small values for the volume paramenter n and fit well to clinical data whilst endpoints with a more complex dose response are poorly suited to the histogram reduction methodology. There is larger variation in the value of D_{50} but most publications are within a few Gy of the QUANTEC value of 76.9 Gy. Generally models that can include clinical factors have been shown to result in better models in recent publications. Since almost all data are based on treatment planning CT there may be scope for further optimisation when accumulated data is available. The following, extracted from studies reviewed above, are recommended as dose-volume constraints to minimise rectal toxicity or identified as a cut-point threshold:

- QUANTEC

  - (late toxicity) rectum $V_{50Gy} < 50\%$, $V_{60Gy} < 35\%$, $V_{65\ Gy} < 25\%$, $V_{70Gy} < 20\%$, $V_{75Gy} < 15\%$.

- Subsequent

  Two articles provide comprehensive summaries of the dose response of the rectum following prostate cancer (Fonteyne et al., 2014; Landoni et al., 2016). Both include graphical summaries overlaying dose-volume constraints from a review of publications. Generally the results show reasonable agreement and remain in broad agreement with QUANTEC. Of note are the CHHiP trial constraints (Dearnaley et al., 2016) which were applied to a trial of over 3000 patients. The trial compared conventional (2 Gy per fraction) with moderate (3 Gy per fraction) hypofractionation, constraints were scaled numerically. No statistically significant differences in late GI toxicity were reported between trial arms indicating that the constraints were equally effective for both conventional and moderate hypofractionation. The trial constraints were more generous than QUANTEC: $V_{30Gy} < 80\%$, $V_{40Gy} < 70\%$, $V_{50Gy} < 60\%$, $V_{60Gy} < 50\%$, $V_{65Gy} < 30\%$, $V_{70Gy} < 15\%$ and $V_{74Gy} < 3\%$ for 2 Gy per fraction. The constraints derived by Fonteyne for a ≤10% probability of developing grade ≥2 toxicity are in good agreement with QUANTEC. Lower constraints were derived for 5% probability. A comprehensive set of constraints at 1 Gy (EQD2Gy) resolution spanning anorectal structures and individual toxicity endpoints can be found in the supplement to Ebert et al. (2015).

Reported risk factors: whole-pelvis/nodal irradiation, use of 3DCRT vs. IMRT, use of calcium channel blockers, use of statins, use of anticoagulants, age, presence of cardiovascular disease, prior abdominal surgery, smoking, diabetes, presence of haemorrhoids.

Bowel

The following, extracted from studies reviewed above, are recommended as dose-volume constraints to minimise bowel toxicity (to limit the risk of grade ≥3 toxicity < 10%, and the risk of grade ≥2 toxicity ≤15%):

- QUANTEC

  - (acute toxicity) small bowel $V_{15Gy} < 120$ cm³; bowel bag $V_{45Gy} < 195$ cm³.

- Subsequent

  - Derived cut-points for small bowel (acute toxicity) for conventional fractionation are summarized in Figure 4.3. Note that these results span treatment sites and grading schemes, as detailed in the sections above.

  - Small bowel, hypofractionation $V_{39Gy} < 158$ cm³, $V_{43Gy} < 110$ cm³, $V_{47Gy} < 28$ cm³, $V_{51Gy} < 6$ cm³, $V_{55Gy} < 0$ cm³.

  - Hypofractionation (per fraction), small bowel $V_{12.5Gy} < 30$ cm³.

Bowel bag (acute toxicity) $V_{15Gy} < 830$ cm³, $V_{25Gy} < 650$ cm³, $V_{30Gy} < 500$ cm³, $V_{40Gy} < 250$ cm³.

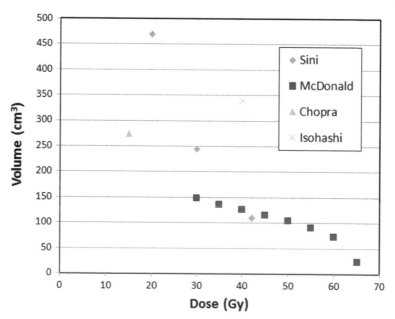

FIGURE 4.3   Published derived small bowel dose constraints for conventionally-fractionated treatments. Studies are Sini et al. (2017), McDonald et al. (2015), Chopra et al. (2014, 2015), and Isohashi et al. (2016).

Reported risk factors: High-dose cisplatin, reduced treatment time, whole-pelvis/nodal irradiation, age, acute toxicity, larger tumour size, smoking, low BMI, previous abdominal/pelvic surgery, bowel bag $V_{30Gy} > 900$ cm^3, small bowel $V_{40Gy} > 340$ cm^3, $V_{45Gy} > 50$ cm^3, $V_{50Gy} > 13$ cm^3, $V_{51.8\,Gy} > 51$ cm^3, $V_{55Gy} > 3$ cm^3.

OPEN ISSUES AND FUTURE DIRECTIONS OF INVESTIGATION

Non-Conventional Anatomical Association

Dosimetric indices for GI toxicities in pelvic radiotherapy have conventionally been derived from rectum and bowel due to an assumed underlying etiology. Alternative anatomical origins have been suggested due to stronger correlations with indices derived from other structures. Smeenk et al. (2012b), similar to Schaake et al. (2016) as discussed in Section "NTCP models," found that mean dose to pelvic floor muscles was associated with incidence of urgency, identifying the following mean dose constraints: ≤30 Gy for the internal anal sphincter muscle, ≤10 Gy for the external anal sphincter muscle, ≤50 Gy for the puborectalis muscle, and the ≤40 Gy for levator ani muscles (see examples of structures in Figure 4.1). Gulliford et al. (2017) found that control related toxicities but not bleeding were predicted by dose-volume parameters related to the region of fat adjacent to the prostate and rectum (the "peri-rectal space") rather than the anorectum.

Planned vs. Accumulated Dose

Until now, almost all studies where models have been generated relating pelvic dosimetry to GI toxicity have utilised parameters extracted from the planned dose distribution. For rectum and bowel, delivery can be perturbed from that planned due to patient setup variations or the daily variation in internal anatomy due to shifts and deformations resulting from organ filling. In prostate radiotherapy for example, studies with repeat daily volumetric imaging at treatment, with dose recalculated on the images or using deformable image registration back to the original plan images, have demonstrated changes across most dosimetric measures for the rectum for the majority of patients (Scaife et al., 2015), with high dose values sometimes varying by factors greater than two. Most recently, Shelley et al. (2017) have demonstrated an improvement in rectal toxicity modelling using dosimetric features (based on DSMs) determined from delivered as opposed to planned dose.

Attempts can be made to minimise motion by specifying rectal and bladder filling requirements, though such approaches have limited efficacy. Balloons or other endorectal devices (Murray et al., 2016) provide stability and impact the dose distribution across the rectum, typically reducing dose-volume indices by pushing tissue further from the treatment area (Serrano et al., 2017).

Molecular Factors Impacting Toxicity

Several analyses of patient cohorts have been undertaken which have incorporated establishing genomic predictors of toxicity in pelvic treatment, via examination of single nucleotide polymorphisms (SNPs) in particular genes or via genome-wide association studies. Barnett et al. (2017) undertook one of the largest genome-wide studies so far and found a large number of genetic variants associated with relative risk of rectal incontinence up to almost 10 in alleles occurring in between 1% and 10% of the population following prostate cancer radiotherapy. Even more common variants have been associated with toxicity in meta-analyses (Kerns et al., 2016). Such associations will become increasingly important for both identifying the etiology of toxicity and for patient-specific prediction of risk of toxicity.

Improved Outcomes Reporting

Electronic reporting of patient-reported outcomes in health research and practice has risen during the past few years (Holch et al., 2017). The International Consortium for Health Outcomes Measurement (ICHOM) is working toward creating a standardised set of patient-reported outcome measures for clinical practice to ensure compatibility across countries (Martin et al., 2015). To create some consistency in reporting trials there needs to be a development of a core set of measured outcomes in clinical practice, with equal weighting to patient and physician reported scales in the consensus process (MacLennan et al., 2015).

REFERENCES

Alsadius, David, Maria Hedelin, Dan Lundstedt, Niclas Pettersson, Ulrica Wilderäng, and Gunnar Steineck. 2012. "Mean Absorbed Dose to the Anal-Sphincter Region and Fecal Leakage among Irradiated Prostate Cancer Survivors." *International Journal of Radiation Oncology Biology Physics* 84 (2): e181-5. doi:10.1016/j.ijrobp.2012.03.065.

Aluwini, S, F Pos, E Schimmel, S Krol, P P van der Toorn, H de Jager, W G Alemayehu, W Heemsbergen, B Heijmen, and L Incrocci. 2016. "Hypofractionated versus Conventionally Fractionated Radiotherapy for Patients with Prostate Cancer (HYPRO): Late Toxicity Results from a Randomised, Non-Inferiority, Phase 3 Trial." *Lancet Oncol* 17 (4): 464–74. doi:10.1016/s1470-2045(15)00567-7.

Andreyev, H., N Jervoise. 2016. "GI Consequences of Cancer Treatment: A Clinical Perspective." *Radiation Research* 185 (4): 341–48. doi:10.1667/RR14272.1.

Andreyev, H J, A Wotherspoon, J W Denham, and M Hauer-Jensen. 2010. "Defining Pelvic-Radiation Disease for the Survivorship Era." *Lancet Oncology* 11 (4): 310–12. doi:10.1016/S1470-2045(10)70026-7.

Arcangeli, S, L Strigari, G Soete, G De Meerleer, S Gomellini, V Fonteyne, G Storme, and G Arcangeli. 2009. "Clinical and Dosimetric Predictors of Acute Toxicity after a 4-Week Hypofractionated External Beam Radiotherapy Regimen for Prostate Cancer: Results from a Multicentric Prospective Trial." *International Journal of Radiation Oncology Biology Physics* 73 (1): 39–45. doi:10.1016/j.ijrobp.2008.04.005.

Arcangeli, G, J Fowler, S Gomellini, S Arcangeli, B Saracino, M G Petrongari, M Benassi, and L Strigari. 2011. "Acute and Late Toxicity in a Randomized Trial of Conventional versus Hypofractionated Three-Dimensional Conformal Radiotherapy for Prostate Cancer." *International Journal of Radiation Oncology Biology Physics* 79 (4): 1013–21. doi:10.1016/j.ijrobp.2009.12.045.

Arcangeli, Giorgio, Biancamaria Saracino, Stefano Arcangeli, Sara Gomellini, Maria Grazia Petrongari, Giuseppe Sanguineti, and Lidia Strigari. 2017. "Moderate Hypofractionation in High-Risk, Organ-Confined Prostate Cancer: Final Results of a Phase III Randomized Trial." *Journal of Clinical Oncology : Official Journal of the American Society of Clinical Oncology* 35 (17): 1891–97. doi:10.1200/JCO.2016.70.4189.

Arunsingh, Moses, Indranil Mallick, Sriram Prasath, B Arun, Sandip Sarkar, Raj Kumar Shrimali, Sanjoy Chatterjee, and Rimpa Achari. 2017. "Acute Toxicity and Its Dosimetric Correlates for High-Risk Prostate Cancer Treated with Moderately Hypofractionated Radiotherapy." *Medical Dosimetry* 42 (1). Elsevier: 18–23. doi:10.1016/j.meddos.2016.10.002.

Baglan, Kathy L, Robert C Frazier, Di Yan, Raywin R Huang, Alvaro A Martinez, and John M Robertson. 2002. "The Dose-Volume Relationship of Acute Small Bowel Toxicity from Concurrent 5-FU-Based Chemotherapy and Radiation Therapy for Rectal Cancer." *International Journal of Radiation Oncology Biology Physics* 52 (1): 176–83. doi:10.1016/S0360-3016(01)01820-X.

Barnett, Gillian C, Deborah Thompson, Laura Fachal, Sarah Kerns, Chris Talbot, Rebecca M Elliott, Leila Dorling, et al. 2017. "A Genome Wide Association Study (GWAS) Providing Evidence of an Association between Common Genetic Variants and Late Radiotherapy Toxicity." *Radiotherapy and Oncology* 111 (2). Elsevier: 178–85. doi:10.1016/j.radonc.2014.02.012.

Bentzen, Soren M, Wolfgang Dörr, Reinhard Gahbauer, Roger W Howell, Michael C Joiner, Bleddyn Jones, Dan T L Jones, Albert J Van Der Kogel, André Wambersie, and Gordon Whitmore. 2012. "Bioeffect Modeling and Equieffective Dose Concepts in Radiation Oncology-Terminology, Quantities and Units." *Radiotherapy and Oncology* 105 (2): 266–68. doi:10.1016/j.radonc.2012.10.006.

Brenner, David J. 2004. "Fractionation and Late Rectal Toxicity." *International Journal of Radiation Oncology Biology Physics* 60 (4): 1013–15. doi:http://dx.doi.org/10.1016/j.ijrobp.2004.04.014.

Buettner, F, S L Gulliford, S Webb, M R Sydes, D P Dearnaley, and M Partridge. 2009. "Assessing Correlations between the Spatial Distribution of the Dose to the Rectal Wall and Late Rectal Toxicity after Prostate Radiotherapy: An Analysis of Data from the MRC RT01 Trial (ISRCTN 47772397)." *Physics in Medicine and Biology* 54 (21): 6535–48. doi:10.1088/0031-9155/54/21/006.

Buettner, F, S L Gulliford, S Webb, and M Partridge. 2011. "Modeling Late Rectal Toxicities Based on a Parameterized Representation of the 3D Dose Distribution." *Physics in Medicine and Biology* 56 (7): 2103–18. doi:10.1088/0031-9155/56/7/013.

Buettner, Florian, Sarah L Gulliford, Steve Webb, Matthew R Sydes, David P Dearnaley, and Mike Partridge. 2012. "The Dose–response of the Anal Sphincter Region – An Analysis of Data from the MRC RT01 Trial." *Radiotherapy and Oncology : Journal of the European Society for Therapeutic Radiology and Oncology* 103 (3). Elsevier Scientific Publishers: 347–52. http://linkinghub.elsevier.com/retrieve/pii/S0167814012001156?showall=true.

Burman, C, G J Kutcher, B Emami, and M Goitein. 1991. "Fitting of Normal Tissue Tolerance Data to an Analytic Function." *International Journal of Radiation Oncology Biology Physics* 21 (1): 123–35.

Capp, A, M Inostroza-Ponta, D Bill, P Moscato, C Lai, D Christie, D Lamb, et al. 2009. "Is There More than One Proctitis Syndrome? A Revisitation Using Data from the TROG 96.01 Trial." *Radiotherapy and Oncology* 90 (3): 400–407. doi:10.1016/j.radonc.2008.09.019.

Casares-Magaz, Oscar, Ludvig Paul Muren, Vitali Moiseenko, Stine E Petersen, Niclas Johan Pettersson, Morten Høyer, Joseph O Deasy, and Maria Thor. 2017. "Spatial Rectal Dose/volume Metrics Predict Patient-Reported Gastro-Intestinal Symptoms after Radiotherapy for Prostate Cancer." *Acta Oncologica* 56 (11): 1507–1513. doi:10.1080/0284186X.2017.1370130.

Catton, C N, H Lukka, C S Gu, J M Martin, S Supiot, P W M Chung, G S Bauman, et al. 2017. "Randomized Trial of a Hypofractionated Radiation Regimen for the Treatment of Localized Prostate Cancer." *International Journal of Clinical Oncology* 35 (17): 1884–90. doi:10.1200/jco.2016.71.7397.

Chang, Daniel T, Devin Schellenberg, John Shen, Jeff Kim, Karyn A Goodman, George A Fisher, James M Ford, Terry Desser, Andrew Quon, and Albert C Koong. 2009. "Stereotactic Radiotherapy for Unresectable Adenocarcinoma of the Pancreas." *Cancer* 115 (3): 665–72. doi:10.1002/cncr.24059.

Chennupati, S K, C A Pelizzari, R Kunnavakkam, and S L Liauw. 2014. "Late Toxicity and Quality of Life after Definitive Treatment of Prostate Cancer: Redefining Optimal Rectal Sparing Constraints for Intensity-Modulated Radiation Therapy." *Cancer Medicine* 3 (4): 954–61. doi:10.1002/cam4.261.

Chopra, Supriya, Tapas Dora, Anand N Chinnachamy, Biji Thomas, Sadhna Kannan, Reena Engineer, Umesh Mahantshetty, Reena Phurailatpam, Siji N Paul, and Shyam Kishore Shrivastava. 2014. "Predictors of Grade 3 or Higher Late Bowel Toxicity in Patients Undergoing Pelvic Radiation for Cervical Cancer: Results from a Prospective Study." *International Journal of Radiation Oncology Biology Physics* 88 (3): 630–5. doi:10.1016/j.ijrobp.2013.11.214.

Chopra, Supriya, Rahul Krishnatry, Tapas Dora, Sadhna Kannan, Biji Thomas, Supriya Sonawone, Reena Engineer, et al. 2015. "Predictors of Late Bowel Toxicity Using Three Different Methods of Contouring in Patients Undergoing Postoperative Radiation for Cervical Cancer." *British Journal of Radiology* 88 (1055). doi:10.1259/bjr.20150054.

Cicchetti, A, T Rancati, M Ebert, C Fiorino, F Palorini, A Kennedy, D J Joseph, et al. 2017. "Modelling Late Stool Frequency and Rectal Pain after Radical Radiotherapy in Prostate Cancer Patients: Results from a Large Pooled Population." *Physica Medica: European Journal of Medical Physics* 32 (12). 1690–7. doi:10.1016/j.ejmp.2016.09.018.

Coen, John J, Kyounghwa Bae, Anthony L Zietman, Baldev Patel, William U Shipley, Jerry D Slater, and Carl J Rossi. 2011. "Acute and Late Toxicity after Dose Escalation to 82 GyE Using Conformal Proton Radiation for Localized Prostate Cancer: Initial Report of American College of Radiology Phase II Study 03-12." *International Journal of Radiation Oncology Biology Physics* 81 (4): 1005–9. doi:10.1016/j.ijrobp.2010.06.047.

Colaco, Rovel J, Bradford S. Hoppe, Stella Flampouri, Brian T McKibben, Randal H Henderson, Curtis Bryant, Romaine C Nichols, et al. 2015. "Rectal Toxicity after Proton Therapy for Prostate Cancer: An Analysis of Outcomes of Prospective Studies Conducted at the University of Florida Proton Therapy Institute." *International Journal of Radiation Oncology Biology Physics* 91 (1): 172–81. doi:10.1016/j.ijrobp.2014.08.353.

Cox, James D, JoAnn Stetz, and Thomas F Pajak. 1995. "Toxicity Criteria of the Radiation Therapy Oncology Group (RTOG) and the European Organization for Research and Treatment of Cancer (EORTC)." *International Journal of Radiation Oncology Biology Physics* 31 (5). Elsevier: 1341–46. doi:10.1016/0360-3016(95)00060-C.

D'Avino, V, G Palma, R Liuzzi, M Conson, F Doria, M Salvatore, R Pacelli, and L Cella. 2015. "Prediction of Gastrointestinal Toxicity after External Beam Radiotherapy for Localized Prostate Cancer." *Radiation Oncology* 10: 80. doi:10.1186/s13014-015-0389-5.

Dearnaley, David, Isabel Syndikus, Helen Mossop, Vincent Khoo, Alison Birtle, David Bloomfield, John Graham, et al. 2016. "Conventional versus Hypofractionated High-Dose Intensity-Modulated Radiotherapy for Prostate Cancer: 5-Year Outcomes of the Randomised, Non-Inferiority, Phase 3 CHHiP Trial." *The Lancet Oncology* 17 (8): 1047–60. doi:10.1016/S1470-2045(16)30102-4.

Defraene, Gilles, Laura Van Den Bergh, Abrahim Al-Mamgani, Karin Haustermans, Wilma Heemsbergen, Frank Van Den Heuvel, and Joos V Lebesque. 2012. "The Benefits of Including Clinical Factors in Rectal Normal Tissue Complication Probability Modeling after Radiotherapy for Prostate Cancer." *International Journal of Radiation Oncology Biology Physics* 82 (3): 1233–42. doi:10.1016/j.ijrobp.2011.03.056.

Delobel, Jean Bernard, Khemara Gnep, Juan David Ospina, Véronique Beckendorf, Ciprian Chira, Jian Zhu, Alberto Bossi, et al. 2017. "Nomogram to Predict Rectal Toxicity Following Prostate Cancer Radiotherapy." *PLoS ONE* 12 (6). doi:10.1371/journal.pone.0179845.

Deville, Curtiland, Stefan Both, Wei Ting Hwang, Zelig Tochner, and Neha Vapiwala. 2010. "Clinical Toxicities and Dosimetric Parameters after Whole-Pelvis versus Prostate-Only Intensity-Modulated Radiation Therapy for Prostate Cancer." *International Journal of Radiation Oncology Biology Physics* 78 (3): 763–72. doi:10.1016/j.ijrobp.2009.08.043.

Di Franco, Rossella, Valentina Borzillo, Vincenzo Ravo, Gianluca Ametrano, Sara Falivene, Fabrizio Cammarota, Sabrina Rossetti, et al. 2017. "Rectal/urinary Toxicity after Hypofractionated vs Conventional Radiotherapy in Low/intermediate Risk Localized Prostate Cancer: Systematic Review and Meta Analysis." *Oncotarget* 8 (10): 17383–95. doi:10.18632/oncotarget.14798.

Di Maio, Massimo, Ethan Basch, Jane Bryce, and Francesco Perrone. 2016. "Patient-Reported Outcomes in the Evaluation of Toxicity of Anticancer Treatments." *Nature Reviews Clinical Oncology* 13 (5): 319–25. doi: 10.1038/nrclinonc.2015.222

Droge, Leif Hendrik, Hanne Elisabeth Weber, Manuel Guhlich, Martin Leu, Lena-Christin Conradi, Jochen Gaedcke, Steffen Hennies, Markus Karl Herrmann, Margret Rave-Frank, and Hendrik Andreas Wolff. 2015. "Reduced Toxicity in the Treatment of Locally Advanced Rectal Cancer: A Comparison of Volumetric Modulated Arc Therapy and 3D Conformal Radiotherapy." *BMC Cancer* 15: 750. doi:10.1186/s12885-015-1812-x.

Ebert, Martin A, Kerwyn Foo, Annette Haworth, Sarah L Gulliford, Angel Kennedy, David J Joseph, and James W Denham. 2015. "Gastrointestinal Dose-Histogram Effects in the Context of Dose-Volume-Constrained Prostate Radiation Therapy: Analysis of Data from the Radar Prostate Radiation Therapy Trial." *International Journal of Radiation Oncology Biology Physics* 91 (3): 595–603. doi:10.1016/j.ijrobp.2014.11.015.

Efstathiou, J A, K Bae, W U Shipley, D S Kaufman, M P Hagan, N M Heney, and H M Sandler. 2009. "Late Pelvic Toxicity After Bladder-Sparing Therapy in Patients With Invasive Bladder Cancer: RTOG 89-03, 95-06, 97-06, 99-06." *Journal of Clinical Oncology* 27 (25): 4055–61. doi:10.1200/jco.2008.19.5776.

Elias, Evelyn, Joelle Helou, Liying Zhang, Patrick Cheung, Andrea Deabreu, Laura D'Alimonte, Perakaa Sethukavalan, Alexandre Mamedov, Marlene Cardoso, and Andrew Loblaw. 2014. "Dosimetric and Patient Correlates of Quality of Life after Prostate Stereotactic Ablative Radiotherapy." *Radiotherapy and Oncology : Journal of the European Society for Therapeutic Radiology and Oncology* 112 (1): 83–88. doi:10.1016/j.radonc.2014.06.009.

Emami, B, J Lyman, A Brown, L Coia, M Goitein, J E Munzenrider, B Shank, L J Solin, and M Wesson. 1991. "Tolerance of Normal Tissue to Therapeutic Irradiation." *International Journal of Radiation Oncology Biology Physics* 21 (1): 109–22.

Engels, Benedikt, Nele Platteaux, Robbe Van Den Begin, Thierry Gevaert, Alexandra Sermeus, Guy Storme, Dirk Verellen, and Mark De Ridder. 2014. "Preoperative Intensity-Modulated and Image-Guided Radiotherapy with a Simultaneous Integrated Boost in Locally Advanced Rectal Cancer: Report on Late Toxicity and Outcome." *Radiotherapy and Oncology* 110 (1): 155–59. doi:10.1016/j.radonc.2013.10.026.

Fellin, G, C Fiorino, T Rancati, V Vavassori, M Baccolini, C Bianchi, E Cagna, et al. 2009. "Clinical and Dosimetric Predictors of Late Rectal Toxicity after Conformal Radiation for Localized Prostate Cancer: Results of a Large Multicenter Observational Study." *Radiotherapy and Oncology* 93 (2): 197–202. doi:10.1016/j.radonc.2009.09.004.

Fiorino, Claudio, Stefano Gianolini, and Alan E Nahum. 2003. "A Cylindrical Model of the Rectum: Comparing Dose-Volume, Dose-Surface and Dose-Wall Histograms in the Radiotherapy of Prostate Cancer." *Physics in Medicine and Biology* 48 (16): 2603–16.

Fiorino, Claudio, Filippo Alongi, Lucia Perna, Sara Broggi, Giovanni Mauro Cattaneo, Cesare Cozzarini, Nadia Di Muzio, Ferruccio Fazio, and Riccardo Calandrino. 2009. "Dose-Volume Relationships for Acute Bowel Toxicity in Patients Treated With Pelvic Nodal Irradiation for Prostate Cancer." *International Journal of Radiation Oncology Biology Physics* 75 (1): 29–35. doi:10.1016/j.ijrobp.2008.10.086.

Fiorino, Claudio, Tiziana Rancati, Gianni Fellin, Vittorio Vavassori, Emanuela Cagna, Valeria Casanova Borca, Giuseppe Girelli, et al. 2012. "Late Fecal Incontinence After High-Dose Radiotherapy for Prostate Cancer: Better Prediction Using Longitudinal Definitions." *International Journal of Radiation Oncology Biology Physics* 83 (1): 38–45. doi:http://dx.doi.org/10.1016/j.ijrobp.2011.06.1953.

Fonteyne, V, G Soete, S Arcangeli, W De Neve, B Rappe, G Storme, L Strigari, G Arcangeli, and G De Meerleer. 2012. "Hypofractionated High-Dose Radiation Therapy for Prostate Cancer: Long-Term Results of a Multi-Institutional Phase II Trial." *International Journal of Radiation Oncology Biology Physics* 84 (4): e483–90. doi:10.1016/j.ijrobp.2012.04.012.

Fonteyne, Valérie, Piet Ost, Frank Vanpachtenbeke, Roos Colman, Simin Sadeghi, Geert Villeirs, Karel Decaestecker, and Gert De Meerleer. 2014. "Rectal Toxicity after Intensity Modulated Radiotherapy for Prostate Cancer: Which Rectal Dose-volume Constraints Should We Use?" *Radiotherapy and Oncology* 113 (3): 398–403. doi:https://doi.org/10.1016/j.radonc.2014.10.014.

Gay, Hiram A, H Joseph Barthold, Elizabeth O'Meara, Walter R Bosch, Issam El Naqa, Rawan Al-Lozi, Seth A Rosenthal, et al. 2012. "Pelvic Normal Tissue Contouring Guidelines for Radiation Therapy: A Radiation Therapy Oncology Group Consensus Panel Atlas." *International Journal of Radiation Oncology Biology Physics* 83 (3): e353–62. doi:10.1016/j.ijrobp.2012.01.023.

Gomez, Caitlin L, Xiaoqing Xu, X Sharon Qi, Pin-Chieh Wang, Patrick Kupelian, Michael Steinberg, and Christopher R King. 2015. "Dosimetric Parameters Predict Short-Term Quality-of-Life Outcomes for Patients Receiving Stereotactic Body Radiation Therapy for Prostate Cancer." *Practical Radiation Oncology* 5 (4). United States: 257–62. doi:10.1016/j.prro.2015.01.006.

Gulliford, S L, K Foo, R C Morgan, E G Aird, A M Bidmead, H Critchley, P M Evans, et al. 2010. "Dose-Volume Constraints to Reduce Rectal Side Effects from Prostate Radiotherapy: Evidence from MRC RT01 Trial ISRCTN 47772397." *International Journal of Radiation Oncology, Biology, Physics* 76 (3): 747–54.

Gulliford, Sarah L, Mike Partridge, Matthew R Sydes, Steve Webb, Philip M Evans, and David P Dearnaley. 2012. "Parameters for the Lyman Kutcher Burman (LKB) Model of Normal Tissue Complication Probability (NTCP) for Specific Rectal Complications Observed in Clinical Practise." *Radiotherapy and Oncology* 102 (3): 347–51. doi:10.1016/j.radonc.2011.10.022.

Gulliford, S L, Soumya Ghose, M A Ebert, Angel Kennedy, Jason A Dowling, Jhimli Mitra, David J Joseph, and James W Denham. 2017. "Radiotherapy Dose-Distribution to the Perirectal Fat Space (PRS) is Related to Gastrointestinal Control-Related Complications." *Clinical and Translational Oncology* 7:62-70.

Hafeez, S, F McDonald, S Lalondrelle, H McNair, K Warren-Oseni, K Jones, V Harris, et al. 2017. "Clinical Outcomes of Image Guided Adaptive Hypofractionated Weekly Radiation Therapy for Bladder Cancer in Patients Unsuitable for Radical Treatment." *International Journal of Radiation Oncology Biology Physics* 98 (1): 115–22. doi:10.1016/j.ijrobp.2017.01.239.

Hoffman, K E, K R Voong, T J Pugh, H Skinner, L B Levy, V Takiar, S Choi, et al. 2014. "Risk of Late Toxicity in Men Receiving Dose-Escalated Hypofractionated Intensity Modulated Prostate Radiation Therapy: Results from a Randomized Trial." *International Journal of Radiation Oncology Biology Physics* 88 (5): 1074–84. doi:10.1016/j.ijrobp.2014.01.015.

Holch, Patricia, Ann M Henry, Susan Davidson, Alexandra Gilbert, Jacqueline Routledge, Leanne Shearsmith, Kevin Franks, Emma Ingleson, Abigail Albutt, and Galina Velikova. 2017. "Acute and Late Adverse Events Associated with Radical Radiation Therapy Prostate Cancer Treatment: A Systematic Review of Clinician and Patient Toxicity Reporting in Randomized Controlled Trials." *International Journal of Radiation Oncology, Biology, Physics* 97 (3): 495–510. doi:10.1016/j.ijrobp.2016.11.008.

Hong, Theodore S, Jennifer Moughan, Michael C Garofalo, Johanna Bendell, Adam C Berger, Nicklas B E Oldenburg, Pramila Rani Anne, et al. 2015. "NRG Oncology Radiation Therapy Oncology Group 0822: A Phase 2 Study of Preoperative Chemoradiation Therapy Using Intensity Modulated Radiation Therapy in Combination With Capecitabine and Oxaliplatin for Patients With Locally Advanced Rectal Cancer." *International Journal of Radiation Oncology, Biology, Physics* 93 (1): 29–36. doi:10.1016/j.ijrobp.2015.05.005.

NCI 2010. "Common Terminology Criteria for Adverse Events (CTCAE) and Common Toxicity Criteria (CTC)." USA: National Cancer Institute.

Incrocci, L, R C Wortel, W G Alemayehu, S Aluwini, E Schimmel, S Krol, P P van der Toorn, et al. 2016. "Hypofractionated versus Conventionally Fractionated Radiotherapy for Patients with Localised Prostate Cancer (HYPRO): Final Efficacy Results from a Randomised, Multicentre, Open-Label, Phase 3 Trial." *Lancet Oncol* 17 (8): 1061–69. doi:10.1016/s1470-2045(16)30070-5.

Isohashi, Fumiaki, Yasuo Yoshioka, Seiji Mabuchi, Koji Konishi, Masahiko Koizumi, Yutaka Takahashi, Toshiyuki Ogata, Shintaroh Maruoka, Tadashi Kimura, and Kazuhiko Ogawa. 2013. "Dose-Volume Histogram Predictors of Chronic Gastrointestinal Complications after Radical Hysterectomy and Postoperative Concurrent Nedaplatin-Based Chemoradiation Therapy for Early-Stage Cervical Cancer." *International Journal of Radiation Oncology Biology Physics* 85 (3): 728–34. doi:10.1016/j.ijrobp.2012.05.021.

Isohashi, Fumiaki, Seiji Mabuchi, Yuichi Akino, Yasuo Yoshioka, Yuji Seo, Osamu Suzuki, Keisuke Tamari, et al. 2016. "Dose-Volume Analysis of Predictors for Chronic Gastrointestinal Complications in Patients with Cervical Cancer Treated with Postoperative Concurrent Chemotherapy and Whole-Pelvic Radiation Therapy." *Journal of Radiation Research* 57 (6): 668–76. doi:10.1093/jrr/rrw037.

Ishikawa, Hitoshi, Hiroshi Tsuji, Tadashi Kamada, Takeshi Yanagi, Jun-Etsu Mizoe, Tatsuaki Kanai, Shinroku Morita, Masaru Wakatsuki, Jun Shimazaki, and Hirohiko Tsujii. 2017. "Carbon Ion Radiation Therapy for Prostate Cancer: Results of a Prospective Phase II Study." *Radiotherapy and Oncology* 81 (1): 57–64. doi:10.1016/j.radonc.2006.08.015.

Katz, Alan J, Michael Santoro, Fred Diblasio, and Richard Ashley. 2013. "Stereotactic Body Radiotherapy for Localized Prostate Cancer: Disease Control and Quality of Life at 6 Years." *Radiation Oncology (London, England)* 8 : 118. doi:10.1186/1748-717X-8-118.

Kavanagh, B D, C C Pan, L A Dawson, S K Das, X A Li, R K Ten Haken, and M Miften. 2010. "Radiation Dose-Volume Effects in the Stomach and Small Bowel." *International Journal of Radiation Oncology Biology Physics* 76 (3): S101–7. doi:10.1016/j.ijrobp.2009.05.071.

Kerns, Sarah L, Leila Dorling, Laura Fachal, Soren Bentzen, Paul D P Pharoah, Daniel R Barnes, Antonio Gomez-Caamano, et al. 2016. "Meta-Analysis of Genome Wide Association Studies Identifies Genetic Markers of Late Toxicity Following Radiotherapy for Prostate Cancer." *EBioMedicine* 10: 150–63. doi:10.1016/j.ebiom.2016.07.022.

Kim, D W Nathan, L Chinsoo Cho, Christopher Straka, Alana Christie, Yair Lotan, David Pistenmaa, Brian D Kavanagh, et al. 2014. "Predictors of Rectal Tolerance Observed in a Dose-Escalated Phase 1-2 Trial of Stereotactic Body Radiation Therapy for Prostate Cancer." *International Journal of Radiation Oncology, Biology, Physics* 89 (3): 509–17. doi:10.1016/j.ijrobp.2014.03.012.

Kishan, Amar U, and Christopher R King. 2017. "Stereotactic Body Radiotherapy for Low- and Intermediate-Risk Prostate Cancer." *Seminars in Radiation Oncology* 27 (3): 268–78. doi:10.1016/j.semradonc.2017.02.006.

Kong, Moonkyoo, Seong Eon Hong, and Sung-Goo Chang. 2014. "Hypofractionated Helical Tomotherapy (75 Gy at 2.5 Gy per Fraction) for Localized Prostate Cancer: Long-Term Analysis of Gastrointestinal and Genitourinary Toxicity." *OncoTargets and Therapy* 7: 553–66. doi:10.2147/OTT.S61465.

Koontz, Bridget F, Alberto Bossi, Cesare Cozzarini, Thomas Wiegel, and Anthony D'Amico. 2017. "A Systematic Review of Hypofractionation for Primary Management of Prostate Cancer." *European Urology* 68 (4): 683–91. doi:10.1016/j.eururo.2014.08.009.

Kouloulias, V, M Tolia, N Kolliarakis, A Siatelis, and N Kelekis. 2013. "Evaluation of Acute Toxicity and Symptoms Palliation in a Hypofractionated Weekly Schedule of External Radiotherapy for Elderly Patients with Muscular Invasive Bladder Cancer." *International Brazilian Journal of Urology* 39 (1): 77–82. doi:10.1590/s1677-5538.ibju.2013.01.10.

Landoni, V, C Fiorino, C Cozzarini, G Sanguineti, R Valdagni, and T Rancati. 2016. "Predicting Toxicity in Radiotherapy for Prostate Cancer." *Medical physics* 32 (3): 521–32. doi:10.1016/j.ejmp.2016.03.003.

Lee, W Robert, James J Dignam, Mahul B Amin, Deborah W Bruner, Daniel Low, Gregory P Swanson, Amit B Shah, et al. 2016. "Randomized Phase III Noninferiority Study Comparing Two Radiotherapy Fractionation Schedules in Patients with Low-Risk Prostate Cancer." *Journal of Clinical Oncology* 34 (20): 2325–32. doi:10.1200/jco.2016.67.0448.

1995. "Lent Soma Scales for All Anatomic Sites." *International Journal of Radiation Oncology Biology Physics* 31 (5): 1049–91. doi:http://dx.doi.org/10.1016/0360-3016(95)90159-0.

Letschert, Joke G J, Joos V Lebesque, Roel W de Boer, Augustinus A M Hart, and Harry Bartelink. 1990. "Dose-Volume Correlation in Radiation-Related Late Small-Bowel Complications: A Clinical Study." *Radiotherapy and Oncology* 18 (4): 307–20. doi:10.1016/0167-8140(90)90111-9.

Letschert, J G, J V Lebesque, B M Aleman, J F Bosset, J C Horiot, H Bartelink, L Cionini, J P Hamers, J W Leer, and M van Glabbeke. 1994. "The Volume Effect in Radiation-Related Late Small Bowel Complications: Results of a Clinical Study of the EORTC Radiotherapy Cooperative Group in Patients Treated for Rectal Carcinoma." *Radiotherapy and Oncology: Journal of the European Society for Therapeutic Radiology and Oncology* 32 (2): 116–23.

Longobardi, Barbara, Genoveffa Berardi, Claudio Fiorino, Filippo Alongi, Cesare Cozzarini, Aniko Deli, Mariangela La MacChia, Lucia Perna, Nadia Gisella Di Muzio, and Riccardo Calandrino. 2011. "Anatomical and Clinical Predictors of Acute Bowel Toxicity in Whole Pelvis Irradiation for Prostate Cancer with Tomotherapy." *Radiotherapy and Oncology* 101 (3): 460–64. doi:10.1016/j.radonc.2011.07.014.

Lutkenhaus, L J, R M van Os, A Bel, and M C Hulshof. 2016. "Clinical Results of Conformal versus Intensity-Modulated Radiotherapy Using a Focal Simultaneous Boost for Muscle-Invasive Bladder Cancer in Elderly or Medically Unfit Patients." *Radiotherapy and Oncology* 11: 45. doi:10.1186/s13014-016-0618-6.

MacLennan, Steven, Paula R Williamson, and Thomas B Lam. 2015. "Re: Neil E. Martin, Laura Massey, Caleb Stowell, et al. Defining a Standard Set of Patient-Centered Outcomes for Men with Localized Prostate Cancer." *European Urology* 67:460–7. doi:10.1016/j.eururo.2015.08.015.

Majewski, W, and R Tarnawski. 2009. "Acute and Late Toxicity in Radical Radiotherapy for Bladder Cancer." *Clinical Oncology (Royal College of Radiologists)* 21 (8): 598–609. doi:10.1016/j.clon.2009.04.008.

Mak, A C, T A Rich, T E Schultheiss, B Kavanagh, D M Ota, and M M Romsdahl. 1994. "Late Complications of Postoperative Radiation Therapy for Cancer of the Rectum and Rectosigmoid." *International Journal of Radiation Oncology, Biology, Physics* 28 (3): 597–603. doi:10.1016/0360-3016(94)90184-8.

Martin, Neil E, Laura Massey, Caleb Stowell, Chris Bangma, Alberto Briganti, Anna Bill-Axelson, Michael Blute, et al. 2015. "Defining a Standard Set of Patient-Centered Outcomes for Men with Localized Prostate Cancer." *European Urology* 67 (3): 460–7. doi:10.1016/j.eururo.2014.08.075.

McDonald, Andrew M, Christopher B Baker, Richard A Popple, Kiran Shekar, Eddy S Yang, Rojymon Jacob, Rex Cardan, Robert Y Kim, and John B Fiveash. 2014. "Different Rectal Toxicity Tolerance with and without Simultaneous Conventionally-Fractionated Pelvic Lymph Node Treatment in Patients Receiving Hypofractionated Prostate Radiotherapy." *Radiation Oncology* 9 (1): 129. doi:10.1186/1748-717X-9-129.

McDonald, F, R Waters, S Gulliford, E Hall, N James, and R A Huddart. 2015. "Defining Bowel Dose-volume Constraints for Bladder Radiotherapy Treatment Planning." *Clinical Oncology (Royal College of Radiologists)* 27 (1): 22–29. doi:10.1016/j.clon.2014.09.016.

Meijer, Gert J, Mandy van den Brink, Mischa S Hoogeman, Jan Meinders, and Joos V Lebesque. 1999. "Dose–Wall Histograms and Normalized Dose–surface Histograms for the Rectum: A New Method to Analyze the Dose Distribution over the Rectum in Conformal Radiotherapy." *International Journal of Radiation Oncology Biology Physics* 45 (4): 1073–80. doi:http://dx.doi.org/10.1016/S0360-3016(99)00270-9.

Mendenhall, Nancy P, Zuofeng Li Bradford S Hoppe, Robert B Marcus, William M Mendenhall, R Charles Nichols, Christopher G Morris, Christopher R Williams, Joseph Costa, and Randal Henderson. 2012. "Early Outcomes from Three Prospective Trials of Image-Guided Proton Therapy for Prostate Cancer." *International Journal of Radiation Oncology Biology Physics* 82 (1): 213–21. doi:10.1016/j.ijrobp.2010.09.024.

Michalski, Jeff M, Hiram Gay, Andrew Jackson, Susan L Tucker, and Joseph O Deasy. 2010. "Radiation Dose-Volume Effects in Radiation-Induced Rectal Injury." *International Journal of Radiation Oncology Biology Physics* 76 (3, Supplement 1): S123–29. doi: 10.1016/j.ijrobp.2009.03.078.

Mirjolet, Céline, Paul M Walker, Mélanie Gauthier, Cécile Dalban, Suzanne Naudy, Frédéric Mazoyer, Etienne Martin, Philippe Maingon, and Gilles Créhange. 2016. "Absolute Volume of the Rectum and AUC from Rectal DVH between 25Gy and 50Gy Predict Acute Gastrointestinal Toxicity with IG-IMRT in Prostate Cancer." *Radiation Oncology* 11 (1): 145. doi:10.1186/s13014-016-0721-8.

Mok, Henry, Christopher H Crane, Matthew B Palmer, Tina M Briere, Sam Beddar, Marc E Delclos, Sunil Krishnan, and Prajnan Das. 2011. "Intensity Modulated Radiation Therapy (IMRT): Differences in Target Volumes and Improvement in Clinically Relevant Doses to Small Bowel in Rectal Carcinoma." *Radiation Oncology (London, England)* 6 (1): 63. doi:10.1186/1748-717X-6-63.

Murray, Julia, Helen McNair, Emma Alexander, Tuathan O'Shea, Karen Thomas, Simeon Nill, Sarah Gulliford, and David P Dearnaley. 2016. "Effect of ProSpare (PS), a Rectal Obturator, on Inter- and Intrafraction Prostate Motion and Anorectal Doses in Prostate Radiotherapy (RT)." *Journal of Clinical Oncology* 34 (2_suppl): e633–e633. doi:10.1200/jco.2016.34.2_suppl.e633.

Niemierko, Andrzej. 1997. "Reporting and Analyzing Dose Distributions: A Concept of Equivalent Uniform Dose." *Medical Physics* 24 (1). American Association of Physicists in Medicine: 103–10. doi:10.1118/1.598063.

Norkus, Darius, Agata Karklelyte, Benedikt Engels, Harijati Versmessen, Romas Griskevicius, Mark De Ridder, Guy Storme, Eduardas Aleknavicius, Ernestas Janulionis, and Konstantinas Povilas Valuckas. 2013. "A Randomized Hypofractionation Dose Escalation Trial for High Risk Prostate Cancer Patients: Interim Analysis of Acute Toxicity and Quality of Life in 124 Patients." *Radiation Oncology (London, England)* 8: 206. doi:10.1186/1748-717X-8-206.

Onal, Cem, Erkan Topkan, Esma Efe, Melek Yavuz, Serhat Sonmez, and Aydin Yavuz. 2009. "Comparison of Rectal Volume Definition Techniques and Their Influence on Rectal Toxicity in Patients with Prostate Cancer Treated with 3D Conformal Radiotherapy: A Dose-Volume Analysis." *Radiation Oncology* 4 (1): 14. doi:10.1186/1748-717X-4-14.

Ospina, Juan D., Jian Zhu, Ciprian Chira, Alberto Bossi, Jean B Delobel, Véronique Beckendorf, Bernard Dubray, et al. 2014. "Random Forests to Predict Rectal Toxicity Following Prostate Cancer Radiation Therapy." *International Journal of Radiation Oncology Biology Physics* 89: 1024–31. doi:10.1016/j.ijrobp.2014.04.027.

Palumbo, Isabella, Fabio Matrone, Giampaolo Montesi, Rita Bellavita, Marco Lupattelli, Simonetta Saldi, Alessandro Frattegiani, et al. 2017. "Statins Protect Against Acute RT-Related Rectal Toxicity in Patients with Prostate Cancer: An Observational Prospective Study." *Anticancer Research* 37 (3): 1453–7. doi:10.21873/anticanres.11469.

Pavy, J -J, J Denekamp, J Letschert, B Littbrand, F Mornex, J Bernier, D Gonzales-Gonzales, J -C Horiot, M Bolla, and H Bartelink. 1995. "Late Effects Toxicity Scoring: The Soma Scale." *International Journal of Radiation Oncology Biology Physics* 31 (5) : 1043–7. doi:10.1016/0360-3016(95)00059-8.

Peach, Matthew Sean, Timothy N Showalter, and Nitin Ohri. 2015. "Systematic Review of the Relationship between Acute and Late Gastrointestinal Toxicity after Radiotherapy for Prostate Cancer." *Prostate Cancer* 2015: 11. doi:10.1155/2015/624736.

Peeters, S T H, M S Hoogeman, W D Heemsbergen, A A M Hart, P C M Koper, and J V Lebesque. 2006. "Rectal Bleeding, Fecal Incontinence, and High Stool Frequency after Conformal Radiotherapy for Prostate Cancer: Normal Tissue Complication Probability Modeling." *International Journal of Radiation Oncology Biology Physics* 66 (1): 11–9. doi:10.1016/j.ijrobp.2006.03.034.

Perna, Lucia, Filippo Alongi, Claudio Fiorino, Sara Broggi, Mauro Cattaneo Giovanni, Cesare Cozzarini, Nadia Di Muzio, and Riccardo Calandrino. 2017. "Predictors of Acute Bowel Toxicity in Patients Treated with IMRT Whole Pelvis Irradiation after Prostatectomy." *Radiotherapy and Oncology* 97 (1): 71–75. doi:10.1016/j.radonc.2010.02.025.

Pervez, N, C Small, M MacKenzie, D Yee, M Parliament, S Ghosh, A Mihai, et al. 2010. "Acute Toxicity in High-Risk Prostate Cancer Patients Treated with Androgen Suppression and Hypofractionated Intensity-Modulated Radiotherapy." *International Journal of Radiation Oncology Biology Physics* 76 (1): 57–64. doi:10.1016/j.ijrobp.2009.01.048.

Peterson, J L, S J Buskirk, M G Heckman, N N Diehl, J R Bernard Jr., K S Tzou, H E Casale, et al. 2014. "Image-Guided Intensity-Modulated Radiotherapy for Prostate Cancer: Dose Constraints for the Anterior Rectal Wall to Minimize Rectal Toxicity." *Medical Dosimetry: Official Journal of the American Association of Medical Dosimetrists* 39 (1): 12–17. doi:10.1016/j.meddos.2013.08.007.

Pollack, Alan, Gail Walker, Eric M Horwitz, Robert Price, Steven Feigenberg, Andre A Konski, Radka Stoyanova, et al. 2013. "Randomized Trial of Hypofractionated External-Beam Radiotherapy for Prostate Cancer." *Journal of Clinical Oncology* 31 (31): 3860–8. doi:10.1200/jco.2013.51.1972.

Quon, Harvey Charles, Aldrich Ong, Patrick Cheung, William Chu, Hans T Chung, Danny Vesprini, Amit Chowdhury, et al. 2015. "PATRIOT Trial: Randomized Phase II Study of Prostate Stereotactic Body Radiotherapy Comparing 11 versus 29 Days Overall Treatment Time." *Journal of Clinical Oncology* 33 (7_suppl): 6. doi:10.1200/jco.2015.33.7_suppl.6.

Rancati, Tiziana, Claudio Fiorino, Gianni Fellin, Vittorio Vavassori, Emanuela Cagna, Valeria Casanova Borca, Giuseppe Girelli, et al. 2011. "Inclusion of Clinical Risk Factors into NTCP Modelling of Late Rectal Toxicity after High Dose Radiotherapy for Prostate Cancer." *Radiotherapy and Oncology* 100 (1): 124–30. doi:10.1016/j.radonc.2011.06.032.

Reis, Tina, Edwin Khazzaka, Grit Welzel, Frederik Wenz, Ralf-Dieter Hofheinz, and Sabine Mai. 2015. "Acute Small-Bowel Toxicity during Neoadjuvant Combined Radiochemotherapy in Locally Advanced Rectal Cancer: Determination of Optimal Dose-Volume Cut-off Value Predicting Grade 2-3 Diarrhoea." *Radiation Oncology (London, England)* 10: 30. doi:10.1186/s13014-015-0336-5.

Reis Ferreira, Miguel, Atia Khan, Karen Thomas, Lesley Truelove, Helen McNair, Annie Gao, Chris C Parker, et al. 2017. "Phase 1/2 Dose-Escalation Study of the Use of Intensity Modulated Radiation Therapy to Treat the Prostate and Pelvic Nodes in Patients With Prostate Cancer." *International Journal of Radiation Oncology, Biology, Physics*, August. United States. doi:10.1016/j.ijrobp.2017.07.041.

Robertson, John M, David Lockman, Di Yan, and Michelle Wallace. 2008. "The Dose-Volume Relationship of Small Bowel Irradiation and Acute Grade 3 Diarrhea During Chemo-radiotherapy for Rectal Cancer." *International Journal of Radiation Oncology Biology Physics* 70 (2): 413–8. doi:10.1016/j.ijrobp.2007.06.066.

Robertson, John M, Matthias Sohn, and Di Yan. 2010. "Predicting Grade 3 Acute Diarrhea during Radiation Therapy for Rectal Cancer Using a Cutoff-Dose Logistic Regression Normal Tissue Complication Probability Model." *International Journal of Radiation Oncology, Biology, Physics* 77 (1): 66–72. doi:10.1016/j.ijrobp.2009.04.048.

Samuelian, Jason M, Matthew D Callister, Jonathan B Ashman, Tonia M Young-Fadok, Mitesh J Borad, and Leonard L Gunderson. 2012. "Reduced Acute Bowel Toxicity in Patients Treated with Intensity-Modulated Radiotherapy for Rectal Cancer." *International Journal of Radiation Oncology Biology Physics* 82 (5): 1981–7. doi:10.1016/j.ijrobp.2011.01.051.

Scaife, Jessica E, Simon J Thomas, Karl Harrison, Marina Romanchikova, Michael P F Sutcliffe, Julia R Forman, Amy M Bates, Raj Jena, M Andrew Parker, and Neil G Burnet. 2015. "Accumulated Dose to the Rectum, Measured Using Dose–volume Histograms and Dose-Surface Maps, is Different from Planned Dose in All Patients Treated with Radiotherapy for Prostate Cancer." *The British Journal of Radiology* 88 (1054): 20150243. doi:10.1259/bjr.20150243.

Schaake, Wouter, Arjen van der Schaaf, Lisanne V van Dijk, Alfons H.H. Bongaerts, Alfons C M van den Bergh, and Johannes A Langendijk. 2016. "Normal Tissue Complication Probability (NTCP) Models for Late Rectal Bleeding, Stool Frequency and Fecal Incontinence after Radiotherapy in Prostate Cancer NTCP Models for Anorectal Side Effects Patients." *Radiotherapy and Oncology* 119 (3): 381–7. doi:10.1016/j.radonc.2016.04.005.

Schellenberg, Devin, Karyn A Goodman, Florence Lee, Stephanie Chang, Timothy Kuo, James M Ford, George A Fisher, et al. 2008. "Gemcitabine Chemotherapy and Single-Fraction Stereotactic Body Radiotherapy for Locally Advanced Pancreatic Cancer." *International Journal of Radiation Oncology Biology Physics* 72 (3): 678–6. doi:10.1016/j.ijrobp.2008.01.051.

Serrano, Nicholas A, Noah S Kalman, and Mitchell S Anscher. 2017. "Reducing Rectal Injury in Men Receiving Prostate Cancer Radiation Therapy: Current Perspectives." *Cancer Management and Research* 9: 339–50. doi:10.2147/CMAR.S118781.

Shelley, L E A, J E Scaife, M Romanchikova, K Harrison, J R Forman, A M Bates, D J Noble, et al. 2017. "Delivered Dose Can Be a Better Predictor of Rectal Toxicity than Planned Dose in Prostate Radiotherapy." *Radiotherapy and Oncology* 123 (3): 466–71. doi:https://doi.org/10.1016/j.radonc.2017.04.008.

Sini, Carla, Barbara Noris Chiorda, Pietro Gabriele, Giuseppe Sanguineti, Sara Morlino, Fabio Badenchini, Domenico Cante, et al. 2017. "Patient-Reported Intestinal Toxicity from Whole Pelvis Intensity-Modulated Radiotherapy: First Quantification of Bowel dose–volume Effects." *Radiotherapy and Oncology* 124 (2): 296–301. doi:10.1016/j.radonc.2017.07.005.

Smeenk, Robert Jan, Wim P M Hopman, Aswin L. Hoffmann, Emile N J Th Van Lin, and Johannes H A M Kaanders. 2012a. "Differences in Radiation Dosimetry and Anorectal Function Testing Imply That Anorectal Symptoms May Arise from Different Anatomic Substrates." *International Journal of Radiation Oncology Biology Physics* 82 (1): 145–52. doi:10.1016/j.ijrobp.2010.08.023.

Smeenk, R J, A L Hoffmann, W P Hopman, E N van Lin, and J H Kaanders. 2012b. "Dose-Effect Relationships for Individual Pelvic Floor Muscles and Anorectal Complaints after Prostate Radiotherapy." *International Journal of Radiation Oncology Biology Physics* 83 (2): 636–44. doi:10.1016/j.ijrobp.2011.08.007.

Sondergaard, J, M Holmberg, A R Jakobsen, M Agerbaek, L P Muren, and M Hoyer. 2014. "A Comparison of Morbidity Following Conformal versus Intensity-Modulated Radiotherapy for Urinary Bladder Cancer." *Acta Oncologica (Stockholm, Sweden)* 53 (10): 1321–8. doi:10.3109/0284186x.2014.928418.

Sonn, G A, N Sadetsky, J C Presti, and M S Litwin. 2013. "Differing Perceptions of Quality of Life in Patients with Prostate Cancer and Their Doctors." *Journal of Urology* 189 (1 Suppl): 59–65. doi:10.1016/j.juro.2012.11.032.

Thornqvist, S, L B Hysing, L Tuomikoski, A Vestergaard, K Tanderup, L P Muren, and B J Heijmen. 2016. "Adaptive Radiotherapy Strategies for Pelvic Tumors - a Systematic Review of Clinical Implementations." *Acta Oncologica (Stockholm, Sweden)* 55 (8): 943–58. doi:10.3109/0284186x.2016.1156738.

Troeller, Almut, Di Yan, Ovidiu Marina, Derek Schulze, Markus Alber, Katia Parodi, Claus Belka, and Matthias Söhn. 2015. "Comparison and Limitations of DVH-Based NTCP Models Derived from 3D-CRT and IMRT Data for Prediction of Gastrointestinal Toxicities in Prostate Cancer Patients by Using Propensity Score Matched Pair Analysis." *International Journal of Radiation Oncology Biology Physics* 91 (2): 435–43. doi:10.1016/j.ijrobp.2014.09.046.

Turgeon, G A, L Souhami, F L Cury, S L Faria, M Duclos, J Sturgeon, and W Kassouf. 2014. "Hypofractionated Intensity Modulated Radiation Therapy in Combined Modality Treatment for Bladder Preservation in Elderly Patients with Invasive Bladder Cancer." *International Journal of Radiation Oncology Biology Physics* 88 (2): 326–31. doi:10.1016/j.ijrobp.2013.11.005.

Widmark, A, A Gunnlaugsson, L Beckman, C Thellenberg-Karlsson, M Hoyer, M Lagerlund, P Fransson, et al. 2017. "Extreme Hypofractionation versus Conventionally Fractionated Radiotherapy for Intermediate Risk Prostate Cancer: Early Toxicity Results from the Scandinavian Randomized Phase III Trial "HYPO-RT-PC" *International Journal of Radiation Oncology Biology Physics* 96 (5): 938–39. doi:10.1016/j.ijrobp.2016.09.049.

Xiao, Canhua, Rosemary Polomano, and Deborah Watkins Bruner. 2013. "Comparison between Patient-Reported and Clinician-Observed Symptoms in Oncology." *Cancer Nursing* 36 (6): E1–E16. doi:10.1097/NCC.0b013e318269040f.

Xu, M J, M Kirk, H Zhai, and L L Lin. 2016. "Bag and Loop Small Bowel Contouring Strategies Differentially Estimate Small Bowel Dose for Post-Hysterectomy Women Receiving Pencil Beam Scanning Proton Therapy." *Acta Oncologica (Stockholm, Sweden)* 55 (7): 900–908. doi:1 0.3109/0284186x.2016.1142114.

Yahya, Noorazrul, Martin A Ebert, Max Bulsara, Annette Haworth, Rachel Kearvell, Kerwyn Foo, Angel Kennedy, et al. 2014. "Impact of Treatment Planning and Delivery Factors on Gastrointestinal Toxicity: An Analysis of Data from the RADAR Prostate Radiotherapy Trial." *Radiation Oncology (London, England)* 9: 282. doi:10.1186/s13014-014-0282-7.

Zhang, Shuo, Yong-Hua Yu, Yong Zhang, Wei Qu, and Jia Li. 2015. "Radiotherapy in Muscle-Invasive Bladder Cancer: The Latest Research Progress and Clinical Application." *American Journal of Cancer Research* 5 (2): 854–68.

Zilli, Thomas, Sandra Jorcano, Michel Rouzaud, Giovanna Dipasquale, Philippe Nouet, José Ignacio Toscas, Nathalie Casanova, et al. 2011. "Twice-Weekly Hypofractionated Intensity-Modulated Radiotherapy for Localized Prostate Cancer with Low-Risk Nodal Involvement: Toxicity and Outcome from a Dose Escalation Pilot Study." *International Journal of Radiation Oncology Biology Physics* 81 (2): 382–9. doi:10.1016/j.ijrobp.2010.05.057.

Zilli, T, M Kountouri, M Rouzaud, A Dubouloz, D Linero, S Jorcano, L Escudé, and R Miralbell. 2017. "EP-1365: Dosimetric Predictors for Rectal Toxicity with Two Hypofractionated Schedules for Prostate Cancer." *Radiotherapy and Oncology* 119 : S637. doi:10.1016/S0167-8140(16)32615-9.

CHAPTER 5

Pelvis

Urinary Toxicity

Tiziana Rancati, Cesare Cozzarini, Riccardo Valdagni,
and Claudio Fiorino

CONTENTS

Introduction 113
    Clinical Significance of the Injury 113
    Clinician-Reported Outcomes 114
    Patient-Reported Outcomes 114
    Relevant Toxicities 115
    Relevant Structures and Dosimetry 115
Short Summary of the QUANTEC Recommendations 116
Critical Review of the Literature 2009–2017 (Conventional Fractionation) 116
    Dose-Volume Relationship 116
    Patient-Related Risk Factors 117
Critical Review of the Literature 2009–2017 (Hypofractionation) 124
Suggestion for DVH Constraints 125
Open Issues and Future Directions of Investigation 126
    Refining Dose-Volume Models 126
    Spatial Effects 127
    Pre-Clinical Animal Models 127
    Incorporation/Estimate of Bladder Deformation into Models 129
    Extreme Hypofractionation 130
References 130

INTRODUCTION

Clinical Significance of the Injury

The term "urinary toxicity" comprises a wide variety of symptoms – including urinary frequency, obstruction and stricture, haematuria, dysuria or incontinence – with very different time patterns and different impact on the single patient's quality of life (Rapariz-González et al. 2014).

The response of the urinary bladder to radiotherapy can be classified into acute/subacute reactions, occurring during radiotherapy and within 3–6 months after treatment completion, and late reactions starting to appear 6 months after therapy and often occurring many years later. The pathophysiology of urinary radiation injury is still not completely understood. Mechanisms of radiation damage affect the urothelium, the vasculature, and the detrusor muscles (Rapariz-González et al. 2014). After irradiation, the urothelium exhibits nuclear irregularity and cellular oedema, and the disruption of polysaccharide layer causes contact between hypertonic urine and isotonic tissue, resulting in tissue inflammation and early urinary symptoms (Marks et al. 1995). Vascular ischaemia, oedema, and cellular demolition cause depletion of bladder smooth muscle and proliferation of fibroblasts, thus leading to decreased bladder compliance and capacity, and to haemorrhagic cystitis. Fibrosis leading to occlusion of the urethral lumen is an important factor for the onset of urethral strictures after radiotherapy (Mundy and Andrich 2012), and is likely linked to reduced urinary functionality in terms of urgency and incontinence symptoms.

Clinician-Reported Outcomes

At the clinical level, urinary toxicity is usually graded using Common Terminology Criteria for Adverse Events (CTCAE): version 4.03 is the most recent release of these scoring criteria (http://ctep.cancer.gov/protocolDevelopment/electronic_applications/ctc.htm). Grade 2 toxicities include: (a) moderate urinary frequency/urgency with indication to medical management; (b) symptomatic haematuria requiring the positioning of a urinary catheter or bladder irrigation; (c) urethral obstruction needing dilation and/or the insertion of a urinary or suprapubic catheter; (d) use of pads; (e) urinary retention leading to the placement of urinary/suprapubic or intermittent catheter; and (f) fistula needing non-invasive intervention. Grade 3 effects embrace: (a) gross haematuria requiring transfusion, hospitalization, hyperbaric oxygen therapy, radiologic, or operative intervention; (b) symptomatic urinary tract obstruction with altered organ function needing operative surgical intervention; (c) urinary incontinence necessitating clamp/collagen injections/surgery; (d) urinary retention with indication to elective operative or radiologic intervention; (e) fistula requiring radiologic/endoscopic/surgery or permanent urinary diversion.

Patient-Reported Outcomes

CTCAE physician-based score is recognized to be much less exhaustive (and often largely different in the result) than patient-reported outcomes. This prompted the increasing use of specific patient-reported questionnaires, as it became evident that these instruments can describe and score many different symptoms, allowing nuances, and letting determination of the impact of symptoms on patient perceived quality of life.

Chiefly, the International Prostate Symptoms Score (IPSS) (Barry et al. 1992) became broadly used to score obstructive symptoms and to generate a global assessment of urinary functionality (Jolnerovski et al. 2017; Carillo et al. 2014; Cozzarini et al. 2015; Palorini et al. 2016b; Malik et al. 2011; Ghadjar et al. 2013; Palorini et al. 2016c; Improta et al. 2016; Yahya et al. 2017; Shaikh et al. 2017; Chin et al. 2017; Shahid et al. 2017; Eriguchi et al. 2016; Miyake et al. 2015; Seymour et al. 2015).

Urinary incontinence can be prospectively assessed by the International Consultation on Incontinence Modular Questionnaire Short Form (ICIQ-SF) (Avery et al. 2004), which also includes the patient perception of the impact of incontinence on quality of life (Donovan et al. 2016; Cozzarini et al. 2017).

The International Consortium for Health Outcomes Measurement (ICHOM) (Martin et al. 2015) proposed the Expanded Prostate Cancer Index Composite 26-question short form (EPIC-26) (Wei et al. 2000), which is already widely used (Shaikh et al. 2017; Shahid et al. 2017; Donovan et al. 2016; Barocas et al. 2017; Venderbos et al. 2017; Qi et al. 2016; Lane et al. 2016) and which addresses all pertinent domains of prostate cancer treatment induced side effects, including urinary obstructive symptoms, urinary incontinence, and haematuria.

Relevant Toxicities

The incidence of acute/late radio-induced urinary toxicity in modern series pertaining to patients treated with radical radiotherapy is largely varying, being mainly related to prescription doses, delivery technique, presence and frequency of image guidance protocols, hypofractionation, and concomitant hormonal therapies. Rates of grade 1 and grade 2 symptoms in patients followed up to 10 years are described to be in the range of 20–43% and 10–46%, respectively (Jolnerovski et al. 2017; Donovan et al. 2016; Zelefsky et al. 1999; Zietman et al. 2005; Talcott et al. 2010; Peeters et al. 2005; Harsolia et al. 2007; Cahlon et al. 2008; Fonteyne et al. 2009; Pederson et al. 2012; Wortel et al. 2016; Byrne et al. 2017). Grade 3 urinary toxicity occurs at a rate of 2–16%. Obstruction, incontinence, and radiation cystitis with gross macroscopic haematuria are the most commonly reported grade 3 symptoms (Rancati et al. 2017).

It has to be mentioned that the prevalence of lower urinary tract symptoms increases with age in the general population: moderate/severe symptoms are present in ≈15% of men aged 50–59 with respect to ≈30% of men ≥70 years old. The most frequent symptom in the general population is nocturia (Rohrmann et al. 2016; Malmsten et al. 2010). For this reason, the rates of late radio-induced urinary toxicity could be overestimated.

A further issue is related to the compelling confirmations on the consequential nature of late radio-induced urinary toxicity, which were recently published in large prospectively followed cohorts (Zelefsky et al. 1999; Martin et al. 2015; Wortel et al. 2016; Zelefsky 2008; Pinkawa et al. 2010; Cozzarini et al. 2012; Jereczek-Fossa et al. 2010; Heemsbergen et al. 2010; Ghadjar et al. 2014). These suggest that a relevant fraction of the late urinary events is a "consequence" of the exuberant repair process following the acute inflammatory phase and it means that any effort to reduce acute toxicity may impact the occurrence of late events.

Relevant Structures and Dosimetry

Traditionally, the anatomical bladder, contoured as a solid organ starting from the external bladder surface has been considered as the relevant structure to be spared during treatment optimisation. It is now acknowledged that the urinary bladder comprises several substructures which may have distinct radiobiological sensitivities, causing different impacts on distinct urinary symptoms. Particularly, several published results (Palorini et al. 2016a; Improta et al. 2016; Yahya et al. 2017; Heemsbergen et al. 2010; Ghadjar et al. 2014; Palorini et al. 2016c) converged in the identification of the trigone as highly associated to an increased risk of severe acute and late injury. The exact mechanisms controlling the trigone are still unclear, however, as this muscle is actively involved in sphincter opening, it is realistic to claim that its damage might elicit frequency/urgency/incontinence symptoms. All these findings are pointing out to the opportunity of a refined optimisation of treatment planning, based on explicit definition of critical/sensitive bladder substructures and on specific dose-constraints for each bladder portion.

Of course, also the dose received by the urethra may have a role: current studies cannot permit to distinguish the relative contribution due to bladder and urethra irradiation. A major impact of urethra irradiation (mostly associated to urethral stenosis) is expected for very high-dose schedule, as reported for brachytherapy series (Sullivan et al. 2009).

SHORT SUMMARY OF THE QUANTEC RECOMMENDATIONS

Apart few quantitative suggestions regarding the tolerance of the bladder to whole organ irradiation, no clear suggestions were available at the time of publication of the QUANTEC recommendations for bladder and more generally, urinary symptoms (Viswanathan et al. 2010). The most robust suggestion dealt with whole irradiation of bladder, showing a quite rapid increase of the risk of grade 3 urinary toxicity for 2-Gy equivalent doses to the whole bladder above 50–55 Gy. As a matter of fact, the chapter dedicated to bladder and urinary toxicity claimed the urgent need for getting more valuable data and research to the topic. At that time, very few and controversial quantitative results were published mainly depending on the lack of high-quality, prospectively collected data, lack of baseline situation (recognized to be a highly confounding factor), lack of patient-reported data, and the strong perception of the potentially jeopardizing impact of the variable filling content.

CRITICAL REVIEW OF THE LITERATURE 2009–2017 (CONVENTIONAL FRACTIONATION)

Dose-Volume Relationship

One of the most outstanding results achieved in recent years is the acknowledgement of the existence of a dose-volume effect for several urinary symptoms arising after and as a consequence of radiotherapy for prostate cancer.

Several trials reported significant associations between dose to the urinary bladder and both acute and late urinary injury: a summary of the more relevant studies is reported in the recent review by Landoni and co-workers (Landoni et al. 2016) reporting on the main findings in terms of constraints/relationships. Predominantly, the bladder seems to act as a highly serial organ, i.e., its functional subunits are arranged as a chain and damage to one single subunit causes loss of functionality for the whole organ. An organ with such an architecture is highly sensitive to small volume receiving high doses and, in the particular case of urinary bladder, reducing the volume receiving more than 75–78 Gy or more than 8–12 Gy/week (Jolnerovski et al. 2017; Palorini et al. 2016b; Zelefsky et al. 1999; Harsolia et al. 2007; Fonteyne et al. 2009; Pederson et al. 2012; Heemsbergen et al. 2010;

Ghadjar et al. 2014; Ahmed et al. 2013; Fleming et al. 2011; Thor et al. 2016; Yahya et al. 2015; Inokuchi et al. 2017; Schaake et al. 2018) may significantly decrease the risk of acute and/or late urinary toxicity. Table 5.1 reports some details on trials highlighting a relationship between late urinary toxicity and bladder doses/prescription dose (with prescription dose being a surrogate for dose to small bladder volumes). Figure 5.1 reports a summary of literature-based dose-volume relationships for late urinary toxicity.

An important consequence of the existence of a dose-volume effect is that any attempt to reduce the fraction of bladder neck receiving high doses (>75–78 Gy, 2 Gy Equivalent Dose) appears to be justified. This highlights the pivotal role of image-guided radiotherapy (IGRT) in lowering urinary toxicity risk, given its potential in reducing the fraction of bladder that overlays the planning target volume, corresponding to the portion of bladder which is irradiated at the full prescription dose. The reduction of urinary toxicity with IGRT with respect to non-IGRT, reported in several studies (Kok et al. 2013; Zelefsky et al. 2012; Gill 2011; Jereczek-Fossa et al. 2011), indirectly supports this argument.

Of note, the role of small bladder volume irradiated at high doses (or of bladder maximum dose) was also established in several trials involving post-prostatectomy settings (Cozzarini et al. 2012; Ghadjar et al. 2014; De Langhe et al. 2014; Cozzarini et al. 2014; Mathieu et al. 2014).

Patient-Related Risk Factors

Many recently published studies also reported relevant patient-related features that are significantly associated with an increased risk of urinary toxicity. Extensive review of these studies can be found in (Rancati et al. 2017; Landoni et al. 2016; Fiorino et al. 2009; Budäus et al. 2012). These clinical features are acting as individual dose-response modifying factors, thus making some patients more sensitive/resistant to radiation.

A first essential risk factor for urinary toxicity, consistently described by different trials, is the baseline urinary functioning (Jolnerovski et al. 2017; Cozzarini et al. 2017; Peeters et al. 2005; Wortel et al. 2016; Barnett et al. 2011; Yahya et al. 2015b), with patients having an already impaired functionality being at higher risk of experiencing severe acute and late urinary toxicity. For this reason, an evaluation of the baseline situation should be mandatory before treatment planning, also considering the possibility of stricter dose limits for some patients.

Other patient-related characteristics have been emphasised as being associated with an increased risk of worsening of acute and late symptoms: previous transurethral resection of the prostate (TURP) (Cozzarini et al. 2017; Peeters et al. 2005; Fonteyne et al. 2009; Byrne et al. 2017; Heemsbergen et al. 2010, De Langhe et al. 2014) as well as smoking (Cozzarini et al. 2015; Stankovic et al. 2016; Bagalà et al. 2016; Steinberger et al. 2015; Solanki and Liauw 2013), age (Cozzarini et al. 2015; Palorini et al. 2016; Cozzarini et al. 2017; Wortel et al. 2016; Ahmed et al. 2013; Mathieu et al. 2014; Yahya et al. 2015), vascular comorbidities and use of cardiovascular drugs (Palorini et al. 2016b; Mathieu et al. 2014; Yahya et al. 2015b), diabetes (Jolnerovski et al. 2017; Mathieu et al. 2014; Yahya et al. 2015b; Stankovic et al. 2016), and use of anti-hypertensive medications (Cozzarini et al. 2015; Palorini et al. 2016b). The last patient-related characteristics are all indirect markers of possible damage

TABLE 5.1 Summary of Trials (Modern Series Pertaining to Patients Treated with Radical Radiotherapy) Highlighting a Relationship Between Acute/Late Urinary Toxicity and Bladder Doses/Prescription Dose (with Prescription Dose Being a Surrogate for Dose to Small Bladder Volumes)

Reference	Pts (N)	RT Technique	Prescription Dose (Gy)	Follow-up	Scoring Type	Toxicity Time	Toxicity Type	Bladder Dose Descriptor	Dose Descriptor Cutpoint	Toxicity Rate below/above Cutpoint (%)
					Overall Urinary Toxicity					
Harsolia et al. (2007)	331	3DCRT	70.2–79.2	Median 1.6 yrs	NCI CTC 2.0	Late, actuarial 3 yrs	Grade ≥2	Bladder volume receiving >82 Gy	2.5%	8 vs. 18
Fonteyne et al. (2009)	260	IMRT	74–80	Median 3 yrs	In-house developed scoring system[a]	Late, crude	Grade ≥2	Bladder volume receiving >70 Gy	20%	15 vs. 28
Pederson et al. (2012)	296	IMRT	68.5–76.4	Median 3.4 yrs	CTCAE	Late, actuarial 4 yrs	Grade ≥2	Prescription dose	76 Gy	13 vs. 24
Ahmed et al. (2013)	503	3DCRT	64–78	Median 5.9 yrs	modified RTOG/LENT	Late, crude	Grade ≥2	Mean bladder dose, bladder wall receiving > 60–70 Gy		
Fleming et al. (2011)	180	3DCRT	70–75	Median 5 yrs	RTOG	Late, actuarial 5 yrs	Grade ≥2	Bladder EUD (with n=0.5)	53.4 Gy	10 vs. 33
Jolnerovski et al. (2017)	177	IMRT	74	Median 6.4 yrs	CTCAE	Late, actuarial 5 yrs	Grade ≥2	Dose to 2% of bladder volume	73 Gy	18.3 vs. 31
Palorini et al. (2017)	319	IMRT	74–80 (2 Gy/fr) 65–75.2 (2.2–2.7 Gy/fr)	Minimum 3 yrs	CTCAE + PRO[&]	Late, crude	Any Grade ≥2 endpoint	Bladder EUD (with n=0.01 & α/β=1 Gy)	75 Gy	<15% if below cutoff

(Continued)

TABLE 5.1 (CONTINUED) Summary of Trials (Modern Series Pertaining to Patients Treated with Radical Radiotherapy) Highlighting a Relationship Between Acute/Late Urinary Toxicity and Bladder Doses/Prescription Dose (with Prescription Dose Being a Surrogate for Dose to Small Bladder Volumes)

Reference	Pts (N)	RT Technique	Prescription Dose (Gy)	Follow-up	Scoring Type	Toxicity Time	Toxicity Type	Bladder Dose Descriptor	Dose Descriptor Cutpoint	Toxicity Rate below/above Cutpoint (%)
Obstruction/Stricture										
Heemsbergen et al. (2010)	557	3DCRT	68 vs. 78	Median 5.9 yrs	RTOG/EORTC	Late, actuarial 5 yrs	Grade ≥2 obstruction	Bladder surface receiving ≥80 Gy	0.5 cc	8 vs. 13
Palorini et al. (2017)	319	IMRT	74–80 (2 Gy/fr) 65–75.2 (2.2–2.7 Gy/fr)	Minimum 3 yrs	CTCAE	Late, crude	Grade ≥3 stricture/obstruction	Bladder EUD (with $n=0.01$ & $\alpha/\beta=10$ Gy)	78 Gy	<3% if below cutoff
Dysuria										
Yahya et al. (2015a)	754	3DCRT	60,70,74	Median 5 yrs	LENT-SOMA	Late, crude OR longitudinal	Grade ≥1 dysuria (P[b]/EC[s])	Relative bladder surface receiving > 64–68 Gy	20%	
Yahya et al. (2015a)	754	3DCRT	60,70,74	Median 5 yrs	LENT-SOMA	Late, crude OR longitudinal	Grade ≥2 dysuria (P[b]/EC[s])	Relative bladder surface receiving > 64–67 Gy	15%	
Schaake et al. (2018)	243	IMRT	78	Minimum 3 yrs	CTCAE	Late, crude	Grade ≥2 pain or discomfort during voiding	Relative bladder trigone receiving > 75 Gy	70%	<8% if below cutoff

(Continued)

TABLE 5.1 (CONTINUED) Summary of Trials (Modern Series Pertaining to Patients Treated with Radical Radiotherapy) Highlighting a Relationship Between Acute/Late Urinary Toxicity and Bladder Doses/Prescription Dose (with Prescription Dose Being a Surrogate for Dose to Small Bladder Volumes)

Reference	Pts (N)	RT Technique	Prescription Dose (Gy)	Follow-up	Scoring Type	Toxicity Time	Toxicity Type	Bladder Dose Descriptor	Dose Descriptor Cutpoint	Toxicity Rate below/above Cutpoint (%)
						Haematuria				
Yahya et al. (2015a)	754	3DCRT	60, 70, 74	Median 5 yrs	LENT-SOMA	Late, crude OR longitudinal	Grade ≥1 haematuria (P[b]/ EC[c])	Relative bladder surface receiving > 60–64 Gy	40%	
Schaake et al (2018)	243	IMRT	78	Minimum 3 yrs	CTCAE	Late, crude	Grade ≥1 haematuria	Relative bladder wall receiving > 75 Gy	20%	<5% if below cutoff
Palorini et al. (2017)	319	IMRT	74–80 (2 Gy/fr) 65–75.2 (2.2–2.7 Gy/ fr)	Minimum 3 yrs	CTCAE	Late, crude	Grade ≥1 haematuria	Bladder EUD (with $n=0.01$ & $\alpha/\beta=1$ Gy)	75 Gy	<5% if below cutoff
Inokuchi et al. (2017)	309	IMRT	78	Median 8.7 yrs	RTOG/ CTCAE	Late, actuarial 10 yrs	Grade ≥2 haematuria	Absolute volume of bladder neck receiving > 75 Gy	12 cc	5.2 vs. 19.2
						Incontinence				
Yahya et al. (2015a)	754	3DCRT	60, 70, 74	Median 5 yrs	LENT-SOMA	Late, crude OR longitudinal	Grade ≥1 incontinence (EC[c])	Relative bladder surface receiving > 15–20 Gy	85%	

(Continued)

TABLE 5.1 (CONTINUED) Summary of Trials (Modern Series Pertaining to Patients Treated with Radical Radiotherapy) Highlighting a Relationship Between Acute/Late Urinary Toxicity and Bladder Doses/Prescription Dose (with Prescription Dose Being a Surrogate for Dose to Small Bladder Volumes)

Reference	Pts (N)	RT Technique	Prescription Dose (Gy)	Follow-up	Scoring Type	Toxicity Time	Toxicity Type	Bladder Dose Descriptor	Dose Descriptor Cutpoint	Toxicity Rate below/above Cutpoint (%)
Yahya et al. (2015a)	754	3DCRT	60, 70, 74	Median 5 yrs	LENT-SOMA	Late, crude OR longitudinal	Grade ≥ 2 incontinence (EC[c])	Relative bladder surface receiving > 70 Gy	20%	<5% if below cutoff
Schaake et al. (2018)	243	IMRT	78	Minimum 3 yrs	CTCAE	Late, crude	Grade ≥ 2 incontinence	Mean dose to the trigone	65 Gy	
Palorini et al. (2017)	319	IMRT	74–80 (2 Gy/fr) 65–75.2 (2.2–2.7 Gy/fr)	Minimum 3 yrs	ICIQ-SF	Late, crude	ICIQ3+ICIQ4>5, in patients with ICIQ3+ICIQ4=0 at baseline	Bladder EUD (with $n=0.01$ & $\alpha/\beta=1$ Gy)	75 Gy	<3% if below cutoff
						Frequency				
Yahya et al. (2015a)	754	3DCRT	60, 70, 74	Median 5 yrs	LENT-SOMA	Late, crude OR longitudinal	Grade ≥ 1 frequency (EC[c])	Relative bladder surface receiving > 40 Gy	27%	
Yahya et al. (2015a)	754	3DCRT	60, 70, 74	Median 5 yrs	LENT-SOMA	Late, crude OR longitudinal	Grade ≥ 2 frequency (P[b]/EC[c])	Relative bladder surface > than 40 Gy	40%	

(Continued)

TABLE 5.1 (CONTINUED) Summary of Trials (Modern Series Pertaining to Patients Treated with Radical Radiotherapy) Highlighting a Relationship Between Acute/Late Urinary Toxicity and Bladder Doses/Prescription Dose (with Prescription Dose Being a Surrogate for Dose to Small Bladder Volumes)

Reference	Pts (N)	RT Technique	Prescription Dose (Gy)	Follow-up	Scoring Type	Toxicity Time	Toxicity Type	Bladder Dose Descriptor	Dose Descriptor Cutpoint	Toxicity Rate below/above Cutpoint (%)
International Prostate Symptom Score										
Ghadjar et al. (2014)	268	IMRT	86.4	Median 5 yrs	IPSS	Late, actuarial 5 yrs	DeltaIPSS$\geq$10	Maximum dose to bladder trigone	91 Gy	5 vs. 20
Palorini et al. (2017)	319	IMRT	74–80 (2 Gy/fr) 65–75.2 (2.2–2.7 Gy/fr)	Minimum 3 yrs	IPSS	Late, crude	DeltaIPSS$\geq$10	Bladder EUD (with $n=0.01$ & $\alpha/\beta=1$ Gy)	75 Gy	<10% if below cutoff

Pts, patients; N, number; RT, radiotherapy; yrs, years; 3DCRT, three-Dimensional Conformal Radiation Therapy; IMRT, Intensity Modulated Radiation Therapy; RTOG, Radiation Therapy Oncology Group; EORTC, European Organisation for Research and Treatment of Cancer; NCI CTC 2.0, National Cancer Institute Common Toxicity Criteria 2.0; CTCAE, Common Terminology Criteria for Adverse Events; LENT, Late Effects on Normal Tissue; SOMA, Subjective, Objective, Management, Analytic; IPSS, International Prostate Symptom Score; deltaIPSS, variation in IPSS between baseline and follow-up measurements; EUD, Equivalent Uniform Dose; ICIQ-SF, International Consultation on Incontinence Modular Questionnaire Short Form; ICIQ3, question number 3 in ICIQ-SF; ICIQ4, question number 4 in ICIQ-SF; PRO, patient-reported outcome.

a   In-house developed scoring system based on the RTOG, LENT/SOMA (Late Effects on Normal Tissue/Subjective, Objective, Management, Analytic), and Common. Terminology Criteria Genito-urinary toxicity scorings system & PRO = patient-reported outcome from IPSS and ICIQ-SF.

b   P, peak event, i.e., presence of symptoms at any time-point, see (Yahya et al. 2015a) for details.

c   EC, event count, i.e., total count of events accumulated during follow-up time, see (Yahya et al. 2015a) for details.

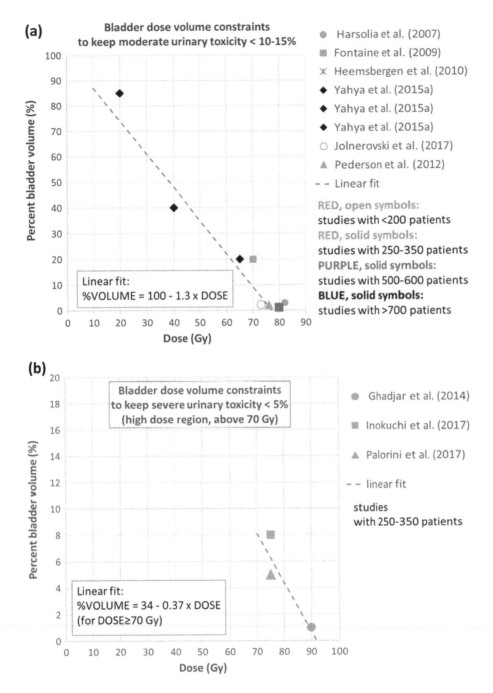

FIGURE 5.1 Summary of literature-based dose-volume relationships for late urinary toxicity (modern series pertaining to patients treated with radical radiotherapy). Panel (a): bladder dose-volume constraints to keep moderate urinary toxicity <10–15%; Panel (b): bladder dose-volume constraints to keep severe urinary toxicity <5%. See Table 5.1 for study details.

of microcirculation leading to a possible impairment of tissue oxygenation, a key step in the repair of radiation-induced tissue damage. The paper by Rancati et al. (2017) reports details on patient-related features which were found to have an association with increased risk of acute/late urinary toxicity.

All these patient-related factors should be combined to dose to critical bladder structures when developing predictive models for urinary toxicity, in order to obtain tools which have the power to individualize treatment planning and optimisation.

CRITICAL REVIEW OF THE LITERATURE 2009–2017 (HYPOFRACTIONATION)

The large majority of patients experiencing urinary toxicity has been treated with conventionally fractionated treatments. However, especially for the case of prostate cancer, the availability of IGRT and some radiobiology motivation, pushed many institutions to implement hypofractionation schedules, even aiming in reducing both patient discomfort and treatment costs.

Different protocols, including moderate to extreme hypofractionated schemes, were suggested and tested in the last 10–15 years; evidence from four large randomized phase III trials were recently published (Dearnaley et al. 2016; Catton et al. 2017; Incrocci et al. 2016; Lee et al. 2016; Pollack et al. 2013), demonstrating that hypofractionation for localized prostate cancer is non-inferior to conventional fractionation. Based on this, a large amount of quantitative information relating the dose received by the bladder and other structures potentially involved to urinary toxicity within hypofractionated delivery schemes was collected in the last years and still more is planned for the coming years as well as hypofractionation will become more and more familiar.

In short, there are some evidences that hypofractionation may have some detrimental impact on urinary morbidity compared to conventional fractionation in prostate cancer patients, both in the acute and in the late settings (Palorini et al. 2016b; Shaikh et al. 2017; Cozzarini et al. 2017; Lee et al. 2016; Pollack et al. 2013; Hoffman et al. 2014; Arcangeli et al. 2012; Aluwini et al. 2016; Sanguineti et al. 2016). Highly interesting, there is increasing evidence of a higher than previously assumed sensitivity to fractionation for bladder (Cozzarini et al. 2017; Sanguineti et al. 2016; Fiorino et al. 2014): historical data and early radiobiology experiments largely supported the conviction that the bladder was relatively insensitive to fractionation with a likely value for α/β (in the linear quadratic approach) around 5 Gy for late toxicity.

Data from a large institutional experience using hypofractionation in post-prostatectomy radiotherapy showed a much worse urinary outcome for late severe incontinence and late severe haematuria, not consistent with such high value (Cozzarini et al. 2014). A best-fit analysis showed that an α/β value slightly below 1 Gy was the most likely (Fiorino et al. 2014). Other quantitative analyses showed results consistent with a very low α/β value for bladder for different symptoms in the prostate cancer radical setting, like late severe haematuria (Sanguineti et al. 2016) and incontinence (Cozzarini et al. 2017); this behaviour was shown to be consistent with the existence of a relevant consequential component of the damage, largely modulated by the daily dose (Dorr and Hendry 2001) and by the baseline situation (Pollack et al. 2013). Interestingly, the only large quantitative study trying to correlate the bladder dose distribution and the risk of urinary toxicity in the context of extreme hypofractionation (Kole et al. 2016) showed a prevalent dose-effect for late "moderate" urinary flare (assessed by IPSS changes) whose entity is more consistent with a low α/β value: In particular, a fraction of bladder larger than 12.7% receiving at least 33.5 Gy in 5 fractions (delivered with robotic linac) was highly predictive of an increased risk of urinary flare. In the same work, a value for volume parameter $n = 0.13$ (with confidence interval: −0.14, 0.41, consistently with the recent findings from Fiorino et al (2014) and Palorini et al. (2017)) was suggested, confirming the limited volume effect reported in conventionally fractionated radiotherapy. Similar results in the moderate hypofractionation context were reported showing a small volume effect for acute symptoms (Carillo et al. 2014; Arunsingh et al. 2017) and a dominant serial behaviour for late toxicity (Palorini et al. 2016b; Improta et al. 2016; Palorini et al. 2017.). Regarding the existence of clinical factors that may exacerbate the risk of urinary toxicity in hypofractionated protocols, to our knowledge there is no any specific evidence, although it may be not excluded that few factors discussed in the previous section may have varying impacts on urinary symptoms depending on the fractionation.

SUGGESTION FOR DVH CONSTRAINTS

In the current era of highly tailored dose distributions, often delivered with IGRT, the dose received by the bladder portion overlapped with Planning Target Volume (PTV) is quite similar to the prescribed dose, excepting the case of Stereotactic Body Radiation Therapy (SBRT, for prostate cancer) where hot spots may also fall in the bladder, if not constrained. The portion of bladder outside PTV receives a "low-medium" dose bath, depending on the PTV shape, delivered dose, delivery, and planning optimisation techniques.

In summary, based on the current knowledge, we may state that:

- The currently available evidence points out that the most relevant toxicities depend on the high-dose delivered in the PTV-bladder overlap and that any reduction of this overlap may be beneficial both in conventionally fractionated and hypofractionated (moderate and extreme) radiotherapy: This means that any effort in reducing the margins with the use of appropriate IGRT techniques may impact on the risk of urinary toxicity.

- Another major point is that, especially for the most severely persistent late toxicities (like incontinence and haematuria), the risk rapidly increases for 2 Gy-equivalent prescribed doses (prescribed EQD2Gy) above 80–85 Gy, using an α/β=1 Gy in case of moderate hypofractionation. Similarly, one may look to the (mean) dose to the trigone as a surrogate of the risk.

- A tentative derivation of "evidence" based constraints from published studies (see Table 5.1) is given by the linear fit presented in Figure 5.1a,b, for moderate and severe late urinary toxicity, respectively. Table 5.2 reports such possible dose-volume histogram constraints in tabular form.

- Concerning SBRT, Table 5.3 summarizes the most popular bladder dose-volume constraints referred to late urinary toxicity. Separated values for one, three, and five fractions (Timmerman 2008; Benedict et al. 2010; Lukka 2016) are reported. The relevant result from Kole et al. (2016) regarding the five fraction scheme was added, as it is, to our knowledge, the only quantitative result from a large prospective trial.

- Several patient-related factors may impact the risk of urinary toxicity, with baseline urinary functioning, previous TURP, smoking, cardio-vascular diseases, and diabetes being the most relevant. Table 5.4 and Figure 5.2 present a summary of trials (modern series pertaining to patients treated with radical radiotherapy) highlighting a relationship between late urinary toxicity and patient-related features. For each patient risk factor a "summary" weighted odds ratio is reported (together with its confidence interval) which was calculated using the inverse variance method.

OPEN ISSUES AND FUTURE DIRECTIONS OF INVESTIGATION

It is clear that our knowledge regarding the quantitative relationship between dose, bladder volume, and urinary toxicity significantly improved in the last decade; however, a lot remains to be explored. Here a tentative summary of the potentially relevant research issues in the field:

Refining Dose-Volume Models

Reliable and validated models of dose-volume effects of bladder are still lacking, efforts in this direction should be warranted. In particular, the evidence of the existence of relevant clinical predictors modulating the dose and dose/volume response should push the research in putting the dose-volume relationship within the context of multi-variable modelling for the different relevant end-points depicting the urinary functionality. In particular, it would be highly relevant to obtain more positive response from independent validation studies (Yahya et al. 2016) quantitatively comparing model predicted toxicity

TABLE 5.2 Suggested Constraints on Bladder Dose-Volume Histogram to Limit Late Urinary Toxicity (Moderate/Severe < 10–15% and Severe < 5%)

DVH Cutpoint	Cutpoint Value
V_{20Gy}	74%
V_{30Gy}	60%
V_{40Gy}	50%
V_{50Gy}	35%
V_{60Gy}	20%
V_{70Gy}	10%
V_{75Gy}	3%
V_{80Gy}	1%
EUD ($n = 0.01$)	80 Gy

Constraints were Derived from Linear Fit of Published Results (see Figure 5.1a, b).
EUD, Equivalent Uniform Dose; DVH, dose-volume histogram; V_{XGy} = percent bladder volume receiving more than X Gy.

TABLE 5.3 Summary of Bladder Dose/Volume Constraints to Limit Late Urinary Toxicity for Bladder in the Case Stereotactic Body Radiation Therapy (SBRT)

Number of Fractions	Dose/Volume/ EUD	Threshold Value	Endpoint	Notes	References
1	D_{max} (Dose to 0.035cc)	22 Gy	Cystitis/fistula (Grade 3)	Limit to wall	Timmerman (2008)
1	$V_{8.7Gy}$	15 cc	Cystitis/fistula (Grade 3)	Limit to wall	Timmerman (2008)
3	D_{max} (Dose to 0.035cc)	30 Gy	Cystitis/fistula (Grade 3)	Limit to wall	Timmerman (2008)
3	V_{15Gy}	15 cc	Cystitis/fistula (Grade 3)	Limit to wall	Timmerman (2008)
5	D_{max} (Dose to 0.035cc)	38 Gy	Cystitis/fistula (Grade 3)	Limit to wall	Timmerman (2008)
5	V_{38Gy}	1 cc	–	Empty bladder	Lukka (2016)
5	V_{37Gy}	10 cc	–		Lukka (2016)
5	$D_{90\%}$	32.6 Gy	–	Empty bladder	Lukka (2016)
5	V_{33Gy}	12%	Urinary flare risk < 10%	Empty bladder	Kole (2016)
5	EUD (Calculated with n=0.13)	30 Gy	Urinary flare risk < 10%	Empty bladder	Kole (2016)
5	V_{18Gy}	15 cc	Cystitis/fistula (Grade 3)	Limit to wall	Timmerman (2008)
5	$D_{50\%}$	18 Gy	–	Empty bladder	Timmerman (2008)
5	$D_{40\%}$	18 Gy	–		Lukka (2016)

Constraints are Reported for One, Three, and Five SBRT Fractions.
EUD, Equivalent Uniform Dose; VXGy, percent/absolute bladder volume receiving more than XGy; DY%, minimum dose to Y% of bladder volume.
[a] Dose to 0.035cc; [b]EUD calculated with n = 0.13.

rates and observed toxicity rates. Stringently related to these developments, the availability of patient-reported, prospectively collected data is of paramount importance as well as the use of more standardized and generalizable methods to score the symptoms, facilitating the pooling of data from different studies.

Spatial Effects

More efforts should be dedicated to the search and confirmation of the existence of substructures more radio-sensitive and/or specifically related to specific symptoms. This issue has been investigated, as previously described by using 2 dimensional (2D) and 3D metrics in addition to dose-volume/dose-surface histograms (Palorini et al. 2016b; Improta et al. 2016; Yahya et al. 2017), with clear indications regarding the higher sensitivity of the trigone. Quantitative analysis of imaging could also be used to assess radiation-induced modifications in normal tissues and their association with changes in urinary functioning, including structures other than bladder, such as the urethra and the muscles involved in the urinary functioning.

Pre-Clinical Animal Models

The patterns of development of urinary toxicities and the relationships with dose, fractionation, and irradiated volumes has never been deeply investigated in pre-clinical studies.

TABLE 5.4 Summary of Trials (Modern Series Pertaining to Patients Treated with Radical Radiotherapy Highlighting a Relationship Between Late Urinary Toxicity and Patient-Related Features

Reference Study	OR	Confidence Interval for OR		Weighted OR	Confidence Interval for Weighted OR	
Baseline Urinary Symptoms						
Peeters et al. (2005)	2.2	1.4	3.5	**2.4**	*2.2*	*2.7*
Heemsbergen et al. (2010)	2.7	1.3	5.6			
Barnett et al. (2011)	2.1	1.5	3.0			
Barnett et al. (2011)	4.2	2.1	8.4			
Yahya et al. (2015b)	2.3	1.7	2.3			
Yahya et al. (2015b)	3.8	2.6	5.5			
Wortel et al. (2016)	2.4	1.5	3.8			
Jolnerovski et al. (2017)	2.4	1.4	4.3			
Cozzarini et al. (2015)	2.4	1.2	5.2			
Trans-Urethral Resection of the Prostate (TURP)						
Peeters et al. (2005)	1.7	1.2	2.5	**1.9**	*1.5*	*2.3*
Fonteyne et al. (2009)	1.4	1.0	1.9			
Heemsbergen et al. (2010)	3.6	1.4	9.5			
Delanghe et al. (2014)	4.1	1.0	16.8			
Byrne et al. (2017)	2.5	1.4	4.6			
Cozzarini et al. (2015)	3.6	1.3	10.0			
Solanki et al. (2013)	3.8	1.4	9.1			
Smoking						
Solanki et al. (2013)	1.5	1.0	2.3	**1.7**	*1.3*	*2.3*
Solanki et al. (2013)	2.7	1.0	7.6			
Steinberger et al. (2015)	1.8	1.1	3.0			
Vascular Comorbidities/use of Cardiovascular Drugs						
Mathieu et al. (2014)	2.4	1.3	4.1	**3.3**	*2.4*	*4.4*
Mathieu et al. (2014)	2.9	1.3	6.5			
Yahya et al. (2015b)	4.0	2.5	5.5			
Diabetes						
Mathieu et al. (2014)	4.0	1.4	11.3	**2.3**	*1.6*	*3.4*
Stankovic et al. (2016)	3.0	1.0	9.0			
Jolnerovski et al. (2017)	2.0	1.1	3.5			
Byrne et al. (2017)	2.0	1.0	4.2			

For Each Patient Risk Factor a "Summary" Weighted Odds Ratio is Reported (Together with its Confidence Interval) which was Calculated using the Inverse Variance Method.
OR, Odds Ratio.

The availability of a new generation of small-animal irradiators, together with in-vivo not invasive imaging modalities, permits to carefully investigate the morphological and functional changes induced in the bladder wall as a function of dose and how they may influence the development of urinary impairment. The results coming from such studies could highlight the processes influencing the dose-volume relationship. On the other hand more reliable models coming from the clinics may inspire new research in the pre-clinical field, hopefully giving rise to a virtuous cycle. An example of the potentials of these systems

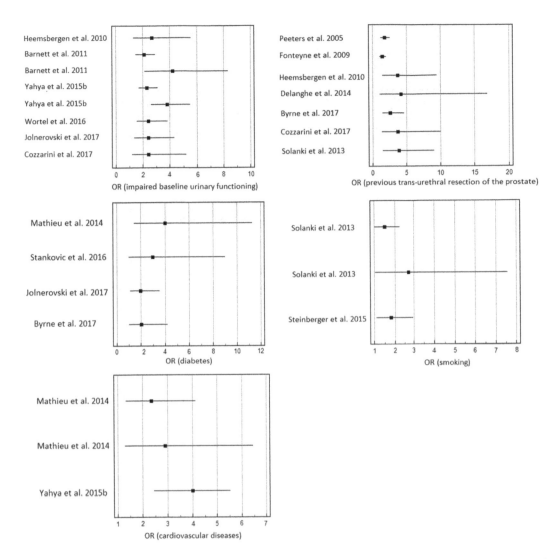

FIGURE 5.2 Summary (forest plot) of trials (modern series pertaining to patients treated with radical radiotherapy) highlighting a relationship between late urinary toxicity and patient-related features.

in localizing (portion of) bladder and selectively delivering tailored dose distributions is shown in Figure 5.3.

Incorporation/Estimate of Bladder Deformation into Models

As already underlined, the bladder is a hollow organ and consequently the dose-volume metrics depend on the degree of its filling. The use of absolute values (especially for bladder surface and wall) has been shown to be more robust compared to relative values (Carillo et al. 2014), once the patients are instructed to keep their bladder empty or full. However, even in the largely prevalent case of patients instructed to keep their bladder full, non-negligible changes in bladder volume may occur. Although dose-surface/volume histograms and

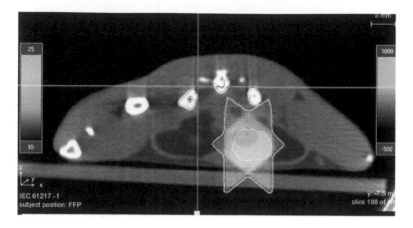

FIGURE 5.3 Highly conformal dose distribution around the bladder of a mouse, obtained with a modern micro-irradiator system and its planning system (single dose, 25 Gy: white, Planning Target Volume (PTV): bladder. *Courtesy of A. Spinelli, Experimental Imaging Research Center, San Raffaele Institute, Milano*).

bladder dose maps have been shown to be sufficiently robust for the assessment of dose-volume/surface relationships in several trials (Palorini et al. 2016a; Hoogeman et al. 2005), the estimate of bladder deformation remains an important issue. The availability of frequent imaging before and during radiotherapy may allow to model bladder deformation, based on non-rigid deformation modelling, and consent in the future to estimate the individually delivered dose distributions, or at least to assess the impact of variable filling in different datasets with different filling protocols, IGRT schedules, and delivery techniques.

Extreme Hypofractionation

The increased use of SBRT in the pelvic region claimed the urgent need of studies dealing with the quantitative relationship between the bladder dose distribution and urinary toxicities in this context. While they are scarce, the presently available findings largely refer to prostate cancer SBRT (mostly with five fractions delivering 35–40 Gy), while the interest toward the use of extreme hypofractionation in other situations is growing. Data pooling from different Institutes would be suitable to speed up the development of dose-volume modelling of urinary toxicity in SBRT.

REFERENCES

Ahmed AA, Egleston B, Alcantara P, Li L, Pollack A, Horwitz EM, Buyyounouski MK. A novel method for predicting late genitourinary toxicity after prostate radiation therapy and the need for age-based risk-adapted dose constraints. *Int J Radiat Oncol Biol Phys.* 2013;86(4):709–15.

Aluwini S, Pos F, Schimmel E, Krol S, van der Toorn PP, de Jager H, et al. Hypofractionated versus conventionally fractionated radiotherapy for patients with prostate cancer (HYPRO): late toxicity results from a randomised, non-inferiority, phase 3 trial. *Lancet Oncol.* 2016;17(4):464–74.

Arcangeli S, Strigari L, Gomellini S, Saracino B, Petrongari MG, Pinnarò P, et al. Updated results and patterns of failure in a randomized hypofractionation trial for high-risk prostate cancer. *Int J Radiat Oncol Biol Phys.* 2012;84(5):1172–8.

Arunsingh M, Mallick I, Prasath S, Arun B, Sarkar S, Shrimali RK, et al. Acute toxicity and its dosimetric correlates for high-risk prostate cancer treated with moderately hypofractionated radiotherapy. *Med Dosim.* 2017;42(1):18–23.

Avery K, Donovan J, Peters TJ, Shaw C, Gotoh M, Abrams P. ICIQ: A brief and robust measure for evaluating the symptoms and impact of urinary incontinence. *Neurourol Urodynamics.* 2004;23:322–330.

Bagalà P, Ingrosso G, Falco MD, Petrichella S, D'Andrea M, Rago M, et al. Predicting genitourinary toxicity in three-dimensional conformal radiotherapy for localized prostate cancer: A dose-volume parameters analysis of the bladder. *J Cancer Res Ther.* 2016;12(2):1018–24.

Barocas DA, Alvarez J, Resnick MJ, Koyama T, Hoffman KE, Tyson MD, et al. Association between radiation therapy, surgery, or observation for localized prostate cancer and patient-reported outcomes after 3 years. *JAMA.* 2017;317(11):1126–40

Barnett GC, De Meerleer G, Gulliford SL, Sydes MR, Elliott RM, Dearnaley DP. The impact of clinical factors on the development of late radiation toxicity: results from the Medical Research Council RT01 trial (ISRCTN47772397). *Clin Oncol (R Coll Radiol).* 2011;23(9):613–24.

Barry MJ, Fowler FJ Jr, O'Leary MP, Bruskewitz RC, Holtgrewe HL, Mebust WK, Cockett AT. The American urological association symptom index for benign prostatic hyperplasia. *J Urol.* 1992;148(5):1549–57.

Benedict SH, Yenice KM, Followill D, Galvin JM, Hinson W, Kavanagh B, et al. Stereotactic body radiation therapy: The report of AAPM Task Group 101. *Med Phys.* 2010;37(8):4078–101.

Budäus L, Bolla M, Bossi A, Cozzarini C, Crook J, Widmark A, Wiegel T. Functional outcomes and complications following radiation therapy for prostate cancer: a critical analysis of the literature. *Eur Urol.* 2012;61(1):112–27.

Byrne K, Hruby G, Kneebone A, Whalley D, Guo L, McCloud P, Eade T. Late genitourinary toxicity outcomes in 300 prostate cancer patients treated with dose-escalated image-guided intensity-modulated radiotherapy. *Clin Oncol (R Coll Radiol).* 2017;29(9):617–25.

Cahlon O, Zelefsky MJ, Shippy A, Chan H, Fuks Z, Yamada Y, et al. Ultra-high dose (86.4 Gy) IMRT for localized prostate cancer: toxicity and biochemical outcomes. *Int J Radiat Oncol Biol Phys* 2008;71(2):330–337.

Carillo V, Cozzarini C, Rancati T, Avuzzi B, Botti A, Borca VC, et al. Relationships between bladder dose–volume/surface histograms and acute urinary toxicity after radiotherapy for prostate cancer. *Radiother Oncol.* 2014;111(1):100–5.

Chin S, Hayden AJ, Gebski V, Cross S, Turner SL. Long term patient reported urinary function following external beam radiotherapy for prostate cancer. *Clin Oncol (R Coll Radiol).* 2017;29(7):421–428.

Catton CN, Lukka H, Gu CS, Martin JM, Supiot S, Chung PWM, et al. Randomized trial of a hypofractionated radiation regimen for the treatment of localized prostate cancer. *J Clin Oncol.* 2017 Jun 10;35(17):1884–1890.

Cozzarini C, Fiorino C, Da Pozzo LF, Alongi F, Berardi G, Bolognesi A, et al. Clinical factors predicting late severe urinary toxicity after postoperative radiotherapy for prostate carcinoma: a single-institute analysis of 742 patients. *Int J Radiat Oncol Biol Phys.* 2012;82(1):191–9.

Cozzarini C, Fiorino C, Deantoni C, Briganti A, Fodor A, La Macchia M, et al. Higher-than-expected Severe (Grade 3–4) late urinary toxicity after postprostatectomy hypofractionated radiotherapy: a single-institution analysis of 1176 patients. *Eur Urol.* 2014;66(:1024–30.

Cozzarini C, Rancati T, Carillo V, Civardi F, Garibaldi E, Franco P, et al. Multivariable models predicting specific patient-reported acute urinary symptoms after radiotherapy for prostate cancer: results of a cohort study. *Radiother Oncol.* 2015;116(2):185–91.

Cozzarini C, Rancati T, Palorini F, Avuzzi B, Garibaldi E, Balestrini D, et al. Patient-reported urinary incontinence after radiotherapy for prostate cancer: quantifying the dose-effect. *Radiother Oncol.* 2017;125(1):101–106.

De Langhe S, De Meerleer G, De Ruyck K, Ost P, Fonteyne V, De Neve W, Thierens H. Integrated models for the prediction of late genitourinary complaints after high-dose intensity modulated radiotherapy for prostate cancer: making informed decisions. *Radiother Oncol.* 2014;112(1):95–9.

Dearnaley D, Syndikus I, Mossop H, Khoo V, Birtle A, Bloomfield D, et al. Conventional versus hypofractionated high-dose intensity-modulated radiotherapy for prostate cancer: 5-year outcomes of the randomised, non-inferiority, phase 3 CHHiP trial. *Lancet Oncol.* 2016;17(8):1047–60.

Donovan JL, Hamdy FC, Lane JA, Mason M, Metcalfe C, Walsh E, et al. Patient-reported outcomes after monitoring, surgery, or radiotherapy for prostate cancer. *N Engl J Med.* 2016;375 (15):1425–1437

Dorr JH, Hendry JH. Consequential late effects in normal tissues. *Radiother Oncol.* 2001;61(3): 223–31.

Eriguchi T, Yorozu A, Kuroiwa N, Yagi Y, Nishiyama T, Saito S et al. Predictive factors for urinary toxicity after iodine-125 prostate brachytherapy with or without supplemental external beam radiotherapy. *Brachytherapy.* 2016;15(3):288–95.

Fleming C, Kelly C, Thirion P, Fitzpatrick K, Armstrong J. A method for the prediction of late organ-at-risk toxicity after radiotherapy of the prostate using equivalent uniform dose. *Int J Radiat Oncol Biol Phys.* 2011;80(2):608–13.

Fiorino C, Valdagni R, Rancati T, Sanguineti G. Dose-volume effects for normal tissues in external radiotherapy: pelvis. *Radiother Oncol.* 2009;93(2):153–67.

Fiorino C, Cozzarini C, Rancati T, Briganti A, Cattaneo GM, Mangili P, et al. Modelling the impact of fractionation on late urinary toxicity after postprostatectomy radiation therapy. *Int J Radiat Oncol Biol Phys.* 2014;90(5):1250–57.

Fonteyne V, Villeirs G, Lumen N, De Meerleer G. Urinary toxicity after high dose intensity modulated radiotherapy as primary therapy for prostate cancer. *Radiothe Oncol.* 2009;92(1):42–7.

Ghadjar P, Jackson A, Spratt DE, Oh JH, Munck af Rosenschöld P, Kollmeier M, et al. Patterns and predictors of amelioration of genitourinary toxicity after high-dose intensity-modulated radiation therapy for localized prostate cancer: implications for defining postradiotherapy urinary toxicity. *Eur Urol.* 2013;64(6):931–8.

Ghadjar P, Zelefsky MJ, Spratt DE, Munck af Rosenschöld P, Oh JH, Hunt M, et al. Impact of dose to the bladder trigone on long-term urinary function after high-dose intensity modulated radiation therapy for localized prostate cancer. *Int J Radiat Oncol Biol Phys.* 2014;88(2):339–44.

Gill S, Thomas J, Fox C, Kron T, Rolfo A, Leahy M, et al. Acute toxicity in prostate cancer patients treated with and without image-guided radiotherapy. *Radiat Oncol.* 2011;6:145.

Harsolia A, Vargas C, Yan D, Brabbins D, Lockman D, Liang J, et al. Predictors for chronic urinary toxicity after the treatment of prostate cancer with adaptive three-dimensional conformal radiotherapy: dose-volume analysis of a phase II dose-escalation study. *Int J Radiat Oncol Biol Phys.* 2007;69(4):1100–9.

Heemsbergen WD, Al-Mamgani A, Witte MG, van Herk M, Pos FJ, Lebesque JV. Urinary obstruction in prostate cancer patients from the Dutch trial (68 Gy vs. 78 Gy): relationships with local dose, acute effects, and baseline characteristics. *Int J Radiat Oncol Biol Phys.* 2010;78(1):19–25.

Hoffman KE, Voong KR, Pugh TJ, Skinner H, Levy LB, Takiar V, et al. Risk of late toxicity in men receiving dose-escalated hypofractionated intensity modulated prostate radiation therapy: results from a randomized trial. *Int J Radiat Oncol Biol Phys.* 2014;88(5):1074–84.

Hoogeman MS, Peeters ST, de Bois J, Lebesque JV. Absolute and relative dose-surface and dose-volume histograms of the bladder: which one is the most representative for the actual treatment? *Phys Med Biol.* 2005;50(15):3589–97.

Improta I, Palorini F, Cozzarini C, Rancati T, Avuzzi B, Franco P, et al. Bladder spatial-dose descriptors correlate with acute urinary toxicity after radiation therapy for prostate cancer. *Phys Med.* 2016;32(12):1681–1689.

Incrocci L, Wortel RC, Alemayehu WG, Aluwini S, Schimmel E, Krol S, et al. Hypofractionated versus conventionally fractionated radiotherapy for patients with localised prostate cancer (HYPRO): final efficacy results from a randomised, multicentre, open-label, phase 3 trial. *Lancet Oncol.* 2016;17(8):1061–9.

Inokuchi H, Mizowaki T, Norihisa Y, Takayama K, Ikeda I, Nakamura K, Hiraoka M. Correlation between urinary dose and delayed radiation cystitis after 78 Gy intensity-modulated radiotherapy for high-risk prostate cancer: a 10-year follow-up study of genitourinary toxicity in clinical practice. *Clin Transl Radiat Oncol.* 2017;6:31–36.

Jereczek-Fossa BA, Zerini D, Fodor C, Santoro L, Serafini F, Cambria R, et al. Correlation between acute and late toxicity in 973 prostate cancer patients treated with three-dimensional conformal external beam radiotherapy. *Int J Radiat Oncol Biol Phys.* 2010;78(1):26–34.

Jereczek-Fossa BA, Zerini D, Fodor C, Santoro L, Cambria R, Garibaldi C, et al. Acute toxicity of image-guided hypofractionated radiotherapy for prostate cancer: nonrandomized comparison with conventional fractionation. *Urol Oncol.* 2011;29(5):523–32.

Jolnerovski M, Salleron J, Beckendorf V, Peiffert D, Baumann AS, Bernier V, et al. Intensity-modulated radiation therapy from 70Gy to 80Gy in prostate cancer: six- year outcomes and predictors of late toxicity. *Radiat Oncol.* 2017;12(1):99.

Kok D, Gill S, Bressel M, Byrne K, Kron T, Fox C, et al. Late toxicity and biochemical control in 554 prostate cancer patients treated with and without dose escalated image guided radiotherapy. *Radiother Oncol.* 2013;107(2):140–6.

Kole TP, Tong M, Wu B, Lei S, Obayomi-Davies O, Chen LN et al. Late urinary toxicity modeling after stereotactic body radiotherapy (SBRT) in the definitive treatment of localized prostate cancer. *Acta Oncologica.* 2016;55(1):52–8.

Landoni V, Fiorino C, Cozzarini C, Sanguineti G, Valdagni R, Rancati T. Predicting toxicity in radiotherapy for prostate cancer. *Phys Med.* 2016;32(3):521–32.

Lane A, Metcalfe C, Young GJ, Peters TJ, Blazeby J, Avery KN, et al. Patient-reported outcomes in the ProtecT randomized trial of clinically localized prostate cancer treatments: study design, and baseline urinary, bowel and sexual function and quality of life. *BJU Int.* 2016;118(6):869–79.

Lee WR, Dignam JJ, Amin MB, Bruner DW, Low D, Swanson GP, et al. Randomized phase III noninferiority study comparing two radiotherapy fractionation schedules in patients with low-risk prostate cancer. *J Clin Oncol.* 2016;34(20):2325–32.

Lukka H, Stephanie P, Bruner D, Bahary JP, Lawton CAF, Efstathiou JA et al. Patient-Reported Outcomes in NRG Oncology/RTOG 0938, a Randomized Phase 2 Study Evaluating 2 Ultrahypofractionated Regimens (UHRs) for Prostate Cancer. *Int J Radiat Oncol Biol Phys.* 2016;(94)1:2.

Malik R, Jani AB, Liauw SL. External beam radiotherapy for prostate cancer: urinary outcomes for men with high International Prostate Symptom Scores (IPSS). *Int J Radiat Oncol Biol Phys.* 2011;80(4):1080–6.

Malmsten UG, Molander U, Peeker R, Irwin DE, Milsom I. Urinary incontinence, overactive bladder, and other lower urinary tract symptoms: a longitudinal population-based survey in men aged 45-103 years. *Eur Urol.* 2010;58(1):149–56.

Marks LB, Carroll PR., Dugan TC, Anscher MS. The response of the urinary bladder, urethra, and ureter to radiation and chemotherapy. *Int J Radiat Oncol Biol Phys.* 1995;31(5):1257–80.

Martin NE, Massey L, Stowell C, Bangma C, Briganti A, Bill-Axelson A, et al. Defining a standard set of patient-centered outcomes for men with localized prostate cancer. *Eur Urol.* 2015;67(3):460–7.

Mathieu R, Arango JDO, Beckendorf V, Delobel JB, Messai T, Chira C. Nomograms to predict late urinary toxicity after prostate cancer radiotherapy. *World J Urol.* 2014;32(3):743–51.

Miyake M, Tanaka N, Asakawa I, Tatsumi Y, Nakai Y, Anai S, et al. Changes in lower urinary tract symptoms and quality of life after salvage radiotherapy for biochemical recurrence of prostate cancer. *Radiother Oncol.* 2015;115(3):321–6

Mundy AR, Andrich DE. Posterior urethral complications of the treatment of prostate cancer. *BJU Int.* 2012;110(3):304–25.

Official website: http://ctep.cancer.gov/protocolDevelopment/electronic_applications/ctc.htm

Palorini F, Botti A, Carillo V, Gianolini S, Improta I, Iotti C, et al. Bladder dose-surface maps and urinary toxicity: robustness with respect to motion in assessing local dose effects. *Phys Med.* 2016a;32(3):506–11.

Palorini F, Rancati T, Cozzarini C, Improta I, Carillo V, Avuzzi B, et al. Multivariable models of large IPSS worsening at the end of therapy in prostate cancer radiotherapy. *Radiother Oncol.* 2016b;118(1):92–8.

Palorini F, Cozzarini C, Gianolini S, Botti A, Carillo V, Iotti C, et al. First application of a pixel-wise analysis on bladder dose-surface maps in prostate cancer radiotherapy. *Radiother Oncol.* 2016c;119(1):123–8.

Palorini F, Cicchetti A, Rancati T, Cozzarini C, Avuzzi B, Botti A, et al. NTCP models of late severe urinary symptoms after radical IMRT for prostate cancer. *Radiother Oncol.* 2017;127(S1):S168.

Pederson AW, Fricano J, Correa D, Pelizzari CA, Liauw SL. Late toxicity after intensity-modulated radiation therapy for localized prostate cancer: an exploration of dose-volume histogram parameters to limit genitourinary and gastrointestinal toxicity. *Int J Radiat Oncol Biol Phys.* 2012;82(1):235–41.

Peeters ST, Heemsbergen WD, van Putten WL, Slot A, Tabak H, Mens JW et al. Acute and late complications after radiotherapy for prostate cancer: results of a multicenter randomized trial comparing 68 Gy to 78 Gy. *Int J Radiat Oncol Biol Phys.* 2005;61(4):1019–34.

Pinkawa M, Holy R, Piroth MD, Fischedick K, Schaar S, Székely-Orbán D, Eble MJ. Consequential late effects after radiotherapy for prostate cancer – a prospective longitudinal quality of life study. *Radiat Oncol.* 2010;5:27–35.

Pollack A, Walker G, Horwitz EM, Price R, Feigenberg S, Konski AA, et al. Randomized trial of hypo-fractionated external-beam radiotherapy for prostate cancer. *J Clin Oncol.* 2013;31(31):3860–8.

Prostate Advances in Comparative Evidence (PACE). Clinical Trials. https://clinicaltrials.gov/ct2/show/NCT-01584258

Qi XS, Wang JP, Gomez CL, Shao W, Xu X, King C, et al. Plan quality and dosimetric association of patient-reported rectal and urinary toxicities for prostate stereotactic body radiotherapy. *Radiother Oncol.* 2016;121(1):113–7.

Rancati T, Palorini F, Cozzarini C, Fiorino C, Valdagni R. Understanding urinary toxicity after radiotherapy for prostate cancer: first steps forward. *Tumori.* 2017;103(5):395–404.

Rapariz-González M, Castro-Díaz D, Mejía-Rendón D. Evaluation of the impact of the urinary symptoms on quality of life of patients with painful bladder syndrome/chronic pelvic pain and radiation cystitis: EURCIS study. *Actas Urologicas Espanolas.* 2014;38(4):224–31.

Rohrmann S, Katzke V, Kaaks R. Prevalence and progression of lower urinary tract symptoms in an aging population. *Urology.* 2016;95:158–63.

Sanguineti G, Arcidiacono F, Landoni V, Saracino BM, Farneti A, Arcangeli S et al. Macroscopic hematuria after conventional or hypofractionated radiation therapy: results from a prospective Phase III study. *Int J Radiat Oncol Biol Phys.* 2016;96(2):304–312.

Schaake W, van der Schaaf A, van Dijk LV, van der Bergh ACM, Lagendijk JA, Development of a prediction model for late urinary incontinence, hematuria, pain and voiding frequency among irradiated prostate cancer patients. *Plos One,* 2018;13(7):e0197757.

Seymour ZA, Chang AJ, Zhang L, Kirby N, Descovich M, Roach M 3rd, et al. Dose-volume analysis and the temporal nature of toxicity with stereotactic body radiation therapy for prostate cancer. *Pract Radiat Oncol.* 2015;5(5):e465–72.

Shaikh T, Li T, Handorf EA, Johnson ME, Wang LS, Hallman MA, et al. Long-term patient-reported outcomes from a phase 3 randomized prospective trial of conventional versus hypo-fractionated radiation therapy for localized prostate cancer. *Int J Radiat Oncol Biol Phys.* 2017;97(4):722–31.

Shahid N, Loblaw A, Chung HT, Cheung P, Szumacher E, Danjoux C, et al. Long-term toxicity and health-related quality of life after single-fraction high dose rate brachytherapy boost and hypofractionated external beam radiotherapy for intermediate-risk prostate cancer. *Clin Oncol (R Coll Radiol).* 2017;29(7):412–420.

Solanki AA, Liauw SL. Tobacco use and external beam radiation therapy for prostate cancer: Influence on biochemical control and late toxicity. *Cancer.* 2013;119(15):2807–14.

Stankovic V, Džamic Z, Pekmezovic T, Tepavcevic DK, Dozic M, Saric M et al. Acute and late genitourinary toxicity after 72 Gy of conventionally fractionated conformal radiotherapy for localised prostate cancer: impact of individual and clinical parameters. *Clin Oncol (R Coll Radiol).* 2016;28(9):577–86.

Steinberger E, Kollmeier M, McBride S, Novak C, Pei X, Zelefsky MJ. Cigarette smoking during external beam radiation therapy for prostate cancer is associated with an increased risk of prostate cancer-specific mortality and treatment-related toxicity. *BJU Int.* 2015;116(4):596–603.

Sullivan L, Williams SG, Tai KH, Foroudi F, Cleeve L, Duchesne GM. Urethral stricture following high dose rate brachytherapy for prostate cancer. *Radiother Oncol.* 2009;91(2): 232–6.

Thor M, Olsson C, Oh JH, Petersen SE, Alsadius D, Bentzen L, et al. Urinary bladder dose-response relationships for patient-reported genitourinary morbidity domains following prostate cancer radiotherapy. *Radiother Oncol.* 2016;119(1):117–22.

Talcott JA, Rossi C, Shipley WU, Clark JA, Slater JD, Niemierko A, Zietman AL. Patient-reported long-term outcomes after conventional and high-dose combined proton and photon radiation for early prostate cancer. *JAMA.* 2010;303(11):1046–53.

Timmerman RD. An overview of hypofractionation and introduction to this issue of seminars in radiation oncology. *Semin Radiat Oncol.* 2008;18(4):215–22.

Viswanathan AN, Yorke ED, Marks LB, Eifel PJ, Shipley WU. Radiation dose-volume effects of the urinary bladder. *Int J Radiat Oncol Biol Phys.* 2010;76(3 Suppl):S116–22.

Venderbos LDF, Aluwini S, Roobol MJ, Bokhorst LP, Oomens EHGM, Bangma CH, Korfage IJ. Long-term follow-up after active surveillance or curative treatment: quality-of-life outcomes of men with low-risk prostate cancer. *Qual Life Res.* 2017;26(6):1635–45.

Wei JT, Dunn RL, Litwin MS, Sandler HM, Sanda MG. Development and validation of the Expanded Prostate Cancer Index Composite (EPIC) for comprehensive assessment of health-related quality of life in men with prostate cancer. *Urology.* 2000;56(6):899–905.

Wortel RC, Incrocci L, Pos FJ, van der Heide UA, Lebesque JV, Aluwini S et al. Late side effects after image guided intensity modulated radiation therapy compared to 3D-conformal radiation therapy for prostate cancer: results from 2 prospective cohorts. *Int J Radiat Oncol Biol Phys.* 2016;95(2):680–9.

Yahya N, Ebert MA, Bulsara M, House MJ, Kennedy A, Joseph DJ, Denham JW. Urinary symptoms following external beam radiotherapy of the prostate: dose-symptom correlates with multiple-event and event-count models. *Radiother Oncol.* 2015a;117(2):277–82.

Yahya N, Ebert MA, Bulsara M, Haworth A, Kennedy A, Joseph DJ, Denham JW. Dosimetry, clinical factors and medication intake influencing urinary symptoms after prostate radiotherapy: An analysis of data from the RADAR prostate radiotherapy trial. *Radiother Oncol.* 2015b;116(1):112–8.

Yahya N, Ebert MA, Bulsara M, Kennedy A, Joseph DJ, Denham JW. Independent external validation of predictive models for urinary dysfunction following external beam radiotherapy of the prostate: Issues in model development and reporting. *Radiother Oncol.* 2016;120(2):339–45.

Yahya N, Ebert MA, House MJ, Kennedy A, Matthews J, Joseph DJ, Denham JW. Modeling urinary dysfunction after external beam radiation therapy of the prostate using bladder dose-surface maps: evidence of spatially variable response of the bladder surface. *Int J Radiat Oncol Biol Phys.* 2017;97(2):420–426.

Zelefsky MJ, Fuks Z, Hunt M, Lee HJ, Lombardi D, Ling CC, et al. Long term tolerance of high dose three-dimensional conformal radiotherapy in patients with localized prostate carcinoma. *Cancer.* 1999;85:2460–8.

Zelefsky MJ, Levin EJ, Hunt M, Yamada Y, Shippy AM, Jackson A, Amols HI. Incidence of late rectal and urinary toxicities after three-dimensional conformal radiotherapy and intensity-modulated radiotherapy for localized prostate cancer. *Int J Radiat Oncol Biol Phys.* 2008;70(4):1124–29.

Zelefsky MJ, Kollmeier M, Cox B, Fidaleo A, Sperling D, Pei X, et al. Improved clinical outcomes with high-dose image guided radiotherapy compared with non-IGRT for the treatment of clinically localized prostate cancer. *Int J Radiat Oncol Biol Phys.* 2012;84(1):125–9.

Zietman AL, DeSilvio ML, Slater JD, Rossi CJ Jr, Miller DW, Adams JA, Shipley WU. Comparison of conventional-dose vs high-dose conformal radiation therapy in clinically localized adenocarcinoma of the prostate: A randomized controlled trial. *JAMA.* 2005;294(10):1233–9.

Stomach, Duodenum, Liver, and Central Hepatobiliary Tract

Giovanni Mauro Cattaneo and Livia Marrazzo

CONTENTS

Introduction 138
Short Summary of the QUANTEC Recommendations 138
    Stomach and Duodenum 138
    Liver 139
Critical Review of the Literature 2009–2017 (Conventional Fractionation) 139
    Stomach and Duodenum 139
    Liver 143
Critical Review of the Literature 2009–2017 (Hypo Fractionation) 144
    Stomach and Duodenum 144
    Liver 146
        Treatment for Primary Liver Cancer 146
    Estimate of Effective Volume 152
    Treatment for Liver Metastasis 152
    Central Hepatobiliary Tract 153
    Mathematical/Biological Models 153
Summary for DVH Constraints/NTCP Model Parameters 156
    Stomach and Duodenum 156
    Liver 157
Open Issues and Future Directions of Investigation 157
    Organ Definition/Delineation 157
    Dose-Surface-Maps 157
    Adaptive Radiotherapy and Replanning 162
    Including Liver Function in Treatment Planning 163
    Effect of Breathing Motion on Liver Doses 163
    Particles 164
References 164

INTRODUCTION

Radiotherapy (RT) of abdominopelvic primary or secondary lesions in conformal or stereotactic techniques is in full development. QUANTEC's (Marks et al. 2010) publication is considered the most accurate assessment of toxicity predictors for conventional fractionation even though the data were scarce/absent for certain organs such as the stomach, duodenum and central hepatobiliary tract. Moreover in QUANTEC's publication, stereotactic body radiotherapy (SBRT) was included in the "special situations," being at that time a relatively new treatment option with little long-term follow-up.

After the QUANTEC publication several studies published detailed analyses of dose-volume data and toxicity; few dosimetric, clinical and treatment related parameters were identified as risk factors associated with toxicity following RT in multivariate analysis; moreover, few authors derived updated normal tissue complication probability (NTCP) model parameters using their own data or available published dose-volume histogram (DVH) data.

This chapter presents a comprehensive update of the dose constraints and available models for the stomach, the duodenum, the liver and the central hepatobiliary tract in conventional and hypofractionated (SBRT) radiotherapy.

SHORT SUMMARY OF THE QUANTEC RECOMMENDATIONS

Stomach and Duodenum

The stomach and small bowel are continuous, hollow visceral digestive organs. The small bowel includes the duodenum, jejunum and ileum. The sphincter pyloric and the jejunum limit the duodenum that is considered largely immobile.

In the QUANTEC publication (Kavanagh et al. 2010) the authors underlined the limited literature at that time. Acute, late and chronic injuries were considered, but very few studies did a separated analysis for stomach and small bowel effects. Volumetric data (volumes in the field) or DVH were rarely reported in studies regarding RT-induced toxicity to the stomach. Concerning duodenum, QUANTEC referred to only one publication reporting specific duodenal toxicity outcomes, with no DVH data available. Among the factors affecting the risk of small bowel/stomach injuries, concurrent chemotherapy has been associated with a higher risk of acute toxicity, and an history of abdominal surgery has been associated with a higher risk of late toxicity.

Most of the studies with quantitative dose-volume analyses of acute RT-induced toxicity referred to small bowel and patients treated for rectal or cervix primary cancer with conventional fractionation, often with concurrent chemotherapy (Baglan et al. 2002; Robertson et al. 2008; Gunnlaugsson et al. 2007; Huang et al. 2007; Roeske et al. 2003).

SBRT studies were included in the "special situation" paragraph: in an analysis of 77 patients treated with 25 Gy single fraction SBRT, the constraints associated with a 9% of crude incidence of late stomach/duodenal toxicity were <5% of duodenal volume receiving >22.5 Gy, <4% of stomach receiving >22.5 Gy and 50% isodose line not reaching the opposite luminal wall (Chang et al. 2009). For three-fraction SBRT, a maximum point dose <30 Gy was suggested for both stomach and duodenum.

Liver

The main endpoints identified in the QUANTEC part on radiation-associated liver injury (Pan et al. 2010) were divided into "classic" and "non-classic" radiation-induced liver disease (RILD). The first involves anicteric hepatomegaly and ascites, typically occurring between 2 weeks to 3 months after therapy (Lawrence et al. 1995) and elevated alkaline phosphatase (more than twice the upper limit of normal or baseline value). This endpoint can occur in patients with well-functioning pretreatment livers (mainly patients treated for liver metastases). Non-classic RILD, typically occurring between 1 week and 3 months after therapy, involves elevated liver transaminases (more than five times the upper limit of normal) or CTCAE v4.0 (Cancer Therapy Evaluation Program, Common Terminology Criteria for Adverse Events, https://evs.nci.nih.gov/ftp1/CTCAE/CTCAE_4.03_2010-06-1 4_QuickReference_5x7.pdf) grade 4 levels in patients with baseline values more than five times the upper limit of normal within 3 months after completion of RT, or a decline in liver function (measured by a worsening of Child-Pugh (CP) score by 2 or more, Table 6.1). This endpoint has been mainly described in hepatocellular carcinoma (HCC) patients who have poor baseline liver function.

Another described endpoint is hepatitis B reactivation (Kim et al. 2007), which can contribute to liver function abnormalities.

At that time, the review of dose-volume data mainly concerned whole liver RT and partial liver RT with conventional fractionation (2 Gy/fraction). Recommended dose-volume limits were separately specified for patients treated for liver metastases or primary liver cancers, since preexisting liver dysfunction may render patients more susceptible to RT-induced liver injury. For palliative whole liver irradiation (2 Gy/fraction), the QUANTEC recommendation for 5% or less risk of RILD was whole-liver dose ≤30 Gy for liver metastases and ≤28 Gy for primary liver cancers (or 21 Gy in 3 Gy/fractions) (Ingold et al. 1965; Russell et al. 1993). For curative partial liver irradiation (2 Gy/fr) the recommendation was mean normal liver dose (liver minus Gross Tumor Volume [GTV]) ≤32 Gy for liver metastases and ≤28 Gy for primary liver cancers, mainly based on the University of Michigan experience (Dawson and Ten Haken 2005).

For SBRT (three to six fractions), the QUANTEC recommendations were: mean normal liver dose <13 Gy (3 fractions) or 18 Gy (6 fractions) for primary liver cancer (<6 Gy for CP B patients in 4–6 Gy/fractions for classic and non-classic RILD) (Hoyer et al. 2006; Tse et al. 2008) and <15 Gy (3 fractions) or 20 Gy (6 fractions) for liver metastases (Lee et al. 2009). Moreover, at least 700 mL of normal liver should receive ≤15 Gy in 3–5 fractions (Kavanagh et al. 2006).

Long-term liver injury or biliary duct system damage was almost uncovered.

CRITICAL REVIEW OF THE LITERATURE 2009–2017 (CONVENTIONAL FRACTIONATION)

Stomach and Duodenum

After the QUANTEC publication, several studies have shown that the risk of toxicity is associated with DVH data.

TABLE 6.1 Child-Pugh Scoring System

Measure	1 Point	2 Points	3 Points
Total bilirubin (mg/dL)	<2	2–3	>3
Serum albumin (g/dL)	>3.5	2.8–3.5	<2.8
Prothrombin time, prolongation (s) or	<4.0	4.0–6.0	> 6.0
INR	<1.7	1.7–2.3	>2.3
Ascites	None	Mild (or suppressed with medication)	Moderate to severe (or refractory)
Hepatic encephalopathy	None	CTCAE Grade I–II	CTCAE Grade III–IV
Points	**Child-Pugh Class**		
5–6	A		
7–9	B		
10–15	C		

Source: Adapted from Pan et al. (2010).
INR, international normalized ratio; CTCAE, Common Terminology Criteria for Adverse Events.

In patients with locally advanced pancreas cancer (LAPC), several papers have reported a high rate of hematologic and gastrointestinal toxicities using gemcitabine-based or 5-fluoracil based concurrent radiochemotherapy; the rather narrow therapeutic window of gemcitabine-based RT, the large irradiated volumes and/or non-IMRT irradiation technique could explain the high rate of grade ≥ 3 toxicity (Loehrer et al. 2011; Zhu 2011).

Huang et al. (2012) retrospectively analyzed data from 46 patients with LAPC treated with definitive concurrent full-dose gemcitabine (1000 mg/cm^2 weekly) and radiotherapy (18 patients received daily erlotinib). The stomach and the duodenum were separately contoured and grade ≥ 3 gastrointestinal (GI) toxicities (CTAE v4.0) were considered. For the entire cohort, considering both acute (22%) and late (17%) toxicities, only the V_{25Gy} of duodenum (percent duodenum volume receiving ≥ 25 Gy) and the concurrent use of erlotibin were independent risk factors on multivariate analysis (MVA) (p=0.006 and p=0.02, respectively). The receiver operating characteristic (ROC) analysis and stepwise discriminant analysis (SDA) showed that duodenum $V_{25Gy} > 45\%$ was the optimal threshold: The 1 year GI toxicity rate was 8% vs. 48% for $V_{25Gy} \leq 45\%$ and $V_{25Gy} > 45\%$, respectively. Excluding the group receiving erlotinib (that increases radiation sensitivity), $V_{35Gy} > 20\%$ was the best threshold (1 year GI toxicity rate of 0% vs. 41% for $V_{35Gy} \leq 20\%$ and $V_{35Gy} > 20\%$, respectively). The high overall rate of grade ≥ 3 toxicity reported in this study is partially related to the use of full-dose gemcitabine.

Nakamura et al. retrospectively analyzed data of 40 LAPC patients (Nakamura et al. 2012). Chemoradiotherapy consisted of conventional fractionated three-dimensional conformal radiotherapy (3DCRT) and concurrent low-dose gemcitabine (250 mg/cm^2 weekly); 95% of the patients received 54 Gy in 38 fractions. Treatment-related acute GI toxicity and upper GI bleeding (UGB) were graded according to the CTCAE v4.0. The crude incidence of acute grade ≥ 2 GI toxicity was 33% and the estimated incidence of UGB at 1 year was 20%. On univariate analysis many dosimetric parameters were significantly correlated with acute GI toxicity such as stomach $V_{50Gy} < 16$ cm^3 (1 year actuarial incidence was 9%

vs. 61%, in patients with V_{50Gy} <16 cm^3 vs. those with V_{50Gy} ≥16 cm^3 respectively, $p = 0.001$). Regarding UGB, for gastroduodenum V_{50Gy} ≤ vs. >33 cm^3, the 1 year rate of bleeding was 0% vs. 44%, respectively ($p = 0.002$).

The risk of gastroduodenal toxicity after RT in cirrhotic patients may be higher than that of non-cirrhotic patients due to the poorer gastric defense mechanism. Kim et al. (2009) retrospectively analyzed DVHs and clinical records of 73 cirrhotic patients treated with 3DCRT for HCC. The median radiation dose was 36 Gy (range, 30–54 Gy) with a daily dose of 3 Gy. The external surface of organs including stomach and duodenum (gastroduodenum) was outlined. The grade of gastroduodenal toxicity was defined by the CTCAE v2.0. Age (Hazard Ratio (HR) = 1.17, $p = 0.03$) and V_{35Gy} (HR = 1.17, $p < 0.01$) were significant predictive parameters of grade 3 gastroduodenal toxicity. The 1 year actuarial rate of grade 3 toxicity in patients with V_{35Gy} <5% was significantly lower than that in patients with V_{35Gy} > 5% (4% vs. 48%, $p < 0.01$). The same group (Yoon et al. 2013) performed a study on 90 HCC patients (87% with cirrhotic liver) considered at risk because the gastroduodenum was in close proximity to the Planning Target Volume (PTV); toxicities were determined by esophagogastroduodenoscopy (EGD) before and after RT. All endoscopic findings related to RT were graded by CTCAE v3.0: grade ≥ 2 toxicity developed in 27 (30%) patients. RT was delivered as 30–50 Gy (median 37.5 Gy) in 2–5 Gy (median 3.5 Gy) per fraction. The planning risk volume (defined to account for organ movement) was chosen for dosimetric analysis and all doses were converted in 2 Gy Equivalent Doses EQD2Gy with linear quadratic model and $\alpha/\beta = 10$ Gy (EQD2Gy^{10}, see chapter 2 for details on EQD2Gy calculation). On MVA, including the clinical factors, stomach V_{25Gy} and duodenum V_{35Gy} were the only significant factors. The gastric toxicity rate at 6 months for V_{25Gy} ≤ vs. > 6.3% was 2.9% and 57.1% respectively; the duodenal toxicity rate at 6 months for V_{35Gy} ≤ vs. > 5.4% was 9.4% and 45.9% respectively.

A group of 119 patients treated for unresectable intrahepatic cancer with 3DCRT were retrospectively analyzed (Feng et al. 2012). A gastric bleed event was defined as a gastric hemorrhage that required transfusion or radiologic/endoscopic/elective operative intervention (grade ≥ 3, CTCAE v4.0); the crude incidence of bleeding was 14%. The median prescribed dose was 54.0 Gy (range 12–90.5 Gy); patients were treated daily, 5 fractions per week (1.8–4.7 Gy per fraction) or twice daily, 11 fractions per week (1.5–1.65 Gy per fraction). Most patients were treated with an empty stomach and active breathing control. The wall of the stomach volume was defined as organ-at-risk (OAR). Three dimensional (3D) dose distribution was converted to EQD2Gy using the linear quadratic model (with $\alpha/\beta = 2.5$ Gy; EQD2Gy$^{2.5}$); the maximum dose was defined as the minimum dose to 1 cm^3 of the organ wall receiving the highest dose (D_{1cm3}). The generalized equivalent uniform dose (EUD) was used for the estimation of Lyman NTCP parameters. In a proportional hazards regression model, only cirrhosis and the NTCP value (or D_{1cm3}) were statistically significant predictors ($p < 0.05$). The mean stomach D_{1cm3} (in terms of EQD2Gy$^{2.5}$) was 43 Gy and 61 Gy for patients with or without gastric bleeds, respectively. Refitting the Lyman NTCP with separate D_{50} for normal and cirrhotic patients (keeping the same m and n values of the entire population), the authors obtained $n = 0.10$, $m = 0.21$, D_{50} (normal) = 56 Gy and D_{50} (cirrhosis) = 22 Gy. The NTCP as a function of EUD for patients with cirrhosis is a very steep curve giving 20% and 50% risk of bleeding at 18 Gy and 22 Gy, respectively.

Concerning gastric bleeding, this study underlined that the stomach has a lower tolerance dose than expected (especially for patients prone to bleeding at baseline), presents a small volume effect and the presence of cirrhosis dramatically increases the risk, consistent with the paper by Kim et al. (2009).

Cattaneo et al. (2013) reported dosimetric-clinical predictors of gastrointestinal toxicity after induction chemotherapy followed by concurrent chemotherapy with moderately hypofractionated IMRT/Image Guided Radiation Therapy (IGRT) (44.25 Gy in 15 fractions) for 61 patients affected by LAPC. The 1 year actuarial rate of acute and late anatomical toxicity was 13%, the rate of acute GI toxicity was 33%. Duodenum V_{40Gy} and V_{45Gy} were significant predictors of grade ≥ 2 anatomical toxicity and their best cut-off values were 16% (or 10 cm^3 if absolute duodenum volume was considered) and 2.6% (or 1.7 cm^3 if absolute duodenum volume was considered) respectively; 40 Gy and 45 Gy delivered in 15 fractions have an EQD2Gy^3 equal to 45.3 Gy and 54 Gy respectively. The stomach V_{20Gy} was a predictor of acute GI grade ≥ 2 toxicity and the best cut-off value assessed by ROC analysis was 31% (76 cm^3, absolute volume).

Another retrospective study (Kelly et al. 2013) analyzed 106 patients with LAPC treated with fractionated chemoradiation. 78 patients were treated to 50.4 Gy in 28 fractions; 28 patients received dose escalated therapy (57.5–75.4 Gy in 28–39 fractions). Duodenal toxicity was scored according to CTCAE v4.0 and confirmed by endoscopic evaluation. V_{55Gy} showed the strongest correlation with grade ≥ 2 duodenal toxicity (HR: 6.7; confidence interval [CI]:2.0–18.8; $p < 0.002$). If more than 1 cm^3 of duodenum received $\geq$55 Gy, the actuarial rate of grade ≥ 2 duodenal toxicity was 47%, compared to 9% for patients with less than 1 cm^3 of the duodenum receiving $\geq$55 Gy. The strong correlation of V_{55Gy} with duodenal toxicity is consistent with the results by Cattaneo et al. (2013).

Verma et al. (2014) investigated late duodenal toxicity in 105 patients receiving IMRT for treatment of paraortic nodes in gynecologic cancer. RT consisted of a nodal Clinical Target Volume (CTV) dose of 45–50.4 Gy with an integrated/sequential boost to a total dose of 60–66 Gy to sites of gross disease; 55% of patients received concurrent platinum based chemotherapy. The ascending portion of duodenum was not delineated due to poor visualization on non-contrast computed tomography (CT); duodenal toxicity was assessed based on endoscopic findings and graded according to the Radiation Therapy Oncology Group (RTOG) scale. The 3 year actuarial rate of grade ≥ 2 late duodenal toxicity was 11.7%; V_{55Gy} was significantly associated with toxicity (V_{55Gy} < vs. > 15 cm^3 correspond to toxicity rates of 48.6% vs 7.4%, respectively, p < 0.01): The predictive value of the duodenum V_{55Gy} was consistent with other investigators (Cattaneo et al. 2013; Kelly et al. 2013), although the suggested cut-off value was higher (15 cm^3). Possibly, this may be due to differences in the type of disease and chemotherapy schedule.

In a phase II clinical trial (Number NCT01391572) 105 patients receiving RT after esophagectomy were randomized into a small-field (only tumor bed area) or large-field (tumor bed + lymphatic drainage area). The prescribed dose to the tumor bed was 60.2–63 Gy (in 28 fractions) and to lymphatic area was 50.4 Gy. The grade of toxicity was defined according to CTCAE v4.0; the crude incidence of grade ≥ 2 acute and late gastric toxicity was 18.1% and 13.3% respectively. V_{50Gy} (of intrathoracic stomach) was the only predicting factor; the optimal cut-off value was V_{50Gy} = 14% and the rate of toxicity was 19.1% vs 34.5% for V_{50Gy}< vs. >14.0% respectively. Looking at the two groups (small vs. large field) the intrathoracic stomach of patients in the large-field group had a larger treatment volume and a different dose distribution, so many DVH values had a statistically significant difference (Liu et al. 2015).

Recently, Holyoake et al. (2017) derived parameters for Lyman-Kutcher-Burman (LKB) (1989) model for duodenal toxicity in conventional fractionated radiotherapy using reconstructed DVH data from a set of publications reporting clinical toxicity outcomes. Data were converted to the 2 Gy equivalent doses using α/β = 4 Gy; observed gastro-intestinal toxicity rates ranged from 0% to 14%. LKB parameters for unconstrained fit to published data were n = 0.07, m = 0.46, D_{50} = 183.8 Gy, while the values for the model incorporating also the individual patient data (from two prospective clinical trials) were n = 0.19, m = 0.51, D_{50} = 299.1 Gy. The low values of the parameter n are consistent with a small duodenum volume effect for grade $\geq$ 3 GI toxicity.

Liver

Very few studies incorporating dose-volume relationship and models on radiation-induced liver toxicity with conventional fractionation (1.5–2.0 Gy/fraction) have been published since 2009 (i.e., in the post-QUANTEC era).

In a dose escalation study (Ren et al. 2011), 40 HCC CP Class (CPC)-A patients were assigned to two subgroups based on tumor diameter (<10 cm or $\geq$10 cm). Radiotherapy was combined with transcatheter arterial chemoembolization (TCA). Escalation was achieved by increments of 4 Gy starting from 46 Gy (2 Gy/fraction). Dose-limiting toxicity was defined as the development of grade $\geq$ 3 acute hepatic or gastrointestinal toxicity or any grade 5 treatment-related adverse event during irradiation or RILD within 4 months after irradiation. In the group of patients with tumor diameter $\geq$10 cm, 1 patient receiving 52 Gy developed RILD. The patient's uninvolved normal liver volume was 827 cm³ and the mean dose to normal liver was 13 Gy. Toxicity depended more on liver baseline function than dosimetry parameters.

In the study by Jung et al. (2014) 20 HCC patients with 33 tumors were treated with helical tomotherapy (mostly at 50 Gy with 2.5 Gy/fraction). Patients were CPC-A or CPC-B and the dose constraints applied for normal liver were: V_{27Gy} $\leq$ 30%, V_{24Gy} $\leq$ 50%, mean normal liver dose was <28 Gy. No patients experienced radiation-induced liver toxicity.

A common limitation to many publications on toxicity is that DVH parameters are derived from dose calculation on planning CT scan, while inter/intra-fraction variations may determine significant differences between planned and actually delivered doses. In the work by Huang and colleagues (Huang et al. 2017), 23 CPC-A patients with HCC received conventional fractionated 3DCRT (48–54 Gy, at 2 Gy/fraction). Seven of them were diagnosed with classic RILD. Mean dose to normal liver was limited to 28 Gy, but when doses were recalculated and accumulated using deformable image registration, adjusted mean doses to normal liver exceeded this level in the group of patients experiencing RILD, indirectly confirming the value of this constraint for primary liver cancers.

The majority of the studies investigating liver toxicity concern liver cancer patients or patients treated for liver metastasis. An analysis on radiation-induced liver toxicity in 20 patients treated with 3DCRT for primary gastric lymphoma (PGL), (Tanaka et al. 2013) 3 with mucosa-associated lymphoid tissue lymphoma (median dose 30 Gy) and 17 with diffuse large B-cell lymphoma (median dose 40 Gy), was published in 2013 (Tanaka et al. 2013). Increased transaminase or alkaline phosphatase (ALP) levels were observed in 19 (95%) patients. A more than 2-fold increase in transaminase or ALP levels, exceeding the upper limit within 4 months of the RT completion was observed in 14 (70%) patients. There were significant differences in dosimetric parameters between patients with and without this ALP increase: Patients with $V_{10Gy} > 60\%$, $V_{15Gy} > 50\%$ or $V_{20Gy} > 30\%$ experienced significantly higher toxicity.

CRITICAL REVIEW OF THE LITERATURE 2009–2017 (HYPO FRACTIONATION)

Stomach and Duodenum

In the last decade the use of SBRT within the upper abdomen has expanded, including primary hepatocellular carcinoma, metastatic liver lesions, spine metastasis, pancreatic malignancies and retroperitoneal tumors; this increasing role of SBRT has the potential for significant gastrointestinal toxicities.

Kopek et al. (2010) explored the dose-volume dependency of GI toxicity in 27 patients with unresectable cholangiocarcinoma (median follow-up of 5.4 years); they received 45 Gy in three fractions prescribed to the isocenter and toxicity was scored using CTCAE v3.0. Although duodenum V_{21Gy}, V_{24Gy}, V_{27Gy}, V_{30Gy} and D_{1cm3} were greater in the group experiencing grade ≥ 2 ulceration or stenosis, the difference reached statistical significance only for D_{1cm3} (37.4 Gy vs. 25.3 Gy, $p = 0.03$). The Stanford group (Murphy et al. 2010) performed a detailed retrospective analysis of duodenal toxicity after pancreas SBRT on 73 patients who received 25 Gy in a single fraction. All patients had 3–5 gold fiducials placed in the tumor. 53 of the patients (73%) were treated on CyberKnife using the Synchrony tracking method, the other patients (27%) were irradiated using a linear accelerator and fluoroscopy was used to ensure that the irradiation occurred when fiducials were within 2–3 mm of their position on the digitally reconstructed radiograph. The superior/inferior border of the duodenum was defined as 1 cm beyond the superior/inferior extent of the PTV; duodenal toxicity was scored according to the CTAE v3.0. 12 (16%) patients experienced grade ≥ 2 duodenal toxicity. In a univariate analysis $V_{10Gy-25 Gy}$ and D_{1cm3} were significant predictors of duodenal toxicity. Keeping $V_{15Gy} < 9.1$ cm³, or $V_{20Gy} < 3.3$ cm³ or $D_{1cm3} < 23$ Gy, the 1 year toxicity rate decreased from 49–52% to 11–12%. NTCP was evaluated with a LKB model and EUD parameter suggesting best-fit values for $m = 0.23$ (95% CI: 0.16–0.35), $n = 0.12$ (95% CI: 0.06–0.26) and $D_{50} = 24.6$ Gy (95% CI: 22.3–27.5 Gy). At MVA, NTCP remained the sole significant ($p = 0.001$) predictor of duodenal toxicity.

Bae et al. (2012) analyzed severe (grade ≥ 3, CTCAE v4.0) gastroduodenal toxicity following SBRT in three fractions for abdominopelvic malignancies. The stomach and duodenum (gastroduodenum) were retrospectively outlined and the superior and inferior borders were defined as 2 cm beyond the superior and inferior extent of the PTV. A group of 40 patients, whose gastroduodenum received at least 20% of the prescribed doses, were considered and severe toxicity was found in 6 (15%). Maximum dose was the best predictor (Area under the ROC curve [AUC] = 0.93) and a complication probability curve was determined fitting a probit function with $m = 0.17$ and $D_{50} = 48.9$ Gy (95% CI: 42.1–58.9 Gy). According to the complication probability curve, maximum doses of 35 and 38 Gy provoked severe gastroduodenal toxicity with probabilities of 5% and 10%, respectively.

In their meta-analysis, Prior et al. (2014) published pooled analysis of duodenum DVH and toxicity data from four studies, two using conventional fractionation and two using SBRT (Kelly et al. 2013; Verma et al. 2014; Kopek et al. 2010; Murphy et al. 2010); the four studies had differences in scoring criteria and in reporting toxicity rates (crude and actuarial rate). A modified linear quadratic model (Guerrero 2004) was used to generate a LKB model in 2 Gy equivalent doses. Dose-volume parameters associated with duodenum/ small bowel toxicity were recalculated using the modified linear quadratic model and the converted data presented some light consistency among the four clinical studies. However, the tolerance dose data available were not sufficient to derive a value of n and this could partially explain the difference between the LKB values they have fitted ($m = 0.21 \pm 0.05$ and $D_{50} = (60.9 \pm 7.9)$ Gy) and the data reported by Holyoake (mainly for $D_{50} > 125$ Gy) (Holyoake et al. 2017).

Bruner et al. (2015) performed a literature search for studies published from 1/2000 to 11/2013 to analyze outcome and toxicity of SBRT in pancreatic ductal carcinoma. A total of twenty studies with information on dose and toxicity grading were analyzed. Acute toxicity was mild in most of the studies but late toxicity, predominantly gastroduodenal, was not negligible with the rate of CTCAE grade ≥ 2 and grade ≥ 3 toxicities ranging between 0 and 23%. The prescribed dose to PTV was considered as a surrogate of dose to OAR, assuming $\alpha/\beta = 3$ Gy: grade ≥ 2 late toxicity was correlated to and target dose expressed as EQD2Gy[3] and as Biological Equivalent Dose (BED[3], see Chapter 2 for details on BED calculation), with a correlation coefficient (R^2) of 0.76 and 0.77 respectively.

Elhammali et al. (2015) pooled severe late gastrointestinal toxicity data from 16 studies after hypofractionated RT for pancreatic cancer; the review included articles published in the period 1981–2013. Serious late GI toxicities (grade ≥ 3) consisted of ulcers, hemorrhages, obstruction, strictures and perforations that occurred after a minimum of 3 months of follow-up; the median rate of toxicity was 7.4% (range 0–32.7%). Among the studies, important differences regarded the treatment techniques: 10 out of 16 studies used SBRT (one to five fractions), 6 out of 16 adopted Intra-Operative Radiotherapy (IORT) and 5 out of 16 combined SBRT/IORT with fractionated external beam radiotherapy (EBRT). In order to compare data with quite different doses per fraction, the Equivalent Dose to 1.8 Gy per fraction (EQD1.8Gy, from the linear quadratic model) was calculated with $\alpha/\beta = 3$ or 4 Gy. Due to the unavailability of DVH data, the authors assumed 1–5% of duodenal volume received the prescription dose while the remaining volume received no dose. The derived LKB NTCP model parameters (n, m, D_{50}) were $n = 0.38$–0.63, $m = 0.48$–0.49, $D_{50} = 35$–95 Gy and were not consistent with other studies (Holyoake et al. 2017; Murphy et al. 2010); as the authors underlined, the use of a linear-quadratic model in hypofractionated treatment is an issue of debate, especially when the fraction size is large, since the mechanism for normal tissue damage/repair in extreme fractionation might follow different pathways than those usually considered in standard fractionation regimens. Moreover, the assumption that a small amount of duodenum receives a uniform dose equal to the prescription dose could be too rough, especially if combining different techniques (SBRT/IORT $\pm$ EBRT).

Goldsmith et al. (2016) following the procedure introduced by Asbell et al. (2016) determined a DVH risk map of duodenal tolerance in the treatment of pancreatic cancer using planning constraints from literature and toxicity data from patients treated with SBRT at their institute. The DVH risk map provided a comparison of radiation tolerance limits as a function of dose, fractionation, volume and risk level for few parameters (D_{5cm3}, D_{1cm3}, maximum dose [D_{max}], minimum dose to 50% of the organ [$D_{50\%}$] and minimum dose to 5% of the organ [$D_{5\%}$]). Applying the logistic model to their duodenal data in three fractions, the risk of grade 3–4 toxicity (hemorrhage and stricture) was 4.7% for $D_{1cm3} = 25.3$ Gy, 10% for $D_{1cm3} = 31.4$ Gy and 20% for $D_{1cm3} = 37.4$ Gy in three fractions. The authors compared the risk of toxicity in patients treated with or without the aid of multiple fiducials, their analysis demonstrated that the use of multiple fiducials for tracking pancreas SBRT significantly reduced the risk of toxicity; these evaluations were based on few data points with large confidence intervals, so a larger dataset is required to confirm the differences in outcome.

In a prospective phase I trial for pancreatic cancer using chemotherapy followed by SBRT and concurrent administration of the human immunodeficiency virus (HIV) protease inhibitor nelfinavir, 13 patients who received pancreaticoduodenectomy were considered for a correlation analysis between dosimetric parameters and histopathologic/clinical duodenal toxicity (Verma et al. 2017). SBRT consisted of five daily fractions of 5–8 Gy; patients were treated within a gating window of $\pm$15% centered on complete exhalation. Histopathologic duodenal damage was classified following predetermined criteria from no (score 1) to marked changes (score 3) and clinical toxicities were assessed according to CTCAE. PTV mean/maximum dose, mean duodenal dose and duodenal $V_{20Gy-35Gy}$ had a statistically significant correlation with pathologic duodenal damage. In this limited number of patients, no correlation was found between clinical toxicities and histopathologic or dosimetric parameters.

Liver

Since the liver has a parallel functional architecture, the risk of radiation-induced damage is generally proportional to the mean dose of radiation delivered to normal liver tissue. Therefore, it should be possible to safely treat small hepatic lesions with high doses of radiation by using SBRT, provided the mean dose to the normal liver can be limited.

Recent studies started to better characterize additional liver toxicity endpoints, such as CPC progression or hepatic enzymatic toxicity. In the following sections the results of the main studies investigating liver toxicity after SBRT are discussed, and are divided into studies on SBRT for primary liver cancers and liver metastases.

Treatment for Primary Liver Cancer

In Table 6.2 the main studies on hepatic toxicity after SBRT of primary liver tumors (mainly HCC) are reported (Cárdenes et al. 2010; Son et al. 2010; Andolino et al. 2011;

Kang et al. 2012; Bujold et al. 2013; Jung et al. 2013; Culleton et al. 2014; Lasley et al. 2015; Scorsetti et al. 2015; Velec et al. 2017), together with factors affecting toxicity and the possible relationship with dose and volume. The studies are largely inhomogeneous in terms of patients' characteristics, dose and fractionation schemes, dosimetric parameters studied for association with toxicity and toxicity evaluation, thus making the attainment of final conclusions difficult.

Some studies failed to identify any strong dose-response relationship, while demonstrating CPC as the main significant parameter for predicting hepatotoxicity (Cárdenes et al. 2010; Andolino et al. 2011; Jung et al. 2013).

A Phase I dose escalation trial was conducted at Indiana University (Cárdenes et al. 2010) to determine the feasibility and toxicity of SBRT for HCC. Dose escalation started at 36 Gy in three fractions (12 Gy/fraction) with a subsequent planned escalation of 2 Gy/fraction/level. 17 patients with 25 lesions were enrolled. Dose was escalated to 48 Gy (16 Gy/fraction) in CPC-A patients without dose limiting toxicity. Two patients with CPC-B disease developed grade 3 hepatic toxicity at 42 Gy (14 Gy/fraction). The CP score was the only factor related to grade $\geq$ 3 liver toxicity or death within 6 months ($p = 0.03$).

In the study by Jung et al. (Jung et al. 2013) 17 out of 92 patients (18.5%) developed grade $\geq$ 2 toxicity after SBRT (30–60 Gy, median 45 Gy/3–4fraction). At MVA, only CPC was a significant parameter for predicting grade $\geq$ 2 hepatic toxicity.

In other studies, together with the CPC, some dosimetric parameters were identified (Son et al. 2010; Bujold et al. 2013; Lasley et al. 2015; Velec et al. 2017). In the study by Son et al. (2010) on 36 HCC patients treated with 30–39 Gy (median 36 Gy) in three fractions, 12 (33%) patients developed radiation-induced hepatotoxicity. The normal liver volume receiving <10 Gy (rV_{10Gy}, reverse volume) was the only parameter associated with the risk of grade $\geq$ 2 hepatic toxicity ($p = 0.022$). Four (11%) patients developed the progression of CPC from A to B and rV_{18Gy} (reverse volume) was the only significant parameter associated with the risk of progression of CP class ($p = 0.05$). Most patients recovered from hepatic toxicity after a median of two months, except one patient who developed the progression of CP class with an elevated level of hepatic enzymes persisting without recovery. For these reasons, the authors concluded that grade $\geq$ 2 hepatic enzymes were unlikely to lead to severe radiation-induced morbidity or mortality.

The presence or absence of progression of CPC could be a dose-limiting factor and the authors suggested that the total liver rV_{18Gy} (reverse volume) should be >800 cm^3 to reduce the risk of the deterioration of hepatic function.

Bujold et al. (2013) analyzed 102 CPC-A HCC patients treated with doses from 24 to 54 Gy in six fractions. A decline in CPC was seen in 29% of the patients at 3 months and in 6% at 12 months. A significantly higher median liver mean dose was observed in patients who developed grade 5 toxicity compared with those who did not (18.1 Gy vs. 15.4 Gy, $p = 0.02$).

In the study by Lasley et al. (2015), 38 CPC-A (39 lesions) and 21 CPC-B patients (26 lesions) were treated with 48 Gy in three fractions or 40 Gy in five fractions and followed for $\geq$6 months. Four (11%) CPC-A patients and eight (38%) CPC-B patients experienced grade 3-4 liver toxicity. CPC-B patients experiencing grade 3-4 liver toxicity had significantly higher mean liver dose (8.82 Gy vs 13.24 Gy, no toxicity vs toxicity, respectively, $p = 0.026$),

TABLE 6.2   Summary of Main Studies on Liver Toxicity After Stereotactic Body Radiation Therapy for Hepatocellular Carcinoma Patients

Study (Reference)	Number of Patients	Lesions per pt	Total Number of Lesions	Median Follow-up (Range) (Months)	Baseline Child-Pugh Score	Prescription Dose and Fractionation	Toxicity	Factors Associated with Toxicity
Cárdenes et al. (2010)	17	1–3 largest <6 cm	25	18	A (6 pts) B (11 pts)	48 Gy in 3fr (CPC-A) if dose to 1/3 normal liver ≤10 Gy 40 Gy in 5 fr (CPC-B) if dose to 1/3 normal liver ≤15 Gy	Grade ≥3: 35.3%;Classic RILD: 17.6% (grade 3 in 2 pts CPC-B at 42 Gy in 3fr dose level)	CPC
Son et al. (2010)	36	–	–	–	A (32 pts) B (3 pts) C (1 pts)	30–39 Gy (median 36 Gy) in 3 fr	Transient toxicity: 33%; 1 pt (3%) no recovery; Progression of CPC class: 11%	absolute total liver rV$_{10Gy}$ for risk of grade ≥2 hepatic toxicity; Absolute total liver rV$_{18Gy}$ for risk of CPC progression.
Andolino et al. (2011)	60	1–3	–	27	A (36 pts) B (24 pts)	48 Gy in 3 fr (CPC-A) if dose to 1/3 normal liver ≤10 Gy 40 Gy in 5 fr (CPC-B) if dose to 1/3 normal liver ≤18 Gy	grade 3 toxicity ≈30%; Increase in CPC:20%,; progressive liver failure: in 50% of CPC=8 pts	none

(Continued)

TABLE 6.2 (CONTINUED) Summary of Main Studies on Liver Toxicity After Stereotactic Body Radiation Therapy for Hepatocellular Carcinoma Patients

Study (Reference)	Number of Patients	Lesions per pt	Total Number of Lesions	Median Follow-up (Range) (Months)	Baseline Child-Pugh Score	Prescription Dose and Fractionation	Toxicity	Factors Associated with Toxicity
Kang et al. (2012)	47	Largest <10 cm	56	17 (6–38)	A (41 pts) B (6 pts)	42–60 Gy in 3 fr (median dose 57 Gy)	Classic RILD: 0%; Grade ≥3 Hyperbilirubinemia: 4%; grade ≥3 Thrombocytopenia: 11%; grade ≥3 Ascites :4% GI ulcer: 11%.	–
Bujold et al. (2013)	102	No limitations in Trial 1. 1–5 largest <15 cm in Trial 2	Multiple in 60.8% pts	31.4 (24.3– 36.4)	A	24–54 Gy in 6 fr in Trial 1 30–54 Gy in 6 fr in Trial 2 (every other day)	Classic RILD: 0%; Grade>3 toxicity: 30%	Normal liver mean dose: 18.1 Gy (pts with grade 5 toxicity) vs. 15.4 Gy (pts without grade 5 toxicity)
Jung et al. (2013)	92	1–3 largest <6 cm	–	25.7 (1.8–55.4)	A (68 pts) B (24 pts)	30–60 Gy (median 45 Gy) in 3–4 fr	grade ≥3 toxicity: 6.5%	Multivariable analysis: CPC; normal liver volume; normal liver rV$_{20Gy}$[a]
Culleton et al. (2014)	29	1–5, sum of diameters ≤26.6 cm	Median of 2 lesions	–	B and C	30 Gy in 6 fr (median) (every other day)	CPC decline: 63%	not reported

(Continued)

TABLE 6.2 (CONTINUED) Summary of Main Studies on Liver Toxicity After Stereotactic Body Radiation Therapy for Hepatocellular Carcinoma Patients

Study (Reference)	Number of Patients	Lesions per pt	Total Number of Lesions	Median Follow-up (Range) (Months)	Baseline Child-Pugh Score	Prescription Dose and Fractionation	Toxicity	Factors Associated with Toxicity
Lasley et al. (2015)	59	1 <6 cm or up to 3 with diameter sum <6 cm	65	33.3 (2.8–61.1)	A (38 pts) B (21 pts)	48 Gy in 3 fr (CPC-A) if dose to 1/3 normal liver ≤10 Gy and $rV_{7Gy} > 500cc$ 40 Gy in 5 fr (CPC-B) if dose to 1/3 normal liver ≤18 Gy and $rV_{12Gy} > 500cc$	grade≥3 liver toxicity: 11% (CPC-A) and 38% (CPC-B). RILD: 4.6% (CPC-B)	Only for CPC-B pts: Mean liver dose: 8.8 Gy (pts with grade<3 toxicity) vs. 13.2 Gy (pts with grade ≥3 toxicity); dose to 1/3 normal liver: 7.4 Gy (pts with grade<3 toxicity) 12.5 Gy (pts with grade ≥3 toxicity); Liver $rV_{2.5Gy}$ to rV_{15Gy} at 2.5 Gy increments
Scorsetti et al. (2015)	43	1–3 largest <6 cm	63	8 (3–43)	A (23 pts) B (20 pts)	48–75 Gy in 3 fr for 30 lesions <3 cm 36–60 Gy in 6 fr For 33 lesions >3 cm	Classic RILD: 0%; Grade ≥3 toxicity: 11.1%	not reported

(Continued)

TABLE 6.2 (CONTINUED) Summary of Main Studies on Liver Toxicity After Stereotactic Body Radiation Therapy for Hepatocellular Carcinoma Patients

Study (Reference)	Number of Patients	Lesions per pt	Total Number of Lesions	Median Follow-up (Range) (Months)	Baseline Child-Pugh Score	Prescription Dose and Fractionation	Toxicity	Factors Associated with Toxicity
Velec et al. (2017)	114	largest <15 cm (CPC-A) largest <10 cm (CPC-B)	Median of 2 lesions	–	A (101 pts) B (13)	24–54 Gy in 6 fr OR 30–54 Gy in 6 fr 2	Liver toxicity (increase in CP score by ≥2 at 3 months from SBRT): 26% in CPC-A pts and 54% in CPC-B pts	Higher baseline CP scores; Normal liver mean dose; Normal liver V_{eff}; Normal liver D_{800cc}. Proposed cutoffs: mean normal liver dose <20 Gy and D800cc <15 Gy, 7% risk vs 64% outside this zone. 75% risk if mean liver dose <19 Gy and D_{800cm3} <25 Gy,

pts, patients; fr, fraction; CP, Child Pugh; CPC, Child Pugh; CPC, Child Pugh Class; rV_{XGy}, reverse V_{XGy}, i.e., organ volume receiving less than X Gy; D_{Xcm3}, maximum dose to X cm3 of organ volume;Normal liver=liver−Gross Target Volume.

[a] Dosimetric factors are significant predictors of toxicity only to univariate analysis, while to multivariate analysis CPC is the only significant factor.

higher dose to one-third normal liver (7.37 Gy vs 12.5 Gy, no toxicity vs toxicity, respectively, $p = 0.029$), and larger volumes of liver-receiving doses in the range 2.5 to 15 Gy in 2.5-Gy increments (all statistically significant, with the highest significance appreciated at the <2.5-Gy level). For CPC-A patients, there was no critical liver dose or volume constraint correlated with toxicity. Dose delivered to 500 mL of normal liver approached significance ($p = 0.054$), as did volume of uninvolved liver ($p = 0.082$).

Velec et al. (2017) analyzed data from patients with HCC treated on clinical trials of six fraction SBRT (see Table 6.2). Liver toxicity was defined as an increase in CP score ≥2 three months after SBRT. Among CPC-A patients, toxicity was seen in 26 (26%) patients. Thrombus, a baseline CP of A6 and lower platelet count were associated with toxicity, as were several liver DVH-based parameters: mean dose, effective volume, doses to 700–900 cm^3.

Among CP class A5, patients with a mean liver dose <20 Gy and D_{800cm3} < 15 Gy had a 7% risk of toxicity versus 64% outside this zone. A similar cutoff occurs at a mean dose <19 Gy and D_{800cm3} < 25 Gy, above which the risk is 75%. For the fewer evaluable A6 and B7 patients, mean liver dose appeared to be better associated with increasing toxicity risk compared to D_{800cm3}.

All these studies further support liver function as one of the most important factors in radiation-induced liver toxicity, even though the irradiated volume treated with SBRT is much smaller than 3DCRT. SBRT is a well-tolerated therapy in adequately selected HCC patients, while those with worse underlying liver function are more at risk for toxicity.

Estimate of Effective Volume

Several current liver SBRT protocols rely on the calculation of effective volume (V_{eff}) to estimate the risk of liver toxicity, which subsequently defines tumor prescription doses.

V_{eff} of normal liver irradiated is defined as the normal liver volume, which, if irradiated to the prescribed dose, would show the same risk as the delivered non-uniform dose. V_{eff} is often calculated without accounting for the biologic effect of fraction size.

Estimates of V_{eff} were often derived from toxicity studies: (Son et al. 2010; Velec et al. 2017; Culleton et al. 2014).

The study of Murphy and colleagues (2014) compared V_{eff} and liver toxicity predictions with and without correction for fraction size: V_{eff} was calculated for 18 liver SBRT plans with and without biologic normalization using the linear quadratic formula. LKB NTCP model was used for estimating the risk of liver toxicity. Accounting for the biologic difference of larger fraction size reduced V_{eff} in all treatment plans (median V_{eff} 0.21 vs. 0.32), thus substantially reducing the estimated risk of liver toxicity (average risk of toxicity 32% vs. 4.5%).

Treatment for Liver Metastasis

In Table 6.3 the main studies on SBRT for liver metastases are reported (Rusthoven et al. 2009; Ambrosino et al. 2009; Goodman et al. 2010; Rule et al. 2011; Scorsetti et al. 2013; abermehl et al. 2013; Meyer et al. 2016; Méndez et al. 2017). The toxicity profile of these patients, who mainly have an adequate baseline hepatic function, is very different from

HCC patients. No grade ≥ 3 liver toxicities are reported independently of dose and fractionation schemes applied. Only Mendez Romero et al. (2017) reported 3 grade 3 cases in their low dose group (37.5 Gy in three fractions) and 1 grade 3 in their high dose group (50.25 Gy in three fractions).

Central Hepatobiliary Tract

Damage to the central hepatobiliary tract (HB) is a recently recognized toxicity with SBRT (Pollom et al. 2017). Eriguchi et al. (2013) found that SBRT at 40 Gy in five fractions for liver tumors adjacent to the central biliary system was feasible with minimal biliary toxicity. Dosimetric parameters predictive of HB toxicity were identified from Osmundson et al. (2015) who analyzed data from 96 patients treated with SBRT for primary (53%) or metastatic (47%) liver tumors, with 53% of patients receiving five fractions and 30% of patients receiving three fractions (median Biological Equivalent Dose calculated with $\alpha/\beta = 10$ Gy [BED^{10}] equal to 85.5 Gy). In this study, the central HB tree was radiographically defined as a uniform 15 mm expansion of the portal vein as it extends from the splenic confluence to the first bifurcation of the left and right portal veins and including any biliary stents. The majority of patients were CPC-A (68%) or B (28%). Dosimetric parameters most predictive of grade ≥ 3 HB toxicity were $V_{72Gy} \geq 21$ cm^3 (with dose expressed as BED^{10}, relative risk [RR]: 11.6, $p < 0.0001$), $V_{66Gy} \geq 24$ cm^3 (with dose expressed as BED^{10}, RR: 10.5, $p < 0.0001$) and mean BED^{10} to the central HB tract ≥ 14 Gy (RR: 9.2, $p < 0.0001$). In their updated analysis (Toesca and Osmundson 2017), the dose to the central HB tract was found to be associated with an occurrence of toxicity only in primary liver cancer patients. The strongest dose predictors for grade ≥ 3 toxicity were $V_{40Gy} \geq 37$ cm^3 (BED^{10}, $p < 0.0001$) and $V_{30Gy} \geq 45$ cm^3 (BED^{10}, $p < 0.0001$).

Mathematical/Biological Models

The QUANTEC description of the application of the Lyman NTCP model was mainly based on the analyses of conventionally fractionated or hyperfractionated treatment involving fraction doses in the range of 2 Gy or less. The application of these models to SBRT regimens should be done with caution, as the linear-quadratic (LQ) conversion for larger fraction sizes will likely be inadequate. Despite the introduction and wide application of SBRT for the treatment of metastatic and primary liver cancer, NTCP modeling has continued to be lacking in this area compared with other sites, such as the lung.

There were few studies applying the NTCP models to liver toxicity in the post-QUANTEC era and there are no new estimates of the m, n and D_{50} parameters available.

The recent study at University of Michigan (El Naqa et al. 2018) aimed at developing NTCP models for HCC patients who undergo liver RT. Further the study evaluated the potential role of incorporating into the model both functional imaging and measurement of blood-based circulating biological markers (before and during RT) to improve the performance of the models. Data from 192 HCC patients were included (146 patients received SBRT to a median physical tumor dose of 49.8 Gy and 46 patients had received conventional RT to a median physical tumor dose of 50.4 Gy). Two approaches were investigated: (1) a generalized LKB model and (2) a generalization of the parallel architecture

TABLE 6.3 Summary of Main Studies on Liver Toxicity After Stereotactic Body Radiation Therapy for Liver Metastases

Study (Reference)	Number of Patients	Lesions per Patients	Total number of Lesions	Type of Metastases	Median Follow-up (Range) (Months)	Prescription Dose and Fractionation	Toxicity	Liver Constraints Applied in Planning
Rusthoven et al. (2009)	47	1–3 largest < 6 cm	63	CRC (15 pts), lung (10 pts), breast (4 pts), other (18 pts)	16 (6–54)	36–60 Gy in 3 fr (13 pts) 60 Gy (36 pts) in 3 fr	grade 3 toxicity (soft tissue): 2.1%; Grade≥3 toxicity: 0%.	$rV_{15Gy} \geq 700cc$
Ambrosino et al. (2009)	27	1–3 largest < 6 cm	49	CRC (11 pts), pancreatic (10 pts), breast (2 pts), other (4 pts)	13 (6–16)	25–60 Gy in 3 fr	Significant toxicity: 0%	not reported
Goodman et al. (2010)	19	1–5 largest < 5 cm	40	CRC (6 pts), pancreatic (3 pts), gastric (2 pts), ovarian (2 pts), other (6 pts)	17.3 (2–55)	18 Gy (3 pts), 22 Gy (6 pts), 26 Gy (9 pts), 30 Gy (8 pts) single fraction in all cases	Grade≥3 toxicity: 0%; RILD: 0%.	$rV_{15Gy} \geq 700cc$
Rule et al. ()	27	1–5	37	CRC (12 pts), carcinoid (3 pts), other (12 pts)	20 (4–53)	30 Gy in 3 fr (9 pts), 50 Gy in 5 fr (9 pts), 60 Gy in 5 fr (9 pts)	Grade≥3 toxicity: 0%; RILD: 0%.	$rV_{15Gy} \geq 700cc$ (in 3 fr) $rV_{21Gy} \geq 700cc$ (in 5 fr)

(Continued)

TABLE 6.3 (CONTINUED) Summary of Main Studies on Liver Toxicity After Stereotactic Body Radiation Therapy for Liver Metastases

Study (Reference)	Number of Patients	Lesions per Patients	Total number of Lesions	Type of Metastases	Median Follow-up (Range) (Months)	Prescription Dose and Fractionation	Toxicity	Liver Constraints Applied in Planning
Scorsetti et al. (2013)	61	1–3 largest <6 cm	76	CRC (29 pts), breast (11 pts), gynecologic (7 pts), other (14 pts)	12 (2–26)	52.5–75 Gy in 3 fr	Grade≥3 toxicity: 0%; RILD: 0%.	rV$_{15Gy}$ ≥700cc
Habermehl et al. (2011)	90	1–4	138	CRC (70 pts), breast (27 pts), pancreatic (11 pts), ovarian (7 pts), other (22 pts)	21.7 (1.6–151.8)	17–20 Gy (43 pts), 21–25 Gy (60 pts), 26–30 Gy (35 pts) single fraction in all cases	Grade≥3 toxicity: 0%; RILD: 0%.	not reported
Meyer et al. ()	14	1–5	17	renal (5 pts), CRC (3 pts), other (6 pts)	30 (6–42)	35 Gy (7 pts), 40 Gy (7 pts) single fraction in all cases	Grade≥3 toxicity: 0%; RILD: 0%.	V$_{9.1Gy}$ <700cc
Mendez et al. (2017)	40	1–3 largest <6 cm	55	CRC	26 (1.5–80) LD and 25 (4.6–64) HD	37.5 Gy in 3 fr (LD 32 lesions), 49.5 Gy in 3 fr (HD, 23 lesions)	Grade 3: 9% in LD and 4.3% in in HD.	–

pts, patients; fr, fractions; RILD, radiation-induced liver disease; LD, low dose, HD, high dose, CRC, colorectal cancer, rV$_{XGy}$, reverse V$_{XGy}$, i.e., organ volume receiving less than X Gy, V$_{XGy}$, absolute or relative organ volume receiving more than X Gy model. Three clinical endpoints were considered: the change in albumin-bilirubin (ALBI score), change in CP score, and grade 3 liver enzymatic changes. Local dynamic contrast-enhanced magnetic resonance imaging portal venous perfusion information was used as an imaging biomarker for local liver function. Four inflammatory cytokines were considered as biological markers: transforming growth factor beta 1 (TGF-β1), eotaxin (C-C Motif Chemokine Ligand 11 [CCL11]), hepatocyte growth factor and Cluster of differentiation 40 (CD40) ligand.

When dosimetric parameters are used in the NTCP modeling, ALBI and CP score are found to be more radiosensitive endpoints and could potentially act as precursors to subsequent enzymatic changes occurring at higher doses. For a mean liver dose (MLD) of 13 Gy in three fractions (0.01% RILD risk from the QUANTEC HCC liver model) the estimated risk is 40.8% for ALBI, 34.1% for CP score and 2.1% for enzymatic changes. For a MLD of 18 Gy in five fractions (0.48% of RILD from the QUANTEC HCC liver model) the estimated risk is 50.2% for ALBI, 42.1% for CP score and 4.15% for enzymatic changes.

Moreover, incorporation of imaging findings and biological markers into NTCP modeling of liver toxicity improved the estimates of expected NTCP risk compared with using dose-only models.

SUMMARY FOR DVH CONSTRAINTS/NTCP MODEL PARAMETERS
Stomach and Duodenum

In most published analyses, the authors not only determined the association between dosimetric variables and toxicity but also suggested optimal dose/volume constraints that separate patients according to the risk of toxicity. Few studies also determined a dose-response curve, a more explicit and complete way to estimate the actual risk level for each OAR as a function of dose, fractionation, volume and other parameters deriving the NTCP parameters using LKB model. However, inconsistencies still permeate the literature due to some variations in:

- endpoint and the grade of toxicity

- organ definition/delineation and related dosimetric parameters (absolute vs. relative)

- dose-fractionation and (the accuracy of) their conversion in a reference dose per fraction

- drugs and their combination regimens

For these reasons, the analysis of published dose tolerance limits presents some difficulties. In Tables 6.4 and 6.5 we summarized dose-volume constraints and NTCP model parameters from literature reports both for conventional and stereotactic body RT.

Timmerman (2008) published one of the first collection of normal tissue dose constraints for SBRT and in 2010 the American Association of Physicist in Medicine TG-101 (Benedict et al. 2010) summarized tolerance doses from the University of Texas Southwestern and the University of Virginia (both studies underlined that mostly constraints were, at that time, clinically unvalidated). A summary of suggested dose constraints for stomach and duodenum in case of SBRT is included in Table 6.6; two studies explicitly reported the low/high risk of toxicity (Bae et al. 2012; Goldsmith et al. 2016).

Liver

As previously described, in the post-QUANTEC studies, baseline liver function was better characterized, new endpoints were introduced, and the quality of the dosimetric data was improved. Despite these steps forward, a summary of dose-volume constraints and NTCP parameters for radiation-induced liver injury can be hardly derived from the data published in the last years. The picture that emerges is heterogeneous: Pre-treatment liver function, large individual radiosensitivity, association with other therapies (such as chemotherapy or thermochemical ablation) and uncertainties in the actually delivered doses (mainly due to intra-fraction motion) lead to high uncertainty. Anyway, based on the available data, we may suggest the dose-volume constraints summarized in Table 6.7.

OPEN ISSUES AND FUTURE DIRECTIONS OF INVESTIGATION

Organ Definition/Delineation

The use of high precision delivery techniques (IMRT, SBRT) for treating abdomen malignancies has increased in the past decade. The main purpose of reducing the dose to healthy tissue to preserve the integrity and function of these structures requires the accurate definition and delineation of the organs. Quite recently, detailed instruction for upper abdominal normal organ delineation guidelines have been published (Jabbour et al. 2014) in the recent RTOG upper GI atlas. The agreement in organ delineation among the 12 experts was excellent for liver and stomach, while it was less accurate for the duodenum as the contributing clinicians frequently missed its fourth portion. Existing data regarding duodenum may be affected by anatomical/delineation inconsistency: For instance, in the paper by Huang et al. (2012) the entire duodenum was delineated, while the duodenum contour was defined according to the PTV extension in the paper by Murphy et al. (2010) and the ascending portion of duodenum was not included in the paper by Verma et al. (2014).

Dose-Surface-Maps

In hollow or tubular organs, the tissue of interest is only a thin layer surrounding a variable amount of contents, then dose-surface-maps may be more appropriate for modeling toxicity in the era of highly tailored dose deposition. Dose-surface maps provide information on dose distribution and spatial dose metrics can be extracted to create models that may improve toxicity predictions (Palorini et al. 2016; Buettner et al. 2009; Zhou et al. 2000). Compared to others tubular organs, the duodenum has a complex geometry so classical methods for dose-surface map generation are inadequate in this case, as these methods can be only used when the CT transaxial slices can be unwrapped individually. Witzum et al. (2016) presented a new methodology to convert the 3D surface dose to a 2D dose-surface map with consistent spatial information. As transaxial CT slices may contain a segment

TABLE 6.4 Summary of Dose Constraints for Stomach and Duodenum from Selected Trials Using Conventional Fractionated or Stereotactic Body Radiotherapy

Study	Cancer Site	Treatment Technique	N. Pts	Dose (Gy) Median (Range)	Dose/fr (Gy) Median (Range)	Concurrent CHT	OAR	Endpoints	Toxicity Rate	DVH Parameters	Risk Comparison	Analysis	Note
Kelly et al. (2013)	LAPC	3DCRT (71%) IMRT (29%)	106	50.4 (50.4–70.4)	1.8 (1.8–1.9)	Gemcitabine (18%)5-FU/ Capecitabine (82%)	Duodenum	CTCAE v4.0 Grade ≥2	14% 1y actuarial	$V_{55Gy} \leq$ vs. > 1 cm³	9% vs. 47%	MVA	Duodenal toxicity. Endoscopic findings.
Verma et al. (2014)	PA nodal metastases	IMRT	105	63 (30.6–72.2)	1.90	Platinum agents (54%)	Duodenum	RTOG Grade≥ 2	11.7% 3y actuarial	$V_{55Gy} \leq$ vs. > 15 cm³	7% vs. 49%	UVA	Duodenal toxicity. Endoscopic findings.
Cattaneo et al. (2013)	LAPC	IMRT	61	44.25 (44.25– 55.0)	2.95 (2.95– 3.7)	Capecitabine (89%) 5FU (8%)	Duodenum	CTCAE v3.0 Grade ≥2	13% 1y actuarial (gastroduodenal)	$V_{45Gy} \leq$ vs. > 16%	3% vs. 28%	MVA	Anatomical Toxicity
							Duodenum	CTCAE v3.0 Grade≥ 2	13% 1y actuarial (gastroduodenal)	$V_{40Gy} \leq$ vs. > 2.6%	NA	MVA	Anatomical Toxicity
							Stomach	CTCAE v3.0 Grade≥ 2	33% crude incidence (gastrointestinal)	$V_{20Gy} \leq$ vs. > 31%	NA	MVA	Acute GI toxicity
Huang et al. (2012)	LAPC	3DCRT (87%) IMRT (19%)	46	36 (22–42)	2.4 (2–2.8)	Full dose gemcitabine (61%) Full dose gemcitabine + erlotinib (39%)	Duodenum	CTCAE v3.0 Grade≥ 2	37% 1y actuarial	$V_{25Gy} \leq$ vs. >45%	8% vs. 48%	MVA	Entire cohort
							Duodenum	CTCAE v3.0 Grade≥ 3	32% Crude incidence	$V_{35Gy} \leq$ vs. >20%	0 vs. 41%		No erlotinib
Nakamura et al. (2012)	LAPC	3DCRT	40	54	1.8	Gemcitabine	Stomach	CTCAE v4.0 Grade≥ 2	33% Crude incidence	$V_{50Gy} \leq$ vs. >16 cm³	9% vs. 61%	UVA	Acute GI toxicity
							Gastroduodenum	CTCAEv4.0 Grade≥ 3	20% 1y actuarial	$V_{50Gy} \leq$ vs. > 33 cm³	0 vs. 44%	UVA	Upper GI bleeding Endoscopic findings
Kim et al. (2009)	HCC	3DCRT	73(All cirrhotic pts)	36 (30–54)	3		Gastroduodenum	CTCAE v4.0 Grade≥ 3	17.5% 1y actuarial	$V_{35Gy} \leq$ vs. >5%	4% vs. 48%	MVA	Gastroduodenal toxicity
Liu et al. (2015)	Esophagus	IMRT	105	60.2 (60.2–63)	2.15 (2.15– 2.25)		Stomach (intrathoracic)	CTCAE v4.0. Grade≥ 2	18.1% acute, crude incidence 13.3% late, crude incidence	$V_{50Gy} \leq$ vs. >14%	19% vs. 34.5%	MVA	Small/large field
Yoon et al. (2013)	HCC	3DCRT	90	35.5 (30–50)	3.5 (2–5)		Stomach	CTCAE v4.0 Grade≥ 2		$V_{25Gy} \leq$ vs. > 6.3%	2.9% vs. 57.1%	MVA	EGD before/after RT EUD_2

(Continued)

TABLE 6.4 (CONTINUED) Summary of Dose Constraints for Stomach and Duodenum from Selected Trials Using Conventional Fractionated or Stereotactic Body Radiotherapy

Study	Cancer Site	Treatment Technique	N. PTs	Dose (Gy) Median (Range)	Dose/fr (Gy) Median (Range)	Concurrent CHT	OAR	Endpoints	Toxicity Rate	DVH Parameters	Risk Comparison	Analysis	Note
Kopek et al. (2010)	Cholangiocarcinoma	Linac SBRT	27	45	15		Duodenum	CTCAE v3.0 Grade ≥ 2	33% Crude incidence	$V_{35Gy} \leq$ vs. >5.4%	9.4% vs. 45.9%	MVA	Duodenal ulceration or stenosis
							Duodenum (delineated segment one and two)			$D_{1cm3} \leq$ vs. > 37.4 Gy PTs with toxicity	NA	UVA	
Murphy et al. (2010)	Pancreatic carcinoma	CyberKnife (73%) Trilogy Linac (27%)	73	25	25		Duodenum	CTCAE v3.0 Grade≥ 2	29% 1y Actuarial	$V_{5Gy} <$ vs. ≥ 25 cm^3	28% vs. 31%	UVA	
										$V_{10Gy} <$ vs. ≥16 cm^3	15% vs. 46%		
										$V_{15Gy} <$ vs. ≥9.1 cm^3	11% vs. 52%		
										$V_{20Gy} <$ vs. ≥3.3 cm^3	11% vs. 52%		
										$V_{25Gy} <$ vs. ≥0.21 cm^3	12% vs. 45%		
										$D_{1cm3} <$ vs. ≥23 Gy	12% vs. 49%		
Bae et al. (2012)	Abdominopelvis malignancies	CyberKnife RapidArc	40	45 (33–60)	15 (11–20)		Gastroduodenum (±2 cm superior/ inferior PTV)	CTCAE v3.0 Grade≥ 3	15% 1y Actuarial	$V_{15Gy} \leq$ >14 cm3	6% vs. 45%	UVA	Gastroduodenal toxicity, excluding nausea and vomiting
										$V_{25Gy} \leq$ vs. >7 cm^3	6% vs. 50%		
										$V_{30Gy} \leq$ vs. >5 cm^3	8% vs. 50%		
										$V_{35Gy} \leq$ vs. >1 cm^3	6% vs. 50%		
										$D_{max} \leq$ vs. >45 Gy	6% vs. 50%		

N, number; fr, fraction; CHT, Chemotherapy; OAR, Organ at risk; DVH, Dose-volume histogram; y, year; LAPC, Locally advanced pancreatic carcinoma; 3DCRT, Three dimensional conformal radiotherapy; IMRT, Intensity modulated radiation therapy; SBRT, Stereotactic Body Radiation Therapy; UVA, Univariate analysis; MVA, Multivariate analysis; CTCAE, Common Terminology Criteria for Adverse Events; RTOG, Radiation Therapy Oncology Group; NA, Not available; 5FU, 5-fluoro.uracil; GI, gastrointestinal; PA, Para-aortic; Gastroduodenum, Stomach+Duodenum; EGD, esophagogastroduodenoscopy; V_{XGy}, (Percent/absolute) organ volume receiving at least X Gy; PT, Patient; D_{1cm3}, maximum dose to 1 cm^3 of organ; D_{max}, maximum dose to organ; PTV, Planning Target Volume.

TABLE 6.5 Summary of NTCP Model Parameters for Stomach and Duodenum for Conventional Fractionated and Stereotactic Body Radiotherapy

Study	Cancer Site	Conventional radiotherapy / SBRT	N. PTs	Dose (Gy) Mean (Range)	Dose/fr (Gy) Mean (Range)	OAR	Endpoints	Toxicity Rate (%)	DVH Physical Dose/ EQD$_{XGy}^Y$	Model	D$_{50}$ (Gy)	m	n	Note
Feng et al. (2012)	Intrahepatic Malignancies	Conventional	116 (6.9% with cirrhosis)	54 (12.0–90.5)	(1.5–4.67)	Stomach (wall)	CTCAE v 4.0 Grade ≥ 3	14% Crude incidence	EQD$_{2Gy}^{2.5}$	LKB	56 (normal) 22 (cirrhotic)	0.21	0.10	Gastric bleeding
Holyoake et al. (2017)	Upper Abdominal tumours	Conventional	Six published DVH data			Duodenum	Grade ≥ 2 GI or duodenal toxicities.		Physical dose	LKB	125.9 (95% CI: 63.1–188.7)	0.36 (95% CI: 0.30–0.43)	0.068 (0.060–0.076)	Model fitted using only 6 published DVH data.
Murphy et al. (2010)	Pancreatic carcinoma	SBRT	73	25	25	Duodenum	CTCAE v3.0 Grade ≥ 2	29% 1y Actuarial	Physical dose	LKB	24.6 (95% CI, 22.3–27.5)	0.23 (95% CI, 0.16–0.35)	0.12 (95% CI,0.06–0.26)	
Bae et al. (2012)	Abdominopelvis malignancies	SBRT	40	45 (33–60)	15 (11–20)	Gastroduodenum (±2 cm superior/inferior PTV)	CTCAE v3.0 Grade ≥ 3	15% 1y Actuarial	Physical dose	Error function D$_{max}$	48.9 (95% CI: 42.1–59.9)	0.170		Late GI toxicities
Elhammali et al. (2015)	Pancreatic carcinoma	SBRT Conventional	Pooled data from 16 studies	125 (EQD$_{1.8Gy}^4$) (50–209)		Duodenum	Grade ≥ 3		EQD$_{1.8Gy}^4$	LKB	95.0 (95% CI:84.4–109.1)	0.48 (95% CI: 0.45–0.51)	0.43 (95% CI: 0.40–0.48)	Late GI toxicities. Assuming that across all studies 5% of the duodenum received the prescription dose.
Prior et al. (2014)	Various	SBRT Conventional	Pooled data from 5 studies	EQD$_{2.0Gy}^4$ 48.3–77.2		Duodenum Small Bowell	CTCAE RTOG		EQD$_{2.0Gy}^4$ (modified LQ model)	LKB	60.9 (±7.9)	0.21 (±0.05)		Modified LQ model, Parameter introduced to replicate constant slope in the survival curve at high doses: δ=0.09±0.03 Gy^{-1}

N, number; PT, patient; fr, fraction; D$_{max}$, Maximum point dose; SBRT, Stereotactic Body Radiotherapy; OAR, Organ at risk; DVH, Dose-volume histogram; PTV, Planning target volume; EQD$_{XGy}^Y$, Equivalent dose in X Gy per fraction using α/β ratio equal to Y; LKB, Lyman-Kutcher-Burman Normal Tissue Complication Probability model; D$_{50}$, Uniform organ dose giving 50% probability of complication; m, related to the inverse slope of dose-response curve; n, correlated to serial/parallel nature of the tissue; CTCAE, Common Terminology Criteria for Adverse Events. LQ, Linear quadratic model; Gastroduodenum, Stomach+Duodenum; RTOG, Radiation Therapy Oncology Group; GI, gastrointestinal; CI, confidence interval.

TABLE 6.6 Collection of (Suggested) Normal Tissue Dose Constraints for Stomach and Duodenum for Stereotactic Body Radiotherapy

OAR	DVH Parameters	One Fraction		Three Fractions		Five Fractions		Endpoint	References
Stomach	D_{max}	16		24		32		Ulceration, fistula	Timmerman (2008)
	D_{10cm^3}	13		21		28			
	$D_{0.03cm^3}$	12.4		22.2		32		Ulceration, fistula	Benedict et al. (2010)
	D_{10cm^3}	11.2		16.5		18			
Duodenum	D_{max}	16		24		32		Ulceration	Timmerman (2008)
	D_{5cm^3}	8.8		15		18			
	$D_{0.035cm^3}$	12.4		22.2		32		Ulceration (avoid circumferential irradiation)	Benedict et al. (2010)
	D_{5cm^3}	11.2		16.5		18			
	D_{10cm^3}	9		11.4		12.5			
	V_{21Gy} (cm³)			1				Duodenal ulceration or stenosis Grade ≥ 3	Kopec et al. (2010)
	$D_{0.035cm^3}$			30		35		Ulcer, bleeding, perforaation Grade ≥ 3	Pollon et a. (2017)
	D_{mean}					20		Duodenal histologic damage	Verma et al. (2017)
	V_{20Gy} (cm³)					39			
	V_{25Gy} (cm³)					27			
	V_{30Gy} (cm³)					14			
	V_{30Gy} (cm³)					5			
		Low risk (<7%)	High risk (<22%)	Low risk (<7%)	High risk (<27%)	Low risk (<6%)	High risk (<31%)	DVH risk map duodenal hemorrhage or stricture	Goldsmith et al. (2016)
	$D_{0.035cm^3}$	16.0	23	30	37	32	42		
	D_{1cm^3}	17	23	25.3	37.4	28	40		
	D_{5cm^3}	11.2	17	21	30	25.8	38		
Gastroduodenum				Low risk (<5%)	High risk (<10%)			Bleeding, perforation, obstruction (Grade ≥3)	Bae et al.(2012)
	D_{max}			35	38				

OAR, Organ at risk; DVH, Dose-volume histogram; Dmax, Maximum point dose; Dmean, mean dose; DXcm3, maximum dose to X cubic centimeters; VXGy (cm3) = Organ volume (in cm3) receiving at least X Gy.

TABLE 6.7   Collection of (Suggested) Normal Tissue Dose Constraints for Liver for Stereotactic Body Radiotherapy

Site	Fractionation	Proposed Dose Constraint	Source
Liver with good baseline function	Conventional fractionation (2 Gy/fr)	Mean normal liver dose < 32 Gy	Pan et al. (2010)
	SBRT (3 fractions)	Mean normal liver dose < 15 Gy	Pan et al. (2010)
	SBRT (6 fractions)	Mean normal liver dose < 20 Gy	Pan et al. (2010)
Liver with poor baseline function	Conventional fractionation (2Gy/fr)	Mean normal liver dose < 28 Gy	Pan et al. (2010) Jung et al. (2013) Huang et al. (2017)
	SBRT (3 fractions)	$rV_{18Gy} > 800$ cm³	Son et al. (2010)
	SBRT (5 fractions)	Mean normal liver dose < 10 Gy (CPC-B) 1/3 normal liver volume < 10 Gy (CPC-B)	Lasley et al. (2015)
	SBRT (6 fractions)	Mean normal liver dose < 15 Gy (CPC-A e B) $D_{800cm3} < 10$ Gy (CPC-A) and < 8 Gy (CPC-B)	Bujold et al. (2013) Velec et al. (2017)
Central liver Hepatobiliary toxicity	SBRT (5 fractions)	Mean dose < 19 Gy $V_{26Gy} < 37$ cm³ $V_{21Gy} < 45$ cm³	Eriguchi et al. (2013) Osmundson et al. (2015)

from two different parts of the duodenum, the authors defined a new axis that follows the curvature of the duodenum and the created slices are oriented normally to a curved line through the organ, greatly reducing the problem of overlapping planes.

Adaptive Radiotherapy and Replanning

IGRT mostly consists in repositioning the patient after the rigid registration of the planning CT with the daily in-room CT acquired just before the treatment delivery. This procedure is not able to take into account for inter-fractional organ deformation and dislocation so that the CTV-to-PTV margin has to take this into account. Liu et al. (2012) quantified the inter-fractional anatomic variations in pancreatic patients treated at 50.4 Gy in 1.8 Gy perfraction and determined the dosimetric advantages of online adaptive replanning procedure. The maximum overlap between the structure volume from the planning and the volume from the daily CT was considered as a surrogate of organ deformation. Organ deformations were considerable mainly for duodenum and stomach; the maximum overlap ratios varied from 40% to 90% for duodenum, and from 30% to 95% for stomach. For each daily CT scan, two plans were generated: a repositioning plan and an online adaptive plan. Mean target coverage was superior for online adaptive replanning compared to repositioning procedure; moreover, the adaptive plans were able to better spare all OARs; for the duodenum $V_{50.4Gy}$ values were 15.6% and 43.4% for adaptive and repositioning plans, respectively (p < 0.001). Improved OAR sparing for adaptive plans is not only related to the smaller PTV margin; standard IGRT strategies lead to either inadequate PTV coverage or higher OAR doses so the additional margins alone are not adequate to manage organ deformation (Liu et al. 2012; Ahunbay et al. 2012).

The dosimetric advantages of a daily adaptive online replanning were evaluated for 10 patients treated with pancreas SBRT (Li et al. 2015). Tumor and OAR contours were deformed from planning CT and transferred to daily CT before planning re-optimisation. Adaptive replanning improved target coverage, reduced hot-spots in the patient and decreased the duodenum volume within the prescription isodose.

The online adaptive replanning is a labour intensive procedure not yet suitable for routine in RT as it requires the approval of new contours and of modified plans but it could have some applications for specific situation such as SBRT.

Including Liver Function in Treatment Planning

We largely discussed how liver patients differ in both pretreatment liver function and radiosensitivity, thus leading to high variability in potential liver toxicity with similar doses.

These two factors are not explicitly taken into account in the planning phase. An attempt of incorporating pre-treatment liver function into treatment planning decisions was done by Wu et al. (2016) who proposed to minimize objective functions accounting for liver function using: (1) perfusion-weighted EUD (fEUD) that avoids delivering dose to highly-perfused liver by explicitly incorporating voxel-based pre-treatment liver perfusion into a treatment planning model, and (2) post-treatment global liver function (GLF) that explicitly captures global liver function using a model of liver dose-response based on pre- and post-treatment perfusion. The GLF model is proven to be the one producing treatment plans that retain the most liver function by delivering dose exceeding the damage saturation threshold to fewer high-functioning voxels. These models are effective in theoretically preserving GLF. Future work consists of assessing what patients may benefit from a planning treatment with this approach.

Effect of Breathing Motion on Liver Doses

One of the challenging aspects of liver RT is the management of respiratory motion.

As previously mentioned, a common limitation of most studies is that DVH parameters refer to static planning CT scan, without taking into consideration intra-fraction variations (mainly due to breathing) and inter-fraction changes. Many studies investigated the effect of respiratory motion on liver doses and toxicity estimates by using deformable image registration and dose accumulation; each study has its own nuances, but there are commonalities in the results. Relatively small variations in normal liver mean doses are detected when comparing 4D (including the effect of breathing) and 3D dose distributions: about 0.1–0.2 Gy on average ((Velec et al. 2011), only breathing; (Velec et al. 2012), positioning, breathing and deformations) or about 3% difference (Andolino et al. 2011), but with large inter-patient dependence. When looking at NTCP, the average variations are within 3% (Jung et al. 2013; Velec et al. 2011), while individual differences may be >5% in one out of three patients (Velec et al. 2012).

Accurate knowledge of doses planned in the presence of breathing motion would allow more informed tumor dose-toxicity risk tradeoffs to be made, potentially reducing patient risks or improving dose escalation.

Particles

This chapter was intentionally limited to photon RT, for which the majority of published data exist. Yet, particle therapy deserves at least to be mentioned, as it may offer advantages over photon-based radiation treatments thanks to the better sparing of liver volume receiving low to moderate dose thus having the potential to reduce radiation-related hepatotoxicity and allowing for tumor dose escalation (Yeung et al. 2017). Currently no randomized data comparing particles and photons in liver treatment exists and it is difficult to extrapolate liver toxicity outcome by comparing published studies, due to large heterogeneity. In the review of the literature recently published by Yeung and colleagues (Yeung et al. 2017) three possible patient categories who may theoretically benefit the most from the use particle therapy are identified: patients with poor baseline liver function (CPC-B and C), patients who have received prior liver RT and patients with large tumors.

REFERENCES

Ahunbay E, Liu F, Yang C, li X. TU-E-BRA-01: A comparison of various online strategies to account for interfractional variations for pancreatic cancer. *Med Phys* 2012;39(6Part24):3910. doi:10.1118/1.4735961.

Ambrosino G, Polistina F, Costantin G, Francescon P, Guglielmi R, Zanco P, et al. Image-guided robotic stereotactic radiosurgery for unresectable liver metastases: Preliminary results. *Anticancer Res* 2009;29(8):3381–4.

Andolino DL, Johnson CS, Maluccio M, Kwo P, Tector AJ, Zook J, et al. Stereotactic body radiotherapy for primary hepatocellular carcinoma. *Int J Radiat Oncol Biol Phys* 2011;81(1):447–53. doi:10.1016/j.ijrobp.2011.04.011.

Asbell SO, Grimm J, Xue J, Chew MS, LaCouture TA. Introduction and clinical overview of the DVH risk map. *Semin Radiat Oncol* 2016;26(2):89–96. doi:10.1016/j.semradonc.2015.11.005.

Bae SH, Kim MS, Cho CK, Kang JK, Lee SY, Lee KN, et al. Predictor of severe gastroduodenal toxicity after stereotactic body radiotherapy for abdominopelvic malignancies. *Int J Radiat Oncol Biol Phys* 2012;84(4):e469–74. doi:10.1016/j.ijrobp.2012.06.005.

Baglan KL, Frazier RC, Yan D, Huang RR, Martinez AA, Robertson JM. The dose-volume relationship of acute small bowel toxicity from concurrent 5-FU-based chemotherapy and radiation therapy for rectal cancer. *Int J Radiat Oncol Biol Phys* 2002;52(1):176–83. doi:10.1016/S0360-3016(01)01820-X.

Benedict SH, Yenice KM, Followill D, Galvin JM, Hinson W, Kavanagh B, et al. Stereotactic body radiation therapy: The report of AAPM Task Group 101. *Med Phys* 2010;37(8):4078–101. doi:10.1118/1.3438081.

Brunner TB, Nestle U, Grosu AL, Partridge M. SBRT in pancreatic cancer: What is the therapeutic window? *Radiother Oncol* 2015;114(1):109–16. doi:10.1016/j.radonc.2014.10.015.

Buettner F, Gulliford SL, Webb S, Partridge M. Using dose-surface maps to predict radiation-induced rectal bleeding: A neural network approach. *Phys Med Biol* 2009;54(17):5139–53. doi:10.1088/0031-9155/54/17/005.

Bujold A, Massey CA, Kim JJ, Brierley J, Cho C, Wong RKS, et al. Sequential phase I and II trials of stereotactic body radiotherapy for locally advanced hepatocellular carcinoma. *J Clin Oncol* 2013;31(13):1631–9. doi:10.1200/JCO.2012.44.1659.

Cárdenes HR, Price TR, Perkins SM, Maluccio M, Kwo P, Breen TE, et al. Phase i feasibility trial of stereotactic body radiation therapy for primary hepatocellular carcinoma. *Clin Transl Oncol* 2010;12(3):218–25. doi:10.1007/s12094-010-0492-x.

Cattaneo GM, Passoni P, Longobardi B, Slim N, Reni M, Cereda S, et al. Dosimetric and clinical predictors of toxicity following combined chemotherapy and moderately hypofractionated rotational radiotherapy of locally advanced pancreatic adenocarcinoma. *Radiother Oncol* 2013;108(1):66–71. doi:10.1016/j.radonc.2013.05.011.

Chang DT, Schellenberg D, Shen J, Kim J, Goodman KA, Fisher GA, et al. Stereotactic radiotherapy for unresectable adenocarcinoma of the pancreas. *Cancer* 2009;115(3):665–72. doi:10.1002/cncr.24059.

Culleton S, Jiang H, Haddad CR, Kim J, Brierley J, Brade A, et al. Outcomes following definitive stereotactic body radiotherapy for patients with Child-Pugh B or C hepatocellular carcinoma. *Radiother Oncol* 2014;111(3):412–7. doi:10.1016/j.radonc.2014.05.002.

Dawson LA, Ten Haken RK. Partial volume tolerance of the liver to radiation. *Semin Radiat Oncol* 2005;15(4):279–83. doi:10.1016/j.semradonc.2005.04.005.

El Naqa I, Johansson A, Owen D, Cuneo K, Cao Y, Matuszak M, et al. Modeling of normal tissue complications using imaging and biomarkers after radiation therapy for hepatocellular carcinoma. *Int J Radiat Oncol Biol Phys* 2018;100(2):335–43. doi:10.1016/j.ijrobp.2017.10.005.

Elhammali A, Patel M, Weinberg B, Verma V, Liu J, Olsen JR, et al. Late gastrointestinal tissue effects after hypofractionated radiation therapy of the pancreas. *Radiat Oncol* 2015;10:4–11. doi:10.1186/s13014-015-0489-2.

Eriguchi T, Takeda A, Sanuki N, Oku Y, Aoki Y, Shigematsu N, et al. Acceptable toxicity after stereotactic body radiation therapy for liver tumors adjacent to the central biliary system. *Int J Radiat Oncol Biol Phys* 2013;85(4):1006–11. doi:10.1016/j.ijrobp.2012.09.012.

Feng M, Normolle D, Pan CC, Dawson LA, Amarnath S, Ensminger WD, et al. Dosimetric analysis of radiation-induced gastric bleeding. *Int J Radiat Oncol Biol Phys* 2012;84(1):e1–6. doi:10.1016/j.ijrobp.2012.02.029.

Goodman KA, Wiegner EA, Maturen KE, Zhang Z, Mo Q, Yang G, et al. Dose-escalation study of single-fraction stereotactic body radiotherapy for liver malignancies. *Int J Radiat Oncol Biol Phys* 2010;78(2):486–93. doi:10.1016/j.ijrobp.2009.08.020.

Goldsmith C, Price P, Cross T, Loughlin S, Cowley I, Plowman N. Dose-volume histogram analysis of stereotactic body radiotherapy treatment of pancreatic cancer: A focus on duodenal dose constraints. *Semin Radiat Oncol* 2016;26(2):149–56. doi:10.1016/j.semradonc.2015.12.002.

Guerrero M, Li XA. Extending the linear-quadratic model for large fraction doses pertinent to stereotactic radiotherapy. *Phys Med Biol* 2004;49(20):4825–35. doi:10.1088/0031-9155/49/20/012.

Gunnlaugsson A, Kjellén E, Nilsson P, Bendahl PO, Willner J, Johnsson A. Dose-volume relationships between enteritis and irradiated bowel volumes during 5-fluorouracil and oxaliplatin based chemoradiotherapy in locally advanced rectal cancer. *Acta Oncol (Madr)* 2007;46(7):937–44. doi:10.1080/02841860701317873.

Habermehl D, Herfarth KK, Bermejo JL, Hof H, Rieken S, Kuhn S, et al. Single-dose radiosurgical treatment for hepatic metastases – therapeutic outcome of 138 treated lesions from a single institution. *Radiat Oncol* 2013;8:1–9. doi:10.1186/1748-717X-8-175.

Holyoake DLP, Aznar M, Mukherjee S, Partridge M, Hawkins MA. Modelling duodenum radiotherapy toxicity using cohort dose-volume-histogram data. *Radiother Oncol* 2017;123(3):431–7. doi:10.1016/j.radonc.2017.04.024.

Hoyer M, Roed H, Hansen AT, Ohlhuis L, Petersen J, Nellemann H, et al. Phase II study on stereotactic body radiotherapy of colorectal metastases. *Acta Oncol (Madr)* 2006;45(7):823–30. doi:10.1080/02841860600904854.

Huang EY, Sung CC, Ko SF, Wang CJ, Yang KD. The different volume effects of small-bowel toxicity during pelvic irradiation between gynecologic patients with and without abdominal surgery: A prospective study with computed tomography-based dosimetry. *Int J Radiat Oncol Biol Phys* 2007;69(3):732–9. doi:10.1016/j.ijrobp.2007.03.060.

Huang J, Robertson JM, Ye H, Margolis J, Nadeau L, Yan D. Dose-volume analysis of predictors for gastrointestinal toxicity after concurrent full-dose gemcitabine and radiotherapy for locally advanced pancreatic adenocarcinoma. *Int J Radiat Oncol Biol Phys* 2012;83(4):1120–5. doi:10.1016/j.ijrobp.2011.09.022.

Huang P, Yu G, Kapp DS, Bian XF, Ma CS, Li HS, et al. Cumulative dose of radiation therapy of hepatocellular carcinoma patients and its deterministic relation to radiation-induced liver disease. *Med Dosim* 2017,83(3):1–9. doi:10.1016/j.meddos.2017.10.002.

Ingold JA, Reed GB, Kaplan HS, et al. Radiation hepatitis. *Am J Roentgenol Radium Ther Nucl Med* 1965;93:200–8. n.d.

Jabbour SK, Hashem SA, Bosch W, Kim TK, Finkelstein SE, Anderson BM, et al. Upper abdominal normal organ contouring guidelines and atlas: A Radiation Therapy Oncology Group consensus. *Pract Radiat Oncol* 2014;4(2):82–9. doi:10.1016/j.prro.2013.06.004.

Jung J, Yoon S, Kim S, Cho B, Park J, Kim S, et al. Radiation-induced liver disease after stereotactic body radiotherapy for small hepatocellular carcinoma: clinical and dose-volumetric parameters. *Radiat Oncol* 2013;8:249. doi:10.1186/1748-717X-8-249.

Jung J, Kong M, Hong SE. Conventional fractionated helical tomotherapy for patients with small to medium hepatocellular carcinomas without portal vein tumor thrombosis. *Onco Targets Ther* 2014;7:1769–75. doi:10.2147/OTT.S69618.

Kang JK, Kim MS, Cho CK, Yang KM, Yoo HJ, Kim JH, et al. Stereotactic body radiation therapy for inoperable hepatocellular carcinoma as a local salvage treatment after incomplete transarterial chemoembolization. *Cancer* 2012;118(21):5424–31. doi:10.1002/cncr.27533.

Kavanagh BD, Schefter TE, Cardenes HR, Stieber VW, Raben D, Timmerman RD, et al. Interim analysis of a prospective phase I/II trial of SBRT for liver metastases. *Acta Oncol (Madr)* 2006;45(7):848–55. doi:10.1080/02841860600904870.

Kavanagh BD, Pan CC, Dawson LA, Das SK, Li XA, Ten Haken RK, et al. Radiation dose-volume effects in the stomach and small bowel. *Int J Radiat Oncol Biol Phys* 2010;76(S3):101–7. doi:10.1016/j.ijrobp.2009.05.071.

Kelly P, Das P, Pinnix CC, Beddar S, Briere T, Pham M, et al. Duodenal toxicity after fractionated chemoradiation for unresectable pancreatic cancer. *Int J Radiat Oncol* 2013;85(3):e143–9. doi:10.1016/j.ijrobp.2012.09.035.

Kim JH, Park JW, Kim TH, Koh DW, Lee WJ, Kim CM. Hepatitis B virus reactivation after three-dimensional conformal radiotherapy in patients with hepatitis B virus-related hepatocellular Carcinoma. *Int J Radiat Oncol Biol Phys* 2007;69(3):813–9. doi:10.1016/j.ijrobp.2007.04.005.

Kim H, Lim DH, Paik SW, Yoo BC, Koh KG, Lee JH, et al. Predictive factors of gastroduodenal toxicity in cirrhotic patients after three-dimensional conformal radiotherapy for hepatocellular carcinoma. *Radiother Oncol* 2009;93(2):302–6. doi:10.1016/j.radonc.2009.05.017.

Kopek N, Holt MI, Hansen AT, Høyer M. Stereotactic body radiotherapy for unresectable cholangiocarcinoma. *Radiother Oncol* 2010;94(1):47–52. doi:10.1016/j.radonc.2009.11.004.

Kutcher G, Burman C. Calculation of complication probability factors for non-uniform normal tissue irradiation: The effective volume method. *Cancer* 1989;16(6):1623–30.

Lawrence TS, Robertson JM, Anscher MS, Jirtle RL, Ensminger WD, Fajardo LF. Hepatic toxicity resulting from cancer treatment. *Int J Radiat Oncol Biol Phys* 1995;31(5):1237–48. doi:10.1016/0360-3016(94)00418-K.

Lasley FD, Mannina EM, Johnson CS, Perkins SM, Althouse S, Maluccio M, et al. Treatment variables related to liver toxicity in patients with hepatocellular carcinoma, Child-Pugh class A and B enrolled in a phase 1-2 trial of stereotactic body radiation therapy. *Pract Radiat Oncol* 2015;5(5):e443–9. doi:10.1016/j.prro.2015.02.007.

Lee MT, Kim JJ, Dinniwell R, Brierley J, Lockwood G, Wong R, et al. Phase i study of individualized stereotactic body radiotherapy of liver metastases. *J Clin Oncol* 2009;27(10):1585–91. doi:10.1200/JCO.2008.20.0600.

Li Y, Hoisak JDP, Li N, Jiang C, Tian Z, Gautier Q, et al. Dosimetric benefit of adaptive re-planning in pancreatic cancer stereotactic body radiotherapy. *Med Dosim* 2015;40(4):318–24. doi:10.1016/j.meddos.2015.04.002.

Liu F, Erickson B, Peng C, Li XA. Characterization and management of interfractional anatomic changes for pancreatic cancer radiotherapy. *Int J Radiat Oncol Biol Phys* 2012;83(3):e423–9. doi:10.1016/j.ijrobp.2011.12.073.

Liu Q, Cai X-W, Fu X-L, Chen J-C, Xiang J-Q. Tolerance and dose-volume relationship of intrathoracic stomach irradiation after esophagectomy for patients with thoracic esophageal squamous cell carcinoma. *Oncotarget* 2015;6(31):32220–7. doi:10.18632/oncotarget.4730.

Loehrer PJ, Feng Y, Cardenes H, Wagner L, Brell JM, Cella D, et al. Gemcitabine alone versus gemcitabine plus radiotherapy in patients with locally advanced pancreatic cancer: An Eastern Cooperative Oncology Group trial. *J Clin Oncol* 2011;29(31):4105–12. doi:10.1200/JCO.2011.34.8904.

Marks LB, Ten Haken RK, Martel MK. Guest editor's introduction to QUANTEC: A users guide. *Int J Radiat Oncol Biol Phys* 2010;76(S3):2009–10. doi:10.1016/j.ijrobp.2009.08.075.

Méndez Romero A, Keskin-Cambay F, van Os RM, Nuyttens JJ, Heijmen BJM, IJzermans JNM, et al. Institutional experience in the treatment of colorectal liver metastases with stereotactic body radiation therapy. *Reports Pract Oncol Radiother* 2017;22(2):126–31. doi:10.1016/j.rpor.2016.10.003.

Meyer JJ, Foster RD, Lev-Cohain N, Yokoo T, Dong Y, Schwarz RE, et al. A phase I dose-escalation trial of single-fraction stereotactic radiation therapy for liver metastases. *Ann Surg Oncol* 2016;23(1):218–24. doi:10.1245/s10434-015-4579-z.

Murphy JD, Christman-Skieller C, Kim J, Dieterich S, Chang DT, Koong AC. A dosimetric model of duodenal toxicity after stereotactic body radiotherapy for pancreatic cancer. *Int J Radiat Oncol Biol Phys* 2010;78(5):1420–6. doi:10.1016/j.ijrobp.2009.09.075.

Murphy JD, Hattangadi-Gluth J, Song WY, Vollans E, Camborde ML, Kosztyla R, et al. Liver toxicity prediction with stereotactic body radiation therapy: The impact of accounting for fraction size. *Pract Radiat Oncol* 2014;4(6):372–7. doi:10.1016/j.prro.2013.12.004.

Nakamura A, Shibuya K, Matsuo Y, Nakamura M, Shiinoki T, Mizowaki T, et al. Analysis of dosimetric parameters associated with acute gastrointestinal toxicity and upper gastrointestinal bleeding in locally advanced pancreatic cancer patients treated with gemcitabine-based concurrent chemoradiotherapy. *Int J Radiat Oncol Biol Phys* 2012;84(2):369–75. doi:10.1016/j.ijrobp.2011.12.026.

Osmundson EC, Wu Y, Luxton G, Bazan JG, Koong AC, Chang DT. Predictors of toxicity associated with stereotactic body radiation therapy to the central hepatobiliary tract. *Int J Radiat Oncol Biol Phys* 2015;91(5):986–94. doi:10.1016/j.ijrobp.2014.11.028.

Pan CC, Kavanagh BD, Dawson LA, Li XA, Das SK, Miften M, et al. Radiation-associated liver injury. *Int J Radiat Oncol Biol Phys* 2010;76(S3):94–100. doi:10.1016/j.ijrobp.2009.06.092.

Palorini F, Cozzarini C, Gianolini S, Botti A, Carillo V, Iotti C, et al. First application of a pixelwise analysis on bladder dose-surface maps in prostate cancer radiotherapy. *Radiother Oncol* 2016;119(1):123–8. doi:10.1016/j.radonc.2016.02.025.

Prior P, Tai A, Erickson B, Allen Li X. Consolidating duodenal and small bowel toxicity data via isoeffective dose calculations based on compiled clinical data. *Pract Radiat Oncol* 2014;4(2):e125–31. doi:10.1016/j.prro.2013.05.003.

Pollom EL, Chin AL, Diehn M, Loo BW, Chang DT. Normal tissue constraints for abdominal and thoracic stereotactic body radiotherapy. *Semin Radiat Oncol* 2017;27(3):197–208. doi:10.1016/j.semradonc.2017.02.001.

Ren ZG, Zhao JD, Gu K, Chen Z, Lin JH, Xu ZY, et al. Three-dimensional conformal radiation therapy and intensity-modulated radiation therapy combined with transcatheter arterial chemoembolization for locally advanced hepatocellular carcinoma: An irradiation dose escalation study. *Int J Radiat Oncol Biol Phys* 2011;79(2):496–502. doi:10.1016/j.ijrobp.2009.10.070.

Roeske JC, Bonta D, Mell LK, Lujan AE, Mundt AJ. A dosimetric analysis of acute gastrointestinal toxicity in women receiving intensity-modulated whole-pelvic radiation therapy. *Radiother Oncol* 2003;69(2):201–7. doi:10.1016/j.radonc.2003.05.001.

Robertson JM, Lockman D, Yan D, Wallace M. The Dose-Volume Relationship of small bowel irradiation and acute grade 3 diarrhea during chemoradiotherapy for rectal cancer. *Int J Radiat Oncol Biol Phys* 2008;70(2):413–8. doi:10.1016/j.ijrobp.2007.06.066.

Rule W, Timmerman R, Tong L, Abdulrahman R, Meyer J, Boike T, et al. Phase i dose-escalation study of stereotactic body radiotherapy in patients with hepatic metastases. *Ann Surg Oncol* 2011;18(4):1081–7. doi:10.1245/s10434-010-1405-5.

Russell AH, Clyde C, Wasserman TH, Turner SS, Rotman M. Accelerated hyperfractionated hepatic irradiation in the management of patients with liver metastases: results of the RTOG dose escalating protocol. *Int J Radiat Oncol Biol Phys.* 1993 Sep 1;27(1):117-23..

Rusthoven KE, Kavanagh BD, Cardenes H, Stieber VW, Burri SH, Feigenberg SJ, et al. Multi-institutional phase I/II trial of stereotactic body radiation therapy for liver metastases. *J Clin Oncol* 2009;27(10):1572–8. doi:10.1200/JCO.2008.19.6329.

Scorsetti M, Comito T, Cozzi L, Clerici E, Tozzi A, Franzese C, et al. The challenge of inoperable hepatocellular carcinoma (HCC): Results of a single-institutional experience on stereotactic body radiation therapy (SBRT). *J Cancer Res Clin Oncol* 2015;141(7):1301–9. doi:10.1007/s00432-015-1929-y.

Scorsetti M, Arcangeli S, Tozzi A, Comito T, Alongi F, Navarria P, et al. Is stereotactic body radiation therapy an attractive option for unresectable liver metastases? A preliminary report from a phase 2 trial. *Int J Radiat Oncol Biol Phys* 2013;86(2):336–42. doi:10.1016/j.ijrobp.2012.12.021.

Son SH, Choi BO, Ryu MR, Kang YN, Jang JS, Bae SH, et al. Stereotactic body radiotherapy for patients with unresectable primary hepatocellular carcinoma: Dose-volumetric parameters predicting the hepatic complication. *Int J Radiat Oncol Biol Phys* 2010;78(4):1073–80. doi:10.1016/j.ijrobp.2009.09.009.

Tanaka H, Hayashi S, Ohtakara K, Hoshi H. Hepatic dysfunction after radiotherapy for primary gastric lymphoma. *J Radiat Res* 2013;54(1):92–7. doi:10.1093/jrr/rrs062.

Timmerman RD. An overview of hypofractionation and introduction to this issue. *Semin Radiat Oncol* 2008;18(4):215–22. doi:10.1016/j.semradonc.2008.04.001.

Tse RV, Hawkins M, Lockwood G, Kim JJ, Cummings B, Knox J, et al. Phase I study of individualized stereotactic body radiotherapy for hepatocellular carcinoma and intrahepatic cholangiocarcinoma. *J Clin Oncol* 2008;26(4):657–64. doi:10.1200/JCO.2007.14.3529.

Toesca DAS, Osmundson EC, Eyben R von, Shaffer JL, Lu P, Koong AC, et al. Central liver toxicity after SBRT: An expanded analysis and predictive nomogram. *Radiother Oncol* 2017;122(1):130–6. doi:10.1016/j.radonc.2016.10.024.

Velec M, Haddad CR, Craig T, Wang L, Lindsay P, Brierley J, et al. Predictors of liver toxicity following stereotactic body radiation therapy for hepatocellular carcinoma. *Int J Radiat Oncol Biol Phys* 2017;97(5):939–46. doi:10.1016/j.ijrobp.2017.01.221.

Velec M, Moseley JL, Eccles CL, Craig T, Sharpe MB, Dawson LA, et al. Effect of breathing motion on radiotherapy dose accumulation in the abdomen using deformable registration. *Int J Radiat Oncol Biol Phys* 2011;80(1):265–72. doi:10.1016/j.ijrobp.2010.05.023.

Velec M, Moseley JL, Craig T, Dawson LA, Brock KK. Accumulated dose in liver stereotactic body radiotherapy: Positioning, breathing, and deformation effects. *Int J Radiat Oncol Biol Phys* 2012;83(4):1132–40. doi:10.1016/j.ijrobp.2011.09.045.

Verma J, Sulman EP, Jhingran A, Tucker SL, Rauch GM, Eifel PJ, et al. Dosimetric predictors of duodenal toxicity after intensity modulated radiation therapy for treatment of the para-aortic nodes in gynecologic cancer. *Int J Radiat Oncol Biol Phys* 2014;88(2):357–62. doi:10.1016/j.ijrobp.2013.09.053.

Verma V, Lazenby AJ, Zheng D, Bhirud AR, Ly QP, Are C, et al. Dosimetric parameters correlate with duodenal histopathologic damage after stereotactic body radiotherapy for pancreatic cancer: Secondary analysis of a prospective clinical trial. *Radiother Oncol* 2017;122(3):464–9. doi:10.1016/j.radonc.2016.12.030.

Witztum A, George B, Warren S, Partridge M, Hawkins MA. Unwrapping 3D complex hollow organs for spatial dose surface analysis. *Med Phys* 2016;43(11):6009–16. doi:10.1118/1.4964790.

Wu VW, Epelman MA, Wang H, Edwin Romeijn H, Feng M, Cao Y, et al. Optimizing global liver function in radiation therapy treatment planning. *Phys Med Biol* 2016;61(17):6465–84. doi:10.1088/0031-9155/61/17/6465.

Yeung RH, Chapman TR, Bowen SR, Apisarnthanarax S. Proton beam therapy for hepatocellular carcinoma. *Expert Rev Anticancer Ther* 2017;17(10):911–24. doi:10.1080/14737140.2017.1368392.

Yoon H, Oh D, Park HC, Kang SW, Han Y, Lim DH, et al. Predictive factors for gastroduodenal toxicity based on endoscopy following radiotherapy in patients with hepatocellular carcinoma. *Strahlentherapie Und Onkol* 2013;189(7):541–6. doi:10.1007/s00066-013-0343-0.

Zhou SM, Marks LB, Tracton GS, Sibley GS, Light KL, Maguire PD, et al. A new three-dimensional dose distribution reduction scheme for tubular organs A new three-dimensional dose distribution reduction scheme for tubular organs. *Med Phys*. 2000;27(8):1727-31

Zhu CP, Shi J, Chen YX, Xie WF, Lin Y. Gemcitabine in the chemoradiotherapy for locally advanced pancreatic cancer: A meta-analysis. *Radiother Oncol* 2011;99(2):108–13. doi:10.1016/j.radonc.2011.04.001.

CHAPTER 7

Central Nervous System (Brain, Brainstem, Spinal Cord), Ears, Ocular Toxicity

Federica Palorini, Anna Cavallo, Letizia
Ferella, and Ester Orlandi

CONTENTS

Central Nervous System: Introduction.. 172
    Pathogenesis and Clinical Significance of Radiation-Induced CNS (Brain,
    Brainstem, Spinal Cord) Toxicity.. 172
    Evaluation of CNS Toxicity Endpoints 174
    Definition of the Relevant Anatomical Structures (CNS) 175
Short Summary of the QUANTEC Recommendations: CNS (Brain, Brainstem,
Spinal Cord) 175
    Brain 176
    Brainstem... 176
    Spinal Cord 177
Critical Review of the Literature 2009–2017 (Conventional Fractionation, CNS)............ 177
    Brain Temporal Lobe.. 177
    Brainstem.. 179
    Brain-Hippocampus.. 179
    Spinal Cord... 180
Critical Review of the Literature 2009–2017 (Hypofractionation CNS) 180
    Brainstem... 180
    Spinal Cord... 181
Suggestion for DVH Constraints/NTCP Model Parameters (CNS) 181
    Spinal Cord... 181
    Brain 182
Ear: Introduction 183
    Pathogenesis and Clinical Significance of Radiation-Induced Ear Toxicity 183

Evaluation of Toxicity Endpoints 183

Definition of Relevant Anatomical Structures.. 183

Short Summary of QUANTEC Recommendations (Ear)..................................... 183

Critical Review of the Literature 2009–2017 (Conventional Fractionation, Ear)......... 184

Critical Review of the Literature 2009–2017 (Hypofractionation, Ear)......................... 186

Suggestion for DVH Constraints/NTCP Model Parameters (Ear).............................. 188

Open Issues and Future Directions of Investigation (Ear).. 188

    Toxicity Scoring.. 188

    Future Directions 189

Eye: Introduction... 189

Pathogenesis and Clinical Significance of Radiation-Induced Ocular Toxicity........... 189

Short Summary of the QUANTEC Recommendations (Ocular Toxicity) 190

Critical Review of the Literature 2009–2017 (Conventional Fractionation,

Ocular Toxicity) 191

    Photons.. 191

    Protons 196

Critical Review of the Literature 2009–2017 (Hypo Fractionation, Ocular Toxicity) 196

    Photons.. 197

    Protons 198

Summary for DVH Constraints/NTCP Model Parameters (Ocular Toxicity).............. 199

Open Issues and Future Directions of Investigation (Ocular Toxicity) 201

References... 201

CENTRAL NERVOUS SYSTEM: INTRODUCTION

Pathogenesis and Clinical Significance of Radiation-Induced CNS (Brain, Brainstem, Spinal Cord) Toxicity

Treatment-induced Central Nervous System (CNS) toxicity has been acknowledged for a long time. It represents a major cause of morbidity in long-term cancer survivors: The majority of them face the risk of developing different degrees of neurocognitive dysfunction, the most severe being in patients receiving cranial irradiation for the control of primary and metastatic brain tumors. Similar effects have also been observed as a late complication of radiation therapy (RT) of tumors of the paranasal sinuses, nasopharynx, middle ear, parotids and lacrimal glands, in which cases the CNS is incidentally involved in the treatment (De Felice and Blanchard, 2017; Perry and Schmidt, 2006). In addition, radiation-associated cranial neuropathy is a rare but functionally devastating radiation-induced late effect of head and neck tumors.

Over the last few decades, there is an increasing body of scientific literature that provides a better understanding of how radiation causes CNS injury.

Radio-induced neurotoxicity can be ascribed to multiple mechanisms: Ionizing radiations hit dividing cells and can cause direct or indirect injury to neural structures through vascular damage to endothelial cells, endocrine disturbance and fibrosis of neural structures. In particular, vascular changes play a crucial role in the pathogenesis of this side effect. Radiation-induced brain injury is described in three phases: acute (within days to weeks after irradiation), early delayed (within 1-6 months post-irradiation) and late (>6 months post-irradiation).

Early disruptions in blood–brain barrier are likely responsible for the vasogenic edema seen in the acute, albeit mostly transient and reversible forms of radiation toxicity consist of fatigue, dizziness, headache, seizures, vomiting and behavioral change. This acute encephalopathy affects up to 50% of irradiated patients, with a reduced risk by using lower fraction doses (<3 Gy) (Gérard et al., 2017). Early delayed effects from a transient demyelinization, typically reversible, include generalized weakness and somnolence. Transitory cognitive disturbances primarily affecting short-term memory and attention were identified in 36% of patients with CNS tumors receiving 54.0–55.8 Gy in 1.8 Gy-fractions; usually, abnormalities are found at Magnetic Resonance Imaging (MRI) (Ésik et al., 2003).

In contrast, more permanent forms of endothelial damage may account for the typical changes of chronic or delayed long-term radiation injury, including thrombosis, hemorrhage, fibrinous exudates, telangiectasias, vascular fibrosis/hyalinization with luminal stenosis and fibrinoid vascular necrosis (Gérard et al., 2017). These severe vascular lesions are associated with necrosis, which primarily affects the white matter (Perry and Schmidt, 2006), named radionecrosis (RN). The exact incidence of intracranial RN remains undetermined, although it ranges from 5 to 50%, depending on treatment modality, dose delivered and fractionation, length of follow-up period, neuroimaging criteria used, and whether clinical signs and symptoms are present. From a clinical point of view, RN may be symptomatic, and thus responsible of serious clinical manifestations, in 2–32% of cases (Le Rhun et al., 2016). The most frequent signs are seizures, sensorimotor deficits, ataxia, cognitive deficits, headaches, nausea, language impairment or intracranial hypertension. These signs and symptoms are not specific, depend on the irradiated region and can mimic tumor recurrence.

Moreover, different anatomical regions of the CNS may possess varying susceptibilities to this injury, due to different tissue vascularity, glial cell population or repair capacity (Ruben et al., 2006). Areas at risk of developing RN are in ascending order: frontal lobe, temporal lobe, intraventricular zone, parietal lobe, cerebellar, corpus callosum, occipital lobe, medulla, thalamus, basal ganglia and pons. Besides, the susceptibility is improved in older patients and patients with diabetes mellitus. RN appears a median of 1–2 years after RT (Ruben et al., 2006).

Brainstem RN is a rare but severe RT complication. Its clinical manifestations are cognitive disorders without dementia, involvement of the cranial nerves and motor, sensory and cerebellar function disorders (Gérard et al., 2017). The rare frequency, the absence of specific neurological signs, the subjective clinical evaluation, and the hard differential diagnosis with tumor progression make its recognition very difficult.

As for the brain, brainstem and spinal cord, injury could be an acute, early delayed and delayed RT effect. Early delayed brainstem injury is subacute encephalopathy. It is characterized by ataxia, diplopia, nystagmus and hearing loss 3 months after RT but it spontaneously regresses in most cases. Chronic complications are brainstem focal RN, cognitive impairment, cranial nerves palsy and motor, sensory and cerebellar function disorders.

With regard to spinal cord early injury, there is no clinical equivalent of acute CNS syndrome after large single doses. Lhermitte syndrome is considered a self-limiting early delayed injury. It occurs after a latent period of 2–4 months and is characterized by paresthesiae in the back and extremities typically upon neck flexion, followed by complete clinical recovery after a few months. Lhermitte syndrome is observed after doses well below the threshold for myelopathy (see below), and it is not associated with permanent myelopathy. Risk factors reported for Lhermitte syndrome include younger age and longer length of irradiated cord (Mul et al., 2012).

Late injury consists in myelopathy, typically irreversible. The symptoms range from minor motor and sensory deficits to a full-blown Brown–Séquard syndrome. The latent time to myelopathy was 18 months following a single course of treatment, and 11 months after reirradiation.

Although RN of CNS has become very rare following conventional fractionated RT, its clinical importance has been renewed in the recent years with the increasing role of stereotactic radiotherapy (SRT) and reirradiation. In particular, it represents the main complication of SRT for brain metastases. It may be observed in up to 34% of cases at 24 months after treatment and associated with significant morbidity in 10-17% (Mul et al., 2012).

However, the most common current serious delayed complication of cerebral RT is a cognitive dysfunction related to radioinduced leucoencephalopathy. This can cause minimal neurocognitive deficits, whereas extensive demyelination is associated with more severe cognitive dysfunction as well as motor and, possibly, visual deficits. More frequently, cognitive decline develops over many years after RT. Cognitive impairment mainly includes deficits in hippocampal-dependent functions – learning, memory, spatial information processing – and the risk of radiation-induced cognitive dysfunction is mainly recognized among long-term survivors.

It is generally believed that the cranial nerves are relatively resistant to radiation. Depending on the dose per fraction, total dose, and time of follow up, the reported incidence of radio-induced cranial nerve palsy is 0–5% (Au et al., 2018).

In an acute phase the irradiated nerve shows transient electrophysiological and biochemical changes. Delayed effects enhance a disorganised patchwork structure in the irradiated volume including direct axonal injury and demyelination, extensive fibrosis and ischemia by injury to capillary networks supplying the nerves.

Evaluation of CNS Toxicity Endpoints

As physician-reported toxicity, the Common Terminology Criteria for Adverse Events, version 4.0 (CTCAE, reference 14) is used to assess the CNS injury according to the symptoms: grade 1 mild or asymptomatic; grade 2 moderate, not interfering with activities of daily living (ADLs); grade 3 severe interference with ADLs; grade 4 life-threatening or disabling, intervention indicated; grade 5 death.

In addition to the CTCAE system, the cognitive functioning assessment needs to consider a variety of skills and abilities, including intellectual capacity, attention and concentration, processing speed, language and communication, visual-spatial abilities and memory.

Moreover, sensorimotor and psychomotor functioning are often measured in order to clarify the brain basis of certain cognitive impairments and are, therefore, considered one of the domains that may be included within a neuropsychological or neurocognitive evaluation. These important skills could be assessed by many tests such as the Mini Mental State Examination, the Hopkins Verbal Learning Test and the Controlled Oral Word Association Test.

Traditionally, MR contrast-enhanced imaging along with clinical observation and response to medical therapies are the main tools in managing and diagnosing RN in these patients.

Definition of the Relevant Anatomical Structures (CNS)

We briefly describe anatomical and functional characteristics of the main normal structures referred to CNS. It is important to underline that they are not always easily recognizable on the imaging used for RT planning and to date no contouring standardization for planning purpose is completely available, hampering a comparison of dose-volume effect relationships as reported in studies.

The brainstem includes the midbrain, the pons and the medulla oblongata. It provides the main motor and sensory innervation to the face and neck via 10 out of 12 cranial nerves. Though small, this is extremely important because motor and sensory nerve connections pass through it; moreover, it plays an important role in the regulation of cardiac and respiratory function, in maintaining consciousness and regulating the sleep cycle.

The spinal cord extends from the foramen magnum, where it is continuous with the medulla, to the level of the first or second lumbar vertebrae. It is a vital link between the brain and the body.

It is a cylindrical structure of nervous tissue composed of white and grey matter, uniformly organized and divided into four regions: cervical, thoracic, lumbar and sacral. The spinal nerve contains motor and sensory nerve fibres to and from all parts of the body.

Hippocampus sits in the medial temporal lobe and comprises the cornu ammonis and the dentate gyrus.

Temporal lobe is one of the four lobes of the brain and largely occupies the middle cranial fossa. Its medial surface comprises the mesial temporal lobe including hippocampus, amygdala and parahippocampal gyrus.

SHORT SUMMARY OF THE QUANTEC RECOMMENDATIONS: CNS (BRAIN, BRAINSTEM, SPINAL CORD)

It is known that for serial and parallel organs at risk (OARs) dose constraints are established on the maximum dose and the volume receiving a certain dose, respectively. Because of their fiber structure, the different parallel organs of CNS are potentially serial structures. Indeed, each fiber is a portion of an entire parallel CNS organ, but each nerve fascicle represents a serial structure (Gérard et al., 2017).

α/β ratios of cerebral OARs as well as their dose-volume constraints used for planning in Intensity Modulated Radiation Therapy (IMRT), radiosurgery (SRS) and SRT are debated. In fact, they are based on historical cohorts or calculated with old mathematical models. The calculation of equivalent doses (biologically effective dose, BED, or equivalent dose in 2 Gy per fraction, EQD2Gy) by the linear quadratic model is questionable, especially when going beyond doses of 10 Gy per session and with uncertain α/β (Jackson et al., 2010). Current dose constraints have been established from clinical-dosimetric correlations; beyond them the risk of radioinduced toxicity increases. These are mainly based on studies using conventional fractionation (1.8-2 Gy per fraction, 5 times a week) and 2 or 3-dimensional RT.

Several mathematical models claim to predict the Normal Tissue Complication Probability (NTCP) of a certain radio-induced toxicity for a given dose level (Tolerance Dose, TD). The Quantitative Analyses of Normal Tissue Effects in the Clinic (QUANTEC) used the Lyman-Kutcher-Burman model taking into account a partial irradiation of a certain OAR (Jackson et al., 2010). However, as mentioned above, the application of this model is controversial during hypofractionation (SRS, SRT). Moreover, as it is based on a homogeneous irradiation of a certain OAR, it fails to provide a valid estimate when partial irradiation to low doses occurs. In addition, highly conformal RT techniques leading to steep dose gradients around the tumor can provide very heterogeneous doses to OARs (Mayo et al., 2010a).

Finally, it is crucial to underline that QUANTEC reviews for CNS data include papers heterogeneous with regard to OARs definition and contouring, type of endpoints and tools for their evaluation, and length of clinical follow-up. Principal QUANTEC recommendations for cerebral OARs are reported below.

Brain

Due to the temporary pattern and good prognosis of acute RT side effects to the brain, QUANTEC data referred to late sequelae, i.e., RN and cognitive deterioration (Lawrence et al., 2010). RT dose and fractionation, brain volume irradiated and overall treatment time are the major RN predictors. Assuming an $\alpha/\beta = 3$ Gy for the brain, in case of fractionated RT with a fraction size <2.5 Gy, an RN incidence of 5% and 10% was expected for a BED of 120 Gy (range 100-140 Gy) and 150 Gy (range 140–170 Gy), respectively (Lawrence et al., 2010). For twice-daily fractionation (6 hours of interval at least), a steep increase in toxicity appeared to occur when the BED was >80 Gy. For large fraction sizes ($\geq$2.5 Gy), toxicity incidence and severity were unpredictable. For single-fraction SRS, a clear correlation has been demonstrated between the target size and the risk of adverse events (Ruben et al., 2006). A dose-escalation study carried out by the Radiation Therapy Oncology Group (RTOG) reported on the maximal brain-tolerated dose (MBTD) in relation to different-size targets for a series of patients previously undergoing whole-brain irradiation. The MBTD for targets of 31–40 mm in diameter, 21–30 mm and <20 mm was 15 Gy, 18 Gy and 24 Gy, respectively. In addition, toxicity increased rapidly once the brain volume exposed to >12 Gy was >5–10 cc. Thus, it was highly suggested to report the volume receiving 12 Gy as a standard method of reporting the dose to the normal brain in SRS procedures (Lawrence et al., 2010).

There were scanty data about neurocognitive injury. In particular, there was no substantial evidence that RT induces irreversible cognitive decline in adults, whereas in children cognitive dysfunctions were largely seen for whole-brain doses of $\geq$18 Gy (Lawrence et al., 2010).

Brainstem

Few papers reported on quantitative brainstem doses and dose-volume measures related to neurotoxicity in the adult population (Mayo et al., 2010b). Because of the remarkable inter-study variability in reporting doses and outcome, a unifying dose-response curve could not be generated from the available data. Using photon RT with conventional fractionation, the maximum dose to the brainstem was 54 Gy, bringing about a limited risk of severe or permanent neurological sequelae such as permanent cranial neuropathy or necrosis. Studies suggest that this dose level may be occasionally exceeded: volumes smaller than 10 cm^3 located on brainstem periphery can be irradiated at a maximum dose of 59 Gy with fraction size ≤2 Gy (De Felice and Blanchard, 2017).

A maximum dose of 64 Gy in a restricted volume (<1 cc) may be acceptable, but, beside it, the risk of severe neurotoxicity is substantially high (D_{50} = 65 Gy, equivalent uniform dose to whole organ leading to 50% toxicity). Ultimately, there is insufficient information to determine whether a further volume effect is to be expected (Mayo et al., 2010b).

Spinal Cord

The estimated risk of myelopathy is <1% and <10% at maximum doses of 54 Gy and 61 Gy, respectively, with a calculated strong dependence on dose per fraction for an estimated α/β = 0.87 Gy. Although some of these patients received relatively high doses per fraction, none of them were treated using stereotactic techniques to exclude a portion of the circumference of the cord. However, it has been noted that this α/β ratio is lower at the values of 2-4 Gy per fraction, but it predicts a more severe effect at larger doses (Kirkpatrick et al., 2010).

A summary of reirradiation studies suggested that partial repair of RT-induced subclinical damage up to 25% would become evident about 6 months post-RT and increase over the next 2 years (Kirkpatrick et al., 2010). In particular, both α/β values of 1 and 3 Gy were employed to calculate the EQD2Gy. Very low incidence of myelopathy (0.2%) was reported despite large cumulative doses, with essentially no cases observed for cumulative doses of 60 Gy (EQD2Gy). The analysis of the results from 9 reports of SRS to the spinal lesions, including de novo RT alone studies, reirradiation alone and combination of the two (mixed series) reported a myelopathy incidence <1% when the maximum spinal cord dose was limited to the equivalent of 13 Gy in a single fraction or 20 Gy in 3 fractions. However, long-term data are insufficient to calculate a dose-volume relationship for myelopathy when the partial cord is treated with a hypofractionated regimen (Kirkpatrick et al., 2010). Thus, QUANTEC report was essentially unable to make evidence-based recommendations for spinal cord tolerance specific to hypofractionated modern RT and SRT practice.

CRITICAL REVIEW OF THE LITERATURE 2009–2017 (CONVENTIONAL FRACTIONATION, CNS)

Brain Temporal Lobe

Radio-induced temporal lobe necrosis (TLN) can affect cognitive functions, such as memory and verbal ability. It has become a common late damage that adversely affects quality of life and survival of patients after RT. The precise mechanism causing TLN remains unknown, but it is generally believed to be associated with volume and dose of irradiated temporal lobes (Lawrence et al., 2010).

Historic data and literature indicate that the incidence of radiation-induced TLN range from 0% to 18.6% after 2D-RT. Lee et al. reported that 64 Gy (at conventional fractionation

178 ■ Modelling Radiotherapy Side Effects

of 2 Gy daily) would lead to a 5% necrotic rate in temporal lobes at 10 years after 2D-RT in nasopharyngeal carcinoma patients (NPC). The 10 year actuarial incidence of TLN was 4.6% for patients irradiated to 60 Gy with 2.5 Gy per fraction and up to 18.6% for those irradiated to 50.4 Gy with 4.2 Gy per fraction. Therefore, this study revealed that the dose per fraction is an independent significant factor affecting cerebral necrosis. Larger fractional doses and a shorter overall radiation treatment time would increase the TLN risk for NPC patients after RT (Lee et al., 2002).

This tolerance dose was derived from publications concerning 2D-RT for patients with nasopharynx, glioma or brain metastases. However, limited published data regarding TLN after IMRT are available and a minority of them concern prospective evaluation.

In Hsiao et al. (2010), after IMRT for NPC, 77% of patients had statistically significant decline in neurocognitive functions (NCF) across domains of short-term memory, language abilities and list generating fluency. They observed that patients with mean doses >36 Gy had significantly greater NCF declines. Moreover, patients with a temporal lobe V_{60Gy} >10% (percent organ volume receiving more than 60 Gy) also had greater NCF declines than those with V_{60Gy} ≤10%.

Su et al. (2013) reported a retrospective analysis of dosimetric factors associated with TLN development in a series of 259 NPC patients treated to 68 Gy with IMRT. Radio-induced TLN was observed at a crude incidence of 15.4% at 5 years and multivariate analysis showed a strong correlation between the absolute volume of temporal lobe receiving 40 Gy and TLN development. In particular, they found that 5-year TLN incidence for V_{40Gy} <10% or <5 cm³ was less than 5%. The incidence for V_{40Gy} ≥15% or ≥10cm³ increased significantly and exceeded 20% (Su et al., 2013). Sun et al. retrospectively analyzed dose-volume data of 20 NPC patients and suggested a dose limit of 69 Gy for a temporal lobes volume of 0.5 cm³.

Other studies focused on the dose-volume parameters related to temporal lobe injury after particle RT. Pehlivan et al. (2012) reported an analysis of temporal lobe toxicity after spot scanning proton therapy in 62 patients treated for skull base tumors. A planning constraint of 74 Gy to 2 cm³ was suggested. Furthermore, McDonald et al. (2015) retrospectively considered 66 patients treated for skull base tumors: All dose levels from 10 to 70 Gy (RBE, Relative Biological Effectiveness[*]) highly correlated with RN, with a 15% three-year risk of any-grade TLN when the absolute volume of a temporal lobe receiving 60 or 70 Gy (RBE) exceeded 5.5 or 1.7 cc, respectively.

The radiation damage after carbon ion therapy appeared similar to that of proton therapies. Schlampp et al. (2011) calculated the RBE of carbon ion therapy in 118 temporal lobes in 59 patients: The dose to 1 cm³ was predictive for radio-induced TLN. Esteemed D_5 (equivalent dose to whole organ leading to 5% toxicity) and D_{50} for the brain were maximum doses of 68.3 (±3.3) Gy and 87.3 (±2.8) Gy, respectively.

---

[*] All proton doses are expressed with the most recent notation Gy (RBE), with an RBE of 1.1 compared to megavoltage X-ray therapy.

Brainstem

According to RTOG 0539, brainstem volumes <0.03 cm^3 Gy can tolerate a cumulative dose of 60 Gy, even if the risk of radio-induced toxicity seems to depend on the dose-volume effect (Xue et al., 2012). The brainstem is a complex serial organ: It is better to consider its topography than simply its volume. In this respect, certain authors suggested to determine two different constraints: one for the central (53–54 Gy) and one for the peripheral part (63–64 Gy) (Weber et al., 2005). This regional variation in radiosensitivity across the brainstem has been accordingly explained by differences in vascularity and migration of oligodendrocyte progenitor cells. Several studies using proton therapy considered this theory as well: Maximum doses of 63-64 Gy (RBE) for the brainstem surface and 53–54 Gy (RBE) for the central part were identified (Weber et al., 2005).

In paediatric patients, Indelicato et al. identified V_{55Gy} <17.7%, maximum dose <56.6 Gy and age <5 years as important predictive factors of brainstem toxicity, considering the higher sensitivity of paediatric nervous structures (Indelicato et al., 2014). However, in this setting tolerance appears to depend not only on maximum doses to brainstem and brainstem volume included in high-dose regions, but it also relates to existing conditions that need to be accounted for.

Indeed, Debus's study reported on 367 skull base tumor patients treated with a combination of photon and proton conformal RT. It suggested that V_{60Gy} >0.9 cc, two or more skull base surgeries and diabetes may relate to higher incidence of brainstem toxicity. Likely, surgeries might lead to subclinical devascularization or neurological injury, making the brainstem less resilient to higher radiation doses (Debus et al., 1997).

Brain-Hippocampus

New memory formation has been associated with a lifelong mitotically active and radiosensitive compartment of neural stem cells (NSCs) located in the subgranular zone of the hippocampal dentate gyrus. Injury to this NSC compartment has been hypothesized to be central to the pathogenesis of radiation-induced cognitive decline.

Several clinical studies confirmed the association between cranial irradiation and cognitive impairment (Gondi et al., 2014) and, consequently, some of them focused on safety and technical feasibility of hippocampus-sparing RT. In particular, these studies showed that the risk of metastasis in the hippocampal area is approximately 0.5% for oligometastatic disease, i.e., up to 3 brain metastases, and approximately 1.5% in case of multiple brain metastasis (Gondi et al., 2010).

From the radiobiological point of view, the effect of ionizing radiation on the hippocampus is not completely understood and remains still debatable. According to the QUANTEC, the brain α/β ratio is 2.9 Gy but this is not explicitly specified for the hippocampus. Some investigators take an α/β ratio between 2 and 3 Gy for the hippocampus, whilst others use a value of 10 Gy for hippocampal NSCs, i.e., the same as for stem cells.

The constraints used for the hippocampus vary a lot in the literature and, at this time, there is no evidence to conclusively support a particular recommendation.

Gondi et al. (2013) conducted a prospective controlled observational study in adult patients with benign or low-grade brain tumors and observed a dose-response relationship between radiation dose to the hippocampus and long-term memory impairment. Specifically, EQD2Gy (assuming $\alpha/\beta = 2$ Gy) to 40% of the bilateral hippocampi >7.3 Gy was associated with long-term impairment in list-learning delayed verbal recall, as measured by the Wechsler Memory Scale-III Word Lists delayed recall test. The radiotoxicity seemed to grow as the dose increased.

RTOG 0933 study suggested a total dose of 37.5 Gy in 15 fractions for whole-brain RT using a $D_{100\%}$ (minimum dose to 100% volume) <10 Gy and a D_{max} (maximum organ dose) ≤17 Gy as bilateral hippocampi constraints (Gondi et al., 2014).

Spinal Cord

In the recent years no papers have been published adding new findings for conventional treatment and spinal cord injury. However, different dose-volume constraints have been identified for spinal cord injury after proton therapy. Some authors accounted 50 Gy (RBE) as limit for the maximum dose; others found a higher tolerance for this structure. In particular, dose constraints of 64 and 54 Gy (RBE) at 2% of the spinal cord surface and central part, respectively, were considered safe in a retrospective cohort of patients treated with high-dose proton therapy for extracranial chondromas and chondrosarcomas (Stieb et al., 2018).

Moreover Marucci et al. (2004) found a maximum dose of 55 and 67 Gy (RBE) to the center and surface of the cervical cord, respectively, to be safe.

CRITICAL REVIEW OF THE LITERATURE 2009–2017 (HYPOFRACTIONATION CNS)

Brainstem

It has been demonstrated that small portions of the brainstem can tolerate high doses, without acute sequelae, if most part of the organ is not involved in the irradiation.

Both SRS and SRT are able to relate a tolerance profile comparable to the conventional fractionation. The QUANTEC reports few data: It cites only the SRT of vestibular schwannomas and trigeminal neuralgias. SRS treatments are not as homogeneous as SRT. In case of punctual doses <12.5 Gy on the 50% isodose, without fractionation, it would be less than 5% cases of serious toxicity (Mayo et al., 2010a). The risk of radio-induced effects is slightly increased with doses like 15–20 Gy in patients with prognostic factors. Timmerman et al. (2008) affirmed that the punctual dose for 1, 3 and 5 fractions could not exceed 15, 23 and 31 Gy, respectively. However, a maximum volume of 0.5 cm³ must be respected when the dose is above 10, 15 and 18 Gy in 1, 2 and 3 fractions, respectively, because the risk of radio-induced toxicity is associated to volumes >1 cc. Concomitant chemotherapy should not be taken into account because it could increase the toxicity risk. It is hard to understand the impact of new types of chemotherapy and immunotherapy associated with RT (Gérard et al., 2017).

The maximum dose can have different definitions, i.e., in a pixel or in 0.03 cc. Nowadays, finding a homogenous dose-volume histogram between classical fractionation therapies and SRS/SRT is debatable.

Spinal Cord

Radiation myelitis re-emerged as a direct result of SRT practice, where high-dose radiation is delivered adjacent to the spinal cord to be spared. In the last few years, a number of dose tolerance guidelines for SRT in 1-5 fractions has been published. The most popular dose planning constraints referred to the maximum point dose (D_{max}), minimum dose to 0.1 cm^3 and to 1 cm^3 ($D_{0.1cm3}$, and D_{1cm3}, respectively).

Published reports document cases of myelopathy but do not account for the total number of patients treated at given dose-volume combinations who do not have myelitis.

In a paper by Sahgal based on a pooled multi-institutional analysis (Sahgal et al., 2013), dose-volume histogram (DVH) data of cohort of patients that developed severe myelopathy resulting in paraplegia or quadriplegia were compared with a series of patients treated without myelopathy. The thecal sac was contoured as the surrogate for the spinal cord and for a length spanning at least 1 vertebral body above and below the target volume. Continuous DVH data for the thecal sac were tabulated according to the D_{max} within the thecal sac, and for the following volumes: 0.1 to 1.0 cm^3 in increments of 0.1 cm^3, and 2.0 cm^3. To equate the various fractionation schemes, the BED was calculated using an α/β = 2 Gy. The authors observed a statistical difference between patient subsets in terms of both mean doses and dose distributions for up to 0.8 cm^3 volumes. It implies that the dose to small volumes of the spinal cord must be respected for safe practice. Furthermore, as the greatest significant difference was observed at the D_{max} volume and the dose to this volume represents the highest dose that can be determined within the organ by the treatment planning system, it is reasonable for safe practice to base thresholds on this volume. Recommended limits to D_{max} thecal sac resulting in 1–5% probability of radiation myelopathy (RM) are 12.4 Gy in a single fraction, 17.0 Gy in 2 fractions, 20.3 Gy in 3 fractions, 23.0 Gy in 4 fractions and 25.3 Gy in 5 fractions SRT. These probabilities were defined with a 15 months follow-up time, underlying that longer-term toxicity outcomes are required to validate the estimates.

In a paper by Katsoulakis (Katsoulakis et al., 2017), the largest study evaluating dosimetric data and radiation-induced myelitis in de novo spine SRS, a median spinal cord D_{max} of 13.85 Gy was safe and able to support that a D_{max} limit of 14 Gy carries a rate <1% of myelopathy. No dose-volume thresholds or relationships between spinal cord dose and myelitis were apparent.

A recent literature review of more than 200 SRT spine articles from the past 20 years found 59 dose tolerance limits for the spinal cord in 1–5 fractions. The authors partitioned these constraints into a unified format of high-risk and low-risk dose tolerance limits (Grimm et al., 2016), showed later in the chapter.

SUGGESTION FOR DVH CONSTRAINTS/NTCP MODEL PARAMETERS (CNS)

Spinal Cord

Considering the possible occurrence of severe irreversible RT-induced injury, constraints are necessarily conservative. With conventional fractionation (2 Gy daily, 5 days a week) a maximum cord dose of 46–50 Gy should be respected. Myelopathy risk rates are estimated at 0.2%, 6% and 50% for a total dose of 50 Gy, 60 Gy and 69 Gy, respectively.

If hypofractionated schedules are used, the spinal cord should receive a lower dose; maximum cord doses of 13 Gy in a single fraction, 20 Gy in 3 fractions and 22 Gy in 5 fractions are associated with less than 1% risk of damage.

An even lower risk rate of myelopathy has been observed for D_{1cm3} of 7, 11 and 13.5 Gy (0.1% risk) and $D_{0.1cm3}$ of 8.5, 16.3 and 20 Gy for 1-, 3- and 5-fractions treatments (0.1%, 0.2%, 0.2% risk), respectively. Furthermore, D_{max} of 14 Gy in 1 fraction, 22 Gy in 3 fractions and 30 Gy in 5 fractions relate to 1.6%, 1.3% and 2.6% myelopathy risk, respectively. Considering $D_{1 cm3}$, limits of 8 Gy in 1 fraction, 16 Gy in 3 fractions, 21 Gy in 5 fractions combine with 0.2%, 0.9% and 2% risk rate. For $D_{0.1 cm3}$, 0.2%, 0.4% and 0.4% risk of cord injury were found for 10 Gy in 1 fraction, 18 Gy in 3 fractions and 22.5 Gy in 5 fractions, respectively (Grimm et al., 2016).

Brain

Neurologic impairments include a decline in memory, focus and manual skills. In general, most studies performed after definitive RT, and typically referring to nasopharynx and paranasal sinuses because of their proximity to temporal lobes, show a direct correlation between radiation dose and CNS. A study by Hsiao et al. (2010) defined some cut-off points in 36 Gy as mean dose to the temporal lobes and 10% of the temporal lobe volume that received 60 Gy: Above these thresholds, patients experienced a more severe loss in their cognitive function.

At present, the QUANTEC only considers injury as an endpoint when it comes to assessing the risks of neurological damages. It is worth investigating a number of variables, i.e., the large fraction size and the volume of brain being treated, besides concomitant chemotherapy. We should probably factor in the natural age-related cognitive decline. Evidence has been found as to the maximum brainstem dose to be delivered of 50-54Gy, V_{65Gy} <3 cm³ and V_{60Gy} <5 cm³ (De Felice and Blanchard, 2017). As for the hippocampus, the maximum dose should not exceed 10 Gy and maintain $V_{7.3Gy}$ <40% (Gondi et al., 2010) whereas for the olfactory bulbs, it should be lower than 40 Gy (De Felice and Blanchard, 2017).

In broad terms, the exposure of CNS structures should be as limited as possible. Recent studies have showed that modulated proton beam therapy results in brain-sparing outcomes compared with a traditional IMRT treatment plan (Holliday et al., 2016). Designing new dose-volume constraints is a matter of great importance. Top priority should be given to maximum doses, then proceed to dose-volume parameters. Neurological structures should be as little affected as possible, and yet never let go of coverage of target volumes as it is paramount in NPC. There is a need for further prospective studies to look into the results brought about by lower doses.

Today, most data available come from retrospective studies that make definitive conclusions hard to draw. In particular, it is difficult to tell whether neurocognitive dysfunction is related to high radiation dose to temporal lobe or hippocampus, or both. Ma et al. (2016) have recently highlighted the role of the cerebellum and the cingulo-opercular network as influencing cognitive abnormalities, implying a connection with other encephalic regions.

EAR: INTRODUCTION

Pathogenesis and Clinical Significance of Radiation-Induced Ear Toxicity

Sensorineural hearing loss (SNHL) is a possible RT adverse event involving the cochlea and/or retrocochlear component of the auditory system (Bhandare et al., 2010; Landier, 2016). This side-effect has been characterized as dose-dependent, late, progressive and permanent, and may, thus, have a substantial impact on health-related and social outcomes, with resultant impaired quality of life. The reported SNHL rates based on audiometric evaluation vary between 0% and 54% (Raaijmakers and Engelen, 2002; Yeh et al., 2005).

RT-induced SNHL in adults treated with conventional or hypofractionated RT for head-and-neck cancers (HNC) and vestibular schwannomas (VS) is hereby considered.

Evaluation of Toxicity Endpoints

Audiometric assessment pre- and post-RT is necessary for a quantitative evaluation: SNHL has been classically defined as a significant increase (>10 dB) in the bone conduction threshold (BCT) at the frequencies between 0.5 and 4.0 kHz, measured by means of pure-tone audiometry (Bhandare et al., 2010) – human speech falls in that range, with frequencies ≥2.0 kHz being important for consonants discrimination. Although already pointed out in the QUANTEC, SNHL reports still vary in terms of frequencies evaluated – a single one and/or pure tone average (PTA) of more frequencies – and change in BCT (ΔBCT) defined as clinically significant. In the papers selected, pre-RT BCT of the same ear is used as control/standard for comparison.

Hearing status evaluation in VS patients have consistently been presented according to the Gardner-Robertson (GR) scale (Gardner and Robertson, 1988; Bhandare et al., 2010), based on the PTA of frequencies 0.5–1.0–2.0 kHz and speech discrimination score (SDS): Hearing variation is expressed in terms of pre-treatment serviceable hearing preservation (HP) and/or improvement or loss in hearing.

Definition of Relevant Anatomical Structures

Due to its small size and deep location in the temporal bone, it may be challenging to recognize the cochlea on CT scans: CT bone view settings and fusion with MRI are recommended. An appropriate image thickness (possibly ≤1.0 mm) should also be used as the cochlea has usually an average volume <0.6 cm^3 (Brouwer et al., 2015; Scoccianti et al., 2015). In the VS case, the internal auditory canal can also be contoured to estimate the dose to the cochlear nerve.

Short Summary of QUANTEC Recommendations (Ear)

In the QUANTEC organ-specific paper, Bhandare et al. (2010) evaluated the mean dose received by the cochlea – a more detailed dose-volume analysis was not feasible because of its small volume – together with other treatment- and patient-related factors. They selected several studies attempting to relate cochlear mean dose to SNHL and proposing NTCP models, but did not choose any specific model for clinical practice due to limitations, such as the variability in SNHL assessment criteria or the relatively small number of patients.

They could not determine a specific threshold dose, but suggested some dose-prescription limits while preserving the desired target coverage.

In HNC conventionally fractionated RT, they recommended to limit cochlear mean dose to ≤45 Gy and, in general, to keep it as low as possible to minimize SNHL risk. For HP after stereotactic radiosurgery (SRS) for VS treatment, they suggested a limit of 12–14 Gy to the prescription marginal tumor dose, which reflects the dose received by the cochlear nerve, and a prescription dose of 21–30 Gy in 3–7 Gy per fraction over 3–10 days as a hypo-fractionation schedule, to provide tumor control and preserve hearing.

Older age, better pre-RT hearing level and post-RT otitis media were some of the factors affecting SNHL risk; the synergistic effect of concurrent cisplatin chemotherapy on ototoxicity had been broadly studied (Low et al., 2006). Finally, they made specific recommendations for evaluating and scoring RT-induced ototoxicity.

Critical Review of the Literature 2009–2017 (Conventional Fractionation, Ear)

A review of the literature published after 2009 was performed to find studies incorporating dose-volume relationship and/or NTCP models. RT conventionally fractionated treatment for HNC were reviewed first (2–2.2 Gy/fraction for a prescribed total dose of 66–74 Gy, 5–6 fractions weekly). Selected studies are listed in Table 7.1. The results mostly confirmed QUANTEC suggestions, i.e., mean dose to the cochlea ≤45 Gy.

TABLE 7.1   Selected Studies for Sensorineural Hearing Loss After Conventional Fractionated Radiotherapy

Author (ref)	Patients, Disease	Mean Dose[a] (Gy)	Median Follow-up[b] (months)	Endpoint (ΔBCT)	Frequency (kHz)	Incidence (%) [Ears]
Petsukiri et al. (2011)	27, NPC	51.0 [25.1–75.5]	13 [6–29]	15 dB	Avg 0.5–1–2 4	7.4 [4/54] 37 [20/54]
Wei et al. (2014)	72, NPC	–	24 [0.3–60]	30 dB	Avg: 0.5–1–2; 4	–
Chan et al. (2009)	87, NPC	48.9 [33–71.7]	24 [6–30]	15 dB (persistent) 30 dB (severe)	Avg 0.5–1–2 4	9.4 [16/170] 51.2 [87/170]
Theunissen et al. (2015)	100, HNC	13.6 [1.1–70.9]	14 [3–31] week	35 dB	Avg 1–2–4	–
Wang et al. (2015)	51, NPC	43.7 [30.0–58.0]	60 [28–84]	15 dB	Avg 0.5–1–2 4	12.7 [13/102] 42.2 [43/102]
Cheraghi et al. (2017)	35, HNC	33 [23–43]	12	CTCAE v.4 G>1		25.7 [9/35]

ΔBCT, change in bone conduction threshold; NPC, nasopharyngeal carcinoma; HNC, head-and-neck cancer; CTCAE, Common Terminology Criteria for Adverse Events.

[a] mean dose to the cochlea.

[b] audiological follow-up, with range expressed in [].

Two retrospective studies reviewed mean cochlear doses in 27 (Petsuksiri et al., 2011) and 72 (Wei et al., 2014) patients treated with IMRT with or without concurrent chemotherapy for NPC. They evaluated ΔBCT post-RT at high and low frequencies: Petsuksiri et al. observed a higher incidence of SNHL (ΔBCT ≥15 dB) for doses >50 Gy to the cochlea (relative risk 1.77, 95% Confidence Interval [CI] 0.82-4.24), while Wei et al. found severe SNHL (ΔBCT ≥30 dB) when the dose was ≥46 Gy ($R^2 = 0.7$). In a longitudinal study on 87 NPC patients, Chan et al. (2009) noted a dose-dependent relationship between the mean cochlear dose and the probability of persistent high-frequency SNHL, with a combined detrimental effect due to concurrent cisplatin addition as the dose-response curve shifted to the left (p = 0.05). At uni- and multivariate analysis, concurrent cisplatin dosage (p <0.01) and mean cochlear dose (p = 0.01) were important determinants of persistent high-frequency SNHL, while age (p = 0.03) was significant for persistent low-frequency SNHL. Doses <47 Gy would result in a severe SNHL incidence <15% (p = 0.02).

In a retrospective study on 100 HNC patients, Theunissen et al. (2015) tried to predict treatment-induced SNHL after cisplatin infusion, aiming to improve individual counseling of patients undergoing chemoradiotherapy. A multilevel mixed-effects linear regression model to predict a 35-dB deterioration in hearing level at 1–2–4 kHz PTA after treatment was established:

$$-5.56 + (0.02 \times C) + (0.21 \times RT) + (0.05 \times PTAL) + (0.68 \times PTAH) + (0.10 \times PTAU)$$

where C indicates the cisplatin dose (mg); RT the cochlear dose (Gy); PTAL, low pre-treatment PTA of 0.5–1–2 kHz BCT (dB); PTAH, high pre-treatment PTA of 1–2–4 kHz BCT (dB); PTAU, ultrahigh pre-treatment PTA of 8–10–12.5 kHz air conduction threshold (dB). AUC was 0.68, with a 29% sensitivity (95% CI, 13–51%) and a 97% specificity (95% CI, 88–100%), resulting in a positive predictive value of 78%.

Wang et al. (2015) found a different dosimetric result. In 51 NPC patients they retrospectively calculated several dosimetric parameters for the cochlea, including mean and maximum dose, multiple DVH cut-points, equivalent uniform dose, etc. A minimum dose ≥39.8 Gy to 0.1 cm^3 of the cochlea was associated with a higher SNHL incidence. It was also a predictive factor for SNHL at multivariate analysis together with accumulative cisplatin dose (≥200 mg/m^2) and secretary otitis media (p = 0.02, 0.03 and 0.01, respectively).

Recently, Cheraghi et al. (2017) studied the dose response relationship for RT-induced SNHL according to 6 known NTCP models in 35 HNC patients, treated with 3D conformal RT. Three models fitted well with the clinical data: the relative seriality, the critical volume individual and the critical volume population model. Resultant D_{50} values varied in the range 50.4–60 Gy.

Older age (Wang et al., 2015; Chan et al., 2009; Petsuksiri et al., 2011) and a better pre-treatment hearing level were patient-related factors affecting SNHL risk. Since the prevalence of presbycusis is 35–50% in the population aged 65 years and older (Parham et al., 2011), age influence on hearing status should be considered in addition to RT effects. In 92 NPC patients, hearing thresholds without age correction were approximately equal to those with age correction plus age-related hearing deterioration, which was worse at high frequency (Hwang et al., 2015). Adjusting for patient age, RT-induced SNHL was progressive and appeared to affect all frequencies equally, suggesting that high-frequency SNHL might be due to both RT toxicity and presbycusis. Zuur et al. (2009) demonstrated that the relative SNHL in dB was larger in patients with good pre-treatment hearing capability compared to those with worse baseline hearing.

RT delivery technique and cisplatin-based concurrent chemotherapy affected SNHL as treatment-related factors. IMRT (Petsuksiri et al., 2011; Theunissen et al., 2014) had a lower SNHL incidence than conventional RT since the cochlea could be addressed as OAR. Gao et al. (2015) showed an even superior performance of volumetric modulated arc therapy in terms of target coverage and cochlea sparing compared with step-and-shoot IMRT. Finally, the synergistic toxicity of cisplatin-based concurrent chemotherapy has been confirmed, with higher cisplatin doses increasing SNHL risk (Petsuksiri et al., 2011; Wei et al., 2014; Wang et al., 2015; Du et al., 2015).

Critical Review of the Literature 2009–2017 (Hypofractionation, Ear)

Radiation therapy including SRS or hypofractionated SRT is one of the established management options for VS (Murphy and Suh, 2011; Tsao et al., 2017) and frequently associated with post-treatment SNHL. Since most VS patients present with hearing loss (98%), it can be hard to distinguish RT effect on the auditory function – resulting in cochlea and cochlear nerve radiation damage – from VS natural progression (Bhandare et al., 2009). A dose-volume analysis is hardly feasible due to the small nerve diameter and imaging issues, but QUANTEC suggested estimating the dose received by the cochlear nerve based on its location and length involved with the tumor and the prescription/marginal tumor dose (Bhandare et al., 2010).

Several studies tried to correlate cochlear dose with HP rates; Table 7.2 lists those selected. Timmer et al. (2009) studied HP in 69 VS patients treated by Gamma Knife SRS,

TABLE 7.2   Selected Studies on Vestibular Schwannomas (Hypofractionated Radiotherapy)

Author (ref)	Pts	Tumor Margin Dose (Gy)	Median Follow-up[a] (months)	Tumor Control (%)	HP (%) [pts]
Timmer et al. (2009)	69	9.3–12.5	14 [3–56]	–	75 [52/69]
Rashid et al. (2016)	37	21–25, 3–5 fr	51 [1518]	–	89
Kano et al. (2009)	77	12–13	20 [6–40]	–	71.4 [55/77]
Hayden Gephart et al. (2013)	94	18, 3 fr	29 [4–107]	100 (2 yrs); 96 (5 yrs)	74
Tsai et al. (2013)	117	18, 3 fr	64 [21–89]	99.1 (5 yrs)	81.5 [53/65]
Jacob et al. (2014)	59	12–13	25 [6–46]	95	96 (1 yr); 83 (2 yrs); 57 (3yrs)
Brown et al. (2011)	53	12–13	15 [6–136]	96	79 [42/53]

HP, hearing preservation rate; pts, patients; yrs=years; fr, fractions.

[a] Audiological follow-up, with range expressed in [].

measuring PTA and SDS before and after treatment. Average marginal tumor dose was 11 Gy and maximum, 50%, 90% and minimum dose to the cochlea were calculated. They found a clear correlation between maximum cochlear dose and PTA difference before and after SRS (p <0.05).

To estimate clinically relevant dose-tolerance limits for HP in a 1–5 fractions SRT setting, Rashid et al. (2016) generated an NTCP model combining the 69 VS from Timmer et al. (2009) and their 37 patients treated by CyberKnife SRT in 3 or 5 fractions (21 or 25 Gy, respectively). Selecting 50 dB ΔBCT as the endpoint, they generated a probit dose-response model for the two datasets (1- and 5-fraction) and chose a 3-fraction regimen for the aggregate dataset, using $\alpha/\beta = 3$ Gy (as in QUANTEC) and a linear quadratic conversion. The respective D_{50} – in terms of maximum cochlear dose – were 23.98, 37.64 and 34.55 Gy. For a SNHL risk <25%, the limits of 14 Gy in 1 fraction, 22.5 Gy in 3 and 27.5 Gy in 5 fractions correspond to a risk lower than 18% (17.9%, 17.7% and 17.4%, respectively).

In a series of 77 VS treated with SRS, patients receiving a radiation dose <4.2 Gy to the central cochlea were more likely (p = 0.0 2) to preserve pre-treatment Gardner–Robertson (GR) hearing class (Kano et al., 2009). In 94 consecutive VS patients treated with 3-fraction SRT – median marginal dose of 18 Gy, radiobiologically equivalent to 11.3 Gy in single fraction – serviceable hearing (GR class I-II was preserved in 74% of patients (Hayden Gephart et al., 2013). The retrospective dosimetric study on the cochlea showed that larger volume and higher dose level (from 6 to 18 Gy, 2 Gy step size) were associated with lower and higher SNHL risk, respectively (p <0.05). Controlling for volume inter-individual variability, SNHL risk significantly increased by 1 to 6% per each additional mm^3 of cochlea receiving 10 to 14 Gy. In a similar setting, Tsai et al. (2013) also found smaller cochlear volumes and higher cochlear doses to be associated with SNHL.

This finding on the cochlear size seemed to suggest a greater "reserve" for functional hearing after RT, but the authors themselves raised doubts about possible inaccuracies in measuring such a small volume like the cochlear one. Jacob et al. (2014), for example, did not find it significant for HP: They evaluated cochlear doses in 59 patients in association with time to SNHL. Serviceable hearing rates at 1, 2 and 3 years after SRS were 96%, 83% and 57%, with a 95% overall tumor control rate. Better pretreatment hearing (PTA < 30 dB or SDS ≥70 or GR class I vs. II , p < 0.01), lower marginal tumor dose (12 vs. 13 Gy, p = 0.01) and lower mean cochlear dose (<5 Gy, p < 0.01) were prognostic factors for HP at univariate analysis.

In a series of 53 patients treated with SRS, Brown et al. (2011) found a strong correlation between the mean percentage of cochlear volume receiving >5.3 Gy and a ΔBCT >20 dB (p = 0.02). A higher tumor coverage would also adversely affect hearing outcome, maybe due to the higher integral dose to the cochlear nerve segment within the internal auditory canal. But the cochlear nerve is hardly recognizable as a discreet structure on MRI scans. Therefore, both Brown et al. and Jacob et al. strongly recommended that tumor marginal dose should not be reduced to spare cochlear structures, as it could decrease long-term tumor control.

Suggestion for DVH Constraints/NTCP Model Parameters (Ear)

In the post-QUANTEC era, several studies tried to investigate the dose-response relationship between SNHL and dose to the auditory system, and their results were in agreement with QUANTEC recommendations. Still neither specific threshold doses nor new constraints for RT treatment planning optimisation could be identified. Some NTCP models were also proposed, but considerable limitations persist (i.e., variability in SNHL assessment criteria, relatively small number of patients, follow up length, lack of external validation) making it quite hard to suggest one for routine clinical use. Two of the selected studies (Wang et al., 2015; Cheraghi et al., 2017) hypothesized a serial architecture for the cochlea, traditionally regarded as a parallel organ. However, its small volume makes a detailed dose-volume analysis difficult, due to variations in cochlea contouring, differences in calculation algorithms, steep dose gradient typical of IMRT dose distributions, setup errors that may influence the actual dose received. Thus, a mean dose should be less affected by all those uncertainties.

For conventionally fractionated RT, it seems reasonable to stick with the 45 Gy limit to cochlear mean dose, as long as target coverage is not compromised. As a rule, it should be kept as low as possible, especially in cisplatin-based chemoradiotherapy treatments due to the well-documented cisplatin ototoxic effect. Older age, a better pre-treatment hearing level and RT delivery technique also seem to affect SNHL risk.

In VS treatments, no specific dose constraint for the cochlea could be identified (different thresholds and parameters), but only suggestions on the prescription/marginal tumor dose – surrogate of the dose received by the cochlear nerve. Recent series limited the tumor margins prescription dose to 12–13 Gy in single fraction SRS, since doses >13 Gy were found to be associated with severe hearing deterioration (Yang et al., 2010; Combs et al., 2015). This approach is one of the management options for VS smaller than 3 cm in diameter and has so far provided a good compromise between tumor control and HP rates (91–99% vs. 41–79% according to Tsao et al. (2017)).

Already at QUANTEC time, some studies had suggested that hypofractionated SRT can help HP in cases of larger and more irregularly shaped lesions, being equally safe and effective for tumor control. Schedule tested ranged between 21–30 Gy in 7–3 Gy per fraction over 3–10 days, although few and small series are reported and it remains unclear if HP rate is actually better.

Open Issues and Future Directions of Investigation (Ear)
Toxicity Scoring
The significant variability in reporting hearing outcomes, maybe also due to the lack of consensus general guidelines, makes data comparison between studies very difficult.

The CTCAE v.5 expresses ototoxicity incidence as a 4-point scale: grade 1, threshold shift 15–25 dB averaged at 2 contiguous frequencies in at least one ear or subjective change in hearing; grade 2, threshold shift >25 dB averaged at 2 contiguous frequencies in at least one ear; grade 3, threshold shift >25 dB averaged at 3 contiguous frequencies in at least one ear; grade 4, decrease in hearing to profound bilateral loss with absolute threshold >80 dB at 2 kHz and above (non-serviceable hearing). The CTCAE grading system takes into account adverse effects induced by RT, but also by other treatment modalities such as chemotherapy or surgery. It is important, though, to report the range of frequencies at which hearing loss occurs (high or low).

Baseline hearing assessment is fundamental to determine SNHL as a side effect of the received treatment: Pre- and post-RT audiometric evaluations of both ears should always be performed to avoid any selection bias – patients with subjective hearing complaints might be more compliant than those without. Audiological follow-up should start six months post-RT to avoid transient hearing fluctuations and continue through the years – SNHL has been characterized as late and progressive over time: A hearing loss observed in two consecutive tests could be considered as persistent (Bhandare et al., 2010).

To assess HP in VS patients, SDS should also be measured. GR scale presents some limitations concerning the range of frequencies evaluated and the descriptive definition of "serviceable" hearing (Coughlin et al., 2018). An interesting suggestion to improve hearing outcomes resolution comes from the American Academy of Otolaryngology – Head & Neck Surgery, which recommends to report baseline PTA (in dB, at 0.5–1–2–3 kHz) versus SDS (in %) on a scattergram. Any change in post-treatment hearing is reported in 10 dB or 10% intervals for PTA and SDS, respectively (Gurgel et al., 2012).

Future Directions

Larger prospective studies with pre- and post-treatment hearing assessment are needed in order to establish an absolute relationship between SNHL and RT dose to the auditory system, with special attention to cisplatin-based chemoradiation therapy.

Prospective investigation is also required to determine the effects of hypofractionation in VS treatments and when it actually is a preferable option compared with SRS: VS are slow-growing benign tumors with low α/β ratio and not typically hypoxic, so they would not benefit from fractionation. But it could help OARs in repairing sublethal radiation-induced damage and provide a more conformal dose distribution around irregularly shaped targets. As well, a more robust relationship with the dose received by cochlea and acoustic nerve should be sought.

EYE: INTRODUCTION

Pathogenesis and Clinical Significance of Radiation-Induced Ocular Toxicity

The optic nerve is the most commonly affected cranial nerve in patients treated for head–neck benign or malignant tumors. Radiation-induced optic neuropathy (RION) is one of the most disabling late complications in these patients and usually arises as a sudden, profound and irreversible vision loss. RION is a rare complication for conventional treatments, when maximum doses to the optic nerves and chiasm are constrained below 55 Gy, but it becomes a common side effect in the treatment of aggressive eye tumors, such as uveal melanomas. In these cases, the optic disc is often very close or included into the Planning Target Volume (PTV), which is irradiated at extreme hypofractionated regimens (usually at fractions of 12–14 Gy/fr and prescribed doses between 60–70 Gy).

RION can be unilateral, for single optic nerve injury, or bilateral for irradiations very close to the chiasm, with simultaneous or sequential appearance. Partial irradiation of the chiasm or the proximal optic nerves can also result in visual field defects.

A proper diagnosis of RION should be performed with ophthalmologic evaluations both before and after RT, along with long-term ophthalmology follow-up (Bhandare et al., 2011). Clinical manifestations can include pallor of the optic nerve or disc, or hyperemic areas with microaneurism on the optic disc, while intravenous fluorescein angiography can detect filling defects or perfusion delays in the optic disc. However, ophthalmology detection of RION at the early stage is particularly difficult in case of lesions affecting the posterior segment of the optic nerve or the chiasm. That is why the diagnosis is often made by presumption, when vision loss in irradiated patients cannot be related to other causes.

The CTCAE v5.0 defines different grades of optic nerve disorders according to the decrease in visual acuity: grade 1 for asymptomatic disorder which is evident at clinical or diagnostic observations only; grade 2 for moderate decrease in visual acuity (best corrected visual acuity 20/40 and better or 3 lines or less decreased vision from known baseline); grade 3 for marked decrease in visual acuity (best corrected visual acuity worse than 20/40 or more than 3 lines of decreased vision from known baseline, up to 20/200); grade 4 in case of best corrected visual acuity of 20/200 or worse in the affected eye. Note that CTCAE v5.0 only added clarifications to grades 2, 3 and 4 with respect to version 4.0.

The majority of most recent publications defined toxicity endpoints according to or identifiable with the CTCAE v4.0 definitions. A few authors, in addition to visual acuity tests, included also evaluations of visual field defects through campimetry tests (Astradsson et al., 2014; Farzin et al., 2016; Leavitt et al., 2013; Pollock et al., 2014).

As stated above, uncertainties in RION toxicity scores based on visual acuity tests can arise for irradiations of tumors close to the anterior visual pathway, where vision loss can be related to other causes, such as retinopathy, cataract, dry eye. In that case, ophthalmology examinations are highly recommended.

Optic nerves and chiasm are the critical structures for RION and for their volume definition we refer to QUANTEC (Mayo et al., 2010a) and to the more recent publication (Brouwer et al., 2015).

Short Summary of the QUANTEC Recommendations (Ocular Toxicity)

QUANTEC recommendations on RION were mainly based on clinical trials with photon and particle therapies at conventional or hyperfractionated schedules (mainly 1–2 Gy/fr,) at total doses between 50–75 Gy, or with single-fraction stereotactic surgeries at 8–30 Gy. There was a lack of data in the hypofractionated regions, with fraction sizes higher than 3 Gy at any total doses (Mayo et al., 2010a).

They selected more than 20 studies documenting the dose to the optic structures, and reported that the two most important risk factors for the incidence of RION were the total dose to the optic nerves or chiasm and the fraction size. The incidence of toxicity showed a probabilistic component, since some patients receiving greater doses did not show complications, but it had a steep increase above 60 Gy.

In most publications, the maximum dose to the optic structures was the only dosimetric data reported, however some studies indicated a volume effect for the optic structures, finding more toxicities in patients with higher mean doses but not in those with higher maximum doses. Data on volume dependence were scarce, however they suggested that a parallel architecture might be present in the very small volume range (<1–3 mm), where different fibers of the nerves and chiasm are associated with specific parts of the visual field.

Some dose limits were suggested. The authors observed that the older Emami estimate, that 50 Gy provided 5% probability of blindness within 5 years, was too conservative (Emami et al., 1991), and reported that a limit of 50 Gy in conventional fractionations was closer to an almost zero incidence. A maximum dose below 55 Gy gave a risk below 3%, while doses in the region 55–60 Gy provided a slightly increased risk about 3–7%. A more substantial risk, about 70–20% was estimated for doses above 60 Gy. However, there was a lack of data in the regimen of very high doses.

In case of single-fraction treatments, the risk of RION was estimated to be rare for D_{max} <8 Gy, to increase in the range 8–12 Gy and to reach a predicted estimate risk >10% for doses between 12–15 Gy.

For particles, the risk of RION was suggested to be low for D_{max} to the optic structures below 54 Gy (RBE) at conventional or moderately hypofractionated schedules. Note that also for protons there were no data on highly aggressive regimens.

In two studies increasing age was found as a risk factor for RION, whilst data on other possible clinical factors were inconsistent.

Critical Review of the Literature 2009–2017 (Conventional Fractionation, Ocular Toxicity)

Conventional schedules are commonly employed for the treatment of skull base, sinonasal or pituitary tumors (see Table 7.3). The analyzed publications for photon and proton treatments report prescribed doses between 54–70 Gy and 53–74 Gy (RBE), respectively, and limit the maximum dose to the optic structures below 55 Gy. All the selected works were based on median follow-up periods longer than 3 years and reported RION toxicity rates that are substantially in agreement with the very low estimates provided in QUANTEC.

Photons

In 2012 Duprez et al. (2012) did not found any radiation-induced blindness in a cohort of 86 patients, with at least 6 months of follow-up, treated with IMRT at 70 Gy for sinonasal tumors. For this population the dose constraint to the optic nerve and chiasm, with a planning organ at risk volume (PRV) expansion of 2 mm, was set at $V_{60Gy} < 5\%$.

Lower prescription doses (54Gy) were employed in a group of 94 patients treated for benign anterior skull base meningioma or pituitary adenoma reported in (Astradsson et al., 2014). The median maximum doses to the combined optic apparatus was almost equal to the prescribed dose, and direct contact of the tumor with the anterior visual pathway was present in 61.5% and 22% of patients in the meningioma and pituitary adenoma groups, respectively. Mild to moderate RION was observed in 11/94 (11.7%) patients and only one patient (1.1%) displayed marked visual acuity loss. Cox regressions models, with

TABLE 7.3 Selected Studies for Radiation-Induced Optic Neuropathy After Photon or Proton Beam Therapies. All Doses Are Expressed in Gy, for Photon Therapies, or in Gy (RBE) for Proton Therapies

Author (ref.), Patients, Median Follow-up (months)	Disease (Patients)	Technique	Toxicity Endpoint/ Ophthalmologic Evaluation	Prescription Dose [Range] (Patients)	Dose per Fraction	D_{max} to Optic Nerves and Chiasm	Incidence
Duprez 130, 52 [15–121]	Adenocarcinoma (82), SCC (23)	IMRT	CTCAE v.3 blindness Late LENT-SOMA/ Ophthalm	70 Gy [a]	2 Gy	$V_{60Gy} < 5\%$ [b]	0/86 (0%) acute grade 4; 1/86 (1.1%) late grade 3
Astradsson 94, 65 [32–127] 82 [36–161]	Benign anterior skull base meningioma (39) pituitary adenoma (55)	SRT[c] (Linac)	Visual acuity loss of at least two Snellen lines; loss of more than 1/24 of visual fields /Ophthalm	54 Gy	1.8–2 Gy	53.3 Gy [6.6–63.6] 54.6 Gy[17.4–64.2]	4/39 (10%) grade ≥2 meningioma, 7/55 (13%) grade ≥2 Pituitary adenoma, 1/94 (1.1%) grade 4
Farzin 213, 75 [2–159]	Meningioma	SRT[d] (Linac)	Visual acuity loss and/ or visual field defect in the absence of other causes /Ophthalm[e]	54 Gy [15–60] / in 1–31 fr	1.75–17 gy	Chiasm EQD2Gy: 48.6 Gy [0.0–53.8][f]; RON EQD2Gy: 48.4 Gy [0–65.3][f] LON EQD2Gy : 49.1 Gy [0–58.3][f]	2/190 (1.1%) grade ≥2 for patients treated with more than 5 fractions
Askoxylakis 122, 36	Sinonasal tumors: cystic carcinoma (47), SCC(26)	IMRT	Acute/late visual impairment and photophobia (not primary endpoint)	64 Gy	1.8–2 Gy	Chiasm: 42.2 Gy [3.4–66.4]; RON: 51.6 Gy [0.7–75.2]; LON: 49.2 Gy [3.4–74.4]	5/122 (4%), acute 8/122 (6.6%), late grade ≥2
Ares 64, 38 [14–92]	Skull-base chordomas (42) chondrosarcoma (22)	Protons	CTCAE v3.0	73.5 Gy [67–74] (42) 68.4 Gy [63–74] (22)	1.8–2 Gy	<60 Gy	1/64 (1.6%) grade 3 1/64 (1.6%) grade 4
Weber 77, 69 [5–191]	Skull-base chondrosarcoma	Protons	CTCAE v4.0	70 gy [64.0–76.0]	1.8–2 Gy	<60 Gy	1/77 (1.3%) grade 4
Weber 222, 50 [4–176]	chordoma (151) chondrosarcoma (71)	Protons	CTCAE v4.0	72.5±2.2 Gy	1.8–2 Gy	<60 Gy	5/222 (2.3%) grade 3 2/222 (0.9%) grade 4
Murray 96, 57 [12–207]	Intracranial meningiomas	Protons	CTCAE v4.0	WHO grade 1: 54 Gy [50-64] WHO grade 2–3: 62 Gy [54–68]	1.8–2 Gy [g]	<60 Gy	2/96 (2.1%) grade 3 1/96 (1.0%) grade 4

(Continued)

TABLE 7.3 (CONTINUED) Selected Studies for Radiation-Induced Optic Neuropathy After Photon or Proton Beam Therapies. All Doses Are Expressed in Gy, for Photon Therapies, or in Gy (RBE) for Proton Therapies

Author (ref.) Patients, Median Follow-up (months)	Disease (Patients)	Technique	Toxicity Endpoint/ Ophthalmologic Evaluation	Prescription Dose [Range] (Patients)	Dose per Fraction	D_{max} to Optic Nerves and Chiasm	Incidence
Iwata 43, 40 [12–92]	Craniopharyngioma	SRT (CyberKnife)	CTCAE v4.0 / Ophthalm	14 Gy in1 fr; 16 Gy in 2 fr; 21 Gy in 3 fr; 25 Gy in5 fr	5–14 Gy	<21/ Gy in 3 fr; <25 Gy in 5 fr	0/43 (0%) grade 4
Leavitt 222, 83 [4–238]	Benign skull base tumor adjacent to the anterior visual pathway	SRT (Gamma Knife)	VA loss > 2 lines and/or worsening VF mean deviation by 2 dB after SRT in the absence of tumor growth /Ophthalm	18 Gy [12–30][h]	18 Gy [12-30]	≤8 Gy (126)[i]; 8.1–10 Gy (39)[i]; 10.1–12 Gy (47)[i]; >12 Gy (10)[i]	0/126 (0%), grade ≥2; 0/39 (0%), grade ≥2; 0/47 (0%), grade ≥2; 1/10 (10%) grade ≥2
Pollock 133, 32 [19–43][1]	Pituitary adenoma	SRT (Gamma Knife)	VA loss > 2 lines and/or worsening VF mean deviation by 2 dB after SRT in the absence of tumor growth /Ophthalm	17.0 Gy [15–25][l,m]	17.0 Gy[15–25]	9.2 [6.9-10.8][l]	0/266[a] (0%) grade ≥2
Hiniker 262, 37 [2–142]	Perioptic tumors	SRT (CyberKnife)	CTCAE/ Ophthalm	18 Gy [12–25] in 1 fr (47); 24 Gy [18–33] in 3 fr (111); 25 Gy [18–40] in 5 fr (66)	18 Gy (47); 8 Gy (111); 5 Gy (66)	Optic nerve: 7.6 Gy [1.9–12.4] in 1 fr; 13.4 Gy [2.7–23.3] in 3 fr; 19.6 Gy [3.8–29.4] in 5 fr	2/262 (0.8%) grade 4
Dunavoelgyi 73, 90 [68–98]	choroidal melanoma	SRT (Linac)	CTCAE v4.0 / Ophthalm	70 Gy (57); 60 Gy (16)	14 Gy; 12 Gy	Not reported[o]	24 (32.9%) grade 0; 1 (1.4%) grade 1; 22 (30.1%) grade 2; 21 (28.9%) grade 3; 5/73 (6.8%) grade 4
Zenda 90, 58 [12–163]	nasal cavity, para-nasal sinus, skull base tumors	Protons	CTCAE v4.0	65 Gy[p]	2.5 Gy	<54 Gy[q]	9/90 (10%) grade ≥2; 4/90 (4.4%) grade 4

(Continued)

TABLE 7.3 (CONTINUED) Selected Studies for Radiation-Induced Optic Neuropathy After Photon or Proton Beam Therapies. All Doses Are Expressed in Gy, for Photon Therapies, or in Gy (RBE) for Proton Therapies

Author (ref.), Patients, Median Follow-up (months)	Disease (Patients)	Technique	Toxicity Endpoint/ Ophthalmologic Evaluation	Prescription Dose [Range] (Patients)	Dose per Fraction	D_{max} to Optic Nerves and Chiasm	Incidence
Kim 93, 66 [6–156]	Parapapillary choroidal melanoma	Protons	Radiation papillopathy[f], visual acuity <20/200/ Ophthalm	70 Gy	14 Gy[g]	>50 Gy[t]	63/93 (67.7%) 77% grade 4 at 3 years in patients with papillopathy
Tran 59, 63 [4–131]	Peripapillary choroidal melanoma,	Protons	Best-corrected visual acuity <20/200 / Ophthalm	54 Gy (19 pts) 60 Gy (40 pts)	13.5 Gy 15 Gy	Not reported	Optic neuropathy: 64% at 5 years Visual loss: 83% grade 4 at 5 years in patients with baseline >20/200
Riechardt 147, 78 [4–140]	Parapapillary choroidal melanoma	Protons	Optic neuropathy[u], visual acuity loss <20/200 /Ophthalm	60 Gy	15 Gy[v]	60 Gy [50-60][s]	Optic neuropathy: 78.1% at 3 years 89.6% at 5 years Visual loss: 86% grade 4 at 10 years
Seibel 1127, 46 [12–170]	Choroidal melanoma	Protons	Optic neuropathy[z] / Ophthalm	60 Gy	15 Gy	6 Gy [0-60]	463/1127 (41%) 39% at 3 years 52% at 5 years 74% at 10 years

(Continued)

TABLE 7.3 (CONTINUED) Selected Studies for Radiation-Induced Optic Neuropathy After Photon or Proton Beam Therapies. All Doses Are Expressed in Gy, for Photon Therapies, or in Gy (RBE) for Proton Therapies

Author (ref.), Patients, Median Follow-up (months)	Disease (Patients)	Technique	Toxicity Endpoint/ Ophthalmologic Evaluation	Prescription Dose [Range] (Patients)	Dose per Fraction	D_{max} to Optic Nerves and Chiasm	Incidence
Thariat 865/53 [6–240]	Parapapillary uveal melanoma within 2mm from the optic disc	Protons	Optic neuropathy, visual acuity loss of at least 0.3 logMAR / Ophthalm	57.2 Gy	14.3 Gy	Not reported	Optic neuropathy: 47.5% Visual loss: 61.7% grade ≥3

RON, Right Optic Nerve; LON, Left Optic Nerve; EQD2Gy, Equivalent Dose in 2Gy per fraction with α/β = 3Gy; SRT, Stereotactic Radiation Therapy; fr, fraction; CTCAE, Common Terminology Criteria for Adverse Events; LENT-SOMA Late Effects of Normal Tissues - Subjective Objective Management Analysis; IMRT, Intensity Modulated Radiation Therapy; V_{xGy}, organ volume receiving more than X Gy; WHO, World Health Organization: VA, visual acuity, VF, visual field.

a Prescribed to 117 patients, in a small group the prescribed dose was 60–66 Gy.
b Dose prescribed to the PRV with an expansion of 2mm.
c 75 patients were treated with 3-dimensional radiotherapy and 19 with Intensity Modulated Radiation Therapy.
d Stereotactic Radiation Therapy was employed for most patients (205).
e Ophthalmological examination only after visual problems.
f In 140 patients at least one of the optic structures was inside the Planning Target Volume, with maximum doses >50Gy (EQD2Gy).
g Hypothesized from previous publications.
h Median tumor margin dose.
i Doses to the anterior visual pathway.
l Interquartile range.
m Median margin dose.
n 266 sides were considered in the analysis.
o The authors analyze the effects of minimum doses enclosing the optic disc.
p Most common schedule. Note that the grade 4 toxicities refer to 3/61 (4.9%) patients treated at 65 GyE in 26 fractions and to 1/14 (7%) patient treated at 60GyE.
q Constraint declared for schedules at 2.5Gy/fraction. However priority was given to target coverage.
r Radiation papillopathy defined as nerve fiber layer infarcts, hemorrhage, exudates, edema or pallor involving the disc.
s Five fractions on 7–10 days.
t Maximum dose to the optic disc.
u Optic neuropathy defined as edema, pallor or hemorrhage of the optic disc.
v Three patients were treated with 8 fractions of 7.5CGE, one patient at 50CGE in 8 fractions of 6.25 CGE.
z Optic neuropathy defined as disc edema, radiation-induced disc pallor, parapapillary hemorrhages or exudates or optice nerve fiber layer infarction.

Bonferroni corrected p-values, did not find any association of RION events with maximum radiation dose to the combined optic structure (COS: optic nerves, chiasm and tracts), dose per fraction, number of operations performed pre-RT, age, diagnosis or tumor size.

Farzin et al. (2016) reported a retrospective analysis on 232 patients with meningioma mainly treated with stereotactic radiotherapy. Various schedules were employed (from 1 to 31 fractions), however median maximum doses to neuro-optic structures were about 48–49 Gy, expressed in EQD2Gy with $\alpha/\beta = 3$ Gy, and in 140 patients at least one of the neuro-optic structures was inside the target volume, and thus received more than 50 Gy in EQD2Gy. A visual impairment, in the absence of other causes, was found in only 2/190 patients (1.1%), considering only the 190 patients treated with more than 5 fractions. The two cases had both optic nerve and chiasm inside the PTV, but this condition was present in all the patients with OARs included in the PTV, suggesting that a probabilistic component exists.

Recently, Askoxylakis et al. (2016) reported a retrospective analysis of 122 patients treated with conventional IMRT at a median total dose of 64 Gy for malignant sinonasal tumors. The evaluation of RION was not the primary endpoint, however a mild vision reduction was seen in 8/122 (6.6%) patients.

Protons

Results on conventionally fractionated proton therapies are mainly based on cohorts of patients treated with spot-scanning proton beam therapy at the Paul Scherrer Institute (PSI), in Switzerland.

Ares et al. (2009) analyzed an old cohort of 64 patients with skull-base chordomas and chondrosarcomas treated between 1998 and 2005. Chordomas received a dose of 73.5 (67–74) Gy (RBE) and chondrosarcomas 68.4 (63–74) Gy (RBE), while dose to the optic nerves and chiasm was limited at 60 Gy (RBE). After a mean follow-up of 38 months, they reported the onset of one (1.6%) grade 3 optic neuropathy at 20 months after the treatment and one (1.6%) grade 4 optic neuropathy at 12 months after the treatment.

Two updated analyses on patients treated between 1998–2014 and between 1998–2012 substantially confirmed the previous results (Weber et al. 2016a,b). After a mean follow-up of 69 months, in a cohort of 77 patients treated for low-grade chondrosarcomas, Weber et al. (2016) reported one case of grade 4 optic neuropathy (1.3%). While in a larger cohort of 222 patients, the same authors reported 5/222 (2.3%) grade 3 unilateral optic neuropathies and 2/222 (0.9%) grade 4 bilateral optic neuropathies (Weber et al., 2016b).

Murray et al. (2017) reported the long-term clinical outcomes of 92 patients with non-benign intracranial meningiomas always treated at the PSI center. Median prescribed doses were equal to 54 Gy (RBE) for WHO grade 1 tumors, and to 62 Gy (RBE) for WHO grade 2–3 tumors. After a median follow-up of 57 months, they reported 2/96 (2.1%) grade 3 optic nerve disorders and 1/96 (1.0%) grade 4 optic nerve edema.

Critical Review of the Literature 2009–2017 (Hypo Fractionation, Ocular Toxicity)

With respect to QUANTEC, more results on hypofractionated treatments have been published.

Moderate schedules are usually employed for the treatment of skull base, sinonasal or pituitary tumors, while much more aggressive schedules are reserved for melanomas.

Stereotactic photon therapies have been performed with linear accelerators, Gamma Knife or CyberKnife.

Photons

Iwata et al., did not find any grade 4 optic nerve disorder in a small cohort of 43 patients with craniopharingioma treated with CyberKnife, after a follow-up of 40 months (Iwata et al., 2012). In this population, the majority of patients were treated with hypofractionated RT at 13–25 Gy in 2–5 fractions. Radiation doses were prescribed at the PTV margin and corresponded to the maximum doses allowed for the organs at risk (optic nerve, chiasm, brainstem).

Leavitt et al. (2013) published a retrospective analysis of 222 patients treated between 1991 and 1999 with single-fraction Gamma Knife for benign skull-base tumors adjacent to the anterior visual pathway (AVP). After mean clinical and imaging follow-up periods of 83 and 123 months, none of the 212 patients who received less than 12 Gy to the AVP experienced mild to moderate RION, whilst it was recorded in 1 of the 10 patients receiving more than 12 Gy.

The same group (Pollock et al., 2014) published a dose-volume analysis on a more recent cohort of 133 patients (266 sides) with pituitary adenomas treated with Gamma Knife at median margin doses equal to 17.0 Gy. Median maximum doses to the AVP were equal to 9 Gy and, after a median follow-up of 32 months, no patient had a documented decline in the visual function or other neurological deficit after RT. The authors also evaluated the volumes receiving at least 8, 10 or 12 Gy, that were equal to 15.8 (3.7–36.2) mm^3; 1.6 (0.5–5.3) mm^3 and 0.1 (0.1–0.6) mm^3, respectively.

Very low rates of grade 4 optic nerve disorders were also found in Hiniker (2016). Only two patients (0.8%) resulted to have experienced unilateral vision loss in the absence of other causes after a retrospective analysis on 262 patients with perioptic tumors within 3 mm from the optic nerves or chiasm. Median prescribed doses varied between 18, 24 and 25 Gy given in 1, 3 or 5 fractions, respectively. Median maximum dose to the optic nerve were 8, 13 and 20 Gy in 1, 3 and 5 fractions, respectively. Note that the first patient had a sellar meningioma and received a maximum dose to the optic nerve equal to 23.9 Gy in 5 fractions. He experienced gradual onset of vision loss 3 months after RT. The second patient received radiosurgery after two previous courses of external beam RT and stereotactic radiosurgery (20 Gy in 1 fraction) and surgery to a right parasellar grade I meningioma. Maximum dose to the optic pathway was 27.7 Gy, vision loss became complete seven months after radiosurgery.

Higher toxicity rates were found in (Dunavoelgyi et al., 2013) on a cohort of 73 patients with choroidal melanoma, treated with hypofractionated stereotactic RT at high doses: 70 Gy or 60 Gy in 5 fractions. After 90 months of follow-up, 49/73 (67%) of patients developed an optic neuropathy of any grade, whilst 48/73 (66%) and 5/73 (6.8%) developed at least grade 2 and grade 4 optic neuropathies, respectively. The authors proposed dose-response curves for the risk of any grade and at least grade 2 optic neuropathy, that were established by considering the minimum dose enclosing the optic nerve circumference. The curves showed a steep increase in the risk of developing at least a grade 2 toxicity: 20% at 2.1 Gy and 50% at 14.7 Gy of EQD2Gy doses (calculated with α/β = 3 Gy). Age, length of follow-up and inclusion of the optic nerve in the 30% isodose were also found as independent risk factors for optic neuropathy of any grade.

Protons

Zenda (2015) published a retrospective analysis on a group of 90 patients treated with moderate hypofractionation between 1999 and 2008 at Chiba (Japan) for malignant sinonasal and skull-base tumors. Prescribed doses were equal to 65 Gy (RBE) at 2.5 Gy (RBE)/fr in 76 patients, and to 60 Gy (RBE) at 4 Gy (RBE)/fr in 14 mucosal melanoma patients. Doses to the optic nerve of the healthy side and chiasm were limited at 54 Gy (RBE) at 2.5 Gy (RBE)/fr. After a median observation period of 58 months, 4 (4.4%), 1 (1.1%) and 4 (4.4%) patients reported late grade 2, grade 3 and grade 4 optic nerve disorders. The patients with grade 4 disorder had small chiasm and optic nerve volumes receiving >50 Gy (RBE), that were equal to 0.03 and 0.5 cm^3; the smallest one for the only patient, in this group, treated at 60 Gy (RBE) in 15 fractions. The time to onset of the optic disorder was also shorter for this patient (17 months) versus 51–74 months for the other three patients.

More aggressive hypofractionated regimens are commonly employed in the treatment of eye melanomas, that usually receive 60–70 Gy (RBE) in four/five high dose fractions of 14–15 Gy (RBE). Optic neuropathy is a common side effect of these treatments and it often appears with retinopathy, both of them affecting visual acuity impairment.

Four of the reported analyses focused on parapapillary melanomas located within 2 mm or 1-disc distance from the optic disc (Kim et al., 2010; Tran et al., 2012; Riechardt et al., 2014; Thariat et al., 2016). In these cases, the optic nerve receives very high doses, usually above 50 Gy (RBE) in 4 or 5 fractions, corresponding to more than 130–150 Gy in EQD2Gy with α/β = 3 Gy. In these conditions, RION has been recorded in 50–90% of patients at 5 years of follow-up.

Kim et al. (2010) evaluated 93 patients treated for parapapillary melanoma. Diagnosis of radiation papillopathy was made by ophthalmologic examination if there was evidence of nerve fiber layer infarcts, hemorrhage, exudate, edema or pallor involving the disc. After a median follow-up of 5.5 years, they found 63/93 (67.7%) cases of optic neuropathy and an estimated vision loss lower than 20/200 in 48%, 71% and 77% of patients with optic neuropathy at 1, 3 and 5 years, respectively. The authors concluded that vision loss was not inevitable in the presence of radiation optic neuropathy, since more than 40% of patients retained counting fingers or better vision at 5 years after the diagnosis of papillopathy.

Similar results were found by Tran et al. (2012), who investigated the clinical outcomes of 59 patients treated between 1995 and 2007 in Canada. Nineteen patients received 54 Gy (RBE) and 40 received 60 Gy (RBE) in 4 fractions. After a median follow-up of 63 months, optic neuropathy was a common treatment-related effect. The actuarial 2 year and 5 year rates of optic neuropathy were estimated to be 45.4% and 63.6%, respectively. Among the 31 patients evaluable at 5 years who had pre-treatment best-corrected visual acuity >20/200, only 17% of patients had visual acuity >20/200.

Riechardt et al. (2014) evaluated 147 patients treated in the Helmoholtz Center in Berlin with 60 Gy (RBE) in 4 consecutive days, and who received a minimum of 50 Gy (RBE) to the optic disc. The authors included early signs of optic neuropathy, that was defined as edema, pallor or hemorrhage of the optic disc, and estimated actuarial 5 year and 10 year rates equal to 89.6% and 94.2%, respectively. The onset was early: at 19 months ± 14 months (mean ± standard deviation). Before treatment median visual acuity was equal to 20/50, but visual deterioration took place between the second and the fifth year after proton therapy, with the median acuity level dropping from 20/200 to 20/500. At 10 years of follow-up 86% of patients had visual acuity <20/200.

An updated analysis on a larger cohort, not limited to parapapillary melanomas, was published by Seibel et al. (2016), who evaluated 1127 patients with uveal melanoma treated between 1998 and 2014 at the same center. In this case median dose to the optic nerve was lower and equal to 6.0 (0–60) Gy (RBE). Focusing again on early signs of optic neuropathy, it was recorded in 463/1127 (41.0%) patients and appeared at 19.8 (0.2–170.4) months after proton beam therapy (PTB). The chance of having optic neuropathy at 5 and 10 years was estimated to be 52% and 74%. Vision was better preserved in the group of 127 patients who did not develop any posterior segment complication: These patients had an initial visual acuity of 0.3 logMAR (20/40) and a final acuity of 0.4 logMAR (20/50). While the patients with optic neuropathy had similar visual acuity of 0.3 logMAR (20/40) before PBT but a significantly reduced visual acuity equal to 1.5 logMAR (20/630) at the most recent follow-up. The authors reported also that in a subgroup of 88 patients with tumor located further than 2.5 mm from the fovea and optic disc only 6/88 (6.8%) patients developed radiation optic neuropathy after 21.4 (12.0–55.8) months. Optic neuropathy was found to be the major cause of visual loss, and the major cause of optic neuropathy was irradiation of the optic disc.

A recent analysis of Papakostas et al. (2017), not reported in Table, confirmed a high rate of grade 4 vision loss (about 84% at 5 years) in a cohort of patients treated with PTB for large choroidal melanomas at 70 Gy (RBE) in 5 fractions. They found that decreased visual acuity depended on the proximity of the tumor to the optic nerve and fovea.

Thariat et al. (2016) reported visual outcomes of 865 patients with parapapillary uveal melanomas treated with PBT at Nice. Prescribed doses were equal to 57.2 Gy (RBE) in 4 fractions of 14.3 Gy (RBE). Radiation optic neuropathy was found in 47.5% of patients and was found correlated with a visual acuity loss of at least 0.3 logMAR. The authors found that visual acuity was associated with age, diabetes, initial visual acuity and optic nerve length and suggested that patients with full dose to 100% or ≥80% of papillary surface could be spared some vision through optimisation of the optic nerve length irradiation. Visual acuity was better preserved when 50% isodose (28.6 Gy (RBE)) could be limited to less than 2 mm of nerve length.

Summary for DVH Constraints/NTCP Model Parameters (Ocular Toxicity)

The constraints suggested by QUANTEC are substantially confirmed by the recent publications on conventional treatments. For maximum doses to the optic nerves and chiasm below 55 Gy and fraction sizes below 2 Gy, grade 4 optic nerve disorder is a rare event, with incidence below 1.1%, while the risk of at least grade 2 disorder is about to 4–10%.

Similar results are found for conventional treatments with protons. For maximum doses to the optic nerves and chiasm below 60 Gy (RBE), and fraction sizes below 2 Gy (RBE), the risk of grade 4 and at least grade 3 optic nerve disorders are below 1.6% and 4%, respectively.

Very low toxicity rates are also found in the publications on moderately hypofractionated photon therapies, where all the EQD2Gy doses to the optic structures were maintained below the limit of 54 Gy calculated for any α/β between 2 and 10 Gy.

The results of the only reported publication on moderate hypofractionation in proton therapies provide slightly higher incidence rates for the risk of at least grade 2, at least grade 3 or grade 4 optic nerve disorders, that can be estimated to be 10%, 5.5% and 4%, respectively. But the Authors declared that priority was given to target coverage, so that maximum EQD2Gy doses to the optic nerves could have been reached values equal to 70–85 Gy (RBE) (calculated with α/β = 3 Gy).

New interesting results were provided by the extreme hypofractionated regimens employed in the treatment of uveal melanoma treatments, when the optic disc is often within or close to the PTV.

For EQD2Gy doses to the optic disc between 150–180 Gy (RBE), the risk of RION incidence can be estimated to be about 65–70%, while for EQD2Gy doses above 200 Gy (RBE) an incidence risk of 65–90% can been estimated. Note that higher incidence rates of grade 4 visual acuity loss have been reported: about 50–80% between 150–180 Gy (RBE) in EQD2Gy, and above 80% for doses. This might be due to the presence of other confounding factors, such as retinopathy.

In Figure 7.1, we provide an attempt of NTCP curve for RION and grade 4 visual acuity loss as a function of EQD2Gy to the optic structures, considering both photon and proton therapies. The estimated D_{50} (dose providing the 50% of risk) for RION is about 175 Gy, while the steepness m of the curve at the D_{50} value is about 0.0367 Gy^{-1}. Instead, the estimated D_{50} and m for grade 4 visual acuity loss are equal to 161 Gy and 0.0378 Gy^{-1}, with a slightly higher steepness reflecting the higher incidence rates recorded. Those curves are comparable with the NTCP curves proposed in (Hiniker et al., 2016) for EQD2Gy doses below 80 Gy. Note that in this publication they reported only two toxicity events and no

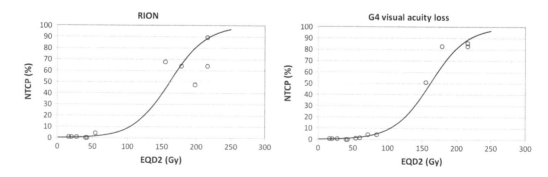

FIGURE 7.1   In bold line, the best-fit Normal Tissue Complication Probability (NTCP) curves for the risk of optic neuropathy, on the left, and grade 4 visual acuity loss, on the right, as a function of maximum dose (D_{max}) to the optic structures expressed in Equivalent Dose to 2 Gy per fraction (EQD2Gy, computed with α/β = 3 Gy). Dots represent the published values of toxicity incidences.

data above the 80 Gy. Interesting dose-response curves were also provided by Dunavoelgyi et al. (2013), but they were not easily comparable with the NTCP curves estimated here, since they were expressed in terms of minimum doses enclosing the optic disc.

Publications showed some evidence that the volume or the length of the optic nerve involved in the medium-high doses might be a risk factor, but more studies are needed to confirm this effect.

Analogously, there is emerging evidence for the risk role of age, diabetes, and initial visual acuity but the reported results are inconsistent.

Open Issues and Future Directions of Investigation (Ocular Toxicity)

More detailed dose-volume analyses are highly desirable, especially in the case of extreme hypofractionated schedules, when high toxicity events are present and when doses to the optic disc span a wide range of values. Particular attention should also be paid to distinguishing visual acuity loss caused by optic nerve (or chiasm) disorders from other eye impairments. The reported publications, show that in the treatment of uveal melanomas the incidence of grade 4 visual losses is often higher than the reported incidence of RION. This might be due to the presence of other eye toxicities but also to the difficulty of detecting early disorders of the optic nerve. An interesting recent work, on a small cohort of patients, suggested the employment of a more objective measure, to establish early damage to the optic nerve (Ozkaya Akagunduz et al., 2017). The authors evaluated the Visual Evoked Potential (VEP) latency, that corresponds to the transmission time of the visual stimulus from the optic nerve to the occipital cortex, which might provide information about optic neuritis before the onset of visual symptoms. In this work, the patients with radiation-induced optic neuropathy showed the worst VEP latency values, and the authors found an interesting correlation between the prolongation of VEP latency and the maximum or median doses received by the optic nerve. In addition, there were some patients with VEP latency values above the normative ones without ophthalmologic evaluation of RION, thus suggesting that VEP latency defects might provide an earlier detection of the optic nerve disorder. Other possible improvements in the definition of RION could come from optical coherence tomography and MRI.

Then NTCP curves provided here were developed by converting all the reported maximum doses to the optic structures in terms of EQD2Gy dose by making use of the linear quadratic model. However Mayo et al. (2010a) observed that other models, such as the Neuret one, might provide a more accurate description of the effects of fraction size on toxicity risk. Future studies might provide a better understanding of the role of fractionation and, possibly, provide new models including also other dose descriptors or significant clinical factors.

REFERENCES

Ares C et al. 2009. "Effectiveness and Safety of Spot Scanning Proton Radiation Therapy for Chordomas and Chondrosarcomas of the Skull Base: First Long-Term Report." *Int J Radiat Oncol Biol Phys*, 75(4):1111–8. doi:10.1016/j.ijrobp.2008.12.055.

Askoxylakis V et al. 2016. "Intensity Modulated Radiation Therapy (IMRT) for Sinonasal Tumors: A Single Center Long-Term Clinical Analysis." *Radiat Oncol*, 11(1):1–9. doi:10.1186/s13014-016-0595-9.

Astradsson A et al. 2014. "Visual Outcome after Fractionated Stereotactic Radiation Therapy of Benign Anterior Skull Base Tumors." *J Neurooncol,* 118(1):101–8. doi:10.1007/s11060-014-1399-0.

Au KH et al. 2018. "Treatment Outcomes of Nasopharyngeal Carcinoma in Modern Era after Intensity Modulated Radiotherapy (IMRT) in Hong Kong: A Report of 3328 Patients (HKNPCSG 1301 Study)." *Oral Oncol,* 77:16–21. doi:10.1016/j.oraloncology.2017.12.004.

Bhandare N et al. 2011. Optic Nerve, Eye, and Ocular Adnexa. In *Human Radiation Injury,* ed. Schrieve DC, Loeffler JS, 190–209. Lippincott Willliams & Wilkins, Philadelphia.

Bhandare N et al. 2009. "Radiation Effects on the Auditory and Vestibular Systems." *Otolaryngol Clin North Am,* 42(4):623–34. doi:10.1016/j.otc.2009.04.002.

Bhandare N et al. 2010. "Radiation Therapy and Hearing Loss." *Int J Radiat Oncol Biol Phys,* 76 (3 Suppl): 50–7. doi:10.1016/j.ijrobp.2009.04.096.

Brouwer CL et al. 2015. "CT-Based Delineation of Organs at Risk in the Head and Neck Region: DAHANCA, EORTC, GORTEC, HKNPCSG, NCIC CTG, NCRI, NRG Oncology and TROG Consensus Guidelines." *Radiother Oncol,* 117(1):83–90. doi:10.1016/j.radonc.2015.07.041.

Brown M et al. 2011. "Predictors of Hearing Loss After Gamma Knife Radiosurgery for Vestibular Schwannomas: Age, Cochlear Dose, and Tumor Coverage." *Neurosurgery,* 69(3):605–14. doi:10.1227/NEU.0b013e31821a42f3.

Chan SH et al. 2009. "Sensorineural Hearing Loss After Treatment of Nasopharyngeal Carcinoma: A Longitudinal Analysis." *Int J Radiat Oncol Biol Phys,* 73(5):1335–42. doi:10.1016/j.ijrobp.2008.07.034.

Cheraghi S et al. 2017. "Normal Tissue Complication Probability Modeling of Radiation-Induced Sensorineural Hearing Loss after Head-and-Neck Radiation Therapy." *Int J Radiat Biol,* 93(12):1327–33. doi:10.1080/09553002.2017.1385872.

Combs SE et al. 2015. "Long-Term Outcome after Highly Advanced Single-Dose or Fractionated Radiotherapy in Patients with Vestibular Schwannomas - Pooled Results from 3 Large German Centers." *Radiother Oncol,* 114(3):378–83. doi:10.1016/j.radonc.2015.01.011.

Coughlin AR et al. 2018. "Systematic Review of Hearing Preservation after Radiotherapy for Vestibular Schwannoma." *Otol Neurotol,* 39(3):273–83. doi:10.1097/MAO.0000000000001672.

CTCAE: National Cancer Institute. Common Terminology Criteria for Adverse Events. Available from: www.ctep.cancer.gov

De Felice F and Blanchard P 2017. "Radiation-Induced Neurocognitive Dysfunction in Head and Neck Cancer Patients." *Tumori,* 103(4): 319–24. doi:10.5301/tj.5000678.

Debus J et al. 1997. "Brainstem Tolerance to Conformal Radiotherapy of Skull Base Tumors." *Int J Radiat Oncol Biol Phys,* 39(5):967–75. doi:https://doi.org/10.1016/S0360-3016(97)00364-7.

Du C et al. 2015. "Concurrent Chemoradiotherapy was Associated with a Higher Severe Late Toxicity Rate in Nasopharyngeal Carcinoma Patients Compared with Radiotherapy Alone: A Meta-Analysis Based on Randomized Controlled Trials." *Radiat Oncol,* 10:70. doi:10.1186/s13014-015-0377-9.

Dunavoelgyi R et al. 2013. "Dose-Response of Critical Structures in the Posterior Eye Segment to Hypofractioned Stereotactic Photon Radiotherapy of Choroidal Melanoma." *Radiother Oncol,* 108(2):348–53. doi:10.1016/j.radonc.2013.08.018.

Duprez F et al. 2012. "IMRT for Sinonasal Tumors Minimizes Severe Late Ocular Toxicity and Preserves Disease Control and Survival." *Int J Radiat Oncol Biol Phys,* 83(1): 252–9. doi:10.1016/j.ijrobp.2011.06.1977.

Emami B et al. 1991. "Tolerance of Normal Tissue to Therapeutic Irradiation." *Int J Radiat Oncol Biol Phys,* 21(1):109–22.

Ésik O et al. 2003. "Increased Metabolic Activity in the Spinal Cord of Patients with Long-Standing Lhermitte's Sign." *Strahlenther Onkol,* 179(10):690–3. doi:10.1007/s00066-003-1115-z.

Farzin M et al. 2016. "Optic Toxicity in Radiation Treatment of Meningioma: A Retrospective Study in 213 Patients." *J Neurooncol,* 127(3):597–606. doi:10.1007/s11060-016-2071-7.

Gao J et al. 2015. "SmartArc-Based Volumetric Modulated Arc Therapy Can Improve the Middle Ear, Vestibule and Cochlea Sparing for Locoregionally Advanced Nasopharyngeal Carcinoma: A Dosimetric Comparison with Step-and-Shoot Intensity Modulated Radiotherapy." *Br J Radiol*, 88(1053):20150052. doi:10.1259/bjr.20150052.

Gardner G and Robertson JH. 1988. "Hearing Preservation in Unilateral Acoustic Neuroma Surgery." *Ann Otol Rhinol Laryngol*, 97(1):55–66. doi:10.1177/000348948809700110.

Gérard M et al. 2017. "Contraintes de Dose En Radiothérapie Conformationnelle Fractionnée et En Radiothérapie Stéréotaxique Dans Les Hippocampes, Le Tronc Cérébral et L'encéphale : Limites et Perspectives." *Cancer Radiother*, 21(6–7):636–47. doi:10.1016/j.canrad.2017.08.108.

Gondi V et al. 2013. "Hippocampal Dosimetry Predicts Neurocognitive Function Impairment after Fractionated Stereotactic Radiotherapy for Benign or Low-Grade Adult Brain Tumors." *Int J Radiat Oncol Biol Phys*, 85(2):345–54. doi:10.1016/j.ijrobp.2012.11.031.

Gondi V et al. 2010. "Estimated Risk of Perihippocampal Disease Progression after Hippocampal Avoidance during Whole-Brain Radiotherapy: Safety Profile for RTOG 0933." *Radiother Oncol*, 95(3):327–31. doi:10.1016/j.radonc.2010.02.030.

Gondi V et al. 2014. "Preservation of Memory with Conformal Avoidance of the Hippocampal Neural Stem-Cell Compartment during Whole-Brain Radiotherapy for Brain Metastases (RTOG 0933): A Phase II Multi-Institutional Trial." *J Clin Oncol*, 32(34):3810–6. doi:10.1200/JCO.2014.57.2909.

Grimm J et al. 2016. "Estimated Risk Level of Unified Stereotactic Body Radiation Therapy Dose Tolerance Limits for Spinal Cord." *Semin Radiat Oncol*, 26(2):165–71. doi:https://doi.org/10.1016/j.semradonc.2015.11.010.

Gurgel RK et al. 2012. "A New Standardized Format for Reporting Hearing Outcome in Clinical Trials." *Otolaryngol Head Neck Surg*, 147(5):803–7. doi:10.1177/0194599812458401.

Hayden Gephart MG et al. 2013. "Cochlea Radiation Dose Correlates with Hearing Loss after Stereotactic Radiosurgery of Vestibular Schwannoma." *World Neurosurg*, 80(3–4):359–63. doi:10.1016/j.wneu.2012.04.001.

Hiniker S et al. 2016. "Dose-Response Modeling of the Visual Pathway Tolerance to Single-Fraction and Hypofractionated Stereotactic Radiosurgery." *Semin Radiat Oncol*, 26(2):97–104. doi:10.1016/j.semradonc.2015.11.008.

Holliday EB et al. 2016. "Dosimetric Advantages of Intensity-Modulated Proton Therapy for Oropharyngeal Cancer Compared with Intensity-Modulated Radiation: A Case-Matched Control Analysis." *Med Dosim*, 41(3):189–94. doi:10.1016/j.meddos.2016.01.002.

Hsiao KY et al. 2010. "Cognitive Function Before and After Intensity-Modulated Radiation Therapy in Patients With Nasopharyngeal Carcinoma: A Prospective Study." *Int J Radiat Oncol Biol Phys*, 77(3):722–6. doi:10.1016/j.ijrobp.2009.06.080.

Hwang CF et al. 2015. "Hearing Assessment after Treatment of Nasopharyngeal Carcinoma with CRT and IMRT Techniques." *Biomed Res Int*, 2015:769806 . doi:10.1155/2015/769806.

Indelicato DJ et al. 2014. "Incidence and Dosimetric Parameters of Pediatric Brainstem Toxicity Following Proton Therapy." *Acta Oncol*, 53(10):1298–304. doi:10.3109/0284186X.2014.957414.

Iwata H et al. 2012. "Single and Hypofractionated Stereotactic Radiotherapy with CyberKnife for Craniopharyngioma." *J Neurooncol*, 106(3):571–7. doi: 10.1007/s11060-011-0693-3.

Jackson A et al. 2010. "The Lessons of QUANTEC: Recommendations for Reporting and Gathering Data on Dose-Volume Dependencies of Treatment Outcome." *Int J Radiat Oncol Biol Phys*, 76(3 Suppl):155–60. doi:10.1016/j.ijrobp.2009.08.074.

Jacob JT et al. 2014. "Significance of Cochlear Dose in the Radiosurgical Treatment of Vestibular Schwannoma: Controversies and Unanswered Questions." *Neurosurgery*, 74(5):466–74. doi:10.1227/NEU.0000000000000299.

Kano H et al. 2009. "Predictors of Hearing Preservation after Stereotactic Radiosurgery for Acoustic Neuroma." *J Neurosurg*, 111(4):863–73. doi:10.3171/2008.12.JNS08611.

Katsoulakis E et al. 2017. "A Detailed Dosimetric Analysis of Spinal Cord Tolerance in High-Dose Spine Radiosurgery." *Int J Radiat Oncol Biol Phys*, 99(3):598–607. doi:10.1016/j.ijrobp.2017.05.053.

Kim IK et al. 2010. "Natural History of Radiation Papillopathy after Proton Beam Irradiation of Parapapillary Melanoma." *Ophthalmology*,117(8):1617–22. doi:https://doi.org/10.1016/j.ophtha.2009.12.015.

Kirkpatrick JP et al. 2010. "Radiation Dose–Volume Effects in the Spinal Cord." *Int J Radiat Oncol Biol Phys*, 76(3): S42–9. doi:https://doi.org/10.1016/j.ijrobp.2009.04.095.

Landier W. 2016. "Ototoxicity and Cancer Therapy." *Cancer*, 122(11):1647–58. doi:10.1002/cncr.29779.

Lawrence YR et al. 2010. "Radiation Dose-Volume Effects in the Brain." *Int J Radiat Oncol Biol Phys*, 76(3 Suppl):20–7. doi:10.1016/j.ijrobp.2009.02.091.

Le Rhun E et al. 2016. "Radionecrosis after Stereotactic Radiotherapy for Brain Metastases." *Expert Rev Neurother*, 16(8):903–14. doi:10.1080/14737175.2016.1184572.

Leavitt JA et al. 2013. "Long-Term Evaluation of Radiation-Induced Optic Neuropathy after Single-Fraction Stereotactic Radiosurgery." *Int J Radiat Oncol Biol Phys*, 87(3):524–7. doi:10.1016/j.ijrobp.2013.06.2047

Lee AWM et al. 2002. "Factors Affecting Risk of Symptomatic Temporal Lobe Necrosis: Significance of Fractional Dose and Treatment Time." *Int J Radiat Oncol Biol Phys*, 53(1):75–85. doi:10.1016/S0360-3016(02)02711-6.

Low WK et al. 2006. "Sensorineural Hearing Loss After Radiotherapy and Chemoradiotherapy: A Single, Blinded, Randomized Study." *J Clin Oncol*, 24(12):1904–9. doi:10.1200/JCO.2005.05.0096.

Ma Q et al. 2016. "Radiation-Induced Functional Connectivity Alterations in Nasopharyngeal Carcinoma Patients with Radiotherapy." *Medicine (Baltimore)*, 95(29):e4275. doi:10.1097/MD.0000000000004275.

Marucci L et al. 2004. "Spinal Cord Tolerance to High-Dose Fractionated 3D Conformal Proton-Photon Irradiation as Evaluated by Equivalent Uniform Dose and Dose Volume Histogram Analysis." *Int J Radiat Oncol Biol Phys*, 59(2):551–5. doi:10.1016/j.ijrobp.2003.10.058.

Mayo C et al. 2010a. "Radiation Dose-Volume Effects of Optic Nerves and Chiasm." *Int J Radiat Oncol Biol Phys*, 76(3 Suppl): 28–35. doi:10.1016/j.ijrobp.2009.07.1753.

Mayo C et al. 2010b. "Radiation Associated Brainstem Injury." *Int J Radiat Oncol Biol Phys*, 76(3 Suppl):36–41. doi:10.1016/j.ijrobp.2009.08.078.

McDonald MW et al. 2015. "Dose-volume Relationships Associated with Temporal Lobe Radiation Necrosis after Skull Base Proton Beam Therapy." *Int J Radiat Oncol Biol Phys*, 91(2):261–7. doi:10.1016/j.ijrobp.2014.10.011.

Mul VEM et al. 2012. "Lhermitte Sign and Myelopathy after Irradiation of the Cervical Spinal Cord in Radiotherapy Treatment of Head and Neck Cancer." *Strahlenther Onkol*, 188(1):71–6. doi:10.1007/s00066-011-0010-2.

Murphy ES and Suh JH. 2011. "Radiotherapy for Vestibular Schwannomas: A Critical Review." *Int J Radiat Oncol Biol Phys*, 79(4):985–97. doi:10.1016/j.ijrobp.2010.10.010.

Murray FR et al. 2017. "Long-Term Clinical Outcomes of Pencil Beam Scanning Proton Therapy for Benign and Non-Benign Intracranial Meningiomas." *Int J Radiat Oncol Biol Phys*, 99(5): 1190–8. doi:10.1016/j.ijrobp.2017.08.005.

Ozkaya Akagunduz O et al. 2017. "Evaluation of the Radiation Dose–Volume Effects of Optic Nerves and Chiasm by Psychophysical, Electrophysiologic Tests, and Optical Coherence Tomography in Nasopharyngeal Carcinoma." *Technol Cancer Res Treat*, 16(6):969–77. doi:10.1177/1533034617711613.

Papakostas TD et al. 2017. "Long-Term Outcomes after Proton Beam Irradiation in Patients with Large Choroidal Melanomas." *JAMA Ophthalmol*, 135(11):1191–6. doi:10.1001/jamaophthalmol.2017.3805.

Parham K et al. 2011. "Challenges and Opportunities in Presbycusis." *Otolaryngol Head Neck Surg*, 144(4):491–5. doi:10.1177/0194599810395079.

Pehlivan B et al. 2012. "Temporal Lobe Toxicity Analysis after Proton Radiation Therapy for Skull Base Tumors." *Int J Radiat Oncol Biol Phys*, 83(5):1432–40. doi:10.1016/j.ijrobp.2011.10.042.

Perry A and Schmidt RE. 2006. "Cancer Therapy-Associated CNS Neuropathology: An Update and Review of the Literature." *Acta Neuropathol*, 111(3):197–212. doi:10.1007/s00401-005-0023-y.

Petsuksiri J et al. 2011. "Sensorineural Hearing Loss after Concurrent Chemoradiotherapy in Nasopharyngeal Cancer Patients." *Radiat Oncol*, 6(1):2–9. doi:10.1186/1748-717X-6-19.

Pollock BE et al. 2014. "Dose-Volume Analysis of Radiation-Induced Optic Neuropathy after Single-Fraction Stereotactic Radiosurgery." *Neurosurgery*, 75(4):456–60. doi:10.1227/NEU.0000000000000457.

Raaijmakers E and Engelen AM. 2002. "Is Sensorineural Hearing Loss a Possible Side Effect of Nasopharyngeal and Parotid Irradiation? A Systematic Review of the Literature." *Radiother Oncol*, 65(1):1–7. doi:10.1016/S0167-8140(02)00211-6.

Rashid A et al. 2016. "Multisession Radiosurgery for Hearing Preservation." *Semin Radiat Oncol*, 26(2):105–11. doi:10.1016/j.semradonc.2015.11.004.

Riechardt AI et al. 2014. "Proton Beam Therapy of Parapapillary Choroidal Melanoma." *Am J Ophthalmol.*, 157(6):1258–65. doi:10.1016/j.ajo.2014.02.032.

Ruben JD et al. 2006. "Cerebral Radiation Necrosis: Incidence, Outcomes, and Risk Factors with Emphasis on Radiation Parameters and Chemotherapy." *Int J Radiat Oncol Biol Phys*, 65(2):499–508. doi:10.1016/j.ijrobp.2005.12.002.

Sahgal A et al. 2013. "Probabilities of Radiation Myelopathy Specific to Stereotactic Body Radiation Therapy to Guide Safe Practice." *Int J Radiat Oncol Biol Phys*, 85(2):341–7. doi:10.1016/j.ijrobp.2012.05.007.

Schlampp I et al. 2011. "Temporal Lobe Reactions after Radiotherapy with Carbon Ions: Incidence and Estimation of the Relative Biological Effectiveness by the Local Effect Model." *Int J Radiat Oncol Biol Phys*, 80(3):815–23. doi:10.1016/j.ijrobp.2010.03.001.

Scoccianti S et al. 2015. "Organs at Risk in the Brain and their Dose-Constraints in Adults and in Children: A Radiation Oncologist's Guide for Delineation in Everyday Practice." *Radiother Oncol*, 114(2):230–8. doi:10.1016/j.radonc.2015.01.016.

Seibel I et al. 2016. "Predictive Risk Factors for Radiation Retinopathy and Optic Neuropathy after Proton Beam Therapy for Uveal Melanoma." *Graefes Arch Clin Exp Ophthalmol*, 254(9):1787–92. doi:10.1007/s00417-016-3429-4.

Stieb S et al. 2018. "Long-Term Clinical Safety of High-Dose Proton Radiation Therapy Delivered With Pencil Beam Scanning Technique for Extracranial Chordomas and Chondrosarcomas in Adult Patients: Clinical Evidence of Spinal Cord Tolerance." *Int J Radiat Oncol Biol Phys*, 100(1):218–25. doi:10.1016/j.ijrobp.2017.08.037.

Su SF et al. 2013. "Analysis of Dosimetric Factors Associated with Temporal Lobe Necrosis (TLN) in Patients with Nasopharyngeal Carcinoma (NPC) after Intensity Modulated Radiotherapy." *Radiat Oncol*, 8(1):1–8. doi:10.1186/1748-717X-8-17.

Sun Y et al. 2013. "Radiation-Induced Temporal Lobe Injury after Intensity Modulated Radiotherapy in Nasopharyngeal Carcinoma Patients: A Dose-Volume-Outcome Analysis." *BMC Cancer*, 13:397. doi:10.1186/1471-2407-13-397.

Thariat J et al. 2016. "Visual Outcomes of Parapapillary Uveal Melanomas Following Proton Beam Therapy." *Int J Radiat Oncol Biol Phys*, 95(1):328–35. doi:10.1016/j.ijrobp.2015.12.011.

Theunissen EAR et al. 2014. "Cochlea Sparing Effects of Intensity Modulated Radiation Therapy in Head and Neck Cancers Patients: A Long-Term Follow-up Study." *J Otolaryngol Head Neck Surg*, 43(1):30. doi:10.1186/s40463-014-0030-x.

Theunissen EAR et al. 2015. "Prediction of Hearing Loss due to Cisplatin Chemoradiotherapy." *JAMA Otolaryngol Head Neck Surg*, 141(9):810–5. doi:10.1001/jamaoto.2015.1515.

Timmer FCA et al. 2009. "Gamma Knife Radiosurgery for Vestibular Schwannomas: Results of Hearing Preservation in Relation to the Cochlear Radiation Dose." *Laryngoscope,* 119(6): 1076–81. doi:10.1002/lary.20245.

Timmerman RD. 2008. "An Overview of Hypofractionation and Introduction to This Issue of Seminars in Radiation Oncology." *Semin Radiat Oncol,* 18(4):215–22. doi:10.1016/j. semradonc.2008.04.001.

Tran E et al. 2012. "Outcomes of Proton Radiation Therapy for Peripapillary Choroidal Melanoma at the BC Cancer Agency." *Int J Radiat Oncol Biol Phys,* 83(5):1425–31. doi:10.1016/j. ijrobp.2011.10.017.

Tsai JT et al. 2013. "Clinical Evaluation of Cyber Knife in the Treatment of Vestibular Schwannomas." *Biomed Res Int,* 2013(2): 297093. doi:10.1155/2013/297093.

Tsao MN et al. 2017. "Stereotactic Radiosurgery for Vestibular Schwannoma : International Stereotactic Radiosurgery Society (ISRS) Practice Guideline." *J Radiosurg SBRT,* 5(1):5–24. PubMed PMID: 29296459

Wang J et al. 2015. "Sensorineural Hearing Loss after Combined Intensity Modulated Radiation Therapy and Cisplatin-Based Chemotherapy for Nasopharyngeal Carcinoma." *Transl Oncol,* 8(6):456–62. doi:10.1016/j.tranon.2015.10.003.

Weber DC et al. 2005. "Results of Spot-Scanning Proton Radiation Therapy for Chordoma and Chondrosarcoma of the Skull Base: The Paul Scherrer Institut Experience." *Int J Radiat Oncol Biol Phys,* 63(2):401–9. doi:10.1016/j.ijrobp.2005.02.023.

Weber DC et al. 2016a. "Long Term Outcomes of Patients with Skull-Base Low-Grade Chondrosarcoma and Chordoma Patients Treated with Pencil Beam Scanning Proton Therapy." *Radiother Oncol,* 120(1):169–74. doi:10.1016/j.radonc.2016.05.011.

Weber DC et al. 2016b. "Long-Term Outcomes and Prognostic Factors of Skull-Base Chondrosarcoma Patients Treated with Pencil-Beam Scanning Proton Therapy at the Paul Scherrer Institute." *Neuro Oncol,* 18(2):236–43. doi:10.1093/neuonc/nov154.

Wei Y et al. 2014. "Long-Term Outcome of Sensorineural Hearing Loss in Nasopharyngeal Carcinoma Patients: Comparison Between Treatment with Radiotherapy Alone and Chemoradiotherapy." *Cell Biochem Biophys,* 69(3):433–7. doi:10.1007/s12013-014-9814-x.

Xue J et al. 2012. "Dose-Volume Effects on Brainstem Dose Tolerance in Radiosurgery." *J Neurosurg,* 117(Suppl): 189–96. doi:10.3171/2012.7.GKS12962.

Yang I et al. 2010. "A Comprehensive Analysis of Hearing Preservation after Radiosurgery for Vestibular Schwannoma." *J Neurosurg,* 112(4):851–9. doi:10.3171/2009.8.JNS0985.

Yeh SA et al. 2005. "Treatment Outcomes and Late Complications of 849 Patients with Nasopharyngeal Carcinoma Treated with Radiotherapy Alone." *Int J Radiat Oncol Biol Phys,* 62(3):672–9. doi:10.1016/j.ijrobp.2004.11.002.

Zenda S et al. 2015. "Late Toxicity of Proton Beam Therapy for Patients with the Nasal Cavity, Para-Nasal Sinuses, or Involving the Skull Base Malignancy: Importance of Long-Term Follow-Up." *Int J Clin Oncol,* 20(3):447–54. doi:10.1007/s10147-014-0737-8.

Zuur CL et al. 2009. "Risk Factors for Hearing Loss in Patients Treated With Intensity-Modulated Radiotherapy for Head-and-Neck Tumors." *Int J Radiat Oncol Biol Phys,* 74(2):490–6. doi:10.1016/j.ijrobp.2008.08.011.

CHAPTER 8

Head and Neck

Parotids

Maria Thor and Joseph O. Deasy

CONTENTS

Introduction and Short Summary of the QUANTEC Recommendations 207
Critical Review of the Literature 2009–2017: Dose-Volume Histogram-Based
Response Relationships 208
    Xerostomia Tolerance Doses after 3DCRT 208
    Xerostomia Tolerance Doses after IMRT 209
Dose-Volume Histogram-Based Response Relationships: Summary and
Perspectives 210
Critical Review of the Literature 2009–2017: Spatial Dose-Response Relationships 210
Spatial Dose-Response Relationships: Summary and Perspectives 212
References 212

INTRODUCTION AND SHORT SUMMARY OF THE QUANTEC RECOMMENDATIONS

The parotid glands are the major salivary glands, and associated injuries following head and neck cancer (HNC) radiotherapy (RT) are primarily attributed to insufficient salivary production, i.e., xerostomia (Deasy, 2010). Xerostomia typically peaks within a one year window after completed RT (Moiseenko, 2012; Lee, 2013; Thor et al., 2017), and is often, though not always, followed by significant recovery of capacity. Based on dose-response studies published up until 2009, the salivary gland-specific Quantitative Analyses of Normal Tissue Effects in the Clinic (QUANTEC) report advised that severe xerostomia (<25% function compared to baseline) is likely to be mitigated if the mean dose to at least one parotid gland is less than 20 Gy, or if the mean dose to both parotids is at least 25 Gy (Deasy, 2010). These recommendations were made largely based on dose-volume histogram (DVH) data, and included RT regimens, such as Intensity Modulated Radiation Therapy (IMRT) treatments, commonly used in the time frame 1992–2009.

The scope of this parotid-focused chapter is to summarize findings on parotid gland response to radiation dating from 2010, i.e., after the publication of the QUANTEC report. The chapter is further stratified into two sections: The first section is a direct continuation of the synthesized QUANTEC report data in that it summarizes dose-response findings based on DVH parameterizations, whereas the second section focuses on spatial dose-response findings. For an overview of image-derived xerostomia models based on either established quantitative imaging, or radiomics, the reader is referred to the associated sections in chapter 16 *"Quantitative imaging for assessing and predicting toxicity"*. In this chapter, a post-QUANTEC study was included if reporting a positive finding, being available in English and in full-text, including ≥50 (ideally 100) patients, and if xerostomia had been assessed within one year after completed treatment either by patients or physicians, or alternatively using salivary flow measurements (ideally accounting for baseline salivary function).

CRITICAL REVIEW OF THE LITERATURE 2009–2017: DOSE-VOLUME HISTOGRAM-BASED RESPONSE RELATIONSHIPS

Xerostomia rates following IMRT are generally significantly lower compared to those observed after three-dimensional conformal RT (3DCRT), e.g., 29% vs. 83% grade ≥2 xerostomia as demonstrated in the phase III randomized control trial reported by Nutting et al. (2011). Given this trend, and the inherently different dose distribution patterns between these two techniques, the following section is, therefore, to the extent possible, further divided into two technique-specific tolerance dose sections.

Xerostomia Tolerance Doses after 3DCRT

In one of the first studies following the parotid-specific QUANTEC report, Dijkema et al. (2010) successfully fitted a dose-response curve (mean dose to both parotids) to xerostomia one year after 3DCRT (<25% flow rate compared to baseline; 3DCRT: the majority of patients) for two institutions including 222 HNC patients. The best-fit model parameters in the combined cohort were a dose, to both parotid glands together, resulting in 50% risk of xerostomia, D_{50}, of 40 Gy and a steepness of the response curve, m, of 0.40. The volume dependence parameter, n, was 1, i.e., the mean dose, and this value was also confirmed within unrestricted n value analysis.

In another series of 165 HNC patients from two institutions, patient-assessed xerostomia (grade ≥ 2) six months after 3DCRT was explained by age, baseline xerostomia, as well as the mean dose to the combined parotids (area under the receiver-operating characteristic curve, AUC = 0.82 (Beetz, 2012a; 105:86–93). The resulting formula was a logistic equation of −5.3 + (0.1*Parotids mean dose) + (Age*0.1) + (Baseline xerostomia*0.9).

In a two-institutional study including 95 HNC patients treated with either 3DCRT or IMRT (similar distributions), Moiseenko et al. (2012) explored the 20 Gy guideline as recommended in the QUANTEC report (Deasy, 2010), but modified it slightly: spare at least one parotid gland to 20 Gy or less. This modified guideline was deemed suitable to prevent xerostomia (<25% flow rate compared to baseline) one year after treatment in both datasets since it resulted in <20% xerostomia rate, with an excellent negative predicted value (NPV = 93%) in the combined data. The associated D_{50} values for the fitted dose-response curves in both datasets were slightly lower than those found in the study by Dijkema et al. (2010) (D_{50} = 32 Gy, 35 Gy vs. 40 Gy).

In another subsequent study by Beetz et al. (2012b; 105:94–100), their earlier proposed 3DCRT-based model (age, baseline xerostomia, and the mean dose to the combined parotids) (Beetz, 2012c; 105:86–93) was explored in an external IMRT cohort consisting of 162 HNC patients investigating the same xerostomia endpoint. Although differences between the two cohorts were accounted for, both discrimination and calibration of the 3DCRT model in the IMRT cohort were significantly worse (3DCRT vs. 3DCRT in IMRT: AUC = 0.82 vs. 0.66; Hosmer-Lemeshow test p-value = 0.47 vs. 0.04). A re-fit of the same variables in the IMRT cohort improved calibration somewhat, but not discrimination (AUC = 0.66; Hosmer-Lemeshow test p-value = 0.79). The rate of xerostomia was similar between the IMRT and the 3DCRT cohort (51% vs. 52%), and the authors stated that lack of generalizability was likely due to the lack of correlation between the paired parotid mean doses in the IMRT, which was not the case in the 3DCRT cohort (Beetz, 2012a; 105:86–93). This likely implies that the dose to each parotid should be studied individually when analyzing xerostomia after IMRT, but to some extent these results are not well understood.

Xerostomia Tolerance Doses after IMRT

Following the unsuccessful generalizability of the 3DCRT model in the IMRT cohort in (Beetz, 2012a; 105:86–93), the authors undertook a re-modeling analysis within a slightly extended IMRT cohort (N = 178) (Beetz, 2012c; 105:101–6). Using the same six-month patient-assessed xerostomia endpoint, but separating the two parotids, generated a best-fit two-variable multivariate model in which mean dose to both parotids as observed in the 3DCRT cohort was now replaced by the mean dose to the contralateral parotid, and baseline xerostomia status still remained (logistic function of −1.4 + [0.05*Contralateral parotid mean dose] + [Baseline xerostomia*0.7]). The authors proposed that this model was considerably better than the previously re-fitted 3DCRT model, but this could be somewhat questioned given a modest increase in AUC from 0.66 to 0.68. Regardless, the majority of dose-response studies succeeding the study by Beetz et al. (2012c; 105:101–6) have focused on individual parotid doses rather than using a paired dose across the two parotids. One such example is the study by Lee et al. (2013), which further explored the efficacy of the 20 Gy/25 Gy QUANTEC guideline in a HNC cohort (N = 212) for xerostomia (grade≥3 patient-assessed; patients with baseline xerostomia excluded) at three months and at 12 months after completed treatment. Best-fit model parameters for both endpoints were D_{50} = 38 Gy, 44 Gy, and m = 0.59, 0.48 implying a recovery of salivary function over time (AUC = 0.68 vs. 0.73; Hosmer-Lemeshow test p-value = 0.67 vs. 0.72; N = 212 vs. 135). Based on this model and tying it back to the QUANTEC 20 Gy/25 Gy guideline the xerostomia rate was 28% and 19% at three months and at 12 months, respectively. The corresponding rates for a modified 20 Gy/20 Gy guideline were 22% and 13%.

Thor et al. (2017) performed a dose-response analysis in a two-institutional cohort of 121 HNC patients with a training/validation data split. The mean dose to the contralateral parotid, and the mean dose to the ipsilateral parotid predicted xerostomia at three months (<25% flow compared to baseline), this model was generalizable between training and validation (AUC = 0.90 and 0.81), and the associated equation was a logistic function of −8.1 + (0.2*Contralateral parotid mean dose) + (0.1*Ipsilateral parotid mean dose). The identified parotid mean dose critical thresholds were somewhat lower than those suggested by QUANTEC: <20 Gy to the contralateral parotid, while ~25 Gy to the ipsilateral parotid (Thor et al., 2017) compared to ~25 Gy for both glands in the QUANTEC guideline (Deasy, 2010).

Consequently, the xerostomia rates were somewhat lower than that at the one- and two-gland QUANTEC guidelines: ≤25% vs. 29% and ≤16% vs. 27%, respectively. Interestingly, in one of the most recently published studies including 252 primarily oropharyngeal patients (73%), the paired parotid mean dose was predictive of xerostomia (mean summary score; baseline status not accounted for) 1–60 months post-IMRT (Hawkins et al., 2018). Also, incorporating the mean dose to the oral cavity (OC), as well as the mean dose to the contralateral subman-dibular gland (cSMG) generated their best model. Hence, and as previously also postulated by others (Eisbruch, 2016), although the parotids are the major salivary glands, xerostomia pathophysiology likely involves the interplay between all major and minor glands. Further, Hawkins et al. (2018) noticed that patients with a bilateral parotid mean dose of 27 Gy, a mean OC dose of 30 Gy, and a mean cSMG dose of 30 Gy had a better xerostomia status after treatment than those with higher mean doses for the corresponding structures, implying a role of submandibular gland irradiation, which has been an area of controversy.

DOSE-VOLUME HISTOGRAM-BASED RESPONSE RELATIONSHIPS: SUMMARY AND PERSPECTIVES

The rate of xerostomia after IMRT may be significantly lower compared to that after 3DCRT (Nutting et al., 2011). This together with dose distribution differences between 3DCRT and IMRT could explain the lack of generalizability of 3DCRT-based models in IMRT cohorts (Beetz, 2012c; 105:101–6). To further reduce xerostomia, there are suggestions that it would be feasible to decrease parotid mean doses beyond the QUANTEC guidelines (Moiseenko, 2012; Lee, 2013; Thor et al., 2017; Gabrys et al., 2017). Xerostomia has primarily been mod-eled using parotid mean doses. Focusing in addition on an inter-organ interplay (Hawkins et al., 2018) while simultaneously incorporating patient and treatment factors, and spatial patterns (cf. next section) should be a goal of future dose-response efforts. An overview of all abovementioned DVH-based dose-response relationships can be found in Table 8.1.

CRITICAL REVIEW OF THE LITERATURE 2009–2017: SPATIAL DOSE-RESPONSE RELATIONSHIPS

In a study from van Luijk et al. (2015), a hypothesis was pursued focused on intra-parotid variability in dose-response. They found, in rat pre-clinical experiments, that the region of the parotid gland most enriched for stem and progenitor cells was that of the major ducts. In the associated clinical analysis, including 73 HNC patients, they found that the dose to the dorsal edge of the mandible, i.e., a region containing the Stensen's duct branching, bet-ter predicted xerostomia (<25% flow compared to baseline [Moiseenko, 2012]) compared to the mean dose of the entire parotid ($R_s = 0.65$ vs. 0.60 [van Luijk, 2015]).

These findings have spurred numerous spatial dose-response efforts. For instance, even pre-dating the study by van Luijk et al. (2015), Buettner et al. (2012) quantified spatial parotid dose distributions along the anterior-posterior, medial-lateral, and the cranio-cau-dal direction using scale-invariant 3D statistical moments. In essence, 10 so-called moments were generated, which describe the variance and the skewness of the dose along the three directions. Incorporating these moments into modeling of xerostomia (physician-assessed grade ≥2) one year after treatment in 111 HNC patients (rate: 48% [Nutting et al., 2011]),

TABLE 8.1 Summary of Dose-Volume-Histogram (DVH)-Based Dose-Response Relationships for Xerostomia Stratified with Respect to Modeling Approach (Lyman Model, Dose-Volume, and Other Variables, or Guideline)

Lyman	Ref	RT Technique	Gland	$n/D_{50}/m$			
	Dijkema et al. (2010)	3DCRT	Bilat	1/40/0.40			
	Lee et al. (2013) 3 m[a]	IMRT	Bilat	1/38/0.59			
Dose/ Volume	**Ref**	**RT Technique**	**Gland**	**Variable (β)**			**Intercept**
	Beetz et al. (2012a)	3DCRT	Bilat	D_{mean} (0.1)	Age (0.1)	Baseline X (0.9)	−5.3
	Beetz et al. (2012c)	IMRT	Contra	D_{mean} (0.05)		Baseline X (0.7)	−1.4
	Thor et al. (2017) 3 m	IMRT	Contra/ Ipsi	D_{mean} (0.2)/ D_{mean} (0.1)			−8.1
Guideline	**Ref**	**RT Technique**	**Gland**	**Guideline**			
	Deasy et al. (2010)	3DCRT/ IMRT	Bilat	$D_{mean} \leq 25$ Gy			
	Deasy et al. (2010)	3DCRT/IMRT	Contra/ Ipsi	$D_{mean} \leq 20$ Gy/ $D_{mean} \leq 25$ Gy			
	Moiseenko et al. (2012)	3DCRT/IMRT	Contra/ Ipsi	$D_{mean} \leq 20$ Gy/ $D_{mean} \leq 20$ Gy			

3 m = Three months; Bilat = Bilateral; Contra = Contralateral; D_{mean} = Mean dose; Ipsi = Ipsilateral; X = Xerostomia; 3DCRT= Three Dimensional Conformal Radiation Therapy; IMRT= Intensity Modulated Radiation Therapy; RT= radiotherapy.
[a] Estimated at the 20 Gy/25 Gy QUANTEC guideline; All β values are logistic regression coefficients.

moment-inclusive models resulted in overall better predictability compared to mean dose only models (Buettner et al., 2012): The best 3DCRT model included one moment that described the skewness of the dose to the deep ipsilateral lobe, plus one describing the skewness of the dose in the cranio-caudal direction, and tumor site (AUC = 0.78). The best IMRT model also included the skewness of the dose in the cranio-caudal direction in addition to the relative concentration of dose in the caudal medial part of the deep ipsilateral lobe, as well as the mean dose to the superficial contralateral lobe and the clinical variable of surgical removal of the submandibular gland (AUC = 0.88).

In an associated effort, Clark et al. (2015) divided the parotid of 102 HNC patients into medial and lateral regions, an inner core, and an outer shell. Models including the mean dose to each of these segments were compared to those including the mean dose to the whole gland both for early and overall loss of parotid function, i.e., <25% flow compared to baseline at three and 12 months, respectively. The mean dose to the medial half of the parotid resulted in the strongest relationship with both endpoints but the improvement over using the mean dose to the whole parotid was marginal (Rs = −0.75, −0.60 vs. −0.73, −0.56).

In a recent study by Gabrys et al. (2018), spatial dose representations were included in a machine learning approach embracing ten different sampling methods, six feature selection algorithms, and seven classification algorithms. Moreover, early, late, and long-term (≤ 6 months, 6–15 months, 15–24 months; patients with baseline xerostomia excluded) grade≥ 2 physician-assessed xerostomia following IMRT in 153 mainly oropharyngeal (65%) patients was modeled using demographics, eight parotid volume 'omics' related to shape, 24 'dosiomics' (including also V_{XGy} [percent organ volume receiving at least X Gy] and $D_{Y\%}$ [minimum dose to Y% of the organ]), the mean dose to nine parotid sub volumes, three dose gradients, and 15 3D dose moments related to variance, correlation, skewness, and coskewness. Models including only mean dose, or morphology had low predictive power (AUC < 0.60), and features were further unable to predict early xerostomia. For late xerostomia, a tendency of contralateral dose-gradients in the left-right direction as well as in the antero-posterior direction was observed, but only for long-term xerostomia a significant relationship was found. Seven variables were included in the seven multivariate models (combination of sampling methods and feature selections), but only four of these variables were included in the majority of the models: the contralateral and the ipsilateral volume, the spread of the contralateral DVH, as well as contralateral parotid gland volume eccentricity (included in 5/7, 7/7, 6/7, and 4/7 models). Given the large number of variables tested, this approach requires a separate validation test.

SPATIAL DOSE-RESPONSE RELATIONSHIPS: SUMMARY AND PERSPECTIVES

A few fairly different attempts to model intra-parotid loss of function have been pursued. The study by van Luijk et al. (2015) suggested that the parotid region including the branching of the duct was the most dose sensitive region, whereas the mean parotid dose in four different segments was not found to be a better predictor compared to the mean dose to the entire gland in the study by Clark et al. (2015). The studies by Buettner et al. (2012) and Garbys et al. (2018), in which intra-parotid loss of function was primarily studied from a dose point-of-view, both indicated spatial dose-responses, but also highlighted the concurrent importance of DVH-based metrics to particular substructures along with patient-specific factors. Taken together, these studies support a spatial dose-response relationship within the parotid glands, but future large and multi-institutional studies are likely to shed further light on the existence of an intra-parotid dose-response, how it relates to a potential inter-organ dose-response, and if there is a co-existing role with DVH metrics, as well as with patient and treatment factors. Hopefully, a reliably predictive model will emerge soon, tested on multiple datasets, and that goes beyond the mean dose guidelines.

REFERENCES

Beetz I, Schilstra C, Bulage FR, Koken PW, Bijl HP, Chouvalova O, Leemans CR, de Bock GH, Christianen ME, van der Laan BF, Vissink A, Steenbakkers RJ, and Langendijk JA. Development of NTCP models for head and neck cancer: Patients treated with three-dimensional conformal radiotherapy for xerostomia and sticky saliva: the role of dosimetric and clinical factors. *Radiother Oncol*, 2012a; 105:86–93.

Beetz I, Schilstra C, van Luijk P, Christianen ME, Doornaert P, Bijl HP, Chouvalova O, van der Heuvel ER, Steenbakkers RJ, and Langendijk JA. External validation of three dimensional conformal radiotherapy based NTCP models for patient-rated xerostomia and sticky saliva among patients treated with intensity modulated radiotherapy. *Radiother Oncol*, 2012b; 105:94–100.

Beetz I, Schilstra C, van der Schaaf A, van der Heuvel ER, Doornaert P, van Luijk P, Vissink A, can der Laan BF, Leemans CR, Bijl HP, Christianen ME, Steenbakkers RJ, and Langendijk JA. NTCP models for patient-rated xerostomia and sticky saliva after treatment with intensity modulated radiotherapy for head and neck cancer: the role of dosimetric and clinical factors. *Radiother Oncol*, 2012c; 105:101–6.

Buettner F, Miah AB, Gulliford SL, Hall E, Harrington KJ, Webb S, Partridge M, and Nutting CM. Novel approaches to improve the therapeutic index of head and neck radiotherapy: an analysis of data from the PARSPORT randomized phase III trial. *Radiother Oncol*, 2012; 103:82–7.

Clark H, Hovan A, Moiseenko V, Thomas S, Wu J, and Reinsberg S. Regional radiation dose susceptibility within the parotid gland: effects on salivary loss and recovery. *Med Phys*, 2015; 42:2064–71.

Deasy JO, Moiseenko V, Marks L, Chao C, Nam J, and Eisbruch A. Radiotherapy dose-volume effects on salivary gland function. *Int J Radiat Oncol Biol Phys*, 2010; 76:58–63.

Dijkema T, Raaijmakers P, Ten Haken RK, Roesink JM, Braam PM, Houewling AC, Moerland MA, Eisbruch A, and Terhaard CHJ. Parotid gland function after radiotherapy: The combined Michigan and Utrecht experience. *Int J Radiat Oncol Biol Phys*, 2010; 78:449–53.

Eisbruch A. Can xerostomia be further reduced by sparing parotid gland cells? *Ann Transl Med*, 2016; 4 (Suppl 1):S16.

Gabryś HS, Buettner F, Sterzing F, Hauswald H, and Bangert M. Parotid gland mean dose as a xerostomia predictor in low-dose domains. *Acta Oncol*, 2017; 56:1197–203.

Gabryś HS, Buettner F, Sterzing F, Hauswald H, and Bangert M. Design and selection of machine learning methods using radiomics and dosiomics for normal tissue complication probability modeling of xerostomia. *Front Oncol*, 2018; 8:35.

Hawkins PG, Lee JY, Mao Y, Li P, Green M, Worden FP, Swiecicki PL, Mierzwa ML, Spector ME, Schipper MJ, and Eisbruch A. Sparing all salivary glands with IMRT for head and neck cancer: Longitudinal study of patient-reported xerostomia and head-and-neck quality of life. *Radiother Oncol*, 2018; 126:68–74.

Lee TF and Fang FM. Quantitative analysis of normal tissue effects in the clinic (QUANTEC) guideline validation using quality of life questionnaire datasets for parotid gland constraint to avoid causing xerostomia during head-and-neck radiotherapy. *Radiother Oncol*, 2013; 106:352–8.

Moiseenko V, Wu JH, Hovan A, Saleh Z, Apte A, Deasy JO, Harrow S, Rabuka C, Muggli A, and Thompson A. Treatment planning constraints to avoid xerostomia in head-and-neck radiotherapy: An independent test of QUANTEC criteria using a prospectively collected dataset. *Int J Radiat Oncol Biol Phys*, 2012; 82:1108–14.

Nutting CM, Morden JP, Harrington KJ, Guerrero Urbano T, Bhide SA, Clark C, Miles EA, Miah AB, Newbold K, Tanay M, Adab F, Jefferies SJ, Scrase C, Yap BK, A'Hern RP, Sydenham MA, Emson M, Hall E, and the PARSPORT trial management group. Parotid-sparing intensity modulated versus conventional radiotherapy in head and neck cancer (PARSPORT): a phase 3 multicentre randomised controlled trial. *Lancet Oncol*, 2011; 12:127–36

Thor M, Owosho A, Clark HD, Oh JH, Riaz N, Hovan A, Tsai J, Thomas SD, Yom SHK, Huryn JM, Moiseenko V, Lee NY, Estilo CL, and Deasy JO. Internal and external generalizability of temporal dose-response relationships for xerostomia following IMRT for head and neck cancer. *Radiother Oncol*, 2017; 122:200–6.

Van Luik P, Pringle S, Deasy JO, Moiseenko V, Faber H, Hovan A, Baanstra M, van der Laan HP, Kierkels RG, van der Schaaf A, Witjes MJ, Schippers JM, Brandenburg S, Langendijk JA, Wu J, and Coppes RP. Sparing the region of the salivary gland containing stem cells preserves saliva production after radiotherapy for head and neck cancer. *Sci Transl Med*, 2015; 7:305ra147.

CHAPTER 9

Head and Neck

*Larynx and Structures Involved
in Swallowing/Nutritional
Problems and Dysphonia*

Giuseppe Sanguineti

CONTENTS

Swallowing/Nutritional Problems: Introduction 216
  Clinical Significance of the Injury..216
Evaluation of Toxicity Endpoints.. 217
  Definition of the Relevant Anatomical Structures 219
Short Summary of the QUANTEC Recommendations with Relationship to Main
Literature on Which They Were Based (Swallowing/Nutritional Problems).....................220
Non-Dosimetric Factors Affecting the Risk of Dysphagia 220
Critical Review of the Literature (Swallowing/Nutritional Problems)............................... 222
  Dosimetric Predictors of Acute and Subacute Toxicity... 222
  Dosimetric Predictors of Late Toxicity 226
Prospective Validation Studies.. 228
Suggestion for DVH Constraints/NTCP Model Parameters (Swallowing/
Nutritional Problems).. 230
Mathematical/Biologic Models.. 230
Special Situations and Future Directions (Swallowing/Nutritional Problems)................. 231
Introduction: Voice Problems (Dysphonia)... 231
  Clinical Significance of the Injury.. 231
  Evaluation of Toxicity Endpoints 232
  Definition of the Relevant Anatomical Structures 232
Short Summary of the QUANTEC Recommendations with Relationship to Main
Literature on Which They Were Based (Dysphonia) 233
  Critical Review of the Literature.. 233
Suggestion for DVH Constraints/NTCP Model Parameters (Dysphonia)................. 234
References.. 234

SWALLOWING/NUTRITIONAL PROBLEMS: INTRODUCTION

Clinical Significance of the Injury

Breakdown of the swallowing function brings both the inability to intake food (either solid or liquid) (dysphagia) and the risk of food penetration in the airways rather than in the upper aerodigestive tract (penetration/aspiration). The former is associated with poor caloric nutrition and dehydration while the latter with *ab ingestis* pneumonia. Both represent potential serious clinical conditions that can be lethal, especially if underestimated.

Swallowing disturbances require parenteral nutrition (PN) or feeding tube (FT) nutrition. Moreover, they cause anxiety, depression and ultimately decrease the quality of life (Nguyen et al. 2005).

In one study, dysphagia was the covariate that mostly affected quality of life (QoL) at 1 year after treatment (Ronis et al. 2008). Dysphagia seems to affect patients' QoL stronger than xerostomia (Ramaekers et al. 2011).

More recently, chronic aspiration has been regarded as a possible cause of long-term survival deterioration of patients undergoing concomitant chemoradiotherapy (cCRT) aiming at laryngeal preservation (Forastiere et al. 2013; Ward et al. 2016). In a consecutive series of 324 patients treated with definitive cCRT for head and neck squamous cell carcinomas (HNSCC) at a single institution in Denmark, severe dysphagia was observed in 32% of patients, 29% patients developed aspiration pneumonia and 3 died from this complication (Mortensen et al. 2013).

While xerostomia rates have been shown to decrease with the introduction of technical ameliorations such as Intensity Modulated Radiation Therapy (IMRT) (Nutting et al. 2011), the prevalence of dysphagia during and after a course of (chemo)radiotherapy remains a major concern due to the incidental irradiation of the swallowing structures (Dirix et al. 2009). In one review, almost 2/3 of the patients being treated with either radiotherapy (RT) alone or in combination with chemotherapy were FT dependent at some point during treatment (Rieger et al. 2006). While only a minority of patients (<18%) remain long-term FT dependent (Rieger et al. 2006), oral feeding restoration may require a long time and up to 41% of patients have been reported to be FT dependent at 1 year (Garden et al. 2008). Another major issue is that FT during treatment is predictive of poorer long-term swallowing function (Langmore et al. 2012; Chen et al. 2010) possibly due to swallowing muscle atrophy caused by prolonged insufficient or absent oral feeding.

Feeding tube rates may be decreasing due to a higher awareness of the problem, which has prompted the use of the FT only when strictly needed or the introduction of prophylactic measures such as swallowing exercises, though their role remains controversial (Mortensen et al. 2015). One study suggested some benefits of IMRT in reducing the duration of FT dependency after treatment (Beadle et al. 2017). In another study, IMRT was found to reduce the risk of the severity of dysphagia over 3DCRT primarily through the decrease in the rates of xerostomia (Mortensen et al. 2013).

While IMRT has the potential advantage of achieving dose distributions that limit the dose to the putative organs at risk (OARs) involved in swallowing, in order to be a successful strategy, several steps (i.e., identification and contouring of appropriate OARs; selection of clinically relevant endpoints) need to be preliminary accomplished.

EVALUATION OF TOXICITY ENDPOINTS

The swallowing function can be assessed as it follows (Russi et al. 2012):

Objective (instrumental) assessment:

- Modified Barium Swallow (MBS)

- Fiberoptic Endoscopic Evaluation of Swallowing (FEES)

Subjective assessment:

- Patient-related scales, i.e., MD Anderson Dysphagia Inventory (MDADI)

- Observer-related scales (i.e., Common Terminology Criteria for Adverse Events [CTCAE], toxicity criteria of the Radiation Therapy Oncology / Group European Organization for Research and Treatment of Cancer [RTOG/EORTC], Late Effects of Normal Tissues Subjective, Objective, Management, Analytic [LENT-SOMA])

Videofluorography includes both a modified barium swallow test and an esophagography to visualize the oral, pharyngeal and esophageal phases of swallowing. FEES lacks the ability to investigate the oral stage of swallowing. The findings of both FEES and MBS can be quantified according to various scales such as the Swallowing Performance Status Scale (SPS) (Karnell et al. 1994) and the 8-point Penetration–Aspiration Scale (8p-PAS) (Rosenbek et al. 1996). However, these tests have been rarely used within dosimetric studies. Among the reasons, the lack of dedicated personnel (i.e., speech therapists) and the lack of baseline information to refer to are the prominent ones.

Therefore, in clinical practice, several studies used the presence of a FT and/or the use of FT/PN as surrogates for swallowing problems. However, one complicating factor is that some institutions have adopted the strategy to place a FT prophylactically (before commencing treatment) in selected patient populations at higher risk of swallowing problems during treatment (i.e., patients undergoing cCRT) rather than when actually needed ('on demand' strategy). In the former subgroup, only a minority of patients end up using the FT regularly for nutritional purposes during treatment. In one study about half of the patients used the FT for 2 weeks or less (Madhoun et al. 2011). Therefore, within this strategy, the endpoints generally used are (1) the use rather than the presence of a FT (even this introduces additional uncertainties) or (2) the time to FT removal (provided that a consistent practice of tube removal is adopted). Within the on-demand strategy, the indication to put in a FT usually follows algorithms or decision trees (Alterio et al. 2017; Wopken et al. 2014), though inconsistently among institutions and possibly confounded by other patient- and tumor-related characteristics.

Finally, among observer-assessed subjective scales, it should be noted that none of the scales currently employed (RTOG/EORTC, CTCAE, LENT-SOMA) has been tested for its validity in measuring dysphagia (Russi et al. 2012).

The MDADI represents the most widely employed patient-reported outcome metric for the assessment of swallowing through questions directed at the emotional, functional, and physical impact of patient-perceived function (Goepfert et al. 2017). Various other instruments have been developed to assess the quality of life of patients with head and neck cancer, all of which include questions about swallowing dysfunction. Although all these tools measure some aspects of HNSCC-related QoL, it is not clear which best applies to the assessment of swallowing dysfunctions. Again, although each instrument as a whole has been tested for validity, similar tests of the specific dysphagia-related questions have not been performed (Russi et al. 2012). Finally, it should be noted that patient assessed symptoms have been found to poorly correlate with objective findings in patients treated with radiotherapy for oropharyngeal cancer (Jensen et al. 2007).

Therefore, clean data detailing the swallowing function deterioration after (chemo) radiotherapy are scarce. Most studies investigating the role of dosimetric factors have used oversimplified endpoints, still prone to interpretation biases. Nevertheless, the bulk of data provide useful and practical information for planning purposes though within the limitations mentioned earlier.

Another issue is how toxicity rates change over time after treatment (Christianen et al. 2015). Several patterns of swallowing toxicity have been described. Examples include the 'severe persistent' one in which moderate or higher swallowing dysfunction at six months after treatment does not improve at any time during follow up; the 'transient' one, in which swallowing dysfunction at six months improves with time; the 'progressive' one, in which patients without symptoms of dysphagia at six months after treatment, progressively deteriorate to develop toxicity during follow-up. The incidence of each pattern after radiotherapy alone or with chemotherapy were 8%, 15% and 8%, respectively (Christianen et al. 2015). It is conceivable that each pattern may have a different pathophysiology and dosimetric correlates (Christianen et al. 2015). Early dysphagia may reflect damage to the superficial lining of the upper aerodigestive tract (mucositis) while the late one may be more strictly related to the development of fibrosis and edema (King et al. 2016). On the other hand, few studies found a correlation between acute and late toxicity rates (King et al. 2016; Popovtzer et al. 2009; van der Laan et al. 2015). While this may just reflect the shift from acute to chronic dysphagia or a prolongation of acute symptoms beyond 3 months, some authors have postulated a causative relationship between acute and late events under what is generally called 'consequential' late toxicity (Popovtzer et al. 2009; van der Laan et al. 2015). In this view, (particularly) intense acute toxicity predisposes to the development of damage to late responding tissues with the breakdown of pharyngeal mucosa leading to inflammation and fibrosis of the underlying constrictor muscles the clearer example (Popovtzer et al. 2009). Within this view, the radiobiology of consequential late reactions would follow the tenets of acute ones (Withers et al. 1995), including the presence of a time factor.

Finally, strictly from a methodological viewpoint, since toxicity may improve over time, a better descriptor of the toxicity burden at a given time would be its prevalence rather than its incidence (Bentzen and 2007), though this has never been implemented in swallowing-related statistics.

Definition of the Relevant Anatomical Structures

Swallowing is an extremely complex function involving voluntary and involuntary stages of several structures (Russi et al. 2012). Because of the complexity, defining the most important anatomic structures whose dose-volume parameters would have a major effect on dysphagia has been difficult and only recently investigated.

One preliminary issue is to assess which OARs are involved in the swallowing process. Eisbruch et al. noted anatomic/functional changes in pharyngeal constrictors and glottic/supraglottic larynx after chemoradiotherapy and correlated radiation-induced changes to objective swallowing assessments (Eisbruch et al. 2004). However, function can be disrupted also in absence of morphological changes of irradiated OARs. Therefore, others have adopted a different approach by taking into account all the various OARs potentially implicated in swallowing regardless their changes after treatment.

The second issue relates to the delineation of the various swallowing-related OARs while maximizing consistency, reproducibility and reliability. To this extent, various guidelines/atlases have been proposed (Eisbruch et al. 2004; Levendag et al. 2007; Christianen et al. 2011; Sanguineti et al. 2006). The reader should be aware that for a given structure the delineation criteria are often slightly different among the different papers and this might impact significantly the dose/volume metrics to a given OAR (Brouwer et al. 2014).

Nevertheless, the OARs that are usually taken into account in this setting are the constrictor muscles ([PC], often divided into Superior [SPC], Middle [MPC] and Inferior [IPC] constrictors), the upper esophageal sphincter (UES), the esophagus (E), the cricopharyngeal muscle (CP), the glottic (GL)/supraglottic (SGL) larynx (L), the base of tongue (BOT), the oral cavity (OC). Sanguineti et al. have introduced a novel structure, named 'oral mucosa', that pools the mucosa of the oral cavity, oropharynx and hypopharynx with the advantage of taking into consideration all the mucosa of the upper gastrointestinal tract at once though spatial localization of the dose is reduced (Sanguineti et al. 2006). Also other authors have introduced novel/artificial structures such as the oropharyngeal lumen expanded by 1 cm in attempt to streamline the process of contouring (Matuschek et al. 2016) and to encompass more generously the mucosa and the muscles in the swallowing pathway over the PCs alone (Matuschek et al. 2016).

Finally, even if the cranial nerves (CN) are usually regarded as relatively radioresistant structures, their damage after (chemo)radiotherapy has been claimed as a potential cause of dysphagia in a recent paper (Hutcheson et al. 2017). Though infrequent, lower cranial nerve damage rate increases with time (estimated cumulative incidence at 8 years: 11%) and precedes swallow deterioration (Hutcheson et al. 2017). Among CNs, those more closely related to dysphagia are the lower motor ones, namely nerves IX (pharyngeal shortening/constriction), X (glottic closure, velar elevation, pharyngeal constriction) and XII (tongue mobility). Although the exact mechanism of damage remains speculative, nerve entrapping by soft tissue fibrosis in the neck seems the most plausible one (Hutcheson et al. 2017), implying that no distinct OAR can be anticipated. However, the location of injury can be determined by the number of nerves affected, suggesting that an isolated XII nerve palsy is presumably related to submandibular damage, isolated X nerve is related to carotid sheath damage and any combination of X, XI and XII nerves reflects skull base pathology (Berger and 1977).

SHORT SUMMARY OF THE QUANTEC RECOMMENDATIONS WITH RELATIONSHIP TO MAIN LITERATURE ON WHICH THEY WERE BASED (SWALLOWING/NUTRITIONAL PROBLEMS)

The QUANTEC report on larynx and pharynx (Rancati et al. 2010) was based on an admittedly limited amount of data available at that time (Eisbruch et al. 2004; Levendag et al. 2007; Dornfeld et al. 2007; Feng et al. 2007; Jensen et al. 2007; Fua et al. 2007; Caglar et al. 2008). Two reports were available only as abstracts (Rancati et al. 2010). Nevertheless, the report ended up providing the recommendations to minimize the volume of the pharyngeal constrictors and larynx receiving 60 + Gy and to reduce it to possibly less than 50 Gy without jeopardizing target coverage (Rancati et al. 2010).

NON-DOSIMETRIC FACTORS AFFECTING THE RISK OF DYSPHAGIA

The risk of dysphagia is related to several non-dosimetric variables that should be taken into consideration when interpreting dose/volume data (confounding factors). It should be preliminarily noted that most of these covariates have been tested only at univariate analysis and their predictive role on dysphagia has been inconsistently found among studies. Moreover, not a single study has evaluated simultaneously all the potential predictors, implying that some degree of correlation may actually exist among covariates. Both lacking factors, comprehensiveness and test for collinearity, help to explain the inconsistency among results.

One major critical issue in evaluating both dysphagia and aspiration rates due to radiotherapy is that a significant proportion of patients present swallowing problems at diagnosis (Mortensen et al. 2013). Baseline swallowing issues should be always ruled out and corrected for, but unfortunately this has not been consistently done in the literature. Pre-existing swallowing problems are linked to primary tumor location, patient age, weight loss, presence of symptoms and smoking (Nguyen et al. 2005; Caglar et al. 2008; Caudell et al. 2010).

Also, supportive measures during treatment can affect long-term dysphagia and confound results. While enteral tube feeding using a gastrostomy is the preferred method to maintain sufficient nourishment of patients over a long period of time, it is an invasive procedure and it is associated with complications. Percutaneous endoscopic gastrostomy (PEG) placement may result in worse diet outcome (Langmore et al. 2012) and delayed oral feeding restoration after treatment (Wopken et al. 2014). Rosenthal et al. (2009) and Mekhail et al. (2001) suggested that a nasogastric FT may decrease the need for esophageal dilation vs. a PEG tube by serving as a stent to prevent stricture formation. Conversely, the role of prophylactic swallowing exercises in swallowing related outcomes is still controversial (Mortensen et al. 2015; Shaw et al. 2015; Hutcheson et al. 2013).

Various tumor-, patient- and treatment-related characteristics, such as male gender (Mekhail et al. 2001; Schweinfurth et al. 2001), advanced age (Mortensen et al. 2013; Schweinfurth et al. 2001; Mangar et al. 2006; Clavel et al. 2012; Cheng et al. 2006; Christianen et al. 2012; Soderstrom et al. 2017; Head and Neck Cancer Symptom Working 2016), poor Eastern Cooperative. Oncology Group (ECOG) score (Matuschek et al. 2016; Poulsen et al. 2008; Munshi et al. 2003), low Body Mass Index (BMI)/weight loss

(Wopken et al. 2014; Mangar et al. 2006; Clavel et al. 2012), nicotine abuse (Goepfert et al. 2017; Mills et al. 2011), high alcohol consumption (Schweinfurth et al. 2001), high TNM stage (Wopken et al. 2014;Goepfert et al. 2017; Mortensen et al. 2013; Mangar et al. 2006; Clavel et al. 2012; Cheng et al. 2006), tumor localization in the nasopharynx and base of tongue (Schweinfurth et al. 2001; Christianen et al. 2012), and tracheotomy (Murray and 1998) as well as treatment characteristics like concurrent chemotherapy (Wopken et al. 2014; Matuschek et al. 2016; Jensen et al. 2007; Mekhail et al. 2001; Sanguineti et al. 2012), concurrent radiotherapy and cetuximab (Wopken et al. 2014), accelerated fractionation of radiation therapy (Wopken et al. 2014; Mortensen et al. 2013; Bhide et al. 2010) have been identified to be associated with an higher risk of swallowing problems. Bilateral irradiation of the neck has been associated with an increased risk of dysphagia compared to unilateral or no irradiation (Langendijk et al. 2009); conversely, a modest decrease in dose from 50 to 40 Gy in the electively treated neck may limit the risk of dysphagia (Nevens et al. 2017). As previously stated, all interventions that increase the risk of acute toxicity on the mucosa, such as concomitant chemotherapy and overall treatment shortening (acceleration) may end up increasing (consequential) swallowing problems as well. Moreover, chemotherapy itself, throughout nausea, vomiting and dehydration may contribute to impaired oral feeding.

It is likely that several of the above mentioned covariates are correlated and dependent on each other. At multivariate logistic regression analysis and after including several clinical but not dosimetric factors, Langendijk et al. found that the following factors to be independently correlated to grade$\geq$2 RTOG dysphagia at six months: T3–T4 stage, bilateral neck irradiation, weight loss prior to radiation, oropharyngeal and nasopharyngeal tumours, accelerated radiotherapy and concomitant chemoradiation (Langendijk et al. 2009; van der Laan et al. 2012). The model was validated on a different subset of patients (Nevens et al. 2016).

Many studies have excluded from the analysis patients developing local recurrent disease, since local(regional) failure may be per se the cause of swallowing symptoms. On the other hand, scarring due to a responding tumor may leave a fibrotic, less functional tissue, that may impact the complexity of the swallowing process; for this reason, some studies have focused only on organs that were not involved by the tumor to begin with (Rancati et al. 2009; Sanguineti et al. 2007). Similarly, most, but not all the studies have excluded patients undergoing upfront major surgery. Interestingly, a correlation between xerostomia and dysphagia has been found (Mortensen et al. 2013; Wopken et al. 2014).

Finally, few studies have investigated also the impact of genetic polymorphism on radiation-induced dysphagia (De Ruyck et al. 2013; Kornguth et al. 2005; Pratesi et al. 2011). Inter-patient variability in susceptibility to radiation effects is a major factor that drives individual response to ionizing radiations. Wopken et al. found that the explained variance, that measures the proportion to which a model accounts for the dispersion/variation of a given dataset, remained relatively low in their final model on FT dependence after (chemo)radiotherapy (Wopken et al. 2014). This would suggest that additional factors beyond dosimetric and clinical ones explain variability, indirectly highlighting the need to account for individual variations in radiosensitivity in such analyses.

CRITICAL REVIEW OF THE LITERATURE (SWALLOWING/NUTRITIONAL PROBLEMS)

In the last decade several papers have investigated dose/volume relationships for swallowing problems both during and after (chemo)radiotherapy for HNSCC. A summary of the available results from selected papers is reported in Table 9.1 (Alterio et al. 2017; Wopken et al. 2014; Matuschek et al. 2016; Mortensen et al. 2013; Christianen et al. 2012; Soderstrom et al. 2017; Head and Neck Cancer Symptom Working 2016; De Ruyck et al. 2013; Sanguineti et al. 2011; Sanguineti et al. 2013; Schwartz et al. 2010; Vlacich et al. 2014). The order reflects the time to (primary endpoint) evaluation; papers can be divided into those who looked at predictors of acute (during treatment and up to three months afterwards) and subacute (up to six months) toxicity versus those who considered later endpoints. As previously stated, in the former subgroup, the degree of acute toxicity (on the mucosa) has likely played a major role in swallowing-related endpoints as well; this translates into slightly different dosimetric predictors compared to those proper of late responding tissues (Sanguineti et al. 2013).

Dosimetric Predictors of Acute and Subacute Toxicity

Sanguineti et al. investigated the use of PEG tube during exclusive IMRT for oropharyngeal cancer (Sanguineti et al. 2011). Out of 59 patients, 22 (37.3%) needed a PEG tube during treatment. Of note, the rate of dose accumulation in the upper gastrointestinal tract (from the lips to the level of the esophageal inlet) was taken into consideration to better reflect the radiobiology of a rapidly responding tissue like the mucosa. In a stepwise multivariate logistic analysis, the amount of mucosa receiving at least 9.5 Gy per week ($\geq$64 vs. <64 cm^3) was the most predictive parameter of PEG tube placement during treatment (Sanguineti et al. 2011).

The analysis was then expanded also to patients receiving chemotherapy and, since these patients typically had undergone a prophylactic tube insertion, the endpoint became the time of PEG dependence after treatment (Sanguineti et al. 2013). One hundred-seventy-one patients, who mostly had oropharyngeal cancer, were analyzed. Dosimetric data on several OARs were extracted and the endpoints were PEG dependent at 3.3 months after treatment completion as well as at seven months (the results on the latter endpoint are reported in the next section) (Table 9.1). At 3.3 months after treatment completion, 43 patients (25%) were PEG dependent (Sanguineti et al. 2013). Independent dosimetric predictors included the volume of OM receiving a least 9.5 Gy per week, the mean dose to the superior pharyngeal muscle and the volume of the larynx receiving at least 50 Gy. Best dosimetry cut-off values (assessed by Receiver Operating Characteristic [ROC] curve analysis) were 35% and 67.4 Gy for larynx V_{50Gy} (percent organ volume receiving at least 50 Gy) and mean dose to SPC, respectively. For the OM, best values for OM $V_{9.5Gy/w}$ (absolute organ volume receiving at least 9.5 Gy per week) were 33 cc and 85 cc for patients receiving combined chemo-IMRT or IMRT alone, respectively (Sanguineti et al. 2013). Alterio et al. found that the risk of grade 3 dysphagia (CTCAE v4.0) during treatment was correlated to the dose to both the CPM and UES (Alterio et al. 2017), thus pointing towards the caudal part of the mucosa as

TABLE 9.1 Selected Studies on Swallowing-Related Organs at Risk Dosimetric Covariates

Authors Year	# pts	Sites	Tech	Surg	Conc Chemo	Scoring	Primary Endpoint	Baseline	OARs Included	OARs Sign	OAR Threshold
Sanguineti et al. (2011)	59	OP	IMRT	None	None	PEG placement	PEG during tmt	Yes	PG, PC, ES, L, OM	OM	$V_{9.5Gy/wk}$ < 64 cc
De Ruyck et al. (2013)	189	Mix	IMRT	22.2% (T) 52.4% (N)	39.7%	CTCAE v3.0	grade≥3 during tmt	Yes	PC, ES	SPC	$D_{2\%}$ AUC: 0.71
Alterio et al. (2017)	42	Mix	IMRT (47%) 3DCRT (53%)	None	81%	CTCAE v4.0	grade≥3 during tmt	Yes	SGL, PC, UES, BOT, SP, CPM	CPM, UES	V_{45Gy} < 14% (UES) Dmean < 35 Gy (CPM) AUC: 0.82
Matuschek et al. (2016)	101	Mix	NA	70%	76%	FT or PN usage	Use of FT or PN during tmt	Yes	SPC, OC, ORO, NPX, Gingiva, Artificial OARs	ORO+1 cm	Dmean (cont)
Sanguineti et al. (2013)	171	Mix (OP)	IMRT	None	66.2%	PEG dep	PEG dep at 3 mths	Yes	PG, PC, ES, L, OM	OM, SPC, L	$V_{9.5Gy/wk}$ < 33 cc[b] (OM) Dmean < 67.4 Gy (SPC) V_{50Gy} = 35% (L) AUC: 0.82
							PEG dep at 7.7 mths	Yes	PG, PC, ES, L, OM	SCP, L	SPC: Dmean (SPC) V_{50Gy} (L) AUC:0.90

(Continued)

TABLE 9.1 (CONTINUED) Selected Studies on Swallowing-Related Organs at Risk Dosimetric Covariates

Authors [Ref]	# pts	Sites	Tech	Surg	Conc Chemo	Scoring	Primary Endpoint	Baseline	OARs Included	OARs Sign	OAR Threshold
Vlacich et al. (2014)	141	Mix	IMRT	None	100%	PEG dep	PEG dep >12 mths	No	BOT, SPC, ES, L, PC	IPC	Dmean (cont)
Christianen et al. (2012)	354	Mix	IMRT/ 3D	None	20%	RTOG EORTC-QLQ35	RTOG grade$\geq$2+ at 6 mths	Yes	PC, UES, BOT, GL, SGL	SPC SGL	Dmean (cont) AUC=0.80
Wopken et al. (2014)	355	Mix	IMRT (49%) 3DCRT (51%)	None	25%	FT dep	FT dep at 6 mths	Yes	PC, UES, BOT, GL, SGL, PG	SPC, IPC, cPG, CPM	Dmean (cont) AUC=0.88
Sodestrom et al. (2017)	114	Mix	3DCRT/IMRT	43.9% (N)	None	MBS EORTC-QLQ30	Aspiration at 60 mths	No	PC, BOT, GL, SGL	MPC	AUC: 0.73
Schwartz et al. (2010)	31	OP	SF-IMRT	None	42%	MBS MDADI, PSS	OPSE at 6–24 mths	Yes	PC, UES, GL, SGL, OC	OC^c SPC	V_{30Gy} <65% (OC) V_{55Gy} <80% V_{65Gy} <30% (SPC)
Mortensen et al. (2013)	259 (65)	Mix	IMRT	None	63%	MBS (subset) EORTC-QLQ30 EORTC H&N35 DAHANCA scale	MBS abnormal at 30 mths	No	PC, CPM, UES, BOT, ES, GL, SGL	SPC (asp), MPC (MBS score) SGL (QoL)	Dmean <60 Gy (SPC, MPC) Dmean <55 Gy (SGL)

(Continued)

TABLE 9.1 (CONTINUED)   Selected Studies on Swallowing-Related Organs at Risk Dosimetric Covariates

Authors [Ref]	# pts	Sites	Tech	Surg	Conc Chemo	Scoring	Primary Endpoint	Baseline	OARs Included	OARs Sign	OAR Threshold
MD Anderson Head and Neck Cancer Symptom Working Group (2016)	300	OP	SF-IMRT	None	65%	Aspiration, stricture, PEG, aspiration pneumonia	Any at 12 mths	No	PC, MPM, LPM, ADM, PDM, ITM, MHM, GGM, MM, BM, PGM, CPM	MHM	$V_{69Gy} < 79.5\%$ AUC 0.835

Mix, mixed Head and Neck Squamous Cell Carcinoma; OP, oropharyngeal Squamous Cell Carcinoma ; OPSE, Oropharyngeal Swallowing Efficiency; Ref, reference; pts, patients; tech, technique; surg, surgery; conc chemo, concurrent chemotherapy; OARs, Organs at Risk; OARs sign, significant Organs at Risk; IMRT, Intensity Modulated Radiation Therapy; SF-IMRT, Split Field Intensity Modulated Radiation Therapy; 3DCRT, three Dimensional Conformal Radiotherapy; mths, months; T, primary tumor; N, involved lymph nodes; AUC, Area Under the Receiver Operating Characteristics Curve; Dmean, mean organ dose; V_{XGy}, organ volume receiving at least X Gy; $V_{XGy/wk}$, organ volume receiving at least X Gy per week; cont, continuous; tmt, treatment; QoL, Quality of Life; PC, pharyngeal constrictor muscles; SPC, MPC, IPC, superior/middle and inferior pharyngeal constrictor muscle; SGL, supraglottic larynx; OC, oral cavity; OM, oral mucosa; L, larynx; GL, glottic larynx; MPM, LPM, medial and lateral pterygoids; ADM, PDM, anterior and posterior digastric muscles; ITM, intrinsic tongue muscles; MHM, mylo/geniohyoid complex; GGM, genioglossus muscle; PGM, palatoglossus muscle; MM, masseter muscle; BM, buccinators muscle; CPM, cricopharyngeus muscle; ES, esophagus; UES, upper esophagus; cPG, contralateral parotid gland; PG, parotid gland; BOT, base of tongue; CTCAE, Common Terminology Criteria for Adverse Events; EORTC H&N35, European Organization for Research and Treatment of Cancer Quality of Life Questionnaire, Head and Neck Module 35 questions; EORTC QLQ30, European Organization for Research and Treatment of Cancer Quality of Life Questionnaire, Core Module 30 questions; DAHANCA scale, the Danish Head and Neck Cancer Group scale; MDADI, MD Anderson Dysphagia Inventory; PEG, percutaneous endoscopic gastrostomy; PN, parenteral nutrition; FT, feeding tube; MBS, Modified Barium Swallow; PSS, Performance Status Scale for Head and Neck Cancer Patients; asp, aspiration.

a No breakdown for individual muscles.

b For pts receiving concomitant chemoradiotherapy.

c Anterior part.

defined by Sanguineti et al. (2011). Similarly, a small study from the Netherlands Cancer Institute showed that the dose to the IPC was correlated to the risk of penetration/aspiration at 10 weeks after completion of chemoradiation (van der Molen et al. 2013). Conversely, according to Matuschek et al. the OAR dosimetry correlating best with the need for enteral nutrition/feeding tube during (chemo) radiotherapy was the mean dose to the oropharyngeal lumen expanded by 1 cm circumferential margin (Matuschek et al. 2016).

Christianen et al. reported on a large prospective cohort study from two Dutch Institutions, University Medical Center Groningen and VU University Medical Center Amsterdam (Christianen et al. 2012), updating & expanding the previous experience reported by Langendjik et al (2009). The sample size consists of 354 consecutive patients treated with (chemo)radiotherapy since 1997. Patients treated with upfront surgery, those failing within six months and those with baseline swallowing dysfunction were excluded. The endpoint of the study was the development of grade≥2 dysphagia according to RTOG/EORTC at six months after treatment completion. Of note, the population consists of highly heterogeneous primary tumors treated with highly different strategies, volumes and techniques. In the multivariate logistic regression analysis, the mean dose to the superior PC and the mean dose to the supraglottic larynx were the only independent predictors of outcome with an Area Under the ROC Curve (AUC) of 0.80 (95% Confidence Interval [CI] 0.75–0.85). Though the authors did not report the actual prevalence of toxicity at six months, from the nomogram a risk of grade≥2 dysphagia at six months of about 35% for a mean dose to both the SGL and SPC of 60 Gy can be estimated; beyond 60 Gy, the risk increases rapidly, with the dose to the SPC having a greater impact than the one to the SGL (Christianen et al. 2012). Two years later, Wopken re-analyzed the same population with a different endpoint, FT dependence at six months (Wopken et al. 2014). The following dosimetric covariates were found to be independent predictors of the endpoint, whose prevalence was 10.7%: the mean dose to the superior and inferior pharyngeal constrictor muscles, to the contralateral parotid gland and to the cricopharyngeal muscle (Wopken et al. 2014). Interestingly the odds ratios (OR) for the pharyngeal constrictors (OR=1.07 and 1.03 for SPC and IPC, respectively) were higher than those for both CPM (OR=1.02) and contralateral parotid gland (OR=1.01). A predictive role of the volume of SPC receiving 55–60 Gy on patient-reported dysphagia deterioration at six months after chemoradiation was also found by Chera et al. (2017; Mavroidis et al. 2017). This study is unique in selecting only favourable Human Papilloma Virus (HPV)-related oropharyngeal cancers (median age: 61 years) and in reducing the prescription dose to the gross tumor volume from 70 Gy to 60 Gy. Interestingly, the average V_{55Gy}/V_{60Gy} were 88.7%/62.5% and 76.2%/37% in patient with (N=9) and without (N=26) dysphagia deterioration, respectively (Chera et al. 2017). Therefore, despite the lower prescription dose a significant proportion of patients still ended up having high values for SPC V_{55Gy} and V_{60Gy}.

Dosimetric Predictors of Late Toxicity

In the paper of Sanguineti et al. on PEG dependence (see previous section), at seven months, only 17 patients (10%) were PEG dependent. Again laryngeal V_{50Gy} and mead dose to SPC were independent predictors of PEG dependence with best cut-off values of 92% and 67.6

Gy, respectively. Of note, contrary to the analysis at 3.3 months, no effect was found for the oral mucosa suggesting that beyond seven months, the impact of acute/consequential late toxicity is irrelevant (Sanguineti et al. 2013). Interestingly, in another study on a partially overlapping patient population, mean laryngeal dose or, alternatively, laryngeal V_{50Gy} were independent predictors of grade 2 laryngeal edema as well (Sanguineti et al. 2007).

Vlacich et al. investigated long-term (>12 months) PEG tube dependence in 88 patients treated with concomitant poly-chemotherapy and IMRT at Vanderbilt. No correction was done for both baseline function and tumor recurrence. Fifteen patients (11%) required PEG for more than 12 months after treatment completion and their survival free of disease was significantly worse than those without a PEG tube at 12 months. The mean dose to the IPC was the only predictor of PEG tube dependence at 12 months (Vlacich et al. 2014).

In an earlier study from Aarhus, Denmark, Jensen et al. retrospectively evaluated 35 cancer-free patients after an average time of five years from radiotherapy alone for pharyngeal cancer (Jensen et al. 2007). Of note, 83% of the patients reported some degree of dysphagia; at FEES, aspiration and penetration were observed in 59% and 18% of patients, respectively (Jensen et al. 2007). Doses less than 60 Gy to supraglottic region, the larynx and the upper esophageal sphincter resulted in a lower risk of aspiration (Jensen et al. 2007). In a more recent study from the same Institution, 259 patients treated with IMRT between 2006 and 2010 answered to both the EORTC QoL-C30 and the EORTC Head & Neck Cancer QoL (H&N35) questionnaires a median of three years after being treated successfully with definitive (chemo)radiotherapy. A smaller subset of patients (N = 65) also underwent modified barium swallow (MBS) testing. This study found a poor correlation between patient-reported and objective recorded outcomes, with the former lower than the latter ones (Mortensen et al. 2013). In general, objective measurements and physician reported endpoints of late dysphagia correlated to the mean dose to the superior and middle constrictor muscles, while the QoL endpoint correlated to the mean dose to the supraglottic larynx. The reason for this mismatch remains unanswered. Thresholds of 60 Gy and 55 Gy have been suggested for the mean dose to the pharyngeal constrictors and the larynx, respectively (Mortensen et al. 2013). Weaknesses of the study include the limited amount of patients, the lack of correction for both baseline function and clinical parameters.

The earlier experience from MD Anderson Cancer Center on a limited number (N = 31) of patients with oropharyngeal cancer undergoing split-field IMRT (SF-IMRT) and evaluating both objective (at MBS) and subjective (MDADI) outcomes found that (1) none of the dosimetric variables predicted for MDADI changes and (2) that both the anterior part of the oral cavity and SPC predicted for long-term objective changes (Table 9.1) (Schwartz et al. 2010). The predictive role of the dose to the anterior part of the oral cavity has been found also by Kumar et al. within a small study on 46 patients undergoing cCRT for oropharyngeal cancer, thought the endpoint (penetration/aspiration score >3) had been investigated at MBS in the majority of the cases within 6 months from treatment end (Kumar et al. 2014).

The latest study from MDACC focused on 300 patients affected by oropharyngeal cancer treated consistently with systemic therapy (65% chemotherapy) and radiotherapy. Chronic radiation associated dysphagia, defined as MBS/FEES detected aspiration or stricture, gastrostomy tube and/or aspiration pneumonia 12 months or later after treatment completion, was observed in 11% of patients (Head 2016). A unique aspect of this study is the fact that both intrinsic and extrinsic muscles of the tongue as well few other 'novel' OARs (Table 9.1) were retrospectively contoured and their dose extracted. Although the number of patients is high, this is a retrospective study with a risk of underreporting, it did not correct for baseline swallowing disorders and it did not include in the analysis some 'standard' OARs, such as the parotid glands and the larynx. Moreover, in the interpretation of the findings the reader should be aware of the fact that SF-IMRT was used and, thus, the dose to the IPC/UES was systematically kept low. Recursive partitioning analysis (RPA) showed that the volume of the mylo/geniohyoid complex (MHM) receiving at least 69 Gy (V_{69Gy}), genioglossus muscle V_{35Gy}, anterior digastric muscle V_{60Gy}, MPC V_{49Gy}, and SPC V_{70Gy} to be associated with chronic dysphagia. A model including age in addition to MHM V_{69Gy} as continuous variables was defined as 'optimal' among tested multivariable models with an AUC of 0.835. The authors conclude that in addition to SPCs, the dose to the MHM should be monitored and constrained, especially in older patients (>62 years), when feasible (Head 2016).

In the recent paper from Sordestrom et al., 124 long-term survivors (minimum and median follow up 25 months and 67 months, respectively) from the ARTSCAN study were investigated at MBS for possible aspiration (Soderstrom et al. 2017). Aspiration was detected in 47% of the patients with the mean dose to the middle pharyngeal constrictor being the strongest dosimetric predictor (Soderstrom et al. 2017). Interestingly, the study found a significant correlation between the volume of the electively treated volume and the dose in the majority of all swallowing related organs, though most the patients were treated with 3 Dimensional Conformal Radiotherapy (3DCRT) (Soderstrom et al. 2017). Another weakness is the lack of baseline information.

PROSPECTIVE VALIDATION STUDIES

Though IMRT has been used for almost two decades, only relatively recently dose-volume objectives have been specifically implemented in order to minimize the risk of dysphagia.

Two groups have tested the potential clinical benefits of 'swallowing-OARs sparing' IMRT.

Having found that the dose to the constrictor muscles was critical in driving the risk of dysphagia (Eisbruch et al. 2004) and that from multiple surgical and radiological series the risk of involvement of the medial retropharyngeal nodes by oropharyngeal SCC is remote, the group at Michigan University postulated that these nodes could be safely excluded from any target volume during IMRT for oropharyngeal SCC, thereby potentially improving function without affecting survival outcomes (Feng et al. 2007). This hypothesis was tested prospectively in a selective group of stage III/IV oropharyngeal carcinoma patients (N = 73) treated with primary chemoradiotherapy (Feng et al. 2007). Parts of PCM, SGL and ES in the region of the uninvolved medial retropharyngeal nodes were spared by setting an optimal dose constraint <50 Gy in the IMRT planning objectives, which subsequently delivered mean doses of 48, 42 and 32 Gy, respectively, to the spared regions. On average, long-term patient-reported, observer-rated, and objective measures of swallowing were only slightly worse than pretreatment measures, while none of the patients failed in the tentatively spared regions (Feng et al. 2007). The strategy of sparing the medial retropharyngeal nodes from target volumes has been endorsed in the recently updated head and neck cancer nodal outlining consensus guidelines (Gregoire et al. 2014). Since this is not an adequately powered study to detect a rare event such as a retropharyngeal nodal failure, patients should be selected cautiously for this approach. Moreover, despite careful planning to minimize the dose to non-target swallow structures including the constrictors and the larynx, up to 20% of survivors developed chronic aspiration (Feng et al. 2010), suggesting the need for further refinement of this strategy.

The Dutch group tested and validated clinically their model predicting RTOG grade≥2 dysphagia at six months after treatment completion (Christianen et al. 2012). Preliminarily, within a comparative dosimetric study, van der Laan et al. (2012) explored the potential benefits of IMRT to spare the swallowing related structures with an estimated 8.9% (range 3–20%) mean reduction in the risk of RTOG grade ≥2 dysphagia over those plans in which IMRT was not stressed to spare swallowing-related OARs. The IMRT plans were further optimized to reduce the dose to SPC, SGL, MPC and UES, in that order of priority. On the other hand (any) Planning Target Volume (PTV) coverage was maintained at no less than V_{95Gy} receiving at least 98% of the dose and allowance was made for a moderate shift of dose to unspecified tissues, such as the neck muscles and the oral cavity (Kierkels et al. 2016). Afterwards, this planning strategy was clinically tested in a prospective/validation cohort of 186 patients (Christianen et al. 2016). Results showed that, by adding dose constraints for swallowing related structures during treatment planning optimisation, at least a 5% improvement in the estimated risk of dysphagia at six months can be obtained in approximately 50% of the patients and that subsequent lower (than expected) prevalence of grade ≥2 dysphagia is observed if the dose to the SPC and SGL can indeed be sufficiently decreased (Christianen et al. 2016). Of note, this approach did not compromise target volume coverage or impact the dose to the major salivary glands. Weaknesses of this strategy include the fact that a six month endpoint may not be representative of the whole spectrum of toxicity (and particularly of later events), that the endpoint is physician-observed rather than patient-reported and that the same group showed that the dose to the SGL may not as relevant for six month grade ≥2 dysphagia as for later events (Christianen et al. 2015) or for other swallowing related endpoints (Wopken et al. 2014). Nevertheless, the study proves the concept that reducing the dose to swallowing related structures is clinically detectable in a selected group of patients.

Finally, the clinical benefits of swallowing sparing IMRT are being investigated within a multicenter phase III clinical trial from the UK that selects patients undergoing radical primary chemoradiation or radiation alone, for primary pharyngeal cancer not involving the retropharyngeal nodes or posterior pharyngeal wall and requiring bilateral neck irradiation (Petkar et al. 2016). In the experimental arm, the mean dose to parts of (superior/middle) pharyngeal constrictors lying outside the high-dose target volume is limited to less than 50 Gy (oropharyngeal primaries) or 40 Gy (hypopharyngeal primaries); likewise, for parts of the IPC outside the high-volume PTV the dose is limited to 20 Gy (oropharyngeal primaries) or 50 Gy (hypopharyngeal primaries)(Petkar et al. 2016). Planning objectives are prioritized in the following order: critical organ constraints (spinal cord and brain-stem); high-dose PTV coverage; constrictor constraints; elective PTV coverage; parotid gland constraints and other non-specified normal tissue (Petkar et al. 2016).

SUGGESTION FOR DVH CONSTRAINTS/NTCP MODEL PARAMETERS (SWALLOWING/NUTRITIONAL PROBLEMS)

Despite an abundance of published literature (Table 9.1), it is challenging to make unequiv-ocal conclusions regarding the optimal swallow-sparing parameters. Duprez et al. have highlighted the many shortcomings associated with these papers (Duprez et al. 2013), that ultimately limit the robustness of the data. Moreover, due to the lack of phase III data on the clinical benefits of the presumed dosimetric improvements, (any) target coverage should always be prioritized over sparing of swallowing-related organs. For contouring OARs we recommend to follow current cooperative group guidelines (Brouwer et al. 2015).

For the portions of OARs that lie outside any PTV, it seems reasonable to attempt to reduce their dose while maintaining the dose limits to other 'named' OARs and checking the dose outside any 'named' structure (unspecified tissue).

The bulk of evidence suggests that:

- The mean dose to each constrictor muscle should be kept below 60 Gy and possibly below 50 Gy; the same dose-volume objective would apply also to the 'extended' ver-sions of the PCs such as those covering also the adjacent mucosa of the oro- and hypo-pharynx (Matuschek et al. 2016).

- The mean dose to the larynx as a whole or to the supraglottic larynx should be kept below 50 Gy and possibly below 40 Gy.

- The mean dose to the anterior oral cavity should not exceed 30 Gy.

- The true vocal cords should be constrained to a mean dose of 20 Gy.

- The upper esophagus/esophageal inlet should be limited to a mean dose of 50Gy and possibly 40 Gy.

- The contralateral parotid should be tentatively spared according to the usual xerosto-mia-related dose-volume objectives.

MATHEMATICAL/BIOLOGIC MODELS

The analysis of dose-volume data relates to the question regarding the most important anatomic structures whose dysfunction after chemo-radiotherapy causes dysphagia. The average shape of the dose-volume distribution of the population under analysis will affect the results, since only OARs that show a high degree of variability among subjects will provide meaningful statistical differences. Because of this, the site of the primary tumor and the extent of neck involvement (and thus the need for unilateral vs. bilateral irradia-tion) as well the technique adopted (3DCRT vs. IMRT/Volumetric Modulated Arc Therapy [VMAT]) will affect the results. If a given OAR is consistently spared at planning, it will likely not be a predictor of toxicity. For instance, in the papers which dealt only on oropharyngeal cancers, the doses to the upper esophageal inlet and to the cricopharyngeous muscle were relatively modest compared to those where laryngo-hypohayngeal tumors were included. Moreover, in the former subgroup a further distinction can be made between those patients who were treated with whole field IMRT (WF-IMRT) in which the larynx is tentatively spared from those treated with SF-IMRT, in which the dose to the larynx (and IPC) can be significantly lower (Dabaja et al. 2005).

It has been noted that the small volume of many structures, such as the pharyngeal constrictors muscles favours the use of mean doses rather than the percent of OAR covered by a given dose (Matuschek et al. 2016). In one study, a very strong correlation between the majority of dose-volume histogram (DVH) parameters within each swallowing OAR and the mean dose of that OAR was found (Christianen et al. 2012). Moreover, the dose to several (adjacent) OARs will be correlated, raising the statistical issue of which covariate should be included in a multivariable model or, in other words, which covariate independently predicts for outcome. Unfortunately only a few (recent) studies have addressed this issue (Alterio et al. 2017; Matuschek et al. 2016; Kumar et al. 2014).

Recent efforts have also been focused on moving away from OAR-based dose response modeling, in favour of voxel-based analyses correlating risk of toxicity with three-dimensional dose maps. Monti et al. performed such an analysis for acute grade ≥ 3 dysphagia and found significantly higher doses in voxels corresponding to the anatomical location of the cricopharyngeus muscle and cervical esophagus (Monti et al. 2017).

SPECIAL SITUATIONS AND FUTURE DIRECTIONS (SWALLOWING/NUTRITIONAL PROBLEMS)

Little is known on the tolerance of swallowing-related OARs to either particles or hypofractionated photon RT. The increase use of protons for HNSCC will elucidate this issue. In a dosimetric study on patients with pharyngeal tumors planned to receive bilateral neck irradiation, intensity-modulated proton therapy aimed at maximizing swallowing-related OARs sparing allowed an about 10% gain in the predicted risk of RTOG grade ≥ 2 dysphagia at six months over swallowing-optimized IMRT (van der Laan et al. 2013). Early clinical data are encouraging (Gunn et al. 2016) and results from phase III studies are awaited.

INTRODUCTION: VOICE PROBLEMS (DYSPHONIA)

Clinical Significance of the Injury

Voice is defined as the sound originating from the vibration of the vocal cords/folds. Voice quality depends on both the myoelastic characteristics of the vocal folds and the resonances and characteristics of the vocal tract. Voice impairment (dysphonia) can have a profound negative impact on patients function and QoL (DeSanto et al. 1995). Dysphonia can be due to the tumor itself (as for laryngeal primary tumors) or its treatment. The latter can be the case for both laryngeal and non-laryngeal primary tumors and will be the topic of the present chapter.

Laryngeal dysfunction after RT for non-laryngeal primary tumors is not a rare event. Fung et al. within a cross-sectional observational study found moderate/severe voice changes at voice handicap index (VHI) score in 27% of patients (Fung et al. 2005). Vainshtein et al. prospectively assessed voice changes in 91 patients with stage III or IV oropharyngeal cancer after chemo-IMRT using a previously-validated communication domain of the Head and Neck Quality of Life instrument (HNQOL-C) (Vainshtein et al. 2014). Voice quality decreased maximally at one month, with 68% of patients reporting worse HNQOL-C compared with baseline (Vainshtein et al. 2014). In the study of Sanguineti et al. voice changes were prospectively assessed using two questions of the Functional Assessment of Cancer Therapy-Head and Neck Scale (FACT-HN)(Cella et al. 1993). 36% of patients reported major (grade 0 or 1) voice changes at any follow up when considering FACT-HN question 4, though this rarely impacted voice communication (FACT-HN question 10)(Sanguineti et al. 2014). Most of the patients recover baseline voice quality within 12 months after treatment (Vainshtein et al. 2014; Stoicheff 1975), though in the study of Sanguineti et al., changes were observed at any follow up to five years after treatment (Sanguineti et al. 2014).

Evaluation of Toxicity Endpoints

Vocal function can be assessed objectively using instruments such as videostroboscopy (Rancati et al. 2010) or subjectively using validated patient-focused questionnaires (Ang 2010). The ones commonly used and validated are the Voice-Related Quality of Life and the Voice Handicap Index (VHI), or its abbreviated version, the VHI-10 (Hogikyan and 1999; Rosen 2004). As previously stated, other authors have used voice specific items/domains of QoL questionnaires, such the HNQOL-C (Vainshtein et al. 2014) or FACT-HN (Sanguineti et al. 2014). Traditionally, late laryngeal toxicity has been scored by RTOG according to both patient symptoms (i.e., cough, hoarseness) or objective morphologic changes (i.e., edema) (Cox et al. 1995), though the overall granularity of this system is low. In an earlier study, Sanguineti et al. have consistently scored laryngeal edema at direct fiberoptic examinations during follow up (Rancati et al. 2009; Sanguineti et al. 2007) and used grade 2 or more laryngeal edema as a surrogate for toxicity. Interestingly, Vainshtein et al. noted disproportionate rates between patient- and physician-reported voice changes after radiotherapy for non-laryngeal primary tumors with the latter ones being less common (Vainshtein et al. 2014). However, a significant correlation was found between patient-reported speech outcomes and evaluation by speech therapist in one study on 62 patients (Karnell et al. 1999). Moreover, two other series indirectly support a close correlation between patient-reported and observer-scored events on voice quality since they share predictors (Sanguineti et al. 2014; Fung et al. 2001).

Definition of the Relevant Anatomical Structures

Sanguineti et al. contoured the larynx as a whole from the tip of the epiglottis superiorly to the bottom of the cricoid inferiorly; the external cartilage framework was excluded from the laryngeal volume (Sanguineti et al. 2007). Eisbruch et al. divided the larynx into the supraglottis (SGL) and the glottis (GL) (Schwartz et al. 2010). The RTOG defines the supraglottic larynx as a "triangular prism-shaped" volume that begins just inferior to the hyoid bone (excluding the suprahyoid epiglottis) and extends to the cricoid cartilage inferiorly.

Both arytenoids and the infrahyoid epiglottis are included (Choi et al. 2014), though other cartilages are not mentioned.

The reader should be aware of the multiple confounding factors when evaluating laryngeal function after RT. First of all, there is level I evidence on the detrimental effect of surgery on vocal cord function (compared to RT) (Aaltonen et al. 2014). Moreover, there is (both direct and indirect) evidence that treatment of the neck contributes to laryngeal dysfunction (Christianen et al. 2012; Soderstrom et al. 2017; Nevens et al. 2017; Sanguineti et al. 2007; Machtay et al. 2008). Radiotherapy alone delivered to small volumes for Stage T1 glottic larynx cancer, usually results in excellent voice quality, even if the whole larynx receives doses in excess of 60 Gy.

SHORT SUMMARY OF THE QUANTEC RECOMMENDATIONS WITH RELATIONSHIP TO MAIN LITERATURE ON WHICH THEY WERE BASED (DYSPHONIA)

Based on the limited amount of literature available at that time (Dornfeld et al. 2007; Rancati et al. 2009; Sanguineti et al. 2007), QUANTEC suggested to limit the mean dose to the noninvolved larynx to 40–45 Gy and the maximal dose to <63–66 Gy (Rancati et al. 2010).

Critical Review of the Literature

Literature data on voice changes after incidental irradiation of the larynx remain scarce.

In the early experience on patients treated with (chemo)radiotherapy for non-laryngeal tumors, mean laryngeal dose or the percentage of laryngeal volume receiving $\geq$50 Gy were independent predictors of grade $\geq$2 edema of the larynx (Sanguineti et al. 2007). The investigators suggested that the percentage of volume receiving $\geq$50 Gy and the mean laryngeal dose should be kept as low as possible, ideally <27% and <43.5 Gy, respectively, to minimize the edema (Sanguineti et al. 2007). Rancati et al. further modeled the data using grade $\geq$2 edema within 15 months after RT as an endpoint (Rancati et al. 2009) finding a significant volume effect for edema, consistent with a parallel architecture of the larynx for this endpoint. More recently the authors focused on a different subgroup of 124 patients, affected mostly by oropharyngeal SCC and undergoing IMRT associated in almost 90% of cases with systemic therapy. Both patient-reported (FACT-HN) and physician-scored (CTCAE) outcomes of laryngeal function were reported (Sanguineti et al. 2014). At multivariate analysis, mean laryngeal dose was an independent predictor of both grade $\geq$2 CTCAE dysphonia (OR = 1.10, 95%CI: $1.01 \div 1.20$, p = 0.025) and grade 0 – grade 1 voice changes scored by FACT-HN question 4 (OR = 1.11, 95% CI: $1.04 \div 1.18$, p = 0.001). Further stratification optimized by a ROC analysis showed that, to minimize the risk of grade 0 – grade 1 voice changes scored by FACT-HN question 4, the mean dose to the larynx should be kept $\leq$49.4 Gy (Sanguineti et al. 2014). It should be noted that in this study, based on the former experience on laryngeal edema (Sanguineti et al. 2007), V_{50Gy} of the larynx had been routinely constrained as a secondary dose objective to be <25%, accepting up to 30% as a minor violation. Therefore, the present toxicity profile should be considered as an 'unavoidable' risk of WF-IMRT for non-laryngeal primary tumors.

Vainshtein et al. found that the mean dose to the glottis independently predicts for both voice quality worsening and speech impairment (Vainshtein et al. 2014). Among patients with mean GL dose ≤20 Gy, >20–30 Gy, >30–40 Gy, >40–50 Gy, and >50 Gy, 10%, 32%, 25%, 30%, and 63%, respectively, reported worse voice quality at 12 months compared with before treatment (p = 0.011) (Vainshtein et al. 2014).

Technical implications, with particular reference to the benefits of WF-IMRT as opposed to SF-IMRT have been extensively debated previously and elsewhere (Dabaja et al. 2005; Lee et al. 2007).

SUGGESTION FOR DVH CONSTRAINTS/NTCP MODEL PARAMETERS (DYSPHONIA)

Whenever feasible, the mean laryngeal dose should be kept as low as possible at planning and possibly less than about 50 Gy or alternatively laryngeal V_{50Gy} should be <33%. Moreover, the mean GL dose should be lowered as much as reasonably achievable, aiming at ≤20 Gy when the larynx is not a target.

REFERENCES

Aaltonen LM, Rautiainen N, Sellman J, Saarilahti K, Makitie A, Rihkanen H, et al. Voice quality after treatment of early vocal cord cancer: A randomized trial comparing laser surgery with radiation therapy. *International Journal of Radiation Oncology, Biology, Physics.* 2014;90:255–60.

Alterio D, Gerardi MA, Cella L, Spoto R, Zurlo V, Sabbatini A, et al. Radiation-induced acute dysphagia: Prospective observational study on 42 head and neck cancer patients. *Strahlentherapie und Onkologie: Organ der Deutschen Rontgengesellschaft [et al].* 2017;193:971–81.

Ang KK. Larynx preservation clinical trial design: Summary of key recommendations of a consensus panel. *The Oncologist.* 2010;15:25–9.

Beadle BM, Liao KP, Giordano SH, Garden AS, Hutcheson KA, Lai SY, et al. Reduced feeding tube duration with intensity-modulated radiation therapy for head and neck cancer: A surveillance, epidemiology, and end results-medicare analysis. *Cancer.* 2017;123:283–93.

Berger PS, Bataini JP. Radiation-induced cranial nerve palsy. *Cancer.* 1977;40:152–5.

Bentzen SM, Trotti A. Evaluation of early and late toxicities in chemoradiation trials. *Journal of Clinical Oncology: Official Journal of the American Society of Clinical Oncology.* 2007;25:4096–103.

Bhide SA, Gulliford S, Fowler J, Rosenfelder N, Newbold K, Harrington KJ, et al. Characteristics of response of oral and pharyngeal mucosa in patients receiving chemo-IMRT for head and neck cancer using hypofractionated accelerated radiotherapy. *Radiotherapy and Oncology: Journal of the European Society for Therapeutic Radiology and Oncology.* 2010;97:86–91.

Brouwer CL, Steenbakkers RJ, Gort E, Kamphuis ME, van der Laan HP, Van't Veld AA, et al. Differences in delineation guidelines for head and neck cancer result in inconsistent reported dose and corresponding NTCP. *Radiotherapy and Oncology: Journal of the European Society for Therapeutic Radiology and Oncology.* 2014;111:148–52.

Brouwer CL, Steenbakkers RJ, Bourhis J, Budach W, Grau C, Gregoire V, et al. CT-based delineation of organs at risk in the head and neck region: DAHANCA, EORTC, GORTEC, HKNPCSG, NCIC CTG, NCRI, NRG Oncology and TROG consensus guidelines. *Radiotherapy and Oncology: Journal of the European Society for Therapeutic Radiology and Oncology.* 2015;117:83–90.

Caglar HB, Tishler RB, Othus M, Burke E, Li Y, Goguen L, et al. Dose to larynx predicts for swallowing complications after intensity-modulated radiotherapy. *International Journal of Radiation Oncology, Biology, Physics.* 2008;72:1110–8.

Cella DF, Tulsky DS, Gray G, Sarafian B, Linn E, Bonomi A, et al. The Functional Assessment of Cancer Therapy scale: Development and validation of the general measure. *Journal of Clinical Oncology: Official Journal of the American Society of Clinical Oncology.* 1993;11:570–9.

Chera BS, Fried D, Price A, Amdur RJ, Mendenhall W, Lu C, et al. Dosimetric predictors of patient-reported Xerostomia and Dysphagia with deintensified chemoradiation therapy for HPV-associated oropharyngeal squamous cell carcinoma. *International Journal of Radiation Oncology, Biology, Physics.* 2017;98:1022–7.

Cheng SS, Terrell JE, Bradford CR, Ronis DL, Fowler KE, Prince ME, et al. Variables associated with feeding tube placement in head and neck cancer. *Archives of Otolaryngology–Head and Neck Surgery.* 2006;132:655–61.

Chen AM, Li BQ, Lau DH, Farwell DG, Luu Q, Stuart K, et al. Evaluating the role of prophylactic gastrostomy tube placement prior to definitive chemoradiotherapy for head and neck cancer. *International Journal of Radiation Oncology, Biology, Physics.* 2010;78:1026–32.

Christianen ME, Langendijk JA, Westerlaan HE, van de Water TA, Bijl HP. Delineation of organs at risk involved in swallowing for radiotherapy treatment planning. *Radiotherapy and Oncology: Journal of the European Society for Therapeutic Radiology and Oncology.* 2011;101:394–402.

Christianen ME, Schilstra C, Beetz I, Muijs CT, Chouvalova O, Burlage FR, et al. Predictive modelling for swallowing dysfunction after primary (chemo)radiation: Results of a prospective observational study. *Radiotherapy and Oncology: Journal of the European Society for Therapeutic Radiology and Oncology.* 2012;105:107–14.

Christianen ME, Verdonck-de Leeuw IM, Doornaert P, Chouvalova O, Steenbakkers RJ, Koken PW, et al. Patterns of long-term swallowing dysfunction after definitive radiotherapy or chemoradiation. *Radiotherapy and Oncology: Journal of the European Society for Therapeutic Radiology and Oncology.* 2015;117:139–44.

Christianen ME, van der Schaaf A, van der Laan HP, Verdonck-de Leeuw IM, Doornaert P, Chouvalova O, et al. Swallowing sparing intensity modulated radiotherapy (SW-IMRT) in head and neck cancer: Clinical validation according to the model-based approach. *Radiotherapy and Oncology: Journal of the European Society for Therapeutic Radiology and Oncology.* 2016;118:298–303.

Choi M, Refaat T, Lester MS, Bacchus I, Rademaker AW, Mittal BB. Development of a standardized method for contouring the larynx and its substructures. *Radiation Oncology.* 2014;9:285.

Clavel S, Charron MP, Belair M, Delouya G, Fortin B, Despres P, et al. The role of computed tomography in the management of the neck after chemoradiotherapy in patients with head-and-neck cancer. *International Journal of Radiation Oncology, Biology, Physics.* 2012;82:567–73.

Cox JD, Stetz J, Pajak TF. Toxicity criteria of the Radiation Therapy Oncology Group (RTOG) and the European Organization for Research and Treatment of Cancer (EORTC). *International Journal of Radiation Oncology, Biology, Physics.* 1995;31:1341–6.

Caudell JJ, Schaner PE, Desmond RA, Meredith RF, Spencer SA, Bonner JA. Dosimetric factors associated with long-term dysphagia after definitive radiotherapy for squamous cell carcinoma of the head and neck. *International Journal of Radiation Oncology, Biology, Physics.* 2010;76:403–9.

Dabaja B, Salehpour MR, Rosen I, Tung S, Morrison WH, Ang KK, et al. Intensity-modulated radiation therapy (IMRT) of cancers of the head and neck: Comparison of split-field and whole-field techniques. *International Journal of Radiation Oncology, Biology, Physics.* 2005;63:1000–5.

De Ruyck K, Duprez F, Werbrouck J, Sabbe N, Sofie de L, Boterberg T, et al. A predictive model for dysphagia following IMRT for head and neck cancer: Introduction of the EMLasso technique. *Radiotherapy and Oncology: Journal of the European Society for Therapeutic Radiology and Oncology.* 2013;107:295–9.

DeSanto LW, Olsen KD, Perry WC, Rohe DE, Keith RL. Quality of life after surgical treatment of cancer of the larynx. *The Annals of Otology, Rhinology, and Laryngology.* 1995;104:763–9.

Dirix P, Abbeel S, Vanstraelen B, Hermans R, Nuyts S. Dysphagia after chemoradiotherapy for head-and-neck squamous cell carcinoma: Dose-effect relationships for the swallowing structures. *International Journal of Radiation Oncology, Biology, Physics.* 2009;75:385–92.

Dornfeld K, Simmons JR, Karnell L, Karnell M, Funk G, Yao M, et al. Radiation doses to structures within and adjacent to the larynx are correlated with long-term diet- and speech-related quality of life. *International Journal of Radiation Oncology, Biology, Physics.* 2007;68:750–7.

Duprez F, Madani I, De Potter B, Boterberg T, De Neve W. Systematic review of dose–volume correlates for structures related to late swallowing disturbances after radiotherapy for head and neck cancer. *Dysphagia.* 2013;28:337–49.

Eisbruch A, Schwartz M, Rasch C, Vineberg K, Damen E, Van As CJ, et al. Dysphagia and aspiration after chemoradiotherapy for head-and-neck cancer: Which anatomic structures are affected and can they be spared by IMRT? *International Journal of Radiation Oncology, Biology, Physics.* 2004;60:1425–39.

Feng FY, Kim HM, Lyden TH, Haxer MJ, Feng M, Worden FP, et al. Intensity-modulated radiotherapy of head and neck cancer aiming to reduce dysphagia: Early dose-effect relationships for the swallowing structures. *International Journal of Radiation Oncology, Biology, Physics.* 2007;68:1289–98.

Feng FY, Kim HM, Lyden TH, Haxer MJ, Worden FP, Feng M, et al. Intensity-modulated chemoradiotherapy aiming to reduce dysphagia in patients with oropharyngeal cancer: clinical and functional results. *Journal of Clinical Oncology: Official Journal of the American Society of Clinical Oncology.* 2010;28:2732–8.

Forastiere AA, Zhang Q, Weber RS, Maor MH, Goepfert H, Pajak TF, et al. Long-term results of RTOG 91-11: A comparison of three nonsurgical treatment strategies to preserve the larynx in patients with locally advanced larynx cancer. *Journal of Clinical Oncology: Official Journal of the American Society of Clinical Oncology.* 2013;31:845–52.

Fua TF, Corry J, Milner AD, Cramb J, Walsham SF, Peters LJ. Intensity-modulated radiotherapy for nasopharyngeal carcinoma: Clinical correlation of dose to the pharyngo-esophageal axis and dysphagia. *International Journal of Radiation Oncology, Biology, Physics.* 2007;67:976–81.

Fung K, Yoo J, Leeper HA, Bogue B, Hawkins S, Hammond JA, et al. Effects of head and neck radiation therapy on vocal function. *The Journal of Otolaryngology.* 2001;30:133–9.

Fung K, Lyden TH, Lee J, Urba SG, Worden F, Eisbruch A, et al. Voice and swallowing outcomes of an organ-preservation trial for advanced laryngeal cancer. *International Journal of Radiation Oncology, Biology, Physics.* 2005;63:1395–9.

Garden AS, Harris J, Trotti A, Jones CU, Carrascosa L, Cheng JD, et al. Long-term results of concomitant boost radiation plus concurrent cisplatin for advanced head and neck carcinomas: a phase II trial of the radiation therapy oncology group (RTOG 99-14). *International Journal of Radiation Oncology, Biology, Physics.* 2008;71:1351–5.

Gregoire V, Ang K, Budach W, Grau C, Hamoir M, Langendijk JA, et al. Delineation of the neck node levels for head and neck tumors: a 2013 update. DAHANCA, EORTC, HKNPCSG, NCIC CTG, NCRI, RTOG, TROG consensus guidelines. *Radiotherapy and Oncology: Journal of the European Society for Therapeutic Radiology and Oncology.* 2014;110:172–81.

Goepfert RP, Lewin JS, Barrow MP, Fuller CD, Lai SY, Song J, et al. Predicting two-year longitudinal MD Anderson Dysphagia Inventory outcomes after intensity modulated radiotherapy for locoregionally advanced oropharyngeal carcinoma. *The Laryngoscope.* 2017;127:842–8.

Gunn GB, Blanchard P, Garden AS, Zhu XR, Fuller CD, Mohamed AS, et al. Clinical outcomes and patterns of disease recurrence after intensity modulated proton therapy for oropharyngeal squamous Carcinoma. *International Journal of Radiation Oncology, Biology, Physics.* 2016;95:360–7.

Head MDA, Neck Cancer Symptom Working G. Beyond mean pharyngeal constrictor dose for beam path toxicity in non-target swallowing muscles: Dose-volume correlates of chronic radiation-associated dysphagia (RAD) after oropharyngeal intensity modulated radiotherapy. *Radiotherapy and Oncology: Journal of the European Society for Therapeutic Radiology and Oncology.* 2016;118:304–14.

Hogikyan ND, Sethuraman G. Validation of an instrument to measure voice-related quality of life (V-RQOL). *Journal of Voice: Official Journal of the Voice Foundation.* 1999;13:557–69.

Hutcheson KA, Bhayani MK, Beadle BM, Gold KA, Shinn EH, Lai SY, et al. Eat and exercise during radiotherapy or chemoradiotherapy for pharyngeal cancers: Use it or lose it. *JAMA Otolaryngology–Head and Neck Surgery.* 2013;139:1127–34.

Hutcheson KA, Yuk M, Hubbard R, Gunn GB, Fuller CD, Lai SY, et al. Delayed lower cranial neuropathy after oropharyngeal intensity-modulated radiotherapy: A cohort analysis and literature review. *Head and Neck.* 2017;39:1516–23.

Jensen K, Lambertsen K, Grau C. Late swallowing dysfunction and dysphagia after radiotherapy for pharynx cancer: Frequency, intensity and correlation with dose and volume parameters. *Radiotherapy and Oncology: Journal of the European Society for Therapeutic Radiology and Oncology.* 2007;85:74–82.

Jensen K, Lambertsen K, Torkov P, Dahl M, Jensen AB, Grau C. Patient assessed symptoms are poor predictors of objective findings. *Results from a Cross Sectional Study in Patients Treated with Radiotherapy for Pharyngeal Cancer. Acta Oncologica.* 2007;46:1159–68.

Karnell M, Maccracken E, Moran W, Vokes E, Haraf D, Panje W. Swallowing function following multispecialty organ preservation treatment of advanced head and neck-cancer. *Oncology Reports.* 1994;1:597–601.

Karnell LH, Funk GF, Tomblin JB, Hoffman HT. Quality of life measurements of speech in the head and neck cancer patient population. *Head and Neck.* 1999;21:229–38.

Kierkels RG, Wopken K, Visser R, Korevaar EW, van der Schaaf A, Bijl HP, et al. Multivariable normal tissue complication probability model-based treatment plan optimisation for grade 2-4 dysphagia and tube feeding dependence in head and neck radiotherapy. *Radiotherapy and Oncology: Journal of the European Society for Therapeutic Radiology and Oncology.* 2016;121:374–80.

King SN, Dunlap NE, Tennant PA, Pitts T. Pathophysiology of radiation-induced Dysphagia in head and neck cancer. *Dysphagia.* 2016;31:339–51.

Kornguth DG, Garden AS, Zheng Y, Dahlstrom KR, Wei Q, Sturgis EM. Gastrostomy in oropharyngeal cancer patients with ERCC4 (XPF) germline variants. *International Journal of Radiation Oncology, Biology, Physics.* 2005;62:665–71.

Kumar R, Madanikia S, Starmer H, Yang W, Murano E, Alcorn S, et al. Radiation dose to the floor of mouth muscles predicts swallowing complications following chemoradiation in oropharyngeal squamous cell carcinoma. *Oral Oncology.* 2014;50:65–70.

Langmore S, Krisciunas GP, Miloro KV, Evans SR, Cheng DM. Does PEG use cause dysphagia in head and neck cancer patients? *Dysphagia.* 2012;27:251–9.

Langendijk JA, Doornaert P, Rietveld DH, Verdonck-de Leeuw IM, Leemans CR, Slotman BJ. A predictive model for swallowing dysfunction after curative radiotherapy in head and neck cancer. *Radiotherapy and Oncology: Journal of the European Society for Therapeutic Radiology and Oncology.* 2009;90:189–95.

Lee N, Mechalakos J, Puri DR, Hunt M. Choosing an intensity-modulated radiation therapy technique in the treatment of head-and-neck cancer. *International Journal of Radiation Oncology, Biology, Physics.* 2007;68:1299–309.

Levendag PC, Teguh DN, Voet P, van der Est H, Noever I, de Kruijf WJ, et al. Dysphagia disorders in patients with cancer of the oropharynx are significantly affected by the radiation therapy dose to the superior and middle constrictor muscle: A dose-effect relationship. *Radiotherapy* and Oncology: Journal of the European Society for Therapeutic Radiology and Oncology. 2007;85:64–73.

Machtay M, Moughan J, Trotti A, Garden AS, Weber RS, Cooper JS, et al. Factors associated with severe late toxicity after concurrent chemoradiation for locally advanced head and neck cancer: An RTOG analysis. Journal of Clinical Oncology: Official Journal of the American Society of Clinical Oncology. 2008;26:3582–9.

Madhoun MF, Blankenship MM, Blankenship DM, Krempl GA, Tierney WM. Prophylactic PEG placement in head and neck cancer: How many feeding tubes are unused (and unnecessary)? World Journal of Gastroenterology. 2011;17:1004–8.

Mangar S, Slevin N, Mais K, Sykes A. Evaluating predictive factors for determining enteral nutrition in patients receiving radical radiotherapy for head and neck cancer: A retrospective review. Radiotherapy and Oncology: Journal of the European Society for Therapeutic Radiology and Oncology. 2006;78:152–8.

Matuschek C, Bolke E, Geigis C, Kammers K, Ganswindt U, Scheckenbach K, et al. Influence of dosimetric and clinical criteria on the requirement of artificial nutrition during radiotherapy of head and neck cancer patients. Radiotherapy and Oncology: Journal of the European Society for Therapeutic Radiology and Oncology. 2016;120:28–35.

Mavroidis P, Price A, Fried D, Kostich M, Amdur R, Mendenhall W, et al. Dose-volume toxicity modeling for de-intensified chemo-radiation therapy for HPV-positive oropharynx cancer. Radiotherapy and Oncology: Journal of the European Society for Therapeutic Radiology and Oncology. 2017;124:240–7.

Mekhail TM, Adelstein DJ, Rybicki LA, Larto MA, Saxton JP, Lavertu P. Enteral nutrition during the treatment of head and neck carcinoma: Is a percutaneous endoscopic gastrostomy tube preferable to a nasogastric tube? Cancer. 2001;91:1785–90.

Mills E, Eyawo O, Lockhart I, Kelly S, Wu P, Ebbert JO. Smoking cessation reduces postoperative complications: A systematic review and meta-analysis. The American Journal of Medicine. 2011;124:144–54 e8.

Mortensen HR, Jensen K, Grau C. Aspiration pneumonia in patients treated with radiotherapy for head and neck cancer. Acta Oncologica. 2013;52:270–6.

Mortensen HR, Jensen K, Aksglaede K, Behrens M, Grau C. Late dysphagia after IMRT for head and neck cancer and correlation with dose-volume parameters. Radiotherapy and Oncology: Journal of the European Society for Therapeutic Radiology and Oncology. 2013;107: 288–94.

Mortensen HR, Overgaard J, Jensen K, Specht L, Overgaard M, Johansen J, et al. Factors associated with acute and late dysphagia in the DAHANCA 6 & 7 randomized trial with accelerated radiotherapy for head and neck cancer. Acta Oncologica. 2013;52:1535–42.

Mortensen HR, Jensen K, Aksglaede K, Lambertsen K, Eriksen E, Grau C. Prophylactic swallowing exercises in head and neck cancer radiotherapy. Dysphagia. 2015;30:304–14.

Monti S, Palma G, D'Avino V, Gerardi M, Marvaso G, Ciardo D, et al. Voxel-based analysis unveils regional dose differences associated with radiation-induced morbidity in head and neck cancer patients. Scientific Reports. 2017;7:7220.

Murray KA, Brzozowski LA. Swallowing in patients with tracheotomies. AACN Clinical Issues. 1998;9:416–26; quiz 56–8.

Munshi A, Pandey MB, Durga T, Pandey KC, Bahadur S, Mohanti BK. Weight loss during radiotherapy for head and neck malignancies: What factors impact it? Nutrition and Cancer. 2003;47:136–40.

Nevens D, Duprez F, Daisne JF, Dok R, Belmans A, Voordeckers M, et al. Reduction of the dose of radiotherapy to the elective neck in head and neck squamous cell carcinoma; a randomized clinical trial. Effect on late toxicity and tumor control. Radiotherapy and Oncology: Journal of the European Society for Therapeutic Radiology and Oncology. 2017;122:171–7.

Nevens D, Deschuymer S, Langendijk JA, Daisne JF, Duprez F, De Neve W, et al. Validation of the total dysphagia risk score (TDRS) in head and neck cancer patients in a conventional and a partially accelerated radiotherapy scheme. *Radiotherapy and Oncology: Journal of the European Society for Therapeutic Radiology and Oncology*. 2016;118:293–7.

Nguyen NP, Frank C, Moltz CC, Vos P, Smith HJ, Karlsson U, et al. Impact of dysphagia on quality of life after treatment of head-and-neck cancer. *International Journal of Radiation Oncology, Biology, Physics*. 2005;61:772–8.

Nutting CM, Morden JP, Harrington KJ, Urbano TG, Bhide SA, Clark C, et al. Parotid-sparing intensity modulated versus conventional radiotherapy in head and neck cancer (PARSPORT): A phase 3 multicentre randomised controlled trial. *The Lancet Oncology*. 2011;12:127–36.

Pratesi N, Mangoni M, Mancini I, Paiar F, Simi L, Livi L, et al. Association between single nucleotide polymorphisms in the XRCC1 and RAD51 genes and clinical radiosensitivity in head and neck cancer. *Radiotherapy and Oncology: Journal of the European Society for Therapeutic Radiology and Oncology*. 2011;99:356–61.

Poulsen MG, Riddle B, Keller J, Porceddu SV, Tripcony L. Predictors of acute grade 4 swallowing toxicity in patients with stages III and IV squamous carcinoma of the head and neck treated with radiotherapy alone. *Radiotherapy and Oncology: Journal of the European Society for Therapeutic Radiology and Oncology*. 2008;87:253–9.

Popovtzer A, Cao Y, Feng FY, Eisbruch A. Anatomical changes in the pharyngeal constrictors after chemo-irradiation of head and neck cancer and their dose-effect relationships: MRI-based study. *Radiotherapy and Oncology: Journal of the European Society for Therapeutic Radiology and Oncology*. 2009;93:510–5.

Petkar I, Rooney K, Roe JW, Patterson JM, Bernstein D, Tyler JM, et al. DARS: A phase III randomised multicentre study of dysphagia- optimised intensity- modulated radiotherapy (Do-IMRT) versus standard intensity- modulated radiotherapy (S-IMRT) in head and neck cancer. *BMC Cancer*. 2016;16:770.

Ramaekers BL, Joore MA, Grutters JP, van den Ende P, Jong J, Houben R, et al. The impact of late treatment-toxicity on generic health-related quality of life in head and neck cancer patients after radiotherapy. *Oral oncology*. 2011;47:768–74.

Rancati T, Fiorino C, Sanguineti G. NTCP modeling of subacute/late laryngeal edema scored by fiberoptic examination. *International Journal of Radiation Oncology, Biology, Physics*. 2009;75:915–23.

Rancati T, Schwarz M, Allen AM, Feng F, Popovtzer A, Mittal B, et al. Radiation dose-volume effects in the larynx and pharynx. *International Journal of Radiation Oncology, Biology, Physics*. 2010;76:S64–9.

Rieger JM, Zalmanowitz JG, Wolfaardt JF. Functional outcomes after organ preservation treatment in head and neck cancer: A critical review of the literature. *International Journal of Oral and Maxillofacial Surgery*. 2006;35:581–7.

Ronis DL, Duffy SA, Fowler KE, Khan MJ, Terrell JE. Changes in quality of life over 1 year in patients with head and neck cancer. *Archives of Otolaryngology-Head and Neck Surgery*. 2008;134:241–8.

Rosen CA, Lee AS, Osborne J, Zullo T, Murry T. Development and validation of the voice handicap index-10. *The Laryngoscope*. 2004;114:1549–56.

Rosenbek JC, Robbins JA, Roecker EB, Coyle JL, Wood JL. A penetration-aspiration scale. *Dysphagia*. 1996;11:93–8.

Rosenthal DI, Trotti A. Strategies for managing radiation-induced mucositis in head and neck cancer. *Seminars in Radiation Oncology*. 2009;19:29–34.

Russi EG, Corvo R, Merlotti A, Alterio D, Franco P, Pergolizzi S, et al. Swallowing dysfunction in head and neck cancer patients treated by radiotherapy: Review and recommendations of the supportive task group of the Italian Association of Radiation Oncology. *Cancer Treatment Reviews*. 2012;38:1033–49.

Sanguineti G, Endres EJ, Gunn BG, Parker B. Is there a "mucosa-sparing" benefit of IMRT for head-and-neck cancer? *International Journal of Radiation Oncology, Biology, Physics.* 2006;66:931–8.

Sanguineti G, Adapala P, Endres EJ, Brack C, Fiorino C, Sormani MP, et al. Dosimetric predictors of laryngeal edema. *International Journal of Radiation Oncology, Biology, Physics.* 2007;68:741–9.

Sanguineti G, Gunn GB, Parker BC, Endres EJ, Zeng J, Fiorino C. Weekly dose-volume parameters of mucosa and constrictor muscles predict the use of percutaneous endoscopic gastrostomy during exclusive intensity-modulated radiotherapy for oropharyngeal cancer. *International Journal of Radiation Oncology, Biology, Physics.* 2011;79:52–9.

Sanguineti G, Sormani MP, Marur S, Gunn GB, Rao N, Cianchetti M, et al. Effect of radiotherapy and chemotherapy on the risk of mucositis during intensity-modulated radiation therapy for oropharyngeal cancer. *International Journal of Radiation Oncology, Biology, Physics.* 2012;83:235–42.

Sanguineti G, Rao N, Gunn B, Ricchetti F, Fiorino C. Predictors of PEG dependence after IMRT+/-chemotherapy for oropharyngeal cancer. *Radiotherapy and Oncology: Journal of the European Society for Therapeutic Radiology and Oncology.* 2013;107:300–4.

Sanguineti G, Ricchetti F, McNutt T, Wu B, Fiorino C. Dosimetric predictors of dysphonia after intensity-modulated radiotherapy for oropharyngeal carcinoma. *Clinical Oncology.* 2014;26:32–8.

Schwartz DL, Hutcheson K, Barringer D, Tucker SL, Kies M, Holsinger FC, et al. Candidate dosimetric predictors of long-term swallowing dysfunction after oropharyngeal intensity-modulated radiotherapy. *International Journal of Radiation Oncology, Biology, Physics.* 2010;78:1356–65.

Schweinfurth JM, Boger GN, Feustel PJ. Preoperative risk assessment for gastrostomy tube placement in head and neck cancer patients. *Head and Neck.* 2001;23:376–82.

Shaw SM, Flowers H, O'Sullivan B, Hope A, Liu LW, Martino R. The effect of prophylactic percutaneous endoscopic gastrostomy (PEG) tube placement on swallowing and swallow-related outcomes in patients undergoing radiotherapy for head and neck cancer: A systematic review. *Dysphagia.* 2015;30:152–75.

Stoicheff ML. Voice following radiotherapy. *The Laryngoscope.* 1975;85:608–18.

Soderstrom K, Nilsson P, Laurell G, Zackrisson B, Jaghagen EL. Dysphagia - results from multivariable predictive modelling on aspiration from a subset of the ARTSCAN trial. *Radiotherapy and Oncology: Journal of the European Society for Therapeutic Radiology and Oncology.* 2017;122:192–9.

Vainshtein JM, Griffith KA, Feng FY, Vineberg KA, Chepeha DB, Eisbruch A. Patient-reported voice and speech outcomes after whole-neck intensity modulated radiation therapy and chemotherapy for oropharyngeal cancer: Prospective longitudinal study. *International Journal of Radiation Oncology, Biology, Physics.* 2014;89:973–80.

van der Laan HP, Christianen ME, Bijl HP, Schilstra C, Langendijk JA. The potential benefit of swallowing sparing intensity modulated radiotherapy to reduce swallowing dysfunction: An in silico planning comparative study. *Radiotherapy and Oncology: Journal of the European Society for Therapeutic Radiology and Oncology.* 2012;103:76–81.

van der Molen L, Heemsbergen WD, de Jong R, van Rossum MA, Smeele LE, Rasch CR, et al. Dysphagia and trismus after concomitant chemo-Intensity-Modulated Radiation Therapy (chemo-IMRT) in advanced head and neck cancer; dose-effect relationships for swallowing and mastication structures. *Radiotherapy and Oncology: Journal of the European Society for Therapeutic Radiology and Oncology.* 2013;106:364–9.

van der Laan HP, van de Water TA, van Herpt HE, Christianen ME, Bijl HP, Korevaar EW, et al. The potential of intensity-modulated proton radiotherapy to reduce swallowing dysfunction in the treatment of head and neck cancer: A planning comparative study. *Acta Oncologica.* 2013;52:561–9.

van der Laan HP, Bijl HP, Steenbakkers RJ, van der Schaaf A, Chouvalova O, Vemer-van den Hoek JG, et al. Acute symptoms during the course of head and neck radiotherapy or chemoradiation are strong predictors of late dysphagia. *Radiotherapy and Oncology: Journal of the European Society for Therapeutic Radiology and Oncology.* 2015;115:56–62.

Vlacich G, Spratt DE, Diaz R, Phillips JG, Crass J, Li CI, et al. Dose to the inferior pharyngeal constrictor predicts prolonged gastrostomy tube dependence with concurrent intensity-modulated radiation therapy and chemotherapy for locally-advanced head and neck cancer. *Radiotherapy and Oncology: Journal of the European Society for Therapeutic Radiology and Oncology.* 2014;110:435–40.

Ward MC, Adelstein DJ, Bhateja P, Nwizu TI, Scharpf J, Houston N, et al. Severe late dysphagia and cause of death after concurrent chemoradiation for larynx cancer in patients eligible for RTOG 91-11. *Oral oncology.* 2016;57:21–6.

Withers HR, Peters LJ, Taylor JM, Owen JB, Morrison WH, Schultheiss TE, et al. Late normal tissue sequelae from radiation therapy for carcinoma of the tonsil: Patterns of fractionation study of radiobiology. *International Journal of Radiation Oncology, Biology, Physics.* 1995;33:563–8.

Wopken K, Bijl HP, van der Schaaf A, van der Laan HP, Chouvalova O, Steenbakkers RJ, et al. Development of a multivariable normal tissue complication probability (NTCP) model for tube feeding dependence after curative radiotherapy/chemo-radiotherapy in head and neck cancer. *Radiotherapy and Oncology: Journal of the European Society for Therapeutic Radiology and Oncology.* 2014;113:95–101.

CHAPTER 10

Thorax

Lungs and Esophagus

Daniel Schanne, Jan Unkelbach, and Matthias Guckenberger

CONTENTS

Introduction 244
  Definition of Relevant Anatomical Structures 244
    Lung 244
    Esophagus 244
    Trachea and Large Bronchi 244
  Evaluation of Toxicity Endpoints 245
    Classification Systems 245
    Diagnostic Approach 245
  Clinical Significance of the Injury 247
  Therapeutic Management and Prevention 247
    Lung 247
    Esophagus 248
    Prevention 248
Short Summary of the QUANTEC Recommendations 249
  The Lyman-Kutcher-Burman Model 249
    Lung: Radiation Pneumonitis 249
    Esophagus: Acute Esophagitis 250
Normal Tissue Complication Probability (NTCP) Models for Standard Fractionation
Beyond the LKB Model 250
    Lung: Radiation Pneumonitis. Literature Review 250
    Esophagus: Acute Esophagitis. Literature Review 251
Non-Dosimetry Predictors 251
NTCP Models for Treatment Plan Optimisation in Intensity-Modulated Radiation
Therapy (IMRT) and Intensity-Modulated Proton Therapy (IMPT) 254
Extrapolation of Dose Constraints to Experimental Fractionation Schemes 256
Toxicities and NTCP Models for Hypofractionated
Treatments and SBRT 258
    Relevant Anatomy and Toxicities 258

Literature Review..259
    Lung...259
    Esophagus and Bronchi...260
Recommendation for Dose-Volume Constraints and NTCP Model Parameters..............262
References..262

INTRODUCTION

Definition of Relevant Anatomical Structures

Lung

The lungs fill the largest share of the thoracic cavity and are the primary site of several carcinomas and a frequent site of metastasis for multiple other tumors. Therefore, parts of the lung will be affected by radiation during virtually any treatment in the thoracic region and the resulting toxicities and changes in organ structure will be observed commonly in these patients. Breathing motions of both healthy tissue and the target volume are of significant importance in the lung because their compensation via increased safety margin results in larger volumes of pulmonary tissue exposed to high-dose irradiation. Considered a parallel organ-at-risk (OAR), pulmonary toxicities following radiation therapy depend largely on the percentage of irradiated healthy lung tissue.

Esophagus

As an anatomical structure that traverses the entire thorax from the pharynx to the gastric sphincter, the esophagus is a major organ-at-risk for most thoracic irradiation plans. Its beginning and end are anatomically defined as spanning from the cricoid cartilage (around vertebra C6) cranially to the cardiac orifice of the stomach (T11) caudally. Within the thoracic cavity, it is situated in the mediastinum between the trachea and the thoracic vertebrae T1-T11 before it pierces through the diaphragm into the abdomen. Its position is somewhat variable in the lateral and dorsoventral direction due to the peristaltic movements of the smooth-muscle layers that comprise its wall. This should be considered upon delineation on planning CTs. As a hollow organ that functions as a conduit for the transport of liquids and food, the esophagus is defined as a serial OAR and therefore perforation is among the most serious of radiation-induced complications. More commonly affected during the course of a radiation treatment is the stratified squamous epithelium that is lining its interior surface and often becomes inflamed.

Trachea and Large Bronchi

Trachea and bronchi connect the pharynx to the lungs and are essential anatomical structures to ensure transport of air into the alveoli and enable exchange of gases. Beginning at the cricoid cartilage, the trachea passes through the mediastinum and bifurcates into the two main bronchi at the level of the manubriosternal junction (or vertebra T4). The multilevel, dichotomous structure of the bronchial tree makes the upper parts of this system a serial OAR with potentially severe toxicity if damaged by perforation. This is especially true for hypo-fractionated treatments of centrally located tumors and has posed a major obstacle for these treatments.

Evaluation of Toxicity Endpoints

Classification Systems

The three most-commonly used classification systems for toxicities (CTCAE, RTOG and LENT/SOMA) are presented and discussed in Chapter 2. Studies trying to evaluate the concordance of these classifications are sparse. One study attempting a comparison between RTOG and CTCAE (version 2) in thoracic irradiation concluded that the two systems are not reliably comparable for higher-grade toxicities and suggested that the CTCAE may be more appropriate to use because it relies solely on clinical appearance and avoids the RTOG's mix of radiographic findings and subjective symptoms. Furthermore, the RTOG system does not consider preexisting clinical symptoms and may therefore falsely identify them as late toxicities (Faria et al. 2009). Attempts to transfer scores from one classification to another should therefore be made with caution (Table 10.1).

Diagnostic Approach

Acute radiation-induced lung disease (RILD), manifesting as pneumonitis, develops approximately 6–12 weeks after irradiation. The most frequent symptoms are nonproductive cough and dyspnea, sometimes accompanied by fever and chest pain. Clinical examination of the patient is often normal but sometimes reveals a crackling sound upon auscultation as well as dullness to percussion. Skin irritation from the recent course of irradiation may still be present but most frequently will not be correlated with the degree of pulmonary toxicity, especially when modern conformal radiation techniques are used. Differential diagnoses include viral, bacterial and fungal pneumonia, exacerbation of underlying diseases (e.g., Chronic obstructive pulmonary disease, COPD, or chronic fibrosis), lymphangitic carcinomatosis, and medication-induced lung injury, e.g., by chemotherapy. The further diagnostic approach should include imaging of the thorax, with CT being preferred over conventional chest radiography because of its higher sensitivity and comparability to pre-treatment scans (Choi et al. 2004). Characteristic for acute RILD are ground-glass opacities with maximum density around the tumor and often not following anatomic boundaries but the pattern of dose distribution (Figure 10.1). However, in some cases consolidation may be seen outside the irradiated area (Bell et al. 1988). Over time, these zones usually progress towards fibrosis unless corticosteroid treatment is initiated (Ikezoe et al. 1988). In addition to imaging, pulmonary function studies can be useful to distinguish RILD from exacerbation of pre-existing conditions like COPD. Carbon monoxide diffusion capacity is often decreased; however, the specificity of this finding is low as multiple other conditions can cause it (Lopez et al. 2012). Invasive diagnostic procedures such as bronchoscopy are usually not necessary and should be limited to situations when non-invasive procedures show equivocal results. Because the differential diagnosis of acute RILD can be difficult and the clinical spectrum may vary from no symptoms to a potentially life-threatening condition, a regular follow-up schedule and close collaboration with the primary-care physician should be ensured for patients after thoracic irradiation.

Esophagitis usually develops within three weeks of radiotherapy and is characterized by dysphagia, odynophagia and sometimes a stinging pain in the posterior chest area. Severe complications like perforation, tracheoesophageal fistulas, and ulceration can occur late

TABLE 10.1 CTCAE Classification for Common Thoracic Toxicities

Adverse Event	1	2	3	4
Esophageal fistula	Asymptomatic; clinical or diagnostic observations only; intervention not indicated	Symptomatic; altered GI function	Severely altered GI function; tube feeding, TPN or hospitalization indicated; elective operative intervention indicated	Life-threatening consequences; urgent intervention indicated
Esophagitis	Asymptomatic; clinical or diagnostic observations only; intervention not indicated	Symptomatic; altered eating/swallowing; oral supplements indicated	Severely altered eating/swallowing; tube feeding, TPN or hospitalization indicated	Life-threatening consequences; urgent operative intervention indicated
Esophageal stenosis	Asymptomatic; clinical or diagnostic observations only; intervention not indicated	Symptomatic; altered GI function	Severely altered GI function; tube feeding; hospitalization indicated; elective operative intervention indicated	Life-threatening consequences; urgent operative intervention indicated
Esophageal ulcer	Asymptomatic; clinical or diagnostic observations only; intervention not indicated	Symptomatic; altered GI function; limiting instrumental ADL	Severely altered GI function; TPN indicated; elective operative or endoscopic intervention indicated; limiting self care ADL; disabling	Life-threatening consequences; urgent operative intervention indicated
Cough	Mild symptoms; nonprescription intervention indicated	Moderate symptoms, medical intervention indicated; limiting instrumental ADL	Severe symptoms; limiting self care ADL	–
Dyspnea	Shortness of breath with moderate exertion	Shortness of breath with minimal exertion; limiting instrumental ADL	Shortness of breath at rest; limiting self care ADL	Life-threatening consequences; urgent intervention indicated
Pneumonitis	Asymptomatic; clinical or diagnostic observations only; intervention not indicated	Symptomatic; medical intervention indicated; limiting instrumental ADL	Severe symptoms; limiting self care ADL; oxygen indicated	Life-threatening respiratory compromise; urgent intervention indicated (e.g., tracheotomy or intubation)
Pulmonary fibrosis	Mild hypoxemia; radiologic pulmonary fibrosis	Moderate hypoxemia; evidence of pulmonary hypertension; radiographic pulmonary fibrosis 25–50%	Severe hypoxemia; evidence of right-sided heart failure; radiographic pulmonary fibrosis >50–75%	Life-threatening consequences (e.g., hemodynamic/pulmonary complications); intubation with ventilatory support indicated; radiographic pulmonary fibrosis >75% with severe honeycombing

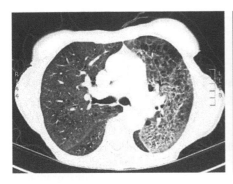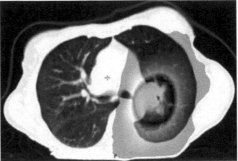

FIGURE 10.1   Computed tomography (CT) of radiation pneumonitis following several weeks after thoracic radiation (left). Pathological changes follow the dose distribution on planning CT (right).

and are rare side effects. In many cases of sever esophageal toxicity, the initial or recurrent cancer involved the esophagus and the severe natural course of the disease therefore contribute to the development of the toxic event. Clinical diagnosis is sufficient in most cases, but diagnostic procedures such as upper esophagoscopy or radiography may be indicated for higher-grade toxicities. Several high-grade toxicities such as bleeding, ulceration, and fistulas can occur years after radiotherapy, so a regular follow-up schedule should be maintained with a low threshold for further diagnostics in case of symptoms.

Clinical Significance of the Injury

The clinical significance of adverse events strongly depends on the parallel or serial nature of the OAR (Table 10.2). Whereas parallel OAR such as the lung consist of multiple units that can function independently from each other, serial OAR like the esophagus, trachea, and large bronchi, undergo significant loss-of-function even in case of a small tissue defect. Dosimetric parameters to predict and prevent toxicity for parallel OAR therefore often focus on the percentage of affected functional units that receive a predefined threshold dose. To evaluate complication risks of serial OAR, more factors have to be considered, as complications like perforation and bleeding additionally depend on maximum point doses to small parts of the organ.

Therapeutic Management and Prevention

Lung

Most treatments for RILD have not been subjected to prospective, randomized clinical studies. Nevertheless, several treatments have emerged as a de-facto standard over time. These can be broadly divided in supportive care, glucocorticoids/anti-inflammatory agents and a combined group of preventive and alternative substances. Supportive treatment can include oxygen, antitussives (e.g., codeine) and treatment of relevant comorbidities. Patients with no or minimal symptoms generally do not need oral medication but should be monitored in regular intervals clinically and with imaging. If cough is present, antitussive medication or inhaled glucocorticoids can be prescribed, e.g., budesonide (Henkenberens et al. 2016).In case of symptoms that interfere with normal activity such

TABLE 10.2   Clinical Significance of Common Toxicities

Organ	Adverse Event	Time of Occurrence (after Onset of Irradiation)	Symptoms/Findings	Clinical Significance
Esophagus	Esophagitis	2–3 weeks	Pain, dysphagia	Mostly self-limiting, proper nutrition has to be ensured, pain management is indicated
	Stricture	Weeks – years	Dysphagia	Impaired nutrition, increased risk for aspiration pneumonia, interventional or surgical treatment may be necessary
	Perforation	Weeks – years	Fistula (pneumonia, dyspnea), abscess, bleeding, dysphagia	Severe toxicity, interventional or surgical treatment necessary
Trachea/Large Bronchi	Stenosis	Weeks – years	Dyspnea, atelectasis, fistula	Depending on symptoms and grade, interventional or surgical treatment may be necessary
	Perforation	Weeks – years	Dyspnea, bleeding	Severe toxicity, interventional or surgical treatment necessary
Lung	Pneumonitis	6–24 weeks	Dyspnea, cough, fever	Variable, from minimal symptoms to potentially fatal complications (e.g., pneumonia)
	Fibrosis	>24 weeks – years	Dyspnea	Supportive treatment depending on the remaining functional lung volume

as dyspnea or severe coughing, oral glucocorticoids (e.g., prednisone 60 mg/day) can be administered for several weeks or until symptoms are controlled and then can be slowly tapered-off. The usual precautions that apply to systemic glucocorticoid treatment should be considered, e.g., proton pump inhibitors and pneumocystis carinii prophylaxis. For corticoid-refractory cases, some data exists for use of azathioprine or cyclosporin A as an alternative treatment (McCarty et al. 1996; Muraoka et al. 2002). Severe cases of RILD will warrant treatment in an in-patient setting with availability of specialist care.

Esophagus

Acute esophagitis is self-limiting in most cases, so treatment is usually symptomatic with topical anesthetics, proton-pump inhibitors and dietary modifications. Stricture is a chronic side effect that can lead to persisting dysphagia and odynophagia and can be treated interventionally with dilation or surgery if necessary. Specialized care may be necessary for severe toxicities such as fistulas, bleeding, and ulceration.

Prevention

Several agents have been proposed to protect normal lung tissue from the effects of radiation. Pentoxifylline is a substance that has been successfully evaluated in a number of other radiation settings to prevent fibrosis and reduce other complications of irradiation (Hille et al. 2005; Delanian et al. 2003; Dion et al. 1990). Regarding pulmonary toxicity, Ozturk et al.

evaluated 40 patients treated for lung or breast cancer and could find a statistically-significant difference in favour of the pentoxifylline group. Carbon monoxide perfusion was increased at 3 and 6 months after therapy and a decrease in higher-grade lung injury on imaging (Ozturk et al. 2004). Amifostine is another agent that has been tested in a clinical trial of Non-small-cell lung cancer (NSCLC) patients. Antonadou et al. included 146 patients and compared radiation monotherapy vs. co-treatment with amifostine in a randomized phase II trial. Occurrence of grade >2 pneumonitis and grade >2 esophagitis were decreased significantly and at 6 months post-RT, lung fibrosis was present in 53% vs. 28% of patients in the monotherapy and experimental group, respectively (Antonadou et al. 2001). Several other agents such as anti-TGFβ1 (Transforming growth factor beta 1) and angiotensin converting enzyme (ACE) inhibitors have been proposed as protective co-treatments of radiation therapy but have not been tested rigorously enough in the clinical setting to warrant treatment recommendations (Kwok et al. 1998; Small et al. 2018). Based on the overall weak evidence, no prophylactic medication is current recommended.

Therefore, the most obvious and effective method to prevent RILD and esophageal toxicities is the reduction of irradiated normal lung volume and the dose that it receives.

SHORT SUMMARY OF THE QUANTEC RECOMMENDATIONS

The Lyman-Kutcher-Burman Model

For standard fractionated treatments, radiation pneumonitis and acute esophagitis are main dose limiting toxicities. The QUANTEC review from 2010 contains organ-specific papers on dose-volume effects in the lung (Marks et al. 2010) and the esophagus (Werner-Wasik et al. 2010) focusing on these two complications. In both cases, the Lyman-Kutcher-Burman (LKB) model has been widely studied. Details on the LKB model can be found in Chapter 2, we here only remind it is characterized by three parameters: D_{50} (the equivalent uniform dose to the whole organ which results in 50% toxicity probability if the organ is irradiated uniformly), m (a parameter to control the slope of the sigmoid curve) and n (the volume effect parameter)

$$\text{NTCP}_{RP}(d) = \frac{\exp\left[b_0 + b_1 \, \text{gEUD}(d;n)\right]}{1 + \exp\left[b_0 + b_1 \, \text{gEUD}(d;n)\right]}$$

$$\text{gEUD}(d;n) = \left(\frac{1}{M} \sum_{i=1}^{M} d_i^{(1/n)}\right)^n$$

Lung: Radiation Pneumonitis

For radiation-induced pneumonitis, the volume parameter n is estimated to be close to 1, reflecting the fact that the lung is a parallel rather than a serial organ. In this case, the generalized-Equivalent Uniform Dose (gEUD) equals the mean lung dose (MLD) and the resulting model is referred to as the MLD model. The MLD model is widely used owing to its simplicity, and due to the lack of evidence that other dose-volume parameters better explain the observed rates of radiation pneumonitis.

The QUANTEC report contains a fit of the LBK model to the pooled data from multiple published studies (containing different grades of radiation pneumonitis as the endpoint). This resulted in model parameters of $D_{50} = 31.4$ Gy [95% confidence interval 29.0–34.7 Gy], and $m = 0.45$ [95% confidence interval 0.39–0.51] (Marks et al. 2010). At a MLD level of 20 Gy, the probability of radiation pneumonitis is approximately 20% based on these parameters and increases by approximately 2% per Gy.

The dose-response relationship for radiation pneumonitis according to the MLD model is shallow. Hence, no clear threshold tolerance dose can be derived. Arguably, treatment planning should therefore follow ALARA ("as low as reasonably achievable") principle. In addition, it appears that patient characteristics other than radiation dose play a significant role in determining the risk of radiation pneumonitis.

Esophagus: Acute Esophagitis

While the lungs are generally considered a parallel organ, the esophagus is considered a rather serial organ. Several studies fitted the LKB model to outcome data for acute esophagitis(Belderbos et al. 2005; Chapet et al. 2005). These fits resulted in $D_{50} \approx 50$ Gy, $m \approx 0.30$ and intermediate values for the volume parameter of $n \approx 0.5$ (Werner-Wasik et al. 2010; Belderbos et al. 2005) (Chapet et al. 2005). Although the confidence intervals for the volume parameter are large, these results are consistent with the general assumption that the esophagus is a partially serial structure.

NORMAL TISSUE COMPLICATION PROBABILITY (NTCP) MODELS FOR STANDARD FRACTIONATION BEYOND THE LKB MODEL

Lung: Radiation Pneumonitis. Literature Review

Models based on MLD remain the most widely used tools to predict radiation-induced toxicities in the lung. Other models have been proposed and will be reviewed here, however, few studies have attempted to compare different models with a sizeable patient cohort so there is a general lack of data in the field. NTCP models based on MLD inherently assume an equal response of the tissue to radiation over the entire organ. In contrast, several preclinical and clinical studies have demonstrated that the NTCP may not always be accurate if derived from the MLD. Preclinical studies in various model organisms (mice, rats, pigs) show limitations of MLD-based prediction algorithms depending on the irradiated region of the lung and dose distribution within it. In line with a parallel organ-at-risk, the percentage of damaged lung seems to determine toxicity in these studies. The irradiated region was also of importance with a lower sensitivity to injury in apical parts of the lung (Liao et al. 1995; Herrmann et al. 2015; Semenenko et al. 2008). Some clinical data confirm the limited generalizability of the MLD and suggest a relationship between the high-dose volume and resulting pulmonary toxicity. This has spurred the concept to redistribute the dose from high-dose areas ("a lot to a little") to larger areas receiving lower doses ("a little to a lot") to minimize toxicity even in case of equal MLD (Choi et al. 2004; Tucker et al. 2013b; Willner et al. 2003). Adapting to these findings, the field has gradually expanded the evaluation toolbox by introducing new parameters, e.g., limiting the dose applied to partial organ volumes (V_{XGy}, relative organ volume receiving more than X Gy).

Multiple studies have been published on the subject in recent years and generally agree that a $V_{20Gy} < 30–35\%$ is a useful cutoff to predict radiation pneumonitis after radiotherapy of NSCLC with or without the addition of chemotherapy.

The increasing use of IMRT and the associated dose redistribution from high-dose to low-dose areas require to reconsider these volumes and their correlation with pulmonary toxicities. Indeed, several studies suggest that dose thresholds well below the established dose levels predict adverse events. In an analysis of 78 NSCLC patients by Yorke et al., the most significant variables were the $V_{5Gy-20Gy}$ in the ipsilateral lung along with somewhat higher p-values for the range of $V_{20Gy-40Gy}$ (Yorke et al. 2005). Seppenwoolde et al., in a study of 382 patients irradiated for breast cancer, lymphoma, and NSCLC, found the best correlation to a fitted NTCP model using V_{13Gy}. Higher dose levels $\geq V_{20Gy}$ lead to a considerably lower predictive power in their model (Seppenwoolde et al. 2003). Another study by Schallenkamp et al. also found the V_{13Gy} as the best predictor of grade >2 radiation pneumonitis in 92 patients with lung cancer (Schallenkamp et al. 2007). Adding to this data, Kong et al. evaluated 109 patients with unresectable NSCLC and identified the MLD, V_{20Gy} and V_{13Gy} as the most significant predictive variables for grade ≥ 2 pneumonitis (Kong et al. 2006). An even lower dose threshold has been proposed by Wang et al. who found all tested levels in the range $V_{5Gy-65Gy}$ to be associated with toxicity. However, the V_{5Gy} was best-suited to discriminate patients with/without radiation pneumonitis: with a cutoff value $V_{5Gy} = 42\%$, the rates of grade ≥ 3 radiation pneumonitis were 3% and 38%, below and above cutoff, respectively (p = 0.001) (Wang et al. 2006). Combined, these data argue for a role of dose thresholds below the V_{20Gy} in the assessment of thoracic irradiation plans. This is especially relevant in case of IMRT treatment planning or large target volumes that lead to extensive low-dose volumes.

Esophagus: Acute Esophagitis. Literature Review

As for the lung, numerous models have been proposed to predict esophageal toxicity after irradiation. Most studies have found a statistically significant relationship between esophagitis and multiple dose-volume cutoffs. A major problem in finding a consensus is the varying reporting of these parameters, different fractionation, treatment schedules, concurrent chemotherapy regimens, and treatment planning. This leads to a disagreement of studies about the appropriate threshold dose-volume parameter, especially in the mid-to-low dose range. More congruence is found in the high-dose range above 50 Gy where multiple publications report a correlation with increased toxicity. Most individual reports choose a dose-volume cutoff based on V_{40Gy} or V_{50Gy}, however, Palma et al. have conducted a meta-analysis of individual patients and found the V_{60Gy} to be the most relevant parameter (Palma et al. 2016). Several other studies have focused on measures apart from simple dose-volume histogram (DVH) cutoffs, such as circumferential irradiation metrics, but have not yet found widespread use in the scientific literature.

NON-DOSIMETRY PREDICTORS

Apart from the dosimetric factors covered in earlier sections, a large number of non-dosimetric predictors have been evaluated over the years. A meta-analysis of 193 studies published between 1990–2010 by Vogelius et al. found pulmonary comorbidities to be the most relevant predictor of pulmonary toxicity with an Odds Ratio (OR) of 2.27 (p=0.007) (Vogelius et al. 2012). This was also the case in two other studies, one reporting a 1 year grade >2 pneumonitis rate of 55% in patients with preexisting interstitial lung disease (ILD) vs. 13.3% in those without ILD (p < 0.001) (Ueki et al. 2015) the other finding an ILD-specific toxicity of 12% (Chen et al. 2017).

Age has been described as a significant risk-factor in a publication by Dang et al. who considered 369 patients with stage III NSCLC and found an overall OR for radiation pneumonitis of 1.99 for patients older than 70 years(Dang et al. 2014). Vogelius et al. also found older age as one of the most relevant predictors for RILD (OR = 1.66, p < 0.0001) (Vogelius et al. 2012). Some other individual studies have failed to demonstrate this relationship, but a large meta-analysis by Zhao et al. confirmed the association in 5708 patients treated with pulmonary Stereotactic Body Radiation Therapy (SBRT) (Zhao et al. 2016).

Vogelius et al. also described mid or inferior tumor location (OR = 1.87, p = 0.002) as a relevant predictor (Vogelius et al. 2012), but Zhao et al. could not find this association in their collective of SBRT patients (Zhao et al. 2016). Tumor location, however, is associated to other factors like the magnitude of breathing motions, so its relationship to RILD may be confounded. Additionally, ventilation and perfusion of the lower lung is higher, causing loss of more functional lung volume when irradiated.

The evaluation of chemotherapy as a predictor of RILD has been suggested, especially for agents that themselves demonstrate pulmonary toxicity such as bleomycin, taxanes, cyclophosphamide, and others. But clinical studies in this field have been hampered by different regimens, dosage, and other factors that complicate the analysis of available data. However, several studies have demonstrated an increased rate of pulmonary toxicity when chemotherapy is administered concurrently with radiation: Palma et al. analyzed 836 patients in a systematic review and identified treatment with carboplatin/paclitaxel (in contrast to etoposide/cisplatin or other chemotherapy) as a risk factor for radiation pneumonitis (OR = 3.33; p < 0.001) (Palma et al. 2013). Focusing on different antineoplastic agent, the PROCLAIM trial reported on 555 patients with stage III NSCLC receiving concurrent radiochemotherapy and subsequent consolidation with pemetrexed (arm A) or etoposide/cisplatin (arm B). Regarding pulmonary toxicity, arm A showed a rate of 11% for grade 2 pneumonitis compared with 5.5% in arm B (p = 0.02) (Senan et al. 2016). Similar results have been reported about gemcitabine in breast cancer (Lingos et al. 1991) and NSCLC (Arrieta et al. 2009) or when chemotherapy was used as an induction treatment in NSCLC (Wang et al. 2008). Interestingly, a meta-study comparing 1205 patients individually did not find an increase of RILD when chemotherapy was combined with radiotherapy, but instead described an increase only in esophageal toxicity with a relative risk of 4.9 (p < 0.001) (Auperin et al. 2010). Vogelius et al. on the other hand surprisingly discovered sequential chemotherapy to be accompanied by higher rates of toxicity than a concomitant treatment (OR = 1.6, p = 0.01) (Vogelius et al. 2012). The authors explain this with a lower intensity chemotherapy when given concomitantly and a bias from older patients preferentially being prescribed a sequential treatment. These data demonstrate the often-conflicting results in this field and the need for a thorough evaluation of available treatment regimens in prospective clinical studies.

Smoking status has been proposed as a predictor for RILD as well: Hildebrandt et al. analyzed 173 patients with NSCLC and found a reduced risk of RILD in patients who were current smokers or recent quitters (Hildebrand et al. 2010). Similarly, Vogelius et al. reported an OR of 0.62 and 0.69 for current and previous smokers, respectively (Vogelius et al. 2012). Confirming these results, smoking was the only significant factor multivariate analysis in a publication by Jin et al. including data from 576 patients with NSCLC (Jin et al. 2009).

Some studies have suggested the involvement of genetic differences in the individual susceptibility to pulmonary toxicity. Pang et al. focused on single-nucleotide polymorphisms (SNP) in heatshock proteins (HSP) which are expressed by cells after the influence of various stressors and confer improved cellular survival via multiple mechanisms. After genotyping 271 NSCLC patients, they selected 2 SNPs in the heat shock protein beta-1 (HSPB1) gene for further analysis and discovered a statistically significant difference in the occurrence of grade ≥ 3 radiation pneumonitis depending on the mean lung dose applied (Pang et al. 2013). Pu et al. chose a wider approach by identifying 11930 SNPs in 201 NSCLC patients. After mathematical modeling, they selected 19 SNPs and further validated them in lymphoblastoid cell lines. Finally, three genes Protein Kinase C Epsilon (PRKCE), DExD/H-Box Helicase 58 (DDX58), and Tumor Necrosis Factor SuperFamily 7 (TNFSF7), involved in cell signaling and immune response, were identified as predictors of pneumonitis and esophagitis with an area under the receiver-operator-characteristics (ROC) curve (AUC) of 0.79 and 0.94, respectively (in combination with clinical data) (Pu et al. 2014). Xiong et al. reported on 362 patients with NSCLC and found an increased risk of radiation pneumonitis for those with a SNP in the Ataxia Teleangiectatica Mutated gene (ATM) [46](Xiong et al. 2013). In another study of 165 patients with NSCLC, SNPs in the DNA repair genes X-Ray Repair Cross Complementing 1 (XRCC1) and Apurinic/Apyrimidinic Endodeoxyribonuclease 1 (APEX1) correlated with grade ≥2 radiation pneumonitis (Yin et al. 2011). Further genes and genetic variants that have been implicated in increased risk for radiation pneumonitis are cytokines like Tumor Necrosis Factor alpha (TNFα), (Tucker et al. 2013b) the DNA-repair gene Ligase IV (LIG4) (Yin et al. 2012) and vascular endothelial growth factor (VEGF) (Guan et al. 2010). Of note, genomic studies associating SNPs with clinical phenotypes should be interpreted with care as the functional consequences of these genetic alterations are often under-explored and a clear cause-effect model is therefore lacking (Emahazion et al. 2001). Extensive preclinical and clinical validation is therefore needed to incorporate these parameters into clinical practice (Beaudet 2010).

Inflammatory cytokines may also play a role in RILD development: By far most data has been collected for TGFβ1 starting with a study by Anscher et al. who found that elevated plasma TGFβ1 levels by the end of radiotherapy increased the risk for RILD in the 73 analyzed NSCLC patients (Anscher et al. 1998). Multiple other reports have repeated this study with very similar designs and most could affirm the relationship between plasma TGFβ1 and RILD (Fu et al. 2001; Zhao et al. 2009; Stenmark et al. 2012; Evans et al. 2006). Other cytokines that have been associated with radiation-induced pulmonary injury are interleukin-8 (IL-8) (Stenmark et al. 2012; Evans et al. 2006; Hart et al. 2005), interleukin-6 (IL-6) (Chen et al. 2001), several molecules in the complement system , (Cai et al. 2010) and alpha 2 macroglobulin (α2M) (Oh et al. 2011). However, more data is needed to develop these findings into reliable clinical tools.

A study by Dehing-Oberije et al. tried to model lung toxicity (defined as dyspnea) combining dosimetric and clinical parameters. After collecting data from 438 lung cancer patients, a univariate model with MLD or V_{20Gy} reached an AUC = 0.47 (Dehing-Oberije et al. 2009). When factoring in clinical parameters including World Health Organization (WHO) performance status, smoking status, Forced Expiratory Volume in 1 second (FEV1), and age, AUC value was increased to 0.62. Other approaches towards complex models have been made using machine-learning methods: Chen et al. reported an AUC of up to 0.76 when feeding a neural network with 93 dosimetric and non-dosimetric variables (Chen et al. 2007).

NTCP MODELS FOR TREATMENT PLAN OPTIMISATION IN INTENSITY-MODULATED RADIATION THERAPY (IMRT) AND INTENSITY-MODULATED PROTON THERAPY (IMPT)

Treatment planning for IMRT/IMPT techniques is based on mathematical algorithms that minimize an objective function that characterizes treatment plan quality. Since the beginning of the IMRT era, the question regarding the adequate objective function persists. Intuitively, one would think that using NTCP models directly in treatment plan optimisation would represent the ideal objective function for OARs. However, this vision of performing inverse planning directly based on NTCP models never became a reality in clinical practice. A frequently claimed reason for that is the uncertainty in the parameter values in NTCP models. However, there is another reason: The commonly used NTCP models are not more powerful than dose-based objective functions, meaning that, inverse planning based on these NTCP models creates the same treatment plans that can as well be generated with the corresponding dose-based objective functions. Below, this is explained for the commonly used NTCP models in the thorax. A more complete and mathematically rigorous presentation of this topic was provided as early as 2004 by Romeijn et al. (2004).

The equivalence of NTCP models and dose-based objective functions can be illustrated using the LKB model for radiation pneumonitis. For simplicity, let us for now assume that the treatment plan optimisation problem is formulated as

$$\text{minimize} \qquad \text{NTCP}_{RP}(d)$$
$$\text{subject to} \qquad c_k(d) \le u_k \qquad k = 1,...,K \tag{10.1}$$

Where $\text{NTCP}_{RP}(d)$ is the probability of radiation pneumonitis as a function of d (or of gEUD) and $c_k(d) \le u_k$ represent constraints on the dose distribution (d) for additional planning criteria such as target coverage, tumor control probability, conformity, and further dose constraints for normal tissues (the index k is running on all K planning criteria). In this case, it is quite easy to recognize that minimizing the gEUD in the lung instead of $\text{NTCP}_{RP}(d)$ is mathematically equivalent. The optimal solution to the treatment plan optimisation problem

$$\text{minimize} \quad \text{gEUD}(d)$$

$$\text{subject to} \quad c_k(d) \le u_k \qquad k = 1, \dots, K \tag{10.2}$$

yields exactly the same treatment plan. This is because the LKB model represents an increasing function of the gEUD, i.e., independent of the parameters D_{50} and m, higher gEUD always leads to higher NTCP. Therefore, the dose distribution that minimizes the gEUD is the same as the dose distribution that minimizes NTCP. Applying the sigmoidal NTCP function to the gEUD represents only a scaling of the objective function that leaves the location of the minimum unaffected. Hence, from an IMRT/IMPT optimisation perspective, minimizing gEUD and NTCP is equivalent and there is no advantage of performing treatment plan optimisation based on the NTCP model directly. In fact, minimizing the corresponding dose-based objective function gEUD would typically be preferable from a computational/numerical perspective.

More generally, this issue can be considered in the context of multi-criteria treatment planning. Let us assume that all planning goals for critical structures are formulated via sigmoidal NTCP models, for example, for radiation pneumonitis in the lung and acute esophagitis. Then, the set of treatment plans that should be considered for treatment are those that cannot lower NTCP for a specific endpoint without increasing NTCP for another complication. These treatment plans form the so-called Pareto-efficient frontier. Formally, all Pareto-efficient treatment plans can be obtained by minimizing a composite objective function, which is a weighted sum of NTCPs for different complications, i.e.,

$$\text{minimize} \quad \sum_s w_s \text{NTCP}_s(d)$$

$$\text{subject to} \quad c_k(d) \le u_k \qquad k = 1, \dots, K \tag{10.3}$$

By varying the importance weights w_s, which control the tradeoff between different NTCP values, the entire Pareto frontier can be obtained. Let us assume now that all NTCP models were sigmoidal models formulated via the gEUD concept, while different normal tissues may have different volume parameters. It can be shown mathematically that optimizing a weighted sum of gEUDs for varying importance weights $\tilde{w}_s$, i.e.,

$$\text{minimize} \quad \sum_s \tilde{w}_s \text{gEUD}_s(d)$$

$$\text{subject to} \quad c_k(d) \le u_k \qquad k = 1, \dots, K \tag{10.4}$$

yields exactly the same set of Pareto-optimal treatment plans. In that sense, using NTCP models for treatment plan optimisation is not more powerful than using the corresponding dose based objective functions. Every treatment plan that is Pareto-optimal regarding the NTCP models, can be generated by minimizing a weighted sum of gEUD objectives. This does not question the value of NTCP models. However, the value of NTCP models lies in treatment plan evaluation and selecting an appropriate treatment plan on the Pareto frontier. There is no additional value in using the NTCP model directly as the objective or constraint function for treatment plan optimisation for intensity-modulated techniques.

Going one step further, other NTCP models that have been suggested that are given by

$$\text{NTCP}(d_i, b_i, n_i) = \frac{\exp(-f(d_i, b_i, n_i))}{1 + \exp(-f(d_i, b_i, n_i))} \tag{10.5}$$

where $f(d_i, b_i, n_i)$ is a function of multiple dose features (d_i), of multiple organ volume parameters (n_i), and of multiple model coefficients (b_i). Commonly, logistic regression is applied, in which case $f(d_i, b_i, n_i)$ is a linear function of dose features (d_i), and (b_i) are the regression coefficients. As an example, let us consider an NTCP model including two dose features $(d_1$ and $d_2)$ expressed with their gEUDs:

$$f(d_i, b_i, n_i) = b_1 \, \text{gEUD}(d_1, n_1) + b_2 \, \text{gEUD}(d_2, n_2) \tag{10.6}$$

where n_1 and n_2 are two different volume parameters. In this case, minimizing the function $f(d_i, b_i, n_i)$ in treatment plan optimisation is equivalent to minimizing NTCP. In this example, this amounts to defining two gEUD objectives with different volume parameters for the same structure. The regression coefficients resulting from NTCP modeling correspond to the relative importance of the two gEUD objectives. Although treatment plan optimisation can still be performed based on dosimetric objectives, the NTCP model now has the advantage that it can guide the treatment planner in specifying the relative importance of different dose features.

In summary, NTCP models are essential for treatment planning in IMRT/IMPT as they guide the selection of dosimetric planning criteria. However, the commonly used NTCP models for radiation pneumonitis and acute esophagitis are not truly more powerful than dose-based objective functions in the context of IMRT and IMPT optimisation, meaning that the treatment plans that are Pareto-optimal with respect to NTCP models can be obtained using traditional dose-based objective functions.

EXTRAPOLATION OF DOSE CONSTRAINTS TO EXPERIMENTAL FRACTIONATION SCHEMES

In the treatment of lung cancer, a large variety of fractionation schemes are being used. The biologically effective dose (BED, discussed with some details in Chapter 2) model (Fowler 2010) is the most commonly used model to compare the effectiveness or toxicity across different fractionation schemes. Two fractionation schemes are assumed to be isotoxic if both schemes correspond to the same BED. Therefore, the BED model can be used to extrapolate the tolerance doses from an established fractionation scheme to an experimental fractionation scheme. If (N_1, D_1) represents an established fractionation scheme consisting of N_1 fractions where for which the tolerance dose is D_1. Then, the BED model can be used to estimate the tolerance dose D_2 for an experimental fractionation scheme consisting of N_2 fractions via the equation

$$D_1\left(1+\frac{D_1/N_1}{\alpha/\beta}\right)=D_2\left(1+\frac{D_2/N_2}{\alpha/\beta}\right) \tag{10.7}$$

Where the parameters α/β is related to the fractionation sensitivity for the specific considered tissue and their values can emphasize the difference between early- and late-responding tissues.

Solving for D_2 yields

$$D_2=\frac{1}{2}\left[\sqrt{\left(N_2\frac{\alpha}{\beta}\right)^2+4D_1\left(\left(N_2\frac{\alpha}{\beta}\right)^2+D_1\frac{N_2}{N_1}\right)}-N_2\frac{\alpha}{\beta}\right] \tag{10.8}$$

This procedure is established for situations in which a single dose level is considered, such as a maximum dose or a dose-volume constraint for an OAR. However, difficulties arise in the context of mean dose constraints and gEUD constraints, i.e., if a dose-volume effect is present. In this case different parts of the organ receive different doses and it is unclear to which dose level Equation 10.8 should be applied. We first discuss mean dose constraints such as a mean lung dose constraint. A naive approach consists in applying Equation 10.8 to the mean dose constraint. However, this leads to inconsistent results. The difficulty arises since the mean BED in the lung does not equal the BED of the mean lung dose.

Let us denote the dose delivered to voxel i as D_i. The mean BED in an organ is given by

$$\overline{\mathrm{BED}}=\frac{1}{M}\sum_{i=1}^{M}D_i\left(1+\frac{D_i/N}{\alpha/\beta}\right)=\bar{D}\left(1+\frac{\bar{D}/N}{\alpha/\beta}\varphi\right) \tag{10.9}$$

where $\bar{D}=1/M\sum_{i=1}^{M}D_i$ is the mean dose in the organ and

$$\varphi=\frac{M\sum_{i=1}^{M}D_i^2}{\left(\sum_{i=1}^{M}D_i\right)^2} \tag{10.10}$$

This means that the mean BED of an inhomogeneous dose distribution in an organ is not given by the BED equation applied to the mean dose. However, a modified BED equation can be derived which includes a correction factor φ. This correction factor characterizes the shape of the dose distribution. Only for a uniform dose distribution, which would hardly ever be observed for the lung, φ is equal to 1 and the traditional BED equation is recovered. For any inhomogeneous dose distribution, φ is larger than 1, which means that the mean BED is higher than what is obtained by applying the traditional BED formula to the mean dose. Intuitively, the reason for this is that high doses near the target volume contribute over-proportionately to the mean BED due to the quadratic term in the BED equation. For contemporary treatment techniques, typical values for φ are in the order of 2 to 3 (Hoffmann et al. 2013; Perkò et al. 2018).

A consequence of this is that extrapolating mean dose constraints from a standard fractionated schedule towards a hypofractionated schedule using the standard BED formula may overestimate normal tissue tolerances, i.e., the mean dose constraint derived while accounting for the inhomogeneity of a dose distribution via φ is lower than what is obtained by applying the standard BED equation to the mean dose. As an example, we consider a mean lung dose constraint of $D_1 = 20$ Gy in $N_1 = 30$ fractions and we calculate the BED equivalent mean dose D_2 in a hypofractionated schedule with $N_2 = 5$ fractions, assuming a generic α/β of 3 Gy. Applying the standard BED equation yields $D_2 = 13.1$ Gy. If instead the inhomogeneity of the dose distribution is accounted for via $φ = 3$, the mean BED equivalent mean dose in 5 fractions is $D_2 = 10.6$ Gy.

It is worth noting that using the generalized BED equation (Eq. 11.9) for comparing mean dose constraints is consistent with NTCP modeling based on fraction size adjusted dose distributions. Assuming that outcome data for radiation pneumonitis containing multiple fractionation schemes and dose levels is analyzed. A common approach for adjusting for fractionation effects consist in calculating fraction size adjusted doses according to

$$\tilde{D}_i = D_i \frac{\left(1 + \dfrac{D_i/N}{\alpha/\beta}\right)}{\left(1 + \dfrac{X}{\alpha/\beta}\right)} = \frac{BED_i}{\left(1 + \dfrac{X}{\alpha/\beta}\right)} \tag{10.11}$$

where X is a reference dose per fraction. This concept has been introduced under a variety of names including Normalized Total Dose (NTD), Equi-effective Dose to dose per fraction of X Gy (EQDXGy), and Fraction Size Equivalent Dose. A reference dose per fraction of $X=2$ Gy is widely used to relate hypofractionated treatments to standard fractionation, in which case $\tilde{D}_i$ is also referred to as the equivalent dose in 2 Gy fractions (EQD2Gy, see Chapter 2 for details). Since $\tilde{D}_i$ is simply the BED multiplied by a constant factor, this implies that, for a volume parameter $n=1$, two fractionation schemes are isotoxic if their mean BED is equal.

The topic is discussed in detail in the publications by Hoffmann et al. (2013) and Perko et al. (2018). These publications also discuss the extensions of the above considerations to gEUD constraints. It is generally unclear whether the BED model is adequate to compare fractionation schemes over a wide range of doses per fraction ranging from standard fractionation to SBRT. Therefore, dose constraints for experimental fractionation schemes always need clinical validation. Nevertheless, the BED model is widely used to compare fractionation schemes. In that regard, it should be recognized that there is an interdependency between dose-volume effect modeling and modeling of fractionation effects. This issue has only recently attracted attention from researchers.

TOXICITIES AND NTCP MODELS FOR HYPOFRACTIONATED TREATMENTS AND SBRT

Relevant Anatomy and Toxicities

Toxicity of SBRT differs from conventional fractionation in several ways. The high BEDs that are achieved pose a disproportionally larger risk especially for serial OARs if they are located close-by or even within the target volume. Organs such as the esophagus, the trachea and large bronchi, and large blood vessels are especially vulnerable because functional failure can lead to serious, potentially life-threatening complications. For this reason, while SBRT has been a highly effective, relatively low-toxicity treatment to treat peripheral lesions in the lung, central tumors have been a challenging target because vital serial OARs are located in the mediastinum. Multiple published reports have described fatal adverse events after irradiation of central tumors, including acute hemoptysis, infections, and respiratory failure. On the other hand, lesions in the far periphery of the lung come with their own specific set of risks for the skin and bones, the brachial plexus, and the spinal cord. The focus of this chapter, however, will be the central structures within the mediastinum as these take up the largest share of severe complications in thoracic SBRT.

Literature Review

Lung

Compared to conventional fractionation, the percentage of lung that receives low-to-medium doses is typically lower in SBRT, therefore hypofractionated treatments are usually accompanied by relatively few high-grade pulmonary toxicities. A recent meta-analysis by Zhao et al. summarizing 88 studies with 7752 patients treated before 2014 found radiation-induced >grade 2 lung toxicities in 2.9% and >grade 3 in 0.5% of patients, respectively (Zhao et al. 2016). With the increased interest in pulmonary SBRT, various models have been proposed to predict toxicity. A general problem in comparing those reports lies in the different calculation of lung volumes (ipsilateral side only vs. both sides), Gross Tumor Volume/Planning Target Volume (GTV/PTV) definition, dosing (number of fractions, dose per fraction), dose calculation methods, toxicity scoring system, (Tucker et al. 2010) and other factors. This should be considered when trying to derive clinically-relevant dose thresholds.

Just as with conventional fractionation, most of the literature is focused on dosimetric parameters such as mean lung dose and percentage of lung receiving a predefined dose threshold (V_{XGy}). Several studies have found the MLD and the V_{20Gy} to be valuable predictors of toxicity using the crude physical dose derived from the DVH: Barriger et al. report a predictive threshold of the median MLD ≤ 4 Gy and $V_{20Gy} < 4\%$ for grade 2–4 radiation pneumonitis in 251 patients (Barriger et al. 2012). Another study by Matsuo et al. is in line with the findings for the V_{20Gy} with similar cutoff values but found an even lower p-value using the V_{25Gy} (Matsuo et al. 2012). More recently, Chang et al. reported on 100 patients with central tumors and suggested slightly higher cutoffs for MLD at 6 Gy and for V_{20Gy} at 12% (Chang et al. 2014).

Various studies have tried to establish the relationship between MLD and pneumonitis using mathematical models. Borst et al. applied an LKB model in 128 SBRT patients and determined a D_{50} of 19.6 Gy (Borst et al. 2009). Using a similar approach, Guckenberger et al. found a somewhat higher value of $D_{50} = 32.6$ Gy in 59 consecutive patients treated with SBRT for primary NSCLC or pulmonary metastases (Guckenberger et al. 2010). Both studies determined doses based on a voxel-based conversion of physical doses to EQD2Gy using $\alpha/\beta = 3$ Gy. Kyas et al. also modeled a dose-response curve but converted doses to the equivalent parameters for a single-dose treatment using the linear-quadratic model, assuming 30 fractions and $\alpha/\beta = 3$ Gy. In contrast to other studies, they used perifocal normal tissue reaction on CT as the outcome parameter and consequentially reported a comparably low D_{50} of 1.2 Gy (Kyas et al. 2007). Ricardi et al., analyzing 63 patients calculated BEDs derived from an LQ model ($\alpha/\beta = 3$ Gy) and determined a D_{50} of 20.3 Gy (Ricardi et al. 2009), similar to the report of Borst et al. (2009). Summarizing these and other studies, Zhao et al. (2016) did not propose a V_{20Gy} or MLD threshold for a lack of clear evidence, however, the sheer number of publications finding a dose-response relationship between MLD/V_{20Gy} and radiation pneumonitis make them the most well-established parameters available, so far. Of interest, Tucker et al. show in their analysis of 357 patients that the MLD underestimates the risk for radiation-induced lung toxicity and demonstrate a superiority of the LKB model for this purpose (Tucker et al. 2013b).

Among the alternative variables that have been proposed is the V_{5Gy}. Bongers et al. treated 79 consecutive patients with volumetric modulated arc therapy for various lung tumors. They converted doses to EQD2Gy using $\alpha/\beta = 3$Gy and then identified the MLD and $V_{5Gy-15Gy}$ of the contralateral lung as well as the MLD, $V_{5Gy-15Gy}$ and internal target volume (ITV) on the ipsilateral side as predictors of grade ≥ 3 radiation pneumonitis. However, in multivariate analysis only the contralateral MLD and ITV volume remained statistically significant (Bongers et al. 2013).Similarly, Baker et al. included 297 tumors and could predict toxicity using V_{5Gy} on univariate analysis but the correlation disappeared in their multivariate model where only clinical factors of female gender, pack-years smoking, and larger gross internal tumor volume and PTV were retained as predictive (Baker et al. 2013). The same applies to the analysis of Chang et al. who nevertheless recommended a cutoff of $V_{5Gy} \leq 30\%$ (Chang et al. 2014). Tumor size, GTV, and PTV volume are additional parameters that have been shown in a number of studies to influence the risk of pulmonary toxicity, although the results are often contradictory between publications (Kanemoto et al. 2015; Allibhai et al. 2013). In their meta-analysis, Zhao et al. could only confirm tumor size and not GTV or PTV volume as a predictive variable, thereby demonstrating the need to collect more data before a clear recommendation can be made about these variables (Zhao et al. 2016).

Esophagus and Bronchi
The pool of data for esophageal and bronchial toxicity after SBRT is limited compared to the lung. This is because the esophagus only becomes a critical OAR when treating central tumors that are located close to the mediastinum, and that the decision to treat such tumors is made much more conservatively following initial reports of fatal toxicities when using high BEDs. Timmerman et al. treated 70 patients with stage I NSCLC using 60–66 Gy in 3 fractions and reported excessive toxicity in patients with central tumors (Timmerman et al. 2006). Tumor location was a strong predictor of toxicity and freedom of severe toxicity was 83% vs. 54% for patients with peripheral and central tumors, respectively. The applied doses equate a BED of 460–550 Gy ($\alpha/\beta = 3$ Gy) and most toxicity modeling studies thereafter have been performed with doses that result in much lower BEDs. The definition of central tumors varies in the literature; however, most authors agree on tumor location

within 2 cm of the proximal bronchial tree, sometimes including other constraints such as the PTV touching the pleura or tumor location adjacent to mediastinal structures.

As expected for a serial OAR, a large share of toxicity modeling focuses on high doses to small volumes because of the severe risks of organ perforation. A frequently used measure is the maximum point dose to the organ, D_{max}: Nuyttens et al. analyzed 56 patients that were treated 45–60 Gy in 3–7 fractions. After applying a LKB-NTCP model they reported a $D_{max} = 43.4/61.4$ Gy for a 50% chance of grade 2/3 toxicity, respectively. They also examined the D_{1cc} (median dose to the hottest 1 cm^3) and found a 50% chance of grade 2/3 toxicity at 32.9/50.7 Gy, respectively (Nuyttens et al. 2016). In another study Wu et al. applied a Cox proportional hazard model to data from 91 patients irradiated with 30–60 Gy in 3–5 fractions. A complication rate <20% was reached if $D_{5cc} \leq 26.3$ Gy (with D_{5cc} expressed as BED calculated with $\alpha/\beta = 10$ Gy) (Wu et al. 2014). They then chose a cutoff value of $D_{5cc} = 14.4$ Gy and described a 2 year complication rate of 24% (above cutoff) compared to 1.6% (below cutoff). Cox et al. evaluated 182 patients treated for spinal metastases abutting the esophagus with a median dose of 24 Gy in one fraction; they found a statistically significant relationship between $D_{2.5cc}$ and grade >3 esophageal toxicity. In addition, they determined a median split for grade >3 esophageal toxicity at the following cutoff values: $V_{12Gy} = 3.8$ cm^3, $V_{15Gy} = 1.9$ cm^3, and $V_{20Gy} = 0.1$ cm^3 (of note V_{XGy} are here expressed as absolute organ volumes in cubic centimeters). Toxicity rates below cutoff values are $\approx$1-4%, while they are $\approx$10–13% above cutoff values (Cox et al. 2012). A recent study by Tekatli et al. has focused on so-called "ultra-central" NSCLC in which the PTV overlaps with the large central OAR, the main bronchus, trachea or esophagus. Patients in this study were treated with 12×5 Gy over 3 weeks resulting in a BED ($\alpha/\beta = 10$ Gy) of 90 Gy to 95% of the PTV. 38% of patients had grade $\geq$ 3 toxicities and in 21% treatment-related death was deemed possible or likely. When modeling toxicity, the authors could not find a significant relationship of any dosimetric parameter (D_{max} to PTV, D_{max} to OARs, $D_{0.1cc}$, $D_{0.5cc}$, D_{1cc}, D_{4cc}, D_{5cc}) to the occurrence of grade $\geq$3 toxicities (Tekatli et al. 2016). Chaudhuri et al. presented somewhat contrasting results in their retrospective analysis of 68 patients with peripheral and central pulmonary tumors. None of the patients with ultra-central lesions (defined as GTV overlapping with OAR) had grade $\geq$2 toxicities when being treated with 5×10 Gy or 4×12.5 Gy. Local tumor control and overall survival did not show a significant difference between peripheral, central, or ultra-central tumors (Chaudhuri et al. 2015).

To address the increased risk of patients with central tumors, an approach of risk-adapted fractionation has been proposed frequently, meaning treatment with equivalent or lower BED, distributed over a larger number of fractions, prolonged treatment schedule or a combination of these factors (compared to peripheral tumors). Lagerwaard et al., in a study of 206 patients with stage I NSCLC, introduced a concept of prescribing doses of 3×20 Gy to uncomplicated T1 tumors, 5×12 Gy to lesions with broad contact to the thoracic wall or category T2, and 8×7.5 Gy to tumors adjacent to critical central structures. With this schedule, grade $\geq$3 toxicity occurred in 3% of patients and no radiation-induced grade 5 event was reported (Lagerwaard et al. 2008). Using the same dose prescription of 8×7.5 Gy but focusing on centrally located NSCLC, Haasbeek et al. treated 68 patients and found no statistically significant difference in local control compared to 445 cases with peripheral tumors. Late grade 3 toxicity was limited to 2 cases of chest wall pain and no grade 4 events. No unequivocally treatment-related grade 5 event was observed but some uncertainty remained in nine cases who deceased of cardiovascular events (Haasbeek et al. 2011). Bral et al. chose a more hypofractionated approach in their study of 40 patients with T1–3 N0 M0 NSCLC who were prescribed 3×20 Gy for peripheral and 4×20 Gy for central tumors, respectively. Local control was not significantly different between peripheral and central tumors and no grade ≥3 toxicities were reported (Bral et al. 2011).

Two large clinical trials are currently ongoing to prospectively evaluate efficacy and toxicity of hypofractionated treatment for central tumors: LungTech includes patients with histologically or cytologically proven early stage, centrally located, inoperable NSCLC (T1–3), treating with 8×7.5 Gy (Adebahr et al. 2015). The RTOG-813 trial uses similar inclusion criteria, however, only T1–T2 tumors are included. Being a dose-escalation study, treatment consists of 5×10–12 Gy over 1.5–2 weeks. Phase-II results have revealed that several patients treated with higher dose levels of 5×11.5 Gy and 5×12 Gy suffered from grade ≥3 late toxicities, including 3 fatal events that were attributed to radiotherapy (Bezjak et al. 2016).

RECOMMENDATION FOR DOSE-VOLUME CONSTRAINTS AND NTCP MODEL PARAMETERS

Dosimetric parameters remain the most reliable predictors of pulmonary and esophageal toxicity and are also factors that are feasible to modify without the risk of adverse events or reduced efficacy of the treatment. Clinical factors and other non-dosimetric variables have emerged over recent years as a promising addition to improve the predictive value of existing models. However, broad clinical application and validation in large clinical studies is still absent for many of these parameters, therefore they should be used cautiously. In general, any generalized model or constraint should be regarded as a guideline and adapted to the individual situation of each patient, e.g., pre-existing medical conditions, patient preference, and available techniques at the treatment site.

REFERENCES

Adebahr, S., Collette, S., Shash, E., et al. 2015. LungTech, an EORTC phase II trial of stereotactic body radiotherapy for centrally located lung tumours:a clinical perspective. *Br J Radiology* 88(1051):20150036.

Allibhai, Z., Taremi, M., Bezjak, A., et al. 2013. The impact of tumor size on outcomes after stereotactic body radiation therapy for medically inoperable early-stage non-small cell lung cancer. *Int J Radiat Oncol Biol Phys* 87(5):1064–70.

Anscher M.S., Kong, F.M., Andrews, K., et al. 1998. Plasma transforming growth factor β1 as a predictor of radiation pneumonitis. *Int J Radiat Oncol Biol Phys* 41(5):1029–35.

Antonadou, D., Coliarakis, N., Synodinou, M., et al. 2001. Randomized phase III trial of radiation treatment +/- amifostine in patients with advanced-stage lung cancer. *Int J Radiat Oncol Biol Phys* 51(4):915–22.

Arrieta, O., Gallardo-Rincón, D., Villarreal-Garza, C., et al. 2009. High frequency of radiation pneumonitis in patients with locally advanced non-small cell lung cancer treated with concurrent radiotherapy and gemcitabine after induction with gemcitabine and carboplatin. *J Thorac Oncol* 4(7):845–52.

Aupérin, A., Le Péchoux, C., Rolland, E., et al. 2010. Meta-analysis of concomitant versus sequential radiochemotherapy in locally advanced non-small-cell lung cancer. *J Clin Oncol* 28(13):2181–90.

Baker, R., Han, G., Sarangkasiri, S., et al. 2013. Clinical and dosimetric predictors of radiation pneumonitis in a large series of patients treated with stereotactic body radiation therapy to the lung. *Int J Radiat Oncol Biol Phys* 85(1):190–5.

Barriger, R.B., Forquer, J.A., Brabham, J.G., et al. 2012. A dose-volume analysis of radiation pneumonitis in non-small cell lung cancer patients treated with stereotactic body radiation therapy. *Int J Radiat Oncol Biology Phys* 82(1):457–62.

Beaudet, A.L. 2010. Ethical issues raised by common copy number variants and single nucleotide polymorphisms of certain and uncertain significance in general medical practice. *Genome Med* 2(7):42.

Belderbos, J., Heemsbergen, W., Hoogeman, M., Pengel, K., Rossi, M. and Lebesque, J. 2005. Acute esophageal toxicity in non-small cell lung cancer patients after high dose conformal radiotherapy. *Radiother Oncol* 75(2):157–64.

Bell, J., McGivern, D., Bullimore, J., Hill, J., Davies, E.R., and Goddard, P. 1988. Diagnostic imaging of post-irradiation changes in the chest. *Clin Radiol* 39(2):109–19.

Bezjak, A., Paulus, R., Gaspar, L.E., et al. 2016. Efficacy and Toxicity Analysis of NRG Oncology/RTOG 0813 Trial of Stereotactic Body Radiation Therapy (SBRT) for Centrally Located Non-Small Cell Lung Cancer (NSCLC). *Int J Radiat Oncol Biol Phys* 96(S2):S8.

Bongers, E.M., Botticella, A., Palma, D.A., et al. 2013. Predictive parameters of symptomatic radiation pneumonitis following stereotactic or hypofractionated radiotherapy delivered using volumetric modulated arcs. *Radiother Oncol* 109(1):95–9.

Borst, G., Ishikawa, M., Nijkamp, J., et al. 2009. Radiation pneumonitis in patients treated for malignant pulmonary lesions with hypofractionated radiation therapy. *Radiother Oncol* 91(3):307–13.

Bral, S., Gevaert, T., Linthout, N., et al. 2011. Prospective, risk-adapted strategy of stereotactic body radiotherapy for early-stage non-small-cell lung cancer:Results of a phase II trial. *Int J Radiat Oncol Biology Phys* 80(5):1343–9.

Cai, X.W., Shedden, K., Ao, X., et al. 2010. Plasma proteomic analysis may identify new markers for radiation-induced lung toxicity in patients with non-small-cell lung cancer. *Int J Radiat Oncol Biol Phys* 77(3):867–76.

Chang, J.Y., Li, Q.Q., Xu, Q.Y., et al. 2014. Stereotactic ablative radiation therapy for centrally located early stage or isolated parenchymal recurrences of non-small cell lung cancer:how to fly in a 'no fly zone'. *Int J Radiat Oncol Biol Phys* 88(5):1120–8.

Chapet, O., Kong, F.-M., Lee, J.S., Hayman, J.A., and Ten Haken, R.K. 2005. Normal tissue complication probability modeling for acute esophagitis in patients treated with conformal radiation therapy for non-small cell lung cancer. *Radiother Oncol* 77(2):176–81.

Chaudhuri, A.A., Tang, C., Binkley, M.S., et al. 2015. Stereotactic ablative radiotherapy (SABR) for treatment of central and ultra-central lung tumors. *Lung Cancer* 89(1):50–6.

Chen, H., Senan, S., Nossent, E.J., et al. 2017. Treatment-related toxicity in patients with early-stage non-small cell lung cancer and coexisting interstitial lung disease: A systematic review. *Int J Radiat Oncol Biol Phys* 98(3):622–31.

Chen, S., Zhou, S., Zhang, J., Yin, F.F., Marks L. and Das S.K. 2007.A neural network model to predict lung radiation induced pneumonitis. *Med Phys* 34(9):3420–7.

Chen, Y., Rubin, P., Williams, J., Hernady, E., Smudzin, T. and Okunieff, P. 2001. Circulating IL-6 as a predictor of radiation pneumonitis. *Int J Radiat Oncol Biol Phys* 49(3):641–8.

Choi, Y.W., Munden, R.F., Erasmus, J.J., et al. 2004. Effects of radiation therapy on the lung: Radiologic appearances and differential diagnosis. *Radiographics* 24(4):985–97.

Cox, B.W., Jackson, A., Hunt, M., Bilsky, M. and Yamada, Y. 2012. Esophageal toxicity from high-dose, single-fraction paraspinal stereotactic radiosurgery. *Int J Radiat Oncol Biol Phys* 83(5):e661–7.

Dang, J., Li, G., Zang, S., Zhang, S., and Yao, L. 2014. Risk and predictors for early radiation pneumonitis in patients with stage III non-small cell lung cancer treated with concurrent or sequential chemoradiotherapy. *Radiat Oncol* 9:172.

Dehing-Oberije, C., De Ruysscher, D., van Baardwijk, A., Yu, S., Rao, B. and Lambin, P. 2009. The importance of patient characteristics for the prediction of radiation-induced lung toxicity. *Radiother Oncol* 91(3):421–6.

Delanian, S., Porcher, R., Balla-Mekias, S., and Lefaix, J.L. 2003. Randomized, placebo-controlled trial of combined pentoxifylline and tocopherol for regression of superficial radiation-induced fibrosis. *J Clin Oncol* 21(13):2545–50.

Dion, MW., Hussey, D.H., Doornbos, J.F., Vigliotti, A.P., Wen, B.C. and Anderson, B. 1990. Preliminary results of a pilot study of pentoxifylline in the treatment of late radiation soft tissue necrosis. *Int J Radiat Oncol Biol Phys* 19(2):401–7.

Emahazion,T., Feuk, L., Jobs, M., et al. 2001. SNP association studies in Alzheimer's disease highlight problems for complex disease analysis. *Trends Genet* 17(7):407–13.

Evans E.S., Kocak, Z., Zhou, et al. 2006. Does transforming growth factor-β1 predict for radiation-induced pneumonitis in patients treated for lung cancer? *Cytokine* 35(3–4):186–92.

Faria, S.L., Aslani, M., Tafazoli, F.S., Souhami, L., and Freeman, C.R. 2009. The challenge of scoring radiation-induced lung toxicity. *Clin Oncol (R Coll Radiol)* 21(5):371–5.

Fowler, J.F. 2010. 21 years of biologically effective dose. *Br J Radiol* 83(991):554–68.

Fu X.L., Huang, H., Bentel, G., et al. 2001. Predicting the risk of symptomatic radiation-induced lung injury using both the physical and biologic parameters V(30) and transforming growth factor beta. *Int J Radiat Oncol Biology Phys* 50(4):899–908.

Guan, X., Yin, M., Wei, Q., et al. 2010. Genotypes and haplotypes of the VEGF gene and survival in locally advanced non-small cell lung cancer patients treated with chemoradiotherapy. *BMC Cancer* 10:431.

Guckenberger M., Baier, K., Polat, B., et al. 2010. Dose-response relationship for radiation-induced pneumonitis after pulmonary stereotactic body radiotherapy. *Radiother Oncol* 97(1):65–70.

Haasbeek, C.J., Lagerwaard, F.J., Slotman, B.J., and Senan, S. 2011. Outcomes of stereotactic ablative radiotherapy for centrally located early-stage lung cancer. *J Thorac Oncol* 6(12):2036–43.

Hart J.P., Broadwater, G., Rabbani, Z., et al. 2005. Cytokine profiling for prediction of symptomatic radiation-induced lung injury. *Int J Radiat Oncol Biol Phys* 63(5):1448–54.

Henkenberens, C., Janssen, S., Lavae-Mokhtari, M., et al. 2016. Inhalative steroids as an individual treatment in symptomatic lung cancer patients with radiation pneumonitis grade II after radiotherapy - a single-centre experience. *Radiat Oncol* 11:12.

Herrmann, T., Geyer, P., and Appold, S. 2015. The mean lung dose (MLD): predictive criterion for lung damage? *Strahlenther Onkol* 191(7):557–65.

Hildebrandt, M. A., Komaki, R., Liao, Z., et al. 2010. Genetic variants in inflammation-related genes are associated with radiation-induced toxicity following treatment for non-small cell lung cancer. *PLoS ONE* 5(8):e12402.

Hille, A., Christiansen, H., Pradier, O., et al. 2005. Effect of pentoxifylline and tocopherol on radiation proctitis/enteritis. *Strahlenther und Onkol* 181(9):606–14.

Hoffmann, A.L. and Nahum, A. E. 2013. Fractionation in normal tissues: the (α/β)eff concept can account for dose heterogeneity and volume effects. *Phys Med Biol* 58(19):6897–914.

Ikezoe, J., Takashima, S., Morimoto, S., et al. 1988. CT appearance of acute radiation-induced injury in the lung. *AJR Am J Roentgenol* 150(4):765–70.

Jin, H., Tucker, S.L., Liu, H.H., et al. 2009. Dose-volume thresholds and smoking status for the risk of treatment-related pneumonitis in inoperable non-small cell lung cancer treated with definitive radiotherapy. *Radiother Oncol* 91(3):427–32.

Kanemoto, A., Matsumoto, Y. and Sugita, T. 2015. Timing and characteristics of radiation pneumonitis after stereotactic body radiotherapy for peripherally located stage I lung cancer. *Int J Clin Oncol* 20(4):680–5.

Kong, F.M., Hayman, J.A., Griffith, K.A., et al. 2006. Final toxicity results of a radiation-dose escalation study in patients with non-small-cell lung cancer (NSCLC): Predictors for radiation pneumonitis and fibrosis. *Int J Radiat Oncol Biol Phys* 65(4):1075–86.

Kwok, E. and Chan, C. K. 1998. Corticosteroids and azathioprine do not prevent radiation-induced lung injury. *Can Respir J* 5(3):211–4.

Kyas, I., Hof, H., Debus, J., Schlegel, W., and Karger, C.P. 2007. Prediction of radiation-induced changes in the lung after stereotactic body radiation therapy of non-small-cell lung cancer. *Int J Radiat Oncol Biol Phys* 67(3):768–74.

Lagerwaard, F.J.., Haasbeek, C.J., Smit, E.F., Slotman, B.J., and Senan, S. 2008. Outcomes of risk-adapted fractionated stereotactic radiotherapy for stage I non-small-cell lung cancer. *Int J Radiat Oncol Biology Phys* 70(3):685–92.

Liao, Z. X., Travis, E. L., and Tucker, S. L. 1995. Damage and morbidity from pneumonitis after irradiation of partial volumes of mouse lung. *Int J Radiat Oncol Biol Phys* 32(5):1359–70.

Lingos, T. I., Recht, A., Vicini, F., Abner, A., Silver, B. and Harris, J.R. 1991. Radiation pneumonitis in breast cancer patients treated with conservative surgery and radiation therapy. Int *J Radiat Oncol Biol Phys* 21(2):355–60.

Lopez Guerra, J. L., Gomez, D., Zhuang, Y., et al. 2012. Change in diffusing capacity after radiation as an objective measure for grading radiation pneumonitis in patients treated for non-small-cell lung cancer. *Int J Radiat Oncol Biol Phys* 83(5):1573–9.

Marks, L.B., Bentzen, S.M., Deasy, J.O., et al. 2010. Radiation dose-volume effects in the lung. *Int J Radiat Oncol Biology Phys* 76(S3):S70–6.

Matsuo, Y., Shibuya, K., Nakamura, M., et al. 2012. Dose-volume metrics associated with radiation pneumonitis after stereotactic body radiation therapy for lung cancer. *Int J Radiat Oncol Biol Phys* 83(4):e545–9.

McCarty, M., Lillis, P., and Vukelja, S. 1996. Azathioprine as a steroid-sparing agent in radiation pneumonitis. *Chest* 109(5):1397–400.

Muraoka, T., Bandoh, S., Fujita, J., et al. 2002. Corticosteroid refractory radiation pneumonitis that remarkably responded to cyclosporin A. *Intern Med* 41(9):730–3.

Nuyttens, J.J., Moiseenko, V., McLaughlin, M., Jain, S., Herbert, S. and Grimm J. 2016. Esophageal dose tolerance in patients treated with stereotactic body radiation therapy. *Semin Radiat Oncol* 26(2):120–8.

Oh, J. H., Craft, J.M., Townsend, R., Deasy, J.O., Bradley, J.D. and El Naqa, I. 2011. A bioinformatics approach for biomarker identification in radiation-induced lung inflammation from limited proteomics data. *J Proteome Res* 10(3):1406–15.

Ozturk, B., Egehan, I., Atavci, S., and Kitapci, M. 2004. Pentoxifylline in prevention of radiation-induced lung toxicity in patients with breast and lung cancer: A double-blind randomized trial. *Int J Radiat Oncol Biol Phys* 58(1):213–9.

Palma, D.A., Senan, S., Oberije, C., et al. 2016. Predicting esophagitis after chemoradiation therapy for non-small cell lung cancer:an individual patient data meta-analysis. *Int J Radiat Oncol Biol Phys* 87(4):690–6.

Palma, D.A., Senan, S., Tsujino, K., et al. 2013. Predicting radiation pneumonitis after chemoradiation therapy for lung cancer: an international individual patient data meta-analysis. *Int J Radiat Oncol Biol Phys* 85(2):444–50.

Pang, Q., Wei, Q., Xu, T., et al. 2013. Functional promoter variant rs2868371 of HSPB1 is associated with risk of radiation pneumonitis after chemoradiation for non-small cell lung cancer. *Int J Radiat Oncol Biol Phys* 85(5):1332–9.

Perkó Z., Bortfeld, T., Hong, T., Wolfgang, J. and Unkelbach, J. 2018. Derivation of mean dose tolerances for new fractionation schemes and treatment modalities. *Phys Med Biol* 63(3):035038.

Pu, X., Wang, L., Chang, J.Y., et al. 2014. Inflammation-related genetic variants predict toxicity following definitive radiotherapy for lung cancer. *Clin Pharmacol Ther* 96(5):609–15.

Ricardi, U., Filippi, A.R., Guarneri, A., et al. 2009. Dosimetric predictors of radiation-induced lung injury in stereotactic body radiation therapy. *Acta Oncol* 48(4):571–7.

Romeijn, E.H., Dempsey, J.F., and Li, J.G. 2004. A unifying framework for multi-criteria fluence map optimisation models. *Phys Med Biol* 49(10):1991–2013.

Schallenkamp, J.M. Miller, R.C., Brinkmann, D.H., Foote, T. and Garces, Y.I. 2007. Incidence of radiation pneumonitis after thoracic irradiation:Dose-volume correlates. *Int J Radiat Oncol Biol Phys* 67(2):410–6.

Semenenko, V.A., Molthen, R.C., Li, C., et al. 2008. Irradiation of varying volumes of rat lung to same mean lung dose: a little to a lot or a lot to a little? *Int J Radiat Oncol Biol Phys* 71(3):838–47.

Senan, S., Brade, A., Wang, L.H., et al. 2016. PROCLAIM: Randomized Phase III Trial of Pemetrexed-Cisplatin or Etoposide-Cisplatin plus thoracic radiation therapy followed by consolidation chemotherapy in locally advanced nonsquamous Non-Small-Cell Lung Cancer. *J Clin Oncol* 34(9):953–62.

Seppenwoolde,Y., Lebesque, J.V., de Jaeger, K., et al. 2003. Comparing different NTCP models that predict the incidence of radiation pneumonitis. *Int J Radiat Oncol Biol Phys* 55(3):724–35.

Small, W. Jr, James, J.L., Moore, T.D., et al. 2018. Utility of the ACE inhibitor captopril in mitigating radiation-associated pulmonary toxicity in lung cancer: Results from NRG oncology RTOG 0123. *Am J Clin Oncol* 41(4):396–401.

Stenmark, M.H., Cai, X.W., et al. 2012. Combining physical and biologic parameters to predict radiation-induced lung toxicity in patients with non-small-cell lung cancer treated with definitive radiation therapy. *Int J Radiat Oncol Biology Phys* 84(2):e217–22.

Tekatli, H., Haasbeek, N., Dahele, M., et al. 2016. Outcomes of hypofractionated high-dose radio-therapy in poor-risk patients with 'ultracentral' non-small cell lung cancer. *J Thorac Oncol* 11(7):1081–9.

Timmerman, R., McGarry, R., Yiannoutsos, C., et al. 2006. Excessive toxicity when treating central tumors in a phase II study of stereotactic body radiation therapy for medically inoperable early-stage lung cancer. *J Clin Oncol* 24(30):4833–9.

Tucker, S. L., Li, M., Xu, T., et al. 2013a. Incorporating single-nucleotide polymorphisms into the Lyman model to improve prediction of radiation pneumonitis. *Int J Radiat Oncol Biol Phys* 85(1):251–7.

Tucker, S.L., Jin, H., Wei, X., et al. 2010. Impact of toxicity grade and scoring system on the relation-ship between mean lung dose and risk of radiation pneumonitis in a large cohort of patients with non-small cell lung cancer. *Int J Radiat Oncol Biol Phys* 77(3):691–8.

Tucker, S.L., Mohan, R., Liengsawangwong, R., Martel, M.K., and Liao, Z. 2013b. Predicting pneumonitis risk: A dosimetric alternative to mean lung dose. *Int J Radiat Oncol Biol Phys* 85(2):522–7.

Ueki, N., Matsuo, Y., Togashi, Y., et al. 2015. Impact of pretreatment interstitial lung disease on radiation pneumonitis and survival after stereotactic body radiation therapy for lung cancer. *J Thorac Oncol* 10(1):116–25.

Vogelius, I.R. and Bentzen, S.M. 2012. A literature-based meta-analysis of clinical risk factors for development of radiation induced pneumonitis. *Acta Oncol* 51(8):975–83.

Wang, S., Liao, Z., Wei, X., et al. 2006. Analysis of clinical and dosimetric factors associated with treatment-related pneumonitis (TRP) in patients with non-small-cell lung cancer (NSCLC) treated with concurrent chemotherapy and three-dimensional conformal radiotherapy (3D-CRT). *Int J Radiat Oncol Biol Phys* 66(5):1399–407.

Wang, S., Liao, Z., Wei, X., Liu, H.H., Tucker, S.L. and Hu, C. 2008. Association between systemic chemotherapy before chemoradiation and increased risk of treatment-related pneumonitis in esophageal cancer patients treated with definitive chemoradiotherapy. *J Thorac Oncol* 3(3):277–82.

Werner-Wasik, M., Yorke, E., Deasy, J., Nam, J., and Marks, L. B. 2010. Radiation dose-volume effects in the esophagus. *Int J Radiat Oncol Biol Phys* 76(S3):S86–93.

Willner, J., Jost, A., Baier, K., and Flentje, M. 2003. A little to a lot or a lot to a little? An analysis of pneumonitis risk from dose-volume histogram parameters of the lung in patients with lung cancer treated with 3-D conformal radiotherapy. *Strahlenther Onkol* 179(8):548–56.

Wu, A.J., Williams, E., Modh, A., et al. 2014. Dosimetric predictors of esophageal toxicity after stereotactic body radiotherapy for central lung tumors. *Radiother Oncol* 112(2):267–71.

Xiong, H., Liao, Z., Liu, Z., et al. 2013. ATM polymorphisms predict severe radiation pneumonitis in patients with non-small cell lung cancer treated with definitive radiation therapy. *Int J Radiat Oncol Biol Phys* 85(4):1066–73.

Yin, M., Liao, Z., Liu, Z., et al. 2012. Genetic variants of the nonhomologous end joining gene LIG4 and severe radiation pneumonitis in nonsmall cell lung cancer patients treated with definitive radiotherapy. *Cancer* 118(2):528–35.

Yin, M., Liao, Z., Liu, Z., et al. 2011. Functional polymorphisms of base excision repair genes XRCC1 and APEX1 predict risk of radiation pneumonitis in patients with non-small cell lung cancer treated with definitive radiation therapy. *Int J Radiat Oncol Biol Phys* 81(3):e67–73.

Yorke, E. D., Jackson, A., Rosenzweig, K.E., Braban, L., Leibel, S.A. and Ling C.C. 2005. Correlation of dosimetric factors and radiation pneumonitis for non-small-cell lung cancer patients in a recently completed dose escalation study. *Int J Radiat Oncol Biol Phys* 63(3):672–82.

Zhao L., Wang, L., Ji, W., et al. 2009. Elevation of plasma TGF-β1 during radiation therapy predicts radiation-induced lung toxicity in patients with non-small-cell lung cancer: A combined analysis from Beijing and Michigan. *Int J Radiat Oncol Biol Phys* 74(5):1385–90.

Zhao, J., Yorke, E.D., Li, L., et al. 2016. Simple factors associated with radiation-induced lung toxicity after stereotactic body radiation therapy of the Thorax: A pooled analysis of 88 studies. *Int J Radiat Oncol Biol Phys* 95(5):1357–66.

CHAPTER 11

Heart and Vascular Problems

Laura Cella and Giovanna Gagliardi

CONTENTS

Introduction 269
    Short Summary of the QUANTEC Recommendations 270
    Critical Review of the Literature 2009–2017 (Conventional Fractionation) 270
        Breast Cancer 273
        Hodgkin's Lymphoma 274
        Lung Cancer 275
        Esophageal Cancer 276
    Critical Review of the Literature 2009–2017 (Hypofractionation) 277
        Breast Cancer 277
        Lung Cancer 278
Summary for DVH Constraints/NTCP Model Parameters 279
    Open Issues and Future Directions of Investigation 279
        Lung-Heart Interaction 280
        Non-Thoracic Irradiation 280
        Biological Markers 281
        Cardiac Imaging Techniques 281
        Voxel-Based Approach 281
        Modern RT Technology 282
References 282

INTRODUCTION

Therapeutic thoracic radiotherapy (RT) has been demonstrated to injure the cardiovascular (CV) system and to cause a broad range of cardiac complications with different manifestations and degrees of severity, many of which have a progressive nature. In radiation-associated CV toxicity, absolute risk appears to increase with increasing follow-up time after exposure. The spectrum of radiation-associated cardiac diseases includes accelerated coronary artery atherosclerosis, pericardial disease, myocardial systolic and diastolic dysfunction as well as valve disease. Multivariable models able to robustly predict radiation-induced CV effects may help to identify the optimal plan that minimizes side effects for individual patients and represents the key to maximizing the benefits of technological advances in RT. Modeling radiation-induced heart disease has been hampered both by the relatively low incidence of complications and the lack of long-term results from 3D-based thoracic RT. However, in the last decade we have seen increasing development of studies that are based on both long follow-up and large populations, which are evaluating the determinants of the risk and the radiation dose-response. In this chapter, a review of the available dosimetric constraints, an analysis of the interaction between them and other relevant patient and treatment-related factors and models on CV toxicity following RT is presented. The main criteria used in this review are fractionation patterns and diagnosis.

Short Summary of the QUANTEC Recommendations

In the QUANTEC part on radiation dose-volume effects in the heart (Gagliardi et al. 2010) the main clinical and dose-volume predictors for acute and late radiation-induced heart disease were identified. The available data at that time, relatively scarce in terms of dose metrics and reasonable follow-up, referred to three main endpoints, i.e., ischemic heart disease, pericardial disease and valvular disease. Volume definition of the organs at risk and its rational, dose-volume metrics and other factors affecting risk were discussed and, when possible, incorporated in the modeling.

In the ischemic heart disease setting, studies on breast cancer (BC) and lymphoma patients were considered. For BC patients the general observation that Normal Tissue Complication Probability (NTCP) values >5% could jeopardize the benefit on survival effect of RT was stated, with the implicit recommendation to refer in the clinic work to a steep dose-response curve (Gagliardi et al. 1996). Considering the very limited use of NTCP modeling in clinical settings, a corresponding recommendation in dose-volume metrics was extrapolated, namely that V25 Gy < 10%, in 2 Gy per fraction, would be associated with a <1% probability of cardiac mortality about 15 years after RT. A quality assurance study of the consistency and compliance to these indications in the clinic routine was then performed (Moiseenko et al. 2016).

For lymphoma patients receiving chemotherapy, particularly doxorubicin and RT, the QUANTEC recommendation was to limit the dose to a whole heart to ~15 Gy.

For pericarditis, most quantitative analysis referred to esophageal cancer patients. Particularly, one study had been able to show that risk increased with mean pericardium dose D_{mean} >26 Gy and V_{30Gy} > 46% (Wei et al. 2008). A previous study had instead found a dependence on the fractionation pattern, providing also NTCP parametrization of the data (Martel et al. 1998). Finally, irradiation of the left ventricle had been found to be the predictor of perfusion defects (Das et al. 2005). Nonetheless, the clinical relevance of this endpoint had been questioned.

Critical Review of the Literature 2009–2017 (Conventional Fractionation)

A literature review was performed to identify studies incorporating dose-volume relationship and models on radiation-associated CV toxicity that have been published since 2009, i.e., in the post-QUANTEC era. The search was first focused on conventionally (1.8–2 Gy) fractionated thoracic RT treatments. Diverse thoracic cancers, including breast, lung, esophageal cancers and lymphomas, are characterized by different irradiation geometries

TABLE 11.1 Summary of Predictive Models Specific to Different Thoracic Treatments and for a Spectrum of Cardiovascular Diseases in Post-Radiotherapy Patient Populations (Conventional Fractionation)

Tumor Site	Heart Structure	Toxicity Endpoint	N#	Model	Model Result	Reference
Breast	Whole Heart	Major Coronary Events	2168	Linear	Rate Risk = $B_s*(1 + K*MHD)$; K=0.074/Gy	Darby et al. (2013)
	Whole Heart	Major Coronary Events	910	Cox	HR(MHD) = exp(0.153) for 1 Gy increase in MHD, HR(age)=exp(0.087) for 1 year increase in age, HR(RISKSCORE)=exp(1.821) for 1 point-increase in RISKSCORE	Van den Bogaard et al. (2017)
	Left Ventricle				HR(LV_V$_{5Gy}$)=exp(0.017) for 1% increase in LV_V$_{5Gy}$ HR(age)=exp(0.063) for 1 year increase in age, HR(RISKSCORE)=exp(0.711) for 1 point-increase in RISKSCORE	
HL	Whole Heart	Cardiovascular disease	898	Logistic	NTCP=$1/(1+e^{-S})$, S=$4\gamma_{50}*[1-d/D_{50}]$; $\gamma_{50}=1.36$, $D_{50}=48.2$ Gy	Maraldo et al. (2012) on Schellong et al. (2010)
	Whole Heart	Coronary Heart Disease	1529	Linear	Rate Risk Ratio = 1 + K*MHD; K=0.074/Gy	van Nimwegen et al. 2015
	Left Ventricle	Heart Valvular Defect	56	Logistic	NTCP=$1/(1+e^{-S})$; S=0.035*LV − V$_{30Gy}$ + 0.037*LV Volume (cc) − 0.002*Lung Volume (cc) − 2.26	Cella et al. (2013) (Dose fraction 1.6 Gy)
	Heart valves	Heart Valvular Defect	246	Non-linear	Rate Risk Ratio = 1+exp(γ)*MVD*exp(δ*MVD); γ = −5.02/Gy, δ = 0.075/Gy	Cutter et al. (2015)
	Coronary arteries	Stenosis	33	Logistic	OR(D$_{median}$) = 1.049 for 1 Gy increase in D$_{median}$	Moignier et al. (2015)
Lungs	Whole Heart	Symptomatic cardiac events‡	112	Cox	HR(MHD)=1.04 for 1 Gy increase in MHD, HR(WHO/ISH 10 years risk)=1.34/stratum	Wang et al. (2017)
	Left Ventricle				HR(LV_V$_{30Gy}$) 1.02 for 1% increase in LV_V$_{30Gy}$, HR(WHO/ISH 10 years risk)=1.46/stratum	

(Continued)

TABLE 11.1 (CONTINUED) Summary of Predictive Models Specific to Different Thoracic Treatments and for a Spectrum of Cardiovascular Diseases in Post-Radiotherapy Patient Populations (Conventional Fractionation)

Tumor Site	Heart Structure	Toxicity Endpoint	N*	Model	Model Result	Reference
	Whole heart	Pericardial effusion ≥ G2	201 prospective 301 retrospective	Cox	HR(Heart_V_{35Gy})=2.14 for Heart_V_{35Gy} <10% vs ≥10%	Ning et al. (2017)
Esophagus	Whole Heart	Symptomatic Cardiac Events‡	58	Cox	HR(Heart_V_{55Gy})= 1.33 for 1% increase in Heart_V_{55Gy}	Ongino et al. (2016)

Note: Unless specified, doses are EQD2 Gy corrected.

HL, Hodgkin's Lymphoma; B_S, basal risk of coronary events; MHD, Mean Heart Dose (Gy); V_{XGy}, percentage volume exceeding X Gy; LV, Left Ventricle; RISK SCORE, 0.8*Diabetes + 1.4*Hypertension + 1.8*History of ischemic cardiac events before RT; γ_{50}, dose-response curve slope; D_{50}, uniform dose to the whole organ leading to 50% complication; D_{median}, median dose; OR, Odds Ratio; HR=Hazard Ratio.

‡ Symptomatic Cardiac Events: Symptomatic pericardial effusion, Myocardial infarction, Unstable angina, Pericarditis, Significant arrhythmia, heart failure.

and dose distributions claiming to be addressed separately. Table 11.1 lists and briefly summarizes the dose-volume constraints or NTCP models specific to different thoracic RT treatments and for a spectrum of CV diseases. Since the QUANTEC heart review (Gagliardi et al. 2010) was published in 2010, there has been a progressive evolution in the philosophy of NTCP modeling, and most studies listed in Table 11.1 are based on the indications suggested by QUANTEC authors for the future toxicity studies (e.g., long follow-up and large populations, doses to heart sub-volumes, global physiological effects of thoracic RT).

Breast Cancer

Long-term BC survival rate is increasing thanks to earlier diagnosis and modern treatment techniques. By 2000, the use of CT simulation as well as three-dimensional treatment-planning techniques had become the standard practice with a consequent general reduction of radiation dose to the heart (Moiseenko et al. 2016; Taylor et al. 2015). However, patients successfully treated with RT for BC have still an increased risk of developing subsequent heart disease, although lower than in earlier eras (Brenner et al. 2014). The risk of heart disease is higher for women who received radiotherapy for left-sided BC compared with women who received right-sided RT (Rehammar et al. 2017).

In a case-control study, Darby et al. (2013) analyzed data of about 2000 women treated in Scandinavia between 1958 and 2001. The authors showed a no-threshold relative increase of 7.4% (95% CI, 2.9–14.5) per Gy of mean heart dose (MHD) in the rate of major coronary events (i.e., myocardial infarction, coronary revascularization or death from ischemic heart disease) over the entire follow-up period. The authors also reported that the increase started within the first 5 years after RT and continued for at least 20 years. Of note, dosimetric data were obtained by reconstruction from two-dimensional data on one model patient.

Recently, Van den Bogaard et al. (2017) validated the model published by Darby et al. in an independent cohort of patients treated with contemporary breast RT between 2005 and 2008 and using individual 3D CT-based planning data. Using exactly the same risk factors and end-point as in the previous study, they found an increase of 16.5% (95% confidence interval [CI], 0.6 to 35.0) in the cumulative incidence of coronary events per Gy of radiation to the whole heart in the first 9 years after treatment. These results are consistent with the hazard ratios of 16.3% increase per Gy, as observed by Darby et al. (2010) in the first 4 years of follow-up, and 15.5% increase in the next 5 to 9 years after RT. Furthermore, their study suggests that the NTCP model for coronary events could be improved by using V_{5Gy} to the left ventricle as the dose-volume parameter for the left ventricle instead of MHD. Model discrimination, in terms of the c-statistic, was indeed significantly improved ($p = 0.042$) by replacing the MHD (AUC = 0.79) with the left ventricle V_{5Gy} and using the weighted coronary event risk score (AUC = 0.83). The weighted coronary event risk score per patient was based on the regression coefficient of the significant comorbidities: 0.8 for diabetes, 1.4 for hypertension and 1.8 for history of ischemic cardiac events before RT.

Finally, it is worthwhile to refer to the meta-analyses performed by Taylor et al. (2017), on 40781 women randomly assigned to BC RT versus no radiotherapy in 75 trials. The authors estimated a proportional increase in cardiac mortality of 0.041 per Gy of MHD estimated at trial-level rather than individual-level. In addition, an absolute risk of cardiac mortality from modern RT of approximately 1% for smokers and 0.3% for nonsmokers was reported.

Hodgkin's Lymphoma

The high-cure rate and the prolonged survival obtained in Hodgkin's lymphoma (HL) have been historically offset by RT-induced cardiac disease (Hancock et al. 1993). In the modern era of combined modality approach in HL, with lower doses (20–30 Gy) and reduced irradiated volumes, with consequently different degrees of heart exposure, rates of late toxicity are expected to be lower (Cella et al. 2011, 2013). However, the potential clinical benefits of all these improvements are not yet completely clear. Indeed, the long latency between radiation exposure and the development of symptomatic heart disease prevents the assessment of the benefits of modern HL therapy settings within a practical time scale. Therefore, the estimation of cardiac toxicity risk through NTCP models would be of great value in order to have early clinical answers from patient cohorts treated in the modern era. In addition, the use of surrogate endpoints such as asymptomatic effects that are predictive of the development of late cardiac events would provide useful timely indications (Darby et al. 2010).

In this framework, in order to quantify the cardiovascular risk reductions of involved node vs. mantle field RT for HL, Maraldo et al. (2012) developed a NTCP model of cardiovascular disease deriving their dose-response curve from the 25 year actuarial risk estimated incidences vs. mediastinal RT dose reported by Schellong et al. (2010) in a pediatric (<18 years) HL cohort. The best esimates for the dose response logistic model parameters were slope γ_{50} of 1.36 (95% CI, 0.78–1.94) and a dose at 50% complication (D_{50}) of 48.2 Gy (95% CI, 38.2–58.2) for any cardiovascular disease.

A more recent retrospective study analyzing about 1500 5-years HL survivors treated with RT in the Netherlands from 1965 to 1995 confirms that mediastinal RT, after a median follow-up of 20 years, increased the risk of coronary heart disease (van Nimwegen et al. 2016). Consistently with the result of Darby et al. (2010), despite the different heart irradiation patterns, the risk ratio of heart disease increased linearly with heart mean dose by 7.4%/Gy (95% CI, 3.3%–14.8%).

Another perspective is given by those studies on cardiac dose-volume effects where dose distributions in heart subregions are investigated. The heart indeed consists of different structures with probably different radiosensitivity.

In a small cohort study by Cella et al. (2011, 2013), in addition to the whole heart structure, the cardiac chambers were also individually evaluated in relation to asymptomatic valve defects assessed at echocardiography. The authors propose multivariable logistic NTCP models for asymptomatic valve defects including the volume of left atrium or left ventricle receiving more than 30 Gy as predictors, but also the combined effects of lung and heart chamber volumes. Model details are provided in Table 11.1. The implemented NTCP model was subsequently applied to an extended dataset of 90 HL survivors confirming the good predictive accuracy of such a model (Cella et al. 2014).

The above results are consistent with those recently reported by Cutter et al. (2015) in a case-control study nested in the above mentioned Dutch cohort. By a retrospective radiation dosimetry analysis, a non-linear dose-response relationship was found between mean dose to heart valves and the valvular heart disease rate, with little or no increase for doses below 30 Gy. However the reliability of retrospective dosimetry on the valves has to be taken into account.

Finally, a French study (Moignier et al. 2015) proposed a method to perform 3D coronary artery segments dosimetry using patients coronary CT angiography. Logistic regression with coronary artery segments yielded an Odds Ratio (OR) associated with the risk of coronary stenosis of 1.049 per additional Gray of the median dose to the coronary segments (95% CI, 1.004–1.095).

Lung Cancer

From the seminal paper by Darby et al. (2013), the RT community learned that the increase of RT-associated CV risk in breast cancer patients started already in the first 5 years after RT. This shed new light on heart dose limits also in the RT planning of those patients whose life expectancy was shorter compared to breast cancer or HL, such as lung carcinoma. Lung cancer patients are also more likely to have comorbid risk factors such as pre-existing cardiac diseases (Janssen-Heijnen et al. 1998) and smoking history that may predispose them to CV events occurring at earlier time points than would be seen in a healthier patient population treated with thoracic RT (Simone 2017; Wang et al. 2017).

Furthermore, the Radiation Therapy Oncology Group (RTOG) 0617 trial (Bradley et al. 2015) demonstrated an association between higher values of heart dosimetric variables (V_{5Gy} and V_{30Gy}) and a worse overall survival at a median follow-up of two years, suggesting that radiation to the heart could contribute to early mortality in non-small cell lung cancer population (NSCLC). Despite this, data on predictors of RT-associated cardiac injury for lung cancer patients are still limited. Following the RTOG 0617 trial, a retrospective study by Speirs et al. (2017) showed that increasing heart V_{50Gy} was associated with worse overall survival in a cohort of 322 locally advanced NSCLC patients (prescribed doses of 50 to 85 Gy; median, 66 Gy). In this analysis, the median heart V_{50Gy} was also found to be significantly higher for patients with CTCAE grade ≥ 1 cardiac toxicity (20.8% vs 13.8%).

A recent retrospective study by Wang et al. (2017) on 112 patients treated for stage III NSCLC in six prospective RT dose-escalation trials (doses of 70 to 90 Gy; median, 74 Gy) reports that cardiac toxicity after RT may occur earlier than traditionally assumed (i.e., ≥ 10 years after RT). In this study, 23% of patients had symptomatic cardiac events at a median of 26 months and most within 5 years of treatment. In a multivariable analysis that accounted for baseline cardiac risk (coronary artery disease or WHO/International Society of Hypertension [ISH] risk score), heart or left ventricle dosimetric parameters (mean dose, V_{5Gy}, V_{30Gy}) were significantly associated with cardiac toxicity. Further details are provided in Table 11.1. The WHO/ISH risk score uses age, sex, smoking status, diabetes and systolic blood pressure to estimate 10 year risk of a cardiovascular event in five strata (0 to 10%, 10% to 20%, 20% to 30%, 30% to, 40% and $\geq 40\%$).

The predictors of pericardial effusion after chemoradiation therapy for locally advanced NSCLC were instead investigated by Ning et al. (2017) in the context of a large prospective randomized trial of IMRT versus proton therapy. The authors report that pericardial effusion occurred in nearly half of all patients with a median time of nine months, even after moderate radiation doses to the heart. Adjuvant chemotherapy increased the risk of pericardial effusion, and a heart $V_{35Gy} > 10\%$ predicted for the development of this toxicity. Of note, pericardial effusion with or without constriction is the most commonly observed type of pericardial disease and effusions, that are initially negligible, may eventually become symptomatic or contribute to the later development of other cardiac toxicities (Wang et al. 2017).

Esophageal Cancer

Esophageal cancer patients, as lung cancer patients, are characterized by poorer prognosis and higher radiation doses to the heart when compared to breast cancer and HL patients; and as for lung cancer patients, data on cardiac morbidity following radiation treatment of esophageal cancer are scarce. This may be once again explained by the fact that cardiac toxicity has traditionally been regarded as a late, or very late, side effect compared to life expectancy. However, thanks to improved prognosis, resulting from the addition of chemotherapy to radiotherapy, the number of long-term survivors of esophageal cancer is increasing and therefore risk of RT-associated CV toxicity is now recognized as an issue of major concern (Beukema et al. 2015).

Konski et al. (2012) retrospectively evaluated the dosimetric data of 74 patients treated with chemoradiotherapy for locally advanced esophageal cancer. Their primary endpoint was symptomatic cardiac toxicity (cardiac ischemia, pericarditis and arrhythmia). Six patients resulted symptomatic and a significant difference was noted in mean heart V_{20Gy}, V_{30Gy} and V_{40Gy} between those patients with and without symptomatic cardiac toxicity. Symptomatic cardiac toxicity was not observed in any patient if the heart V_{20Gy}, V_{30Gy} and V_{40Gy} was below 70%, 65% or 60%, respectively. In a subsequent small retrospective cohort study (Ogino et al. 2016) evaluating clinical and dosimetric factors associated with the risk of symptomatic cardiac disease in 46 patients without a history of heart disease, only heart V_{55Gy} was found to be a significant risk factor (Hazard Ratio [HR] = 1.33, 95% CI 1.02–1.71).

One of the most reported CV complications after esophageal cancer treated with RT (median prescription dose of 60 Gy) is pericardial effusion (Beukema et al. 2015) and a wide range of heart or pericardium dosimetric predictors has been reported. Shirai et al. (2011) analyzed 43 esophageal cancer patients treated with definitive chemoradiotherapy. By multivariate analysis, the authors identified the heart V_{50Gy} as the strongest predictive factor for pericardial effusion of any grade. Other Japanese studies on larger patient cohorts report pericardium V_{30Gy} (cut-off value of 41.6%) (Tamari et al. 2014) or heart V_{10Gy} (cut-off value of 72.8%) (Hayashi et al. 2015) to be the most influential factor.

A more recent study (Ogino et al. 2017), with a longer follow-up (at least 24 months) and using CT scan for objective diagnosis, also showed that a wide range of radiation doses to the heart and pericardium were related to the incidence of pericardial effusion in esophageal cancer patients treated with concurrent chemoradiotherapy. On multivariable analysis heart or pericardium V_{50Gy} were predictive for pericardial effusion (HR = 1.043, 95%CI 1.022–1.065 and HR = 1.057 95%CI = 1.024–1.09, respectively).

As a whole, all selected studies described in Table 11.1 suggest that several heart dosimetric parameters rather than one single parameter are important for cardiac toxicity outcome in thoracic cancer RT treatment.

Critical Review of the Literature 2009–2017 (Hypofractionation)

Another relevant factor that deserves consideration is the relationship between hypofractionated thoracic RT and CV effects. Moderate hypofractionated regimes are used in breast cancer patients. High-dose per fraction are instead used in stereotactic body radiation therapy (SBRT) of lung cancer.

Breast Cancer

Different moderate hypofractionated regimes following breast-conserving surgery, such as 42.5 Gy in 16 daily fractions of Canadian trial (Whelan et al. 2010), 41.6 Gy or 39 Gy in 13 fractions over 5 weeks of UK START) A trial, and or 40 Gy in 15 daily fractions over 3 weeks of START B trial (Haviland et al. 2013), following breast-conserving surgery have been introduced. These regimes have proven to be equally effective and to achieve similar or better cosmetic and normal tissue outcomes for both invasive and *in situ* diseases and when treating the regional nodes (Koulis et al. 2015). The relationship between hypofractionation and late CV effects has been investigated in several analyses and no difference in cardiac disease between standard and hypofractionated regimen was reported.

Mahrin et al. (2007) performed an analysis on about 3800 left-sided and 3700 right-sided breast cancer patients treated between 1984 and 2000, comparing the different fractionation schedules and concluded that a statistical increase in overall and cardiac-specific mortality could not be found when comparing left vs. right breast cancer patients. Furthermore, after a median follow-up of 7.9 years, hypofractionated adjuvant RT regimens did not significantly increase the risk of cardiac mortality. In the Canadian trial (Whelan et al. 2010) no difference in cardiac-related deaths between standard and hypofractionated regimen was reported (1.5% vs. 1.9%, $p > 0.05$, 10 years of follow-up). The 10-year follow-up of the START A and B trials (Haviland et al. 2013) reported that ischemic heart disease were rare and occurred in much the same proportions with each treatment schedule thus confirming the 5 years results that "appropriately dosed hypofractionated radiotherapy is safe and effective."

In support to clinical data, Appelt et al. (2013) performed an *in silico* dosimetric study comparing heart dose distributions for different fractionation schedules (START A, START B and Canadian), corrected for fraction size by Linear Quadratic model (so using the Equivalent dose to 2 Gy/fraction [EQD2Gy]), with the standard 50 Gy/25 regimen. They concluded that using hypofractionated regimes results in a lower biological dose to the heart compared with normofractionated schedules, even for very low values of α/β (as long as $\alpha/\beta \geq 1.5$ Gy). In particular, they analyzed the EQD2Gy corrected heart mean dose and V_{40Gy}. However, to the best of our knowledge, no specific dosimetric studies have been performed.

Lung Cancer

In conventionally fractionated lung RT treatments, major limitations to the use of higher doses are represented by the prolonged overall treatment time and the higher risk of deleterious effects on healthy tissues, including the CV system. These considerations, which are associated with the advanced RT technologies that allow larger doses per treatment to be delivered more safely, have led to the design of aggressive hypofractionated Stereotactic Body Radiation Therapy (SBRT) schedules in treating pulmonary tumors (Ricardi et al. 2009). Several clinical trials (e.g., RTOG protocols 0236, 0813, 0915) have been designed to investigate multiple dosing regimens for patients with early stage, medically inoperable NSCLC with 1–10 delivered fractions depending on the size and central/peripheral location of the target (Guckenberger et al. 2014; Chang et al. 2015). Tumor location and size of Gross Tumor Volume (GTV) > 10 cc were found to be strong predictors of toxicity, with hilar or pericentral tumors showing an 11-fold increased risk in grade 3–5 adverse events when compared to more peripheral tumors (Timmerman et al. 2006).

Heart constraints depend on the dose regimens used. Published constraints based on MD Anderson experience (Li et al. 2014; Evans et al. 2013; Chang et al. 2014) along with those recommended by RTOG protocols (https://www.rtog.org/ClinicalTrials/ProtocolTable.aspx) for different dose/fractionation schedules are summarized in Table 11.2. It is important to underline (Guckenberger et al. 2014) that the constraints reported are simple recommendations not yet formally validated by cohort study analysis.

Notably, although biologically effective dose (BED) formulas were generally used to compare different RT dose-(Chang et al. 2014) fractionation schedules in terms of both dose to Planning Target Volume (PTV) and Organs at Risk, these formulas may not be appropriate for the extreme hypofractionation (in conjunction with very small volumes irradiated) (Kavanagh et al. 2003).

TABLE 11.2 Summary of the Heart/Pericardium Dose Constraints Indicated by RTOG Protocols and MD Anderson Studies for Different Lung SBRT Dose/Fractionation Schedules

Lung Tumor Location	Toxicity Endpoint	Fractionation Scheme	Constraint	Reference
Peripheral lesion				
	Pericarditis (≥grade 3)	1 fr × 34 Gy	$D_{max} \leq 22$ Gy; $V_{16Gy} < 15$ cc	RTOG 0915
	Cardiac Disorders	3 fr × 20 Gy	$D_{max} \leq 30$ Gy	RTOG 0236, RTOG 0618
	Pericarditis (≥grade 3)	4 fr × 12 Gy	$D_{max} \leq 34$ Gy; $V_{28Gy} < 15$ cc	RTOG 0915
Central lesion				
	Pericarditis	5 fr × 10–12 Gy	$D_{max} \leq 52.5$ Gy; $V_{32Gy} < 15$ cc	RTOG 0813
	Cardiac Disorders/Pericarditis	4 fr × 12.5 Gy	$D_{max} \leq 45$ Gy; $V_{40Gy} < 1$ cc, $V_{20Gy} < 5$ cc	Chang (2014), Evans et al. (2013)
	Cardiac Disorders/Pericarditis	7 fr × 10 Gy	$D_{max} \leq 60$ Gy; $V_{40Gy} < 1$ cc	Li et al. (2014)

VXGy, absolute organ volume (in cubic centimeters) exceeding X Gy; Dmax = maximum dose in Gy; fr = fraction.

As described in the previous paragraph, several analyses have identified cardiac dose as an important predictor of overall survival after standard fractionated chemo-radiation for locally advanced NSCLC. The survival influence of the cardiac dose after lung SBRT is much less investigated. The association between heart dose and non-cancer death or heart disease has been recently investigated by Stam et al. (2017) in 803 patients with early stage NSCLC treated with SBRT. The authors report that doses to the left atrium and superior vena cava were significantly associated with non-cancer death. Although these kinds of studies cannot prove a direct cause-effect relationship between heart irradiation and cardiac death, dose sparing, in particular to the upper region of the heart, might potentially improve outcome and should be subject of further studies.

SUMMARY FOR DVH CONSTRAINTS/NTCP MODEL PARAMETERS

At the time of QUANTEC, the available knowledge on radiation-induced CV diseases was limited to specific endpoints and in general was hampered by the quality of dosimetric data, size of populations and length of follow-up.

The post-QUANTEC studies have taken us several steps forward, being based on large populations with a long follow-up where several cardiac comorbidities are taken into account as shown in Table 11.1. Furthermore, both the micro- and macro-perspectives have been considered; in some cases the focus is on the doses to heart sub-volumes, in others CV complications are studied looking instead at the global physiological effect of thoracic RT. The quality of the dosimetric data has also evolved in the direction of individual three-dimensional dose distribution. However, a summary of dose-volume constraints and NTCP parametrization for CV injury following radiation therapy could not be identified. The picture is complex; age at the time of irradiation, longer intervals from the time of radiation, coexisting cardiovascular risk factors and smoking habits all seem to predispose to radiation-induced CV injuries. In particular, the recent meta-analysis performed by Taylor et al. (2017) emphasizes the important role of background smoking history in assessing the cardiac risk of RT cancer survivors.

In a dosimetric perspective, the most straightforward result for both breast cancer and HL patients is a no-threshold relative increase of 7.4% per Gy in the rate of major coronary events (Darby et al. 2013; van Nimwegen et al. 2016). Importantly, for breast, the model was validated in an independent cohort of patients treated with contemporary breast RT (van den Bogaard et al. 2017).

For lung and esophageal cancer patients, data are at present still scarce, although analyses are evolving. Indeed, those cancer patients are generally more challenging when compared to healthier patient population such as breast cancer and HL patients.

Open Issues and Future Directions of Investigation

The literature review performed identifies several issues that need to be further addressed in order to improve our understanding on radiation-induced CV effects and accordingly our ability to make robust predictions.

Lung-Heart Interaction

A better understanding of the mechanism underlying cardiac damage may lead to improved selection of risk predictive metrics. The Groningen group performed a number of studies on the mechanism of cardiac toxicity using an irradiated rat as a model showing that heart damage was aggravated if also the lung was irradiated and vice versa (Ghobadi et al. 2012; van Luijk et al. 2007). However, in a clinical setting, conflicting results have been reported regarding heart-lung interaction in the evolution of heart (or lung) late toxicity (Tucker et al. 2014).

By applying improved statistical machine learning method, Cella et al. (2015) suggested the importance of jointly considering lung irradiation and lung volume size in the prediction of subclinical radiation-related heart disease. Further investigations are needed for testing the clinical validity of this statistical result.

Non-Thoracic Irradiation

An increased risk of cardiovascular events and cardiac mortality after non-thoracic irradiation, such as testicular cancer para-aortic RT or neck irradiation, has been reported (Zagars et al. 2004; Gujral et al. 2014), although it is recognized that radiation is not the only cause of late effect induction. Huddart et al. (2003) hypothesized secondary effects of infradiaphragmatic RT as the possible cause of the increased risk of cardiovascular disease. Blood vessels may be damaged by radiation exposure and RT may have a negative impact on the vascular system by promoting accelerated atherosclerotic vascular complications. Of note, the increase in cardiovascular risk may not be directly correlated with the localization and degree of stenosis according to the principle that atherosclerosis is a diffuse disease and detection in one vascular bed implies a high likelihood of association with atherosclerosis in a different bed (Duprez and Cohn 2008). Thus, the atheroma can be considered as a predictor of vascular events and useful for CV risk stratification.

Most reported cases of radiation-induced arterial stenosis in non-thoracic irradiation concern carotid arteries in head and neck cancers (Xu and Cao 2014). However, cases of stenosis affecting the aorta or the coronary, iliac, subclavian and renal arteries at an atypically young age have been described in patients treated for HL or seminoma (Patel et al. 2006). A recent study (Cella et al. 2017) has showed that patients with an infradiaphragmatic RT history can develop asymptomatic vascular lesions in abdominal arteries even at moderate-low radiation doses. Artery mean dose (OR = 1.13, 95% CI 1.01–1.26) along with younger age at irradiation were found to be associated with increased risk of stenosis evaluated by echo-color Doppler. This observation seems to be in agreement with the hypothesis of a secondary effect of the infradiaphragmatic irradiation on the increased risk of cardiovascular events reported in the literature (Huddart et al. 2003). These findings could be generally relevant for the follow-up of patients irradiated in non-thoracic region. Indeed, non-invasive ultrasound-based follow-up may allow for early detection of asymptomatic vascular radiation-induced damage, eventually helping to prevent severe vascular events usually appearing more than 10 years after RT.

Biological Markers

Biological markers of radiation-related side effects can be used to identify patients at increased risk for radiation-associated CV. Indeed, RT patients display different suscepti-bilities in their normal-tissue responses to identical radiation dose distributions, therefore, the inclusion of biological variability in NTCP modeling could improve predictive capa-bility. RT-induced cardiac-cell damage and changes in the left ventricular loading condi-tions have been linked to several biomarkers including N-terminal pro-B–type natriuretic peptide (NT-proBNP) and troponins. In their studies on left-sided breast cancer patients, D'Errico et al. (2012, 2015) analyzed the relationship between NT-proBNP plasma levels and a large group of dosimetric parameters for the heart and left ventricle. Significant correlations between NT-proBNP plasma levels and some dosimetric parameters (high doses in small volumes) of heart and left ventricle were found in those patients whose NT-proBNP values were above the pathological cut-off threshold (>125 pg/ml). More recently, in a prospective study on a relatively small number of breast cancer patients (58), another group (Skytta et al. 2015) demonstrated that cardiac troponin T (cTnT) levels, detected by high-sensitive cardiac troponin T assay, increased during the entire duration of adjuvant breast RT. In addition, the increase in the cTnT release was positively associated with cardiac radiation doses and with minor changes in echocardiographic measurements revealing subclinical myocardial damage. Further studies are needed to determine bio-marker plasma levels as well as the optimal timing of biomarker measurements to predict the severity and progression of cardiac damage after RT.

Cardiac Imaging Techniques

Several imaging approaches can be used to detect, evaluate, and monitor radiation-induced CV disease.

Echocardiography takes a central role in evaluating the morphology and function of the heart. Global Longitudinal Strain (GLS) and strain rate assessed using automated 2D-speckle-tracking echocardiography is a recent technique for detecting and quantifying subtle disturbances in left ventricular systolic function (Erven et al. 2011). Other imaging modalities, including coronary computed tomographic angiography (CCTA) (Girinsky et al. 2014), cardiac magnetic resonance (CMR) (Machann et al. 2011) and nuclear cardiology, can be used to assess and evaluate the presence of radiation-induced CV disease depending on the type of pathological features (Lancellotti et al. 2013).

Detection and prediction of early subclinical cardiac dysfunction and lesions induced by thoracic RT using functional and anatomical cardiac imaging may enhance our knowl-edge on the potentially involved biological mechanisms (Jacob et al. 2016).

Voxel-Based Approach

A crucial aspect of several NTCP modeling studies on CV morbidities is the diffi-culty or inconsistency in the delineation of the heart substructures. In particular, the definition and contouring of heart vessels and coronary arteries can be time consum-ing and prone to inter- and intra-operator variations with concrete risks of structure misclassification.

A possible alternative is represented by a Voxel-Based (VB) approach which evaluate local dose response patterns and goes beyond the organ-based and dose-volume histogram (DVH)-based philosophy of NTCP modeling thus allowing for a blind identification of the involved regions, irrespective of their anatomical classification (Palma et al. 2016; Monti et al. 2017). A VB approach would identify correlations between radiation-induced heart morbidity and local dose release providing a new insight into spatial signature of radiation sensitivity in composite regions.

A recent study by McWilliam et al. (2017) applied the VB approach on a cohort of 1101 lung cancer patients treated with curative intent with a fractionation pattern of 55 Gy in 20 fractions. The authors identified a significant region across the base of the heart where higher doses were associated with worse patient survival.

Modern RT Technology

In recent years, a considerable effort has been made to identify techniques that reduce the dose to the heart in thoracic cancer irradiation. Several dosimetric and clinical studies have shown that both deep inspiration breath hold (DIBH) and proton beam therapy (PBT) further decrease doses to heart in patients receiving mediastinal RT. By breathing control, the distance between the chest wall and heart is maximized at or near the deep inspiration. Treatment delivery only at deep inspiration reduces the high-dose heart volume. Several groups have reported a reduction in the heart and in the descending coronary artery dose using DIBH in both dose planning and clinical studies on breast cancer or HL (Vikstrom et al. 2011; Nissen and Appelt 2013; Rechner et al. 2017; Aznar et al. 2015).

In a systematic review (Taylor et al. 2015) of studies published during 2003 to 2013, typical heart doses from left-sided breast cancer RT have been reported: The lowest average mean heart doses were from tangential radiation therapy with either breathing control (1.3 Gy; range, 0.4–2.5 Gy) or from PBT therapy (0.5 Gy; range, 0.1–0.8 Gy). PBT represents a promising modality for left breast irradiation to minimize the risk of cardiac events (Tommasino et al. 2017) although definitive clinical experiences are nowadays still sparse (Verma et al. 2016).

In patients receiving mediastinal RT for HL, PBT has been shown to further decrease doses to the heart in both dose planning and phase II clinical studies (Cella et al. 2013; Horn et al. 2016; Hoppe et al. 2014, 2017). A significantly lower radiation exposure to the whole heart and cardiac substructures using PBT has been recently demonstrated in patients with mid- to distal esophageal cancer (Shiraishi et al. 2017).

However, the ongoing randomized trials of protons versus photons and long-term follow-up studies may provide an insight into the impact on cardiac sparing and, accordingly, the risk reduction of radiation-associated CV effects.

REFERENCES

Appelt, A. L., et al. 2013. "Modern hypofractionation schedules for tangential whole breast irradiation decrease the fraction size-corrected dose to the heart." *Clin Oncol (R Coll Radiol)* no. 25 (3):147–52. doi: 10.1016/j.clon.2012.07.012.

Aznar, M. C., et al. 2015. "Minimizing late effects for patients with mediastinal Hodgkin lymphoma: Deep inspiration breath-hold, IMRT, or both?" *Int J Radiat Oncol Biol Phys* no. 92 (1):169–74. doi: 10.1016/j.ijrobp.2015.01.013.

Beukema, J. C., et al. 2015. "Is cardiac toxicity a relevant issue in the radiation treatment of esophageal cancer?" *Radiother Oncol* no. 114 (1):85–90. doi: 10.1016/j.radonc.2014.11.037.

Bradley, J. D., et al. 2015. "Standard-dose versus high-dose conformal radiotherapy with concurrent and consolidation carboplatin plus paclitaxel with or without cetuximab for patients with stage IIIA or IIIB non-small-cell lung cancer (RTOG 0617): A randomised, two-by-two factorial phase 3 study." *Lancet Oncol* no. 16 (2):187–99. doi: 10.1016/S1470-2045(14)71207-0.

Brenner, D. J., et al. 2014. "Risk and risk reduction of major coronary events associated with contemporary breast radiotherapy." *JAMA Intern Med* no. 174 (1):158–60. doi: 10.1001/jamainternmed.2013.11790.

Cella, L., et al. 2011. "Dosimetric predictors of asymptomatic heart valvular dysfunction following mediastinal irradiation for Hodgkin's lymphoma." *Radiother Oncol* no. 101 (2):316–21. doi: 10.1016/j.radonc.2011.08.040.

Cella, L., et al. 2013a. "Hodgkin's lymphoma emerging radiation treatment techniques: Trade-offs between late radio-induced toxicities and secondary malignant neoplasms." *Radiat Oncol* no. 8:22. doi: 10.1186/1748-717X-8-22.

Cella, L., et al. 2013b. "Multivariate normal tissue complication probability modeling of heart valve dysfunction in Hodgkin lymphoma survivors." *Int J Radiat Oncol Biol Phys* no. 87 (2):304–10. doi: 10.1016/j.ijrobp.2013.05.049.

Cella, L., et al. 2014. "Complication probability models for radiation-induced heart valvular dysfunction: Do heart-lung interactions play a role?" *PLoS One* no. 9 (10):e111753. doi: 10.1371/journal.pone.0111753.

Cella, L., et al. 2015. "Predicting radiation-induced valvular heart damage." *Acta Oncol* no. 54 (10):1796–804. doi: 10.3109/0284186X.2015.1016624.

Cella, L., et al. 2017. "Predictors of asymptomatic radiation-induced abdominal atherosclerosis." *Clin Oncol (R Coll Radiol)*. doi: 10.1016/j.clon.2017.07.013.

Chang, J.Y., et al. 2014. "Stereotactic ablative radiation therapy for centrally located early stage or isolated parenchymal recurrences of non-small cell lung cancer: How to fly in a "no fly zone." *Int J Radiat Oncol Biol Phys* no. 88 (5):1120–8. doi: 10.1016/j.ijrobp.2014.01.022.

Chang, J.Y., et al. 2015. "Stereotactic ablative radiotherapy for centrally located early stage non-small-cell lung cancer: What we have learned." *J Thorac Oncol* no. 10 (4):577–85. doi: 10.1097/JTO.0000000000000453.

Cutter, D. J., et al. 2015. "Risk of valvular heart disease after treatment for Hodgkin lymphoma." *J Natl Cancer Inst* no. 107 (4). doi: 10.1093/jnci/djv008.

D'Errico, M. P., et al. 2012. "N-terminal pro-B-type natriuretic peptide plasma levels as a potential biomarker for cardiac damage after radiotherapy in patients with left-sided breast cancer." *Int J Radiat Oncol Biol Phys* no. 82 (2):e239–46. doi: 10.1016/j.ijrobp.2011.03.058.

D'Errico, M. P., et al. 2015. "Kinetics of B-type natriuretic peptide plasma levels in patients with left-sided breast cancer treated with radiation therapy: Results after one-year follow-up." *Int J Radiat Biol* no. 91 (10):804–9. doi: 10.3109/09553002.2015.1027421.

Darby, S. C., et al. 2010. "Radiation-related heart disease: Current knowledge and future prospects." *Int J Radiat Oncol Biol Phys* no. 76 (3):656–65. doi: 10.1016/j.ijrobp.2009.09.064.

Darby, S. C., et al. 2013. "Risk of ischemic heart disease in women after radiotherapy for breast cancer." *N Engl J Med* no. 368 (11):987–98. doi: 10.1056/NEJMoa1209825.

Das, S. K., et al. 2005. "Predicting radiotherapy-induced cardiac perfusion defects." *Med Phys* no. 32 (1):19–27. doi: 10.1118/1.1823571.

Duprez, D. A., and J. N. Cohn. 2008. "Identifying early cardiovascular disease to target candidates for treatment." *J Clin Hypertens (Greenwich)* no. 10 (3):226–31.

Erven, K., et al. 2011. "Acute radiation effects on cardiac function detected by strain rate imaging in breast cancer patients." *Int J Radiat Oncol Biol Phys* no. 79 (5):1444–51. doi: 10.1016/j.ijrobp.2010.01.004.

Evans, J. D., et al. 2013. "Cardiac (1)(8)F-fluorodeoxyglucose uptake on positron emission tomography after thoracic stereotactic body radiation therapy." *Radiother Oncol* no. 109 (1):82–8. doi: 10.1016/j.radonc.2013.07.021.

Gagliardi, G., et al. 1996. "Long-term cardiac mortality after radiotherapy of breast cancer – application of the relative seriality model." *Br J Radiol* no. 69 (825):839–46. doi: 10.1259/0007-1285-69-825-839.

Gagliardi, G., et al. 2010. "Radiation dose-volume effects in the heart." *Int J Radiat Oncol Biol Phys* no. 76 (3 Suppl):S77–85. doi: 10.1016/j.ijrobp.2009.04.093.

Ghobadi, G., et al. 2012. "Physiological interaction of heart and lung in thoracic irradiation." *Int J Radiat Oncol Biol Phys* no. 84 (5):e639–46. doi: 10.1016/j.ijrobp.2012.07.2362.

Girinsky, T., et al. 2014. "Prospective coronary heart disease screening in asymptomatic Hodgkin lymphoma patients using coronary computed tomography angiography: Results and risk factor analysis." *Int J Radiat Oncol Biol Phys* no. 89 (1):59–66. doi: 10.1016/j.ijrobp.2014.01.021.

Guckenberger, M., et al. 2014. "Definition of stereotactic body radiotherapy: Principles and practice for the treatment of stage I non-small cell lung cancer." *Strahlenther Onkol* no. 190 (1):26–33. doi: 10.1007/s00066-013-0450-y.

Gujral, D. M., et al. 2014. "Radiation-induced carotid artery atherosclerosis." *Radiother Oncol* no. 110 (1):31–8. doi: 10.1016/j.radonc.2013.08.009.

Hancock, S. L., S. S. Donaldson and R. T. Hoppe. 1993. "Cardiac disease following treatment of Hodgkin's disease in children and adolescents." *J Clin Oncol* no. 11 (7):1208–15. doi: 10.1200/JCO.1993.11.7.1208.

Haviland, J. S., et al. 2013. "The UK Standardisation of Breast Radiotherapy (START) trials of radiotherapy hypofractionation for treatment of early breast cancer: 10-year follow-up results of two randomised controlled trials." *Lancet Oncol* no. 14 (11):1086–94. doi: 10.1016/S1470-2045(13)70386-3.

Hayashi, K., et al. 2015. "Predictive factors for pericardial effusion identified by heart dose-volume histogram analysis in oesophageal cancer patients treated with chemoradiotherapy." *Br J Radiol* no. 88 (1046):20140168. doi: 10.1259/bjr.20140168.

Hoppe, B. S., et al. 2014. "Involved-node proton therapy in combined modality therapy for Hodgkin lymphoma: Results of a phase 2 study." *Int J Radiat Oncol Biol Phys* no. 89 (5):1053–9. doi: 10.1016/j.ijrobp.2014.04.029.

Hoppe, B. S., et al. 2017. "Consolidative proton therapy after chemotherapy for patients with Hodgkin lymphoma." *Ann Oncol* no. 28 (9):2179–84. doi: 10.1093/annonc/mdx287.

Horn, S., et al. 2016. "Comparison of passive-beam proton therapy, helical tomotherapy and 3D conformal radiation therapy in Hodgkin's lymphoma female patients receiving involved-field or involved site radiation therapy." *Cancer Radiother* no. 20 (2):98–103. doi: 10.1016/j.canrad.2015.11.002.

Huddart, R. A., et al. 2003. "Cardiovascular disease as a long-term complication of treatment for testicular cancer." *J Clin Oncol* no. 21 (8):1513–23. doi: 10.1200/JCO.2003.04.173.

Jacob, S., et al. 2016. "Early detection and prediction of cardiotoxicity after radiation therapy for breast cancer: The BACCARAT prospective cohort study." *Radiat Oncol* no. 11:54. doi: 10.1186/s13014-016-0627-5.

Janssen-Heijnen, M. L., et al. 1998. "Prevalence of co-morbidity in lung cancer patients and its relationship with treatment: A population-based study *Lung Cancer* no. 21 (2):105–13.

Kavanagh, B. D., et al. 2003. "How should we describe the radioblologic effect of extracranial stereotactic radlosurgery: Equivalent uniform dose or tumor control probability?" *Med Phys* no. 30 (3):321–4. doi: 10.1118/1.1543571.

Konski, A., et al. 2012. "Symptomatic cardiac toxicity is predicted by dosimetric and patient factors rather than changes in 18F-FDG PET determination of myocardial activity after chemoradiotherapy for esophageal cancer." *Radiother Oncol* no. 104 (1):72–7. doi: 10.1016/j.radonc.2012.04.016.

Koulis, T. A., et al. 2015. "Hypofractionated whole breast radiotherapy: Current perspectives." *Breast Cancer (Dove Med Press)* no. 7:363–70. doi: 10.2147/BCTT.S81710.

Lancellotti, P., Nkomo, V.T., Badano, L.P., et al. 2013. "Expert consensus for multi-modality imaging evaluation of cardiovascular complications of radiotherapy in adults: A report from the European Association of Cardiovascular Imaging and the American Society of Echocardiography." *Eur Heart J Cardiovasc Imaging* no. 14 (8):721–40. doi: 10.1093/ehjci/jet123.

Li, Q., Swanick, C.W., Allen, P.K., et al. 2014. "Stereotactic ablative radiotherapy (SABR) using 70 Gy in 10 fractions for non-small cell lung cancer: Exploration of clinical indications." *Radiother Oncol* no. 112 (2):256–61. doi: 10.1016/j.radonc.2014.07.010.

Machann, W., Beer, M., Breunig, M., et al. 2011. "Cardiac magnetic resonance imaging findings in 20-year survivors of mediastinal radiotherapy for Hodgkin's disease." *Int J Radiat Oncol Biol Phys* no. 79 (4):1117–23. doi: 10.1016/j.ijrobp.2009.12.054.

Maraldo, M.V., Brodin, N.P., Vogelius, I.R., et al. 2012. "Risk of developing cardiovascular disease after involved node radiotherapy versus mantle field for Hodgkin lymphoma." *Int J Radiat Oncol Biol Phys* no. 83 (4):1232–7. doi: 10.1016/j.ijrobp.2011.09.020.

Marhin, W., et al. 2007. "Impact of fraction size on cardiac mortality in women treated with tangential radiotherapy for localized breast cancer." *Int J Radiat Oncol Biol Phys* no. 69 (2):483–9. doi: 10.1016/j.ijrobp.2007.03.033.

Martel, M. K., et al. 1998. "Fraction size and dose parameters related to the incidence of pericardial effusions." *Int J Radiat Oncol Biol Phys* no. 40 (1):155–61.

McWilliam, A. J., et al. 2017. "Radiation dose to heart base linked with poorer survival in lung cancer patients." *Eur J Cancer* no. 85:106–13. doi: 10.1016/j.ejca.2017.07.053.

Moignier, A., et al. 2015. "Coronary stenosis risk analysis following Hodgkin lymphoma radiotherapy: A study based on patient specific artery segments dose calculation." *Radiother Oncol* no. 117 (3):467–72. doi: 10.1016/j.radonc.2015.07.043.

Moiseenko, V., et al. 2016. "Clinical evaluation of QUANTEC guidelines to predict the risk of cardiac mortality in breast cancer patients." *Acta Oncol* no. 55 (12):1506–10. doi: 10.1080/0284186X.2016.1234067.

Monti, S., et al. 2017. "Voxel-based analysis unveils regional dose differences associated with radiation-induced morbidity in head and neck cancer patients." *Sci Rep* no. 7 (1):7220. doi: 10.1038/s41598-017-07586-x.

Ning, M. S., et al. 2017. "Incidence and predictors of pericardial effusion after chemoradiation therapy for locally advanced non-small cell lung cancer." *Int J Radiat Oncol Biol Phys* no. 99 (1):70–9. doi: 10.1016/j.ijrobp.2017.05.022.

Nissen, H. D., and A. L. Appelt. 2013. "Improved heart, lung and target dose with deep inspiration breath hold in a large clinical series of breast cancer patients." *Radiother Oncol* no. 106 (1):28–32. doi: 10.1016/j.radonc.2012.10.016.

Ogino, I. S., et al. 2016. "Symptomatic radiation-induced cardiac disease in long-term survivors of esophageal cancer." *Strahlenther Onkol* no. 192 (6):359–67. doi: 10.1007/s00066-016-0956-1.

Ogino, I., et al. 2017. "Dosimetric predictors of radiation-induced pericardial effusion in esophageal cancer." *Strahlenther Onkol* no. 193 (7):552–60. doi: 10.1007/s00066-017-1127-8.

Palma, G., et al. 2016. "A voxel-based approach to explore local dose differences associated with radiation-induced lung damage." *Int J Radiat Oncol Biol Phys* no. 96 (1):127–33. doi: 10.1016/j.ijrobp.2016.04.033.

Patel, D. A., et al. 2006. "Clinical manifestations of noncoronary atherosclerotic vascular disease after moderate dose irradiation." *Cancer* no. 106 (3):718–25. doi: 10.1002/cncr.21636.

Rechner, L. A., et al. 2017. "Life years lost attributable to late effects after radiotherapy for early stage Hodgkin lymphoma: The impact of proton therapy and/or deep inspiration breath hold." *Radiother Oncol.* doi: 10.1016/j.radonc.2017.07.033.

Rehammar, J. C., et al. 2017. "Risk of heart disease in relation to radiotherapy and chemotherapy with anthracyclines among 19,464 breast cancer patients in Denmark, 1977–2005." *Radiother Oncol* no. 123 (2):299–305. doi: 10.1016/j.radonc.2017.03.012.

Ricardi, U., et al. 2009. "Dosimetric predictors of radiation-induced lung injury in stereotactic body radiation therapy." *Acta Oncol* no. 48 (4):571–7. doi: 10.1080/02841860802520821.

Schellong, G., et al. 2010. "Late valvular and other cardiac diseases after different doses of mediastinal radiotherapy for Hodgkin disease in children and adolescents: Report from the longitudinal GPOH follow-up project of the German-Austrian DAL-HD studies." *Pediatr Blood Cancer* no. 55 (6):1145–52. doi: 10.1002/pbc.22664.

Shirai, K., et al. 2011. "Dose-volume histogram parameters and clinical factors associated with pleural effusion after chemoradiotherapy in esophageal cancer patients." *Int J Radiat Oncol Biol Phys* no. 80 (4):1002–7. doi: 10.1016/j.ijrobp.2010.03.046.

Shiraishi, Y., et al. 2017. "Dosimetric comparison to the heart and cardiac substructure in a large cohort of esophageal cancer patients treated with proton beam therapy or Intensity-modulated radiation therapy." *Radiother Oncol.* doi: 10.1016/j.radonc.2017.07.034.

Simone, C. B., 2nd. 2017. "New era in radiation oncology for lung cancer: Recognizing the importance of cardiac irradiation." *J Clin Oncol* no. 35 (13):1381–3. doi: 10.1200/JCO.2016.71.5581.

Skytta, T., Tuohinen, S., Boman, E., Virtanen, V., Raatikainen, P. & Kellokumpu-Lehtinen P.L. et al. 2015. "Troponin T-release associates with cardiac radiation doses during adjuvant left-sided breast cancer radiotherapy." *Radiat Oncol* no. 10:141. doi: 10.1186/s13014-015-0436-2.

Speirs, C.K., DeWees, T.A., Rehman, S, et al. 2017. "Heart dose is an independent dosimetric predictor of overall survival in locally advanced non-small cell lung cancer." *J Thorac Oncol* no. 12 (2):293–301. doi: 10.1016/j.jtho.2016.09.134.

Stam, B., et al. 2017. "Dose to heart substructures is associated with non-cancer death after SBRT in stage I-II NSCLC patients." *Radiother Oncol* no. 123 (3):370–5. doi: 10.1016/j.radonc.2017.04.017.

Tamari, K. F., et al. 2014. "Risk factors for pericardial effusion in patients with stage I esophageal cancer treated with chemoradiotherapy." *Anticancer Res* no. 34 (12):7389–93.

Taylor, C. W., et al. 2015. "Exposure of the heart in breast cancer radiation therapy: A systematic review of heart doses published during 2003 to 2013." *Int J Radiat Oncol Biol Phys* no. 93 (4):845–53. doi: 10.1016/j.ijrobp.2015.07.2292.

Taylor, C., et al. 2017. "Estimating the risks of breast cancer radiotherapy: Evidence from modern radiation doses to the lungs and heart and from previous randomized trials." *J Clin Oncol* no. 35 (15):1641–9. doi: 10.1200/JCO.2016.72.0722.

Timmerman, R., et al. 2006. "Excessive toxicity when treating central tumors in a phase II study of stereotactic body radiation therapy for medically inoperable early-stage lung cancer." *J Clin Oncol* no. 24 (30):4833–9. doi: 10.1200/JCO.2006.07.5937.

Tommasino, F., et al. 2017. "Model-based approach for quantitative estimates of skin, heart, and lung toxicity risk for left-side photon and proton irradiation after breast-conserving surgery." *Acta Oncol* no. 56 (5):730–6. doi: 10.1080/0284186X.2017.1299218.

Tucker, S. L., et al. 2014. "Is there an impact of heart exposure on the incidence of radiation pneumonitis? Analysis of data from a large clinical cohort." *Acta Oncol* no. 53 (5):590–6. doi: 10.3109/0284186X.2013.831185.

van den Bogaard, V.A., Ta, B.D., van der Schaaf, A., et al. 2017. "Validation and modification of a prediction model for acute cardiac events in patients with breast cancer treated with radiotherapy based on three-dimensional dose distributions to cardiac substructures." *J Clin Oncol* no. 35 (11):1171–8. doi: 10.1200/JCO.2016.69.8480.

van Luijk, P., et al. 2007. "The impact of heart irradiation on dose-volume effects in the rat lung." *Int J Radiat Oncol Biol Phys* no. 69 (2):552–9. doi: 10.1016/j.ijrobp.2007.05.065.

van Nimwegen, F. A., et al. 2016. "Radiation dose-response relationship for risk of coronary heart disease in survivors of hodgkin lymphoma." *J Clin Oncol* no. 34 (3):235–43. doi: 10.1200/JCO.2015.63.4444.

Verma, V., et al. 2016. "Clinical outcomes and toxicity of proton radiotherapy for breast cancer." *Clin Breast Cancer* no. 16 (3):145–54. doi: 10.1016/j.clbc.2016.02.006.

Vikstrom, J., et al. 2011. "Cardiac and pulmonary dose reduction for tangentially irradiated breast cancer, utilizing deep inspiration breath-hold with audio-visual guidance, without compromising target coverage." *Acta Oncol* no. 50 (1):42–50. doi: 10.3109/0284186X.2010.512923.

Wang, K., et al. 2017. "Cardiac toxicity after radiotherapy for stage III non-small-cell lung cancer: Pooled analysis of dose-escalation trials delivering 70 to 90 Gy." *J Clin Oncol* no. 35 (13):1387–94. doi: 10.1200/JCO.2016.70.0229.

Wei, X., et al. 2008. "Risk factors for pericardial effusion in inoperable esophageal cancer patients treated with definitive chemoradiation therapy." *Int J Radiat Oncol Biol Phys* no. 70 (3):707–14. doi: 10.1016/j.ijrobp.2007.10.056.

Whelan, T. J., et al. 2010. "Long-term results of hypofractionated radiation therapy for breast cancer." *N Engl J Med* no. 362 (6):513–20. doi: 10.1056/NEJMoa0906260.

Xu, J., and Y. Cao. 2014. "Radiation-induced carotid artery stenosis: A comprehensive review of the literature." *Interv Neurol* no. 2 (4):183–92. doi: 10.1159/000363068.

Zagars, G. K., et al. 2004. "Mortality after cure of testicular seminoma." *J Clin Oncol* no. 22 (4):640–7. doi: 10.1200/JCO.2004.05.205.

CHAPTER 12

Adverse Effects to the Skin and Subcutaneous Tissue

Michele Avanzo, Joseph Stancanello, and Rajesh Jena

CONTENTS

Introduction 289
    Structure of the Skin 290
    Acute Side Effects to the Skin 290
    Late Adverse Effects 291
    Systems to Grade and Quantify Radiation-Induced Side Effects to the Skin 291
    Non-Dosimetric Factors Related to Skin Toxicity 292
    Dosimetry of the Skin in RT 292
Dose Response Models 293
    Models for Acute Effects in the Frame of Post-Mastectomy Breast Cancer RT
    Using Orthovoltage Photons Beams 293
    NTCP Models for Late Effects in the Frame of Post-Mastectomy Breast
    Cancer RT Using Orthovoltage Photons Beams 296
    NTCP for Skin Effects in the Frame of Post-Quadrantectomy Breast Cancer RT
    Using Megavoltage Photons Beams 297
    Incomplete Repair Between Fractions and Repopulation 299
    Volume/Surface Effect 299
    Considerations for Stereotactic Body Radiation Therapy (SBRT) 301
Recommendations for RT Planning 302
    Contouring of the Skin for RT Planning 302
    Strategies to Reduce Risk for Skin Side Effects 302
    Tolerance Doses for RT 303
Conclusions 305
References 305

INTRODUCTION

The skin is one of the largest organs in the body and its primary function is to provide a protective barrier for the body from physical, chemical, infectious, and thermoregulatory threats, but it also serves sensory and immune functions. Cutaneous skin reactions are common side effects of radiotherapy (RT). Severe radiation-induced skin toxicity is burdensome to patients and sometimes necessitates interruption of radiation treatment, which can impact on locoregional tumor control. Because skin changes are easily observed and followed, knowledge of the normal early stage response of skin to irradiation warns the therapist of impending late changes and gives insight into the patient's total response to irradiation.

The aim of this chapter is to provide a comprehensive review of currently available models for prediction of the risk for skin side effects. As dose responses for skin effects such as moist desquamation, telangiectasia, and severe dermatitis are known for orthovoltage photons but are less described for high-energy photons. We here provide novel models for the description of these side effects. Finally, the reviewed models are used in order to derive tolerance doses for adverse skin effects, and a summary of recommendations regarding skin contouring and planning of the treatment useful for prevention of skin side effects RT is provided.

Structure of the Skin

The two main layers of the skin are the epidermis, which is the outer shell, and the dermis, the inner one. The epidemis is 30 to 300 μm thick, the deepest layer is the stratum basale, a proliferative basal cell monolayer whose cells divide to form the keratinocytes of the epidermal cells. The outermost layer of the epidermis is the stratum corneum, made of flattened dead cells. The epidermal shell is bonded through the basement membrane to the dermal shell, which is 1–3 mm thick. The upper portion is the papillary layer, which contains the microvessels supplying the epidermis. The remaining dermis, the rete dermis, consisting of collagen bundles, scattered fibroblasts, and single microvessels, is bonded to the subcutaneous fat layer, which may be several centimeters thick. The subcutaneous tissue contains predominantly fat cells, but also a network of arteries, veins, and lymphatics (Archambeau et al. 1995).

Acute Side Effects to the Skin

Radiation reactions in the skin can be divided into early or acute changes and delayed or chronic changes. The early, acute changes are frequent side effects of RT, usually occurring within 90 days, and probably resulting from damage of epithelial cells. Their range begins with transient, early erythema, often seen within a few hours from irradiation, which is possibly related to inflammation. The main erythematous reaction, which reflects a loss of epidermal basal cells, takes usually two to three weeks to occur after the start of RT, because of the time required for differentiating cells to move from the basal layer to the keratinized layer of the skin. Dry desquamation, occurring at 3–6 weeks, is an atypical keratinization of the skin due to the reduction in the number of clonogenic cells within the basal layer of the epidermis.

The recovery of the epidermis occurs as a result of the proliferation of surviving clonogenic basal cells within the irradiated area, and migration of viable cells from the edges of the irradiated site. With doses below the threshold for moist desquamation, the repopulation of the basal layer of the epidermis restores the skin back to normal. With larger doses, moist desquamation, the loss of the epidermis due to the destruction of a high proportion of the clonogenic cells within the epidermis, may appear, typically at 3–5 weeks (Schulte et al. 2005). Other effects include pigmentation, epilation, and atrophy (thinning of the dermal tissues associated with the contraction of the previously irradiated area).

The acute changes can be painful and debilitating. Moist desquamation can heal by 50 days following irradiation or not heal and progress to necrosis. With symptomatic treatment aimed at control of pain, necrosis generally will not occur as epithelial remodeling (hyperplasia) and reepithelialization restore the skin to normal, usually within four weeks. Ulceration can be cured by ointments containing gentamicin but, in some cases, can become refractory, requiring surgical procedures such as a skin graft (Schulte et al. 2005).

Irradiation-induced epilation is due to high susceptibility of anagen follicles, which are located approximately 5 mm below the scalp to radiation. Complete hair regrowth generally occurs 2–4 months after irradiation. With higher doses, permanent alopecia may occur (De Puysseleyr et al. 2014).

Late Adverse Effects

The long-term negative effects of radiation include breast pain, fibrosis, scaling, atrophy, discoloration, telangiectasia, and necrosis of the skin, which can appear after a time period of months to years, following the acute reaction, during which the skin appears normal. The time to expression of 90% of moderate to severe complications has been estimated as 3.2 years for fibrosis and 4.7 years for telangiectasia (Bentzen et al. 1989).

Telangiectasia is the atypical dilation of the superficial dermal capillaries and results from the loss of microvascular endothelial cells and damage to the basement membrane, appearing as an area of reddish discoloration displaying multiple, prominent, thin-walled, and dilated vessels ("spider veins"). It may disappear spontaneously in the course of time, but it is often permanent.

Radiation-induced fibrosis of the subcutaneous layer is characterized by a progressive induration, edema formation, and thickening of the dermis and subcutaneous tissues. Fibrosis represents a proliferative response of surviving fibrocytes to growth factors released by injury; e.g., the transforming growth factor β (TGF-β), originated by activation of fibroblasts in subcutaneous tissue in response to tissue injury and their subsequent proliferation and overproduction of connective tissue. It is most severe in those areas where there was an earlier moist reaction. Onset and progression of fibrosis is dose dependent and, once started, slowly progressive (Archambeau et al. 1995).

In general, every acute skin reaction is followed by some degree of permanent damage or late change. There is a growing clinical evidence showing that telangiectasia is a late sequelae of moist desquamation and acute erythema is shown to be a risk factor for poor cosmetic outcome (Lilla et al. 2007).

Systems to Grade and Quantify Radiation-Induced Side Effects to the Skin

Several grading scales exist for reproducible assessment and documentation of radiodermatitis. The Radiation Therapy Oncology Group/European Organization for Research and Treatment of Cancer (RTOG/EORTC) toxicity criteria and National Cancer Institute

Common Toxicity Criteria for Adverse Events version 3 and 4.03 (NCI CTCAE) systems are the most commonly used (Cox et al. 1995).

More objective measures of radiation-induced skin toxicity are needed to overcome the inter-user uncertainty of subjective scales. Statistically significant differences were found in microcirculation index, an index based on blood flow parameters obtained via Doppler fluxometry, at the end of RT among the patients with different CTCAE classification of skin reactions (Gonzalez Sanchis et al. 2017). Evaluation of skin oxygenation and perfusion with hyperspectral imaging has shown a dose-response relationship between radiation exposure and oxygenated hemoglobin after a 6 Gy cumulative dose threshold, and association with visual skin reactions classified according to the RTOG system (Chin et al. 2017).

Non-Dosimetric Factors Related to Skin Toxicity

The presence of comorbidities such as psoriasis (Pastore et al. 2016) and diabetes (Jagsi et al. 2015) increase the risk for acute effects to the skin. Allergy, hypertension and long-term smoking was associated with a significant increase in risk of telangiectasia (Lilla et al. 2007). Moreover, a number of drugs are known to increase radiosensitivity of the skin and subcutaneous tissue. When given in conjunction with RT, mitoxantrone, 5-fluorourcil, cyclophosphamide, paclitaxel, docetaxel, and possibly tamoxifen can result in cutaneous toxicity. A separate form of radiation-related drug toxicity is termed *radiation recall*. This is an inflammatory skin reaction of unknown origin that occurs in a previously irradiated body part after drug administration (Balter et al. 2010). Grade ≥ 3 dermatitis after RT for head and neck cancer is associated with concomitant administration of cetuximab (Studer et al. 2011). The administration of prophylactic skin care is effective in reducing the risk and severity of skin reactions (Chen et al. 2010a).

Dosimetry of the Skin in RT

The measurement or calculation of absorbed dose to the skin from RT treatment is essential for studying its dose response in RT. Since the skin extends from a few millimeters deep to the surface of the patient, the limited accuracy in calculating surface dose by commercial treatment planning systems (TPSs) in regions of electronic disequilibrium, like the build-up region, is a major concern.

The skin is one of the few sites in the human body where dosimeters, such as films or thermoluminescent dosimeters, can be easily placed for in vivo measurements. When in vivo dosimetry is performed on the patient's surface, relatively large differences from the calculated dose are often found. The difference between calculated and measured dose at surface depends on the orientation of beams, and it is larger were normally incident beams are used (Avanzo et al. 2013). As a consequence, the skin dose depends on the treatment delivery modality. For treatment of the head and neck, it was found that static, bilateral fields produce a measured dose of 69% average of the target dose, while for tomotherapy treatment and Intensity Modulated RT (IMRT) it is 71% and 82%, respectively. The discrepancy between calculated and measured was 13, 11, and 10% relative to the prescribed dose for bilateral fields, IMRT with multiple equispaced fields, and helical tomotherapy, respectively (Higgins et al. 2007). The differences in discrepancies among various treatment delivery modalities could be partly due to a change in depth dose. Depth dose measurements, extracted from embedded films, indicated that the depth of dose build-up (>99%) was the shallowest for IMRT (2–5 mm) followed by tomotherapy (5–8 mm) and bilateral fields (10–15 mm). The addition of bolus or immobilizing mask material results in both a significant increase of the dose to the surface, and reduction of the discrepancy from the calculated dose (Higgins et al. 2007).

For breast cancer treatment , surface dose appears lower in tangential 3D-conformal RT (3D-CRT) techniques with respect to IMRT (Rudat et al. 2014). Measurements of superficial dose in phantoms with radiochromic films at different depths (up to 5 mm) showed that the average dose to the skin with wedged 3D-CRT was 89.4% of the prescribed dose and with tangential IMRT it was 87.6% (Almberg et al. 2011).

The accuracy of dose calculation on the patient's surface also depends on the grid size used for dose calculation. Measurements with ultrathin thermoluminescent dosimeters for head and neck IMRT treatments found differences on measured to calculated doses of +21%, +9.5%, $\leq$+2%, −11% for 5, 3, 2, 1 mm^3 voxel sizes, respectively. The dose to the surface therefore was accurately predicted for 2 mm^3 voxels. The same result was confirmed for Volumetric Modulated Arc Therapy (VMAT) (Price et al. 2014). A study comparing radiochromic film readings with calculated dose on cadavers in the region of the scalp showed better accuracy with the use of a 1 mm^3 calculation grid (De Puysseleyr et al. 2014).

Quite surprisingly, there are only few studies correlating measured dose in vivo with appearance of side effects. Clinical tests conducted on six patients revealed that measured dose greater than 49.9 Gy was associated with a grade 3 skin reaction, and skin reactions corresponding to RTOG grade 2 or lower were observed for doses below 40 Gy, when extrapolated to a course of 30 treatment fractions (Fu et al. 2017).

DOSE RESPONSE MODELS

In the following sections we will review the models currently available for the estimation of Normal Tissue Complication Probability (NTCP) for adverse skin reactions and propose new dose response models based on recently published data on the incidence of adverse skin effects on patients treated with megavoltage photon RT for breast cancer. The models and their parameters described in the following sections are summarized in Table 12.1a for acute reactions, and Table 12.1b for late reactions.

Models for Acute Effects in the Frame of Post-Mastectomy Breast Cancer RT Using Orthovoltage Photons Beams

Dose response relationships for various acute and late side effects to the skin after breast RT were extensively studied using linear-quadratic (LQ) based models (Turesson et al. 1989). The clinical assay consisted of 450 breast cancer patients irradiated postoperatively to various dose schedules. The majority of patients received 200-kV orthovoltage radiation. In some patients electron beams of 12, 13 MeV were used and a relative biological effectiveness (RBE) for electrons of 0.83 relative to 200 kV photons was derived in the same patient assay. The acute endpoints studied were the appearance of erythema and moist desquamation. The appearance of erythema was quantified by reflectance spectrophotometry with

TABLE 12.1 Summary of Existing Models for (a) Acute and (b) Late Toxicities to the Skin. In the Last Column, the 2Gy Per Fraction Equivalent Dose (EQD2Gy) Corresponding to 5% and 50% Normal Tissue Complication Probability (EQD2Gy5% and EQD2Gy50%) is Calculated Using Each Model

References	End-Point	Dataset	Model, Parameters	Notes	EQD2Gy5%, EQD2Gy50%
A					
Turesson and Thames (1989)	Erythema	Breast orthovoltage RT	LQ model $\ln(K)=4.81$, $\alpha=0.95\bullet10^{-1}$, $\beta=1.27\bullet10^{-2}$, $\alpha/\beta=7.5$ Gy	No surface effect, electrons RBE=0.83	30.8 Gy, 43.8 Gy
Turesson and Thames (1989)	Moist Desquamation		LQ model, $\ln(K)=5.57$, $\alpha=1.02\bullet10^{-1}$, $\beta=0.91\bullet10^{-2}$, $\alpha/\beta=11.2$ Gy		37.2 Gy, 49.4 Gy
Present study	Grade ≥2 Dermatitis	Published studies on breast RT with	Lyman, $D_{50}=58.3$ Gy, m=0.28, $\alpha/\beta=10$ Gy	Assumed skin dose 90% of prescribed.	29.5 Gy, 58.3 Gy
Present study	Moist Desquamation	MV-photons, Table 12.2	Lyman, $D_{50}=55.8$ Gy, m=0.22, $\alpha/\beta=10$ Gy	Electrons RBE=0.89	33.1 Gy, 55.8 Gy
Burman et al. (1991)	Ulceration/Necrosis	Based on previous model by Von Essen (1973)	Lyman, n=0.1, m=0.12, D50=70 Gy	Reference surface 100 cm², no correction for fractionation	55 Gy, 70 Gy
Pastore et al. (2016)	grade ≥3 RTOG Toxicity	Single institution breast cancer dataset	Lyman, $n=0.38$, m=0.14, $D_{50}=39$ Gy	No correction for fractionation	30.2, 39 Gy
B					
Turesson and Thames (1989)	grade ≥1 Telangiectasia at 3 years	Breast, orthovoltage	LQ model $\ln(K)=3.98$, $\alpha=0.48\bullet10^{-1}$, $\beta=1.73\bullet10^{-2}$; $\alpha/\beta=2.77$ Gy	No surface effect considered, RBE=0.83	34.9 Gy, 52.6 Gy
Turesson and Thames (1989)	grade ≥2 Telangiectasia at 5 years	Breast, orthovoltage	LQ model $\ln(K)=5.36$, $\alpha=0.64\bullet10^{-1}$, $\beta=2.29\bullet10^{-2}$; $\alpha/\beta=2.79$ Gy	RBE=0.83 for electrons	38.8 Gy, 52.2 Gy
Bentzen and Overgaard (1991)	Telangiectasia	Post mastectomy	Logit function and LQ Model with complete repair, $\alpha0=-8.6$, $\alpha=0.099$ Gy -2, $\beta=0.036$		33.1 Gy, 50.3 Gy
Present study	Telangiectasia	Published studies on breast MV-RT; Table 12.2	Lyman, $D_{50}=58.2$ Gy, Lyman, m=0.11, $\alpha/\beta=3$ Gy	Assumed skin dose 90% of prescribed, Electrons RBE=0.89	47.4 Gy, 58.2 Gy
Avanzo et al. (2012)	Subcutaneous Fibrosis	Published data	Lyman, $D_{50}=107.2$ Gy, m=0.22, $n=0.06$, $\alpha/\beta=3$ Gy		47.3 Gy, 73.7 Gy
Mukesh et al. (2013b)	Subcutaneous Fibrosis	Data from 3 trials	Lyman, $D_{50}=132$ Gy, m=0.35, $n=0.012$, $\alpha/\beta=3$ Gy		34.5 Gy, 81.4 Gy

LQ=linear quadratic; MV=megavoltage; RBE=Relative Biological Effectiveness.

light of wavelengths 578 and >660 nm, corresponding to the absorption of oxyhemoglobin and melanin, respectively. A score of more than 50% change of reflectance relative to the baseline value was adopted as the endpoint of relevant erythema.

As the dataset included patients treated with various fractionation schemes, it was possible to study the sensitivity of different endpoints to fractionation, and α/β ratio was determined as 7.5 and 11.2 Gy for erythema and moist desquamation, respectively, for treatments lasting less than 40 days. In Figure 12.1 the Turesson et al. (1989) dose-response relationships without repopulation and repair corrections are shown for erythema and moist desquamation versus dose at 2 Gy per fraction (EQD2y, details for definition in Chapter 2).

Large scale analyses of skin toxicity data are complicated by the subjective nature of clinical assessment, differences in treatment planning techniques, fractionation schedules, and the retrospective nature of data collection. An analysis used prospective data collection from a single institution, with all evaluations of skin toxicity undertaken by the same observers (Pastore et al. 2016). A model for NTCP of severe acute skin toxicity (RTOG grade 3) was then derived using data from 140 breast cancer patients. The dose prescribed to the whole breast was of 50 Gy with 2 Gy daily fractions in five weeks. A boost of 10 Gy with electrons was delivered to the tumor bed in one week. Two models were derived: an NTCP Lyman model (see Chapter 2 for details) recast for inclusion of dose-surface histogram (DSHs) and a multivariate logistic model. DSHs of the body were considered as representative of the irradiation in thin cutaneous layers. Absolute DSH was estimated from dose to the outer 3 mm of the patient body contour. The surface of the patient receiving at least 5 Gy was used to delimit the skin area belonging to the breast region and to normalize

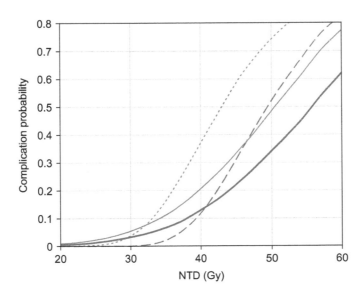

FIGURE 12.1   Dose response models for acute erythema (short-dashed line) and moist desquamation (long-dashed line) by Turesson et al. (1989) and models for grade≥2 dermatitis (thin, solid line) and moist desquamation (thick, solid line) derived by fitting data in Table 12.2 versus 2Gy Equivalent Dose (EQD2Gy) in the present work.

the absolute DSH. The multivariate analysis yielded a two-variable model based on the skin surface receiving more than 30 Gy and the presence of psoriasis, which had high discriminative power (area under the Receiver Operating Characteristics curve, AUC = 0.84).

NTCP Models for Late Effects in the Frame of Post-Mastectomy Breast Cancer RT Using Orthovoltage Photons Beams

In the work of Turesson et al. (1989), the appearance of telangiectasia was scored blindly by two observers from photographs taken weekly/biweekly. Scores ranged from 0 to 3, describing no, minimal, distinct, and very marked telangiectasia The incidence of telangectasia with score ≥1 at three years and score ≥2 at five years post RT, was fitted with a model based on the LQ model (Table 12.1a,b).

Models for late skin necrosis and ulceration, at 5 years from RT were derived by Emami et al. (1991) by using previously published dose response data (von Essen 1963) which were considered consistent with the experience of radiation oncologists in the group. The dose response models were fitted with a probit formula. The dependence on the area irradiated normalized to a reference area of 100 cm² was described using a volume parameter, n, of 0.1 (see Chapter 2 for details on the meaning of organ volume parameter n).

The appearance of fibrosis and telangectasia on post-mastectomy patients was modeled using a logit formula by Bentzen et al. (1989). A correction was applied to account for the latency of the appearance of side effect, where the time at which 90% of the ultimately expected damage is expressed was chosen to characterize the position of the latent-time distribution. The α/β ratios were 1.9 Gy and 3.7 Gy for fibrosis and telangectasia, respectively. The dose response curves obtained with these models for telangiectasia are shown in Figure 12.2.

Alexander et al. (2007) reported that the volume parameter exhibited a fairly parallel effect (n = 0.78) t on breast fibrosis. However, later studies all demonstrated a weaker

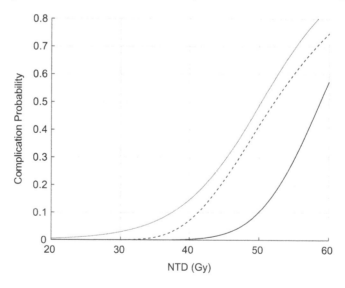

FIGURE 12.2 Dose response for telangiectasia derived for postmastecomy patients (dotted), for whole breast irradiation with orthovoltage radiotherapy (dashed) and derived for whole breast RT with high energy photons in the present study by fitting the data in Table 12.2 (continuous).

dependency of fibrosis on irradiated volume. Avanzo et al. (2012) estimated the best-fit parameters of a model for moderate and severe breast fibrosis using average values of dosimetric parameters (prescription dose, fraction dose, median follow-up, and dose-volume data) from three whole breast irradiation studies without boost and four external beam partial breast irradiation studies. A latency function correction was also included to account for limited follow-up time, according to which 90% of patients develop fibrosis within 8.1 years. The Equivalent Uniform Dose (EUD) corrected for fractionation with $\alpha/\beta = 3$ Gy was used. The fitting process yielded to $n = 0.15$, showing a more serial-like behavior of the skin when compared with the previous studies. Mukesh et al. (2013b) derived ann NCTP model for fibrosis by pooling data of 5856 patients from 2 trials of whole breast irradiation with/without boost. Dose distribution for each patient was approximated by a two compartment model, where the boost volume was the first, and the remaining breast volume was the second compartment, each one receiving a uniform dose (corrected for fractionation with $\alpha/\beta = 3$ Gy). A very small volume effect was found ($n = 0.012$) in this case. The model also proved good predictive ability on a validation dataset of 1410 patients.

NTCP for Skin Effects in the Frame of Post-Quadrantectomy Breast Cancer RT Using Megavoltage Photons Beams

The NTCP models described in the previous sections for erythema, telangiectasia, and moist desquamation cannot be generalised to RT with high-energy photons, because they were derived from patients treated with orthovoltage radiation and post-mastectomy RT, where the type of surgery may impact the risk for radiation-related side effects. It would be tempting to rescale the models by using the RBE of orthovoltage photons with respect to megavoltage photons, but the estimate of the RBE is affected by a large uncertainty, estimated in the range 1.2–1.4 (Marthinsen et al. 2010).

Novel models for grade ≥ 2 dermatitis, moist desquamation, and telangectasia are derived here by fitting incidences of toxicities in the published studies on RT of the whole breast listed in Table 12.3. Since most patients were treated with roughly the same field sizes, it was assumed that the effect of irradiated surface/volume could be ignored. In the analysis it was assumed that overall treatment time played a negligible role in determining the frequencies of these late effects; also, for those studies in which a fraction of patients received the boost, the weighted sum of doses with and without the boost, with the fraction of patients receiving or not receiving the boost as weight, was considered. Following the results of studies on superficial dose in phantoms following tangential irradiation (Almberg et al. 2011), we assumed that the skin received 90% of the prescribed dose. The dose received from the electron boost was corrected for RBE of electrons with respect to high-energy photons, assumed as of 0.89 (Bentzen et al. 1991). The effect of fractionation was included by converting into EQDGy assuming $\alpha/\beta = 10$ Gy for acute effects, and $\alpha/\beta = 3$ Gy for telangiectasia. The parameters of a probit formula minimizing the difference between complication rates in the clinical studies and calculated NTCPs were determined by the weighted least squares method, using the number of patients in each dataset as weights (Table 12.2a and b). The dose-response curves obtained are shown in Figure 12.3.

TABLE 12.2 Summary of Studies Reporting Toxicities to the Skin Used to Derive Dose Response in the Present Chapter

Authors	Technique	PTV Dose/Fractions	Boost Dose/Fractions/% of Patients	Total EQD2Gy	Patients	End-Point
Keller et al. (2012)	Hybrid IMRT	46 Gy/23 fx	14 Gy/7 fx/99%	60. 1 Gy	946	Telangectasia 8.2%
Haviland et al. (2013) START-A	3D-CRT	50 Gy/25 fx, 41.6 Gy/13 fx, 39 Gy/13 fx	10 Gy/5 fx/60%	56.9, 55.0, 53.8 Gy	730, 730, 627	Telangectasia 5.7/5.9/2.5% at 10 years
Haviland et al. (2013); START-B	3D-CRT	50 Gy/25 fx, 40.05 Gy/15 fx	10 Gy/5 fx/60%	56,9/52.4 Gy Gy	1081/1094	Telangectasia 4.8/3.1% at 10 years
Lilla et al. (2007)	3D-CRT	50.4–50 Gy/25–28 fx	12 Gy/6 fx/90%	59.5 Gy	409	Telangectasia 31.8%
Jagsi et al. (2015)	3D-CRT	50 Gy/25, 40.05 Gy/16 fx	10 Gy/5 fx, 93%/60%	54.1/43.5 Gy	1731/578	MD 28/6.6%
Pignol (2008)	IMRT, 3D	50 Gy/25 fx	16 Gy/8 fx/30%	54.8 Gy	170/161	MD 31.2/47.8%
Chen et al (2010)	3D-CRT	50.4 Gy/28 fx		49.56 Gy	158	MD 23%
De Langhe et al. (2014)	IMRT	50 Gy/25 fx, 40.05/15 fx	10 Gy/4 fx/75%	57.8, 50.1 Gy	45/332	MD 51.1, 9.9%
Rudat et al. (2016)	Inverse IMRT/3D-CRT	40.05/15 fx, 50 Gy/25 fx	Evaluated before boost	42.3 Gy, 50, Gy	121/145	AD2 2/19%
De Langhe et al. (2014)	IMRT	50 Gy/25 fx, 40.05/15 fx	10 Gy/4 fx/75% patients	57.8, 50.1 Gy	45/332	AD2 87, 54.5%
Nagai et al. (2017)	IMRT Tomodirect	50 Gy/25 fx		50 Gy	152	AD2 ≥ 2: 15.8%
Freedman et al. (2009)	3D-CRT/IMRT	48 Gy/24 fx	14–20 Gy/95% patients	63.2 Gy	60/73	AD2 75/52%
Harsolia et al. (2007)	IMRT/3D-CRT	45 Gy/25 fx	16 Gy/8 fx	52.7 Gy	93/79	AD2 41/85%
Jagsi et al. (2015)	3D-CRT	50 Gy/25, 40.05 Gy/16 fx	10 Gy/5 fx, 93%/60% patients	54.1/43.5 Gy	1731/578	AD2 61/27%
Monten et al. (2017)	VMAT	28.5 Gy/5 fx, 33.5 Gy/5 fx		32.3 Gy/40.3 Gy	32/63	AD2 0/17.5%
Hijal et al. (2010)	3D-CRT	42.6 Gy/16 fx		39.6 Gy	162	AD2 11.7%

PTV=Planning Target Volume; IMRT=Intensity Modulated Radiation Therapy; 3D-CRT=three-dimensional Conformal Radiotherapy; VMAT=Volumetric Modulated Arc Therapy; fx=fractions; EQD2Gy=2Gy equivalent dose; MD=moist desquamation; AD2=grade ≥ 2 acute dermatitis.

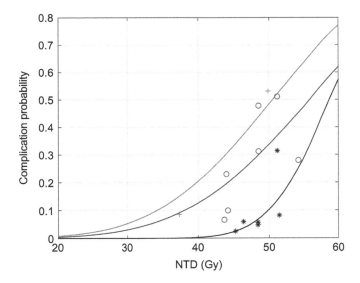

FIGURE 12.3   Plots of fits of dose response relationships (solid lines) for grade≥2 dermatitis (red), moist desquamation (blue), and telangiectasia (black) versus 2Gy Equivalent Dose (EQD2Gy). For calculation of EQD2Gy, α/β of 10 Gy and 3 Gy were used for acute and late effects respectively. The rates of side effects from clinical studies (see Table 12.2) used for fitting are also plotted. For easy reading, the datasets for grade≥2 dermatitis have been condensed into two datasets.

Incomplete Repair Between Fractions and Repopulation

The acutely responding cell population might be spared by cell repair and proliferation. The repair between multiple fractions of RT when the time between fractions is less than 24 hours is modeled using an exponential function, with repair halftime between 1.1 and 1.3 hours for acute effects. There is also a suggestion of a second faster repair component for acute and late effects in the skin, with half-time of the order of 0.3–0.4 hours (Turesson et al. 1989).

For late effects, direct analysis of skin reactions after post-mastectomy breast treatments indicates longer repair half-times, of 3 hours or more, for late telangiectasia (Turesson et al. 1989). In accelerated partial breast irradiation (APBI), good results in terms of side effects have been generally reported (Chen et al. 2010), but in two APBI studies, significant short-term toxicities have also been observed (Hepel et al. 2009; Jagsi et al. 2010). For modeling subcutaneous fibrosis after twice daily APBI, biologically-corrected EUD calculations included a repair half-time of 4.4 hours for RT that using twice-a-day fractionation (Avanzo et al. 2012).

Following irradiation, surviving basal cells, capable of sustained proliferation and scattered over a wide area, can repopulate the basal monolayer through lateral migration and preserve function. The repopulation of epithelial cells seems to start after a time lag of 4 weeks (Turesson et al. 1989). The survival of one cell/cm^2 is sufficient to permit epithelial regeneration of the irradiated area (Archambeau et al. 1995).

Volume/Surface Effect

It can reasonably be assumed that there is an effect of the irradiated surface on the tolerated dose, because with a smaller irradiated area the migration of viable cells from the edges of the irradiated site contributes significantly to the recovery of the epidermis. Results of orthovoltage RT demonstrated that the risk and severity of side effects to the skin depended on the field size. It has been observed that with small field sizes (30–180 cm²) minimal erythema was observed, whereas larger field sizes (150–210 cm²) could result in fibrosis and, when even larger field size were used (230–420 cm²) a markedly increased incidence of intense skin erythema and fibrosis could be observed It was confirmed that also the risk for ulceration (Schulte et al. 2005) and necrosis (Lee et al. 2002), was dependent on the size of the field The dose required to produce necrosis for a field size of 1 cm² was 42.1 Gy while it was reduced to 21.9 Gy (reduced to about 50%) when a larger field size of 10 cm² was used (Lee et al. 2002).

The dependency of side effects to the skin on irradiated area can be explained using the concept of functional subunits (FSU), whose dose response define the response of the skin. It was postulated that an FSU consists of a micro-vessel with associated epidermis and dermis. This element of skin, also called skin functional unit," is about 30 pm in diameter and 350 pm in length (Archambeau et al. 1995).

Another model for the structure of the skin (with regards to its response to irradiation) was proposed by Withers (1988), who considered that the clonogenic cells can be repopulated from adjacent basal layer and hence the effective clonogenic population of the FSU is larger than the FSU itself. Therefore the cell survival fraction consistent with one surviving clonogen per FSU, has been called the tissue rescuing unit" (TRU). For example, in mouse skin, moist desquamation is avoided if about 10 clonogenic cells per cm² survive, so that FSU can be described operationally as a 1/10 cm² area, and the TRU, 1 cell per cm².

Data from the days of the orthovoltage era showed evidence that the maximum tolerance dose for normal tissues is related to the irradiated area A through a power law, $D = kA^{-0.16}$ or $D = kL^{-0.32}$ where L is the diameter of the irradiated area (von Essen 1963).

There is less evidence for dependence of dose response relationship on the irradiated surface for RT with high-energy photons. In Werbrouck (2009) the patients with grade ≤ 2 dermatitis had a mean dose to the skin of 50.45 Gy, while patients with grade 3 dermatitis 56.4 Gy. V_{50Gy} (the percent volume of skin receiving more than 50 Gy) was 62.55% and 73.1%, respectively.

Grade 2 dermatitis is generally related to breast volume (Moody et al. 1994) and Clinical Target Volume (CTV) (Keller et al. 2012) sizes. This correlation could be explained with larger field size needed for larger breast volumes; however, another possible explanation could be the greater dose inhomogeneity (<95% and >105% of prescribed dose) usually found in women with large breasts (Neal et al. 1995). In patients with cetuximab related grade ≥ 3 dermatitis, there was evidence of a surface area effect. Patients who experienced grade 3 dermatitis had mean surface areas of 79 cm² and 35 cm² receiving at least 50 and 60 Gy, respectively, whilst patients with grade 4 dermatitis had mean surface areas of 167 cm² and 107 cm² receiving at the same 50 and 60 Gy levels respectively (Studer et al. 2011).

In a study on skin side effects after RT with carbon ions, the area irradiated with over 60 GyE (dose in Gy corrected for RBE of ions) on DSH was found to be a predictor of late grade 3 skin reactions (Yanagi et al. 2010).

The surface effect for skin necrosis/telangiectasia at 5 years from RT has been modeled by Burman et al. (1991) with n = 0.1 based, on previous results of orthovoltage radiation.

Considerations for Stereotactic Body Radiation Therapy (SBRT)

The data on adverse skin reactions on patients who underwent SBRT is very limited (Hoppe et al. 2008; Welsh et al. 2011). Given the steep dose gradient in SBRT, patients with tumors at more than 1–2 cm from the chest wall and more than 5 cm from the posterior skin are at a very low risk of toxicity in thoracic SBRT delivered by one to seven 6 MV arcs (Stephans et al. 2012). Incidence of grade ≥ 2 dermatitis of or have been observed when the skin receives more than 30 Gy in 3–5 fractions (Hoppe et al. 2008; Welsh et al. 2011). 50 patients with Stage I non-small cell lung cancer (NSCLC) were treated at Memorial Sloan-Kettering Cancer Center with 60 Gy in three fractions or 44–48 Gy in four fractions using IMRT delivered with multiple 6 MV linac coplanar beams (Hoppe et al. 2008). Four patients (8%) exhibited grade 2 skin toxicity, , two patients (4%) grade 3, and one patient (2%) grade 4 (skin necrosis). Factors associated with grade ≥ 2 or higher acute skin toxicity included using only 3 beams (p = 0.0007), a distance from the tumor to the posterior chest wall skin of less than 5 cm (p = 0.006), and a maximum skin dose ≤50% of the prescribed dose (p = 0.02). These results suggest that skin doses of 30 Gy in three fractions (EQD2Gy = 50 Gy for α/β = 10 Gy) or 22–24 Gy in four fractions (N EQD2Gy = 28.4–32 Gy for α/β = 10 Gy) may be related to significant risk for grade ≥ 2 skin toxicity.

Welsh et al. (2011) reported the results on a series of 360 patients with pulmonary tumors who received a total dose of 50 Gy in four daily 12.5 Gy fractions, 265 (268 tumors) had tumors within 2.5 cm of the chest wall. 104/265 (39%) developed skin toxicity (any grade). The incidence of both skin toxicity and chest wall pain was correlated with the volume of the chest wall receiving ≤ 30 Gy among the patients with lesions within 2.5 cm from the chest wall (Welsh et al. 2011).

Other studies reported only grade 1 dermatitis in a minority of patients treated with SBRT with multiple non-coplanar arcs or static beams for liver metastases (54 Gy in three fractions and 60 in three fractions) (Schefter et al. 2005).

In a study on thoracic hepatobiliary malignancies treated with RT doses ≥52.5 Gy in 15 fractions between January 2009 and December 2012, 36% of patients developed skin toxicity (45 grade 1, 3 grade 2) at a median time of 18 days. On multivariate analysis, V_{40Gy} ≥ 117 cm³ was associated with the occurrence of skin toxicity (Swanick et al. 2017).

A fibrotic mass that arises in a noncritical region is likely to have a minimal effect on a patient's quality of life. However, a fibrotic lesion around the brachial plexus might cause forearm neuropathy and lymphedema, which, although not life-threatening, will significantly affect a patient's quality of life. Kawase et al. (2009) focused on the appearance of masses of fibrosis after SBRT. Soft-tissue masses were found in 5/92 (5.4%) patients irradiated in five fractions compared with 4/287 (1.4%) patients irradiated in four fractions. The range of mean and maximum doses to the tissue originating the soft-tissue masses were 25.8–53.9 Gy (median 43.7 Gy) and 47.5–62.5 Gy (median, 50.2 Gy), respectively.

RECOMMENDATIONS FOR RT PLANNING

Contouring of the Skin for RT Planning

The skin is contoured as the outer shell of the body outline with a uniform thickness of 3 mm (Rudat et al. 2014) or, more frequently, of 5 mm thickness (Werbrouck et al. 2009; Saibishkumar et al. 2008; De Rose et al. 2017), in order to include all layers of the skin.

The calculation of dose to the skin requires high spatial resolution because of the steepness of the dose gradient in the build-up region. For this reason, when side effects to the skin are of concern, it is recommended to choose a proper grid size of 2 mm^3 (Price et al. 2014) or 1 mm^3 (De Puysseleyr et al. 2014) during planning.

In order to avoid excessive doses to the skin in the build-up region as a result of IMRT planning, it has been recommended to delineate the Planning Target Volume (PTV) to exclude the skin surface in head and neck treatments (Lee et al. 2002). For breast cancer treatments, the CTV can be expanded to PTV within 3 mm of the skin surface (Chen et al. 2010).

Strategies to Reduce Risk for Skin Side Effects

Dose inhomogeneity both within and beyond the PTV have a significant impact on radiation-induced dermatitis, and, in order to reduce the risk of side effects, it should be minimized. Chen et al. (2010) reported that when A volume of the PTV receiving more than 110% of the prescribed dosewas >5.13% there was an increased incidence of moist desquamation after adjuvant 3D-CRT with 50.4 Gy in 28 fractions. The studies comparing 3D-CRT with forward IMRT or inverse IMRT for breast cancer showed less incidence of acute skin effects with IMRT, possibly because of reduction of the hot spots (Pignol et al. 2008; Harsolia et al. 2007; Mukesh et al. 2013a). This is confirmed by statistical association among the presence of hot spots and incidence of acute effects (Chen et al. 2010; Pignol et al. 2008). A reduction of moist desquamation in the field-in-field IMRT arm (31% versus 48%), associated with improved breast dosimetry (Pignol et al. 2008).

The time factor and the relative long half-time for repair for late effects have important implications for multiple-fraction-per-day treatment, and imply that interfraction intervals of 6 hours or longer are recommended.

IMRT inverse planning can increase skin dose for head and neck patients, and for other patients when the PTV extends to the skin surface. This results from the requirement of PTV-based optimisation to minimize objective functions, and hence to remove regions of low dose from the PTV, even where these are caused by the build-up effect. In cases where the PTV extends to the skin, this results in excessive fluence being delivered to the skin (Thomas et al. 2004; Lee et al. 2002) first suggested to explicitly include the skin as an organ at risk (OAR) in the IMRT optimisation.

By configuring the skin as an OAR, it is possible to achieve skin dose reduction while delivering whole-breast IMRT without compromising dose profiles to PTV and other OARs. In breast RT plans with Helical Tomotherapy, the mean skin dose and volume of skin receiving ≤ 50 Gy were significantly lower with the skin-sparing plan compared with the non-skin-sparing plan (42.3 Gy vs. 47.7 Gy and 12.2% vs. 57.8%, respectively; p < 0.001). The planning objective in these cases was to give a dose of 50 Gy to at least 90% of the PTV while reducing the volume of skin receiving that same dose (Saibishkumar et al. 2008). The inclusion of skin in optimisation also reduces estimated NTCP for cutaneous side effects in proton plans (Tommasino et al. 2017). It must also be considered that, since the inaccuracy in calculated dose to the skin is larger when normally incident beams are used, delivery techniques using multiple equi-spaced beams, arcs, or beamlets may reduce the discrepancy between delivered and calculated dose to the skin.

Hair-sparing whole brain RT has been indicated as a method to decrease dose to hair follicles, thus reducing the severity of alopecia. One method employed 2 arc 6 MV VMAT, with a median dose of 20 Gy (5 fractions) prescribed to the PTV (De Puysseleyr et. al. 2014). The hair follicle volume was defined as the tissue underlying the skin up to the outer table of the skull using an automated script written in Pinnacle, version 9.0 (Philips Medical Systems, Andover, MA, US). During optimisation, the dose to the hair follicle volume was reduced as much as possible, without compromising the dose to the brain.

Tolerance Doses for RT

In order to study conservative tolerance doses for the adverse effects to the skin, the EQD2Gy giving 5% risk for each effect (EQD2Gy$_{5\%}$) calculated using the available models are reported in Table 12.3 for different fractionation schedules and irradiated volumes. The EQD2Gy$_{5\%}$ values for acute reactions for mild reactions are in the range 30.8–31.2 Gy. This can be considered as a threshold for the appearance of any relevant reaction to the skin. 30 Gy was also found to be associated with the appearance of alopecia (Rosenthal et al. 2008). Above this level, both the risk and severity of side effects increases, as the tolerated dose for moist desquamation is in the range 33.1–37.2 Gy. Of note, using an $\alpha/\beta = 10$ Gy, 30 Gy to the skin in 2 Gy per fractions (EQD2Gy$_{5\%}$ for any skin effect) translates into 21 Gy to the skin if dose is given three fractions, 22.8Gy if four fractions are used and 24 Gy in five fractions. For moist desquamation, a dose of 33 Gy to the skin in 2 Gy per fractions translates into 22.5 Gy to the skin if dose is given three fractions, 24.4 Gy and 26 Gy in four and five fractions, respectively. These results are in agreement with the SBRT report of Memorial Sloan Kettering, where skin dose of 30 Gy in three fractions (EQD2Gy = 50 Gy with α/β of 10 Gy) or 22–24 Gy in four fractions (EQD2Gy = 28.4–32 Gy) were related to significant risk for grade ≥ 2 toxicity (Hoppe et al. 2008).

Given the evidence of surface effect for the skin, the tolerance doses for irradiation of small surfaces may be higher. The dose response models used to describe the occurrence of skin side effects were mostly derived from datasets of patients treated with whole breast irradiation, where the irradiated area is around 816.0 cm² (range 319.0–1718.8) (Neal et al. 1995). For an irradiated volume of 1/10 (about 80 cm²) of the reference surface considered in the above models and assuming a value for n=0.1 to describe the surface effect as in Burman et al. (1991), the safety level for moist desquamation increases to 46.8 Gy at 2 Gy per fraction. For SBRT, this is equivalent to to 28.8 Gy in three fractions, 31.4 Gy in four fractions and 33.5 in five fractions.

TABLE 12.3   Summary of Tolerance Doses for the Skin Based on the Models Derived in this Chapter from the Studies Described in Table 12.2

Endpoint	Irradiated Surface	Schedule	Models Used	Tolerance Dose
Any acute effect	No surface effect	2 Gy/fraction	Models shown in Figures 12.1 and 12.2	30 Gy
Any acute effect	No surface effect	3/4/5 fractions	Models in Figures 12.1 and 12.2, $\alpha/\beta = 10$ Gy	21/22.8/24 Gy
Moist desquamation	Surface of whole breast	2 Gy/fraction	Model in Figure 12.1, $\alpha/\beta = 10$ Gy	33 Gy
Moist desquamation	Surface of whole breast	3/4/5 fractions	Model shown in Figure 12.1, $\alpha/\beta = 10$ Gy	22.5/24.4/26 Gy
Moist desquamation	1/10 surface of whole breast, $\approx$ 80 cm^2	2 Gy/fraction	Model shown in Figure 12.1, $\alpha/\beta = 10$ Gy, n = 0.1	46.8 Gy
Moist desquamation	1/10 surface of whole breast, $\approx$ 80 cm^2	3/4/5 5 fractions	Model shown in Figure 12.1, $\alpha/\beta = 10$ Gy, n = 0.1	28.8/31.4/33.5 Gy
RTOG $\geq$ 3	Surface of whole breast	2 Gy/fraction	Model of Pastore et al. (2016)	30.2 Gy
RTOG $\geq$ 3	1/10 surface of whole breast, $\approx$80 cm^2	2 Gy/fraction	Model of Pastore et al. (2016), n = 0.38	72.5 Gy
Telangiectasia	Surface of whole breast	2 Gy/fraction	Model in Figure 12.2	47.4 Gy
Subcutaneous fibrosis	Whole breast volume	2 Gy/fraction	Model by Mukesh et al. (2013a)	33.6 Gy
Subcutaneous fibrosis	1/10 subcutaneous tissue of breast, $\approx$100 cc	2 Gy/fraction	n = 0.012	34.5 Gy
Subcutaneous fibrosis	1/10 subcutaneous tissue of breast, $\approx$100 cc	3/4/5 fractions	n = 0.012, $\alpha/\beta = 3$ Gy	18.7/21/22.8 Gy

A suggested tolerance dose to 1/10 of the breast for RTOG grade $\geq$ 3 skin toxicity could be 72.5 Gy, as it can be derived using a value for n=0.38 from the model of Pastore et al. (2016).

The tolerance doses for telangiectasia are significantly different according to the presence of mastectomy. For whole breast irradiation with high-energy photons, the threshold EQD2$_{Gy}$ is 47.4 Gy with $\alpha/\beta = 3$ Gy (Table 12.3). For postmastectomy patients, the tolerated EQD2$_{Gy}$ decreases to 33 Gy as shown in Figure 12.2. As there is no evidence in published clinical results for a surface effect for telangiectasia, it is not possible to extrapolate tolerance doses for small irradiated surface. Also, there are no sufficient data on late telangiectasia from SBRT.

For subcutaneous fibrosis, we used the model of Mukesh et al. (2013b), in which a weak surface effect was described with n = 0.012 and $\alpha/\beta = 3$ Gy was used for correction for fractionation. Assuming that 1/10 of average breast volume is irradiated, the tolerance doses are 34.5 Gy for 2 Gy per fraction schedule, and 18.7 Gy, 21 Gy and 22.8 Gy for schedules involving three, four or five fractions, respectively. These values are, as expected, lower than the 25.8–53.9 Gy average doses found to be associated to development of soft-tissue masses in the study by Kawase et al. (2009) in thoracic SBRT in 4–5 fractions.

CONCLUSIONS

The skin is the most studied organ in terms of response to radiation, as the risk for acute side effects and its dependence on fractionation schedule, overall treatment time, and irradiated surface/volume are well established. Some open issues remain, in particular, the risk for late effects to the skin following SBRT which needs further investigation.

REFERENCES

Alexander, M.A., Brooks, W.A., and Blake, S.W. 2007. "Normal Tissue Complication Probability Modelling of Tissue Fibrosis Following Breast Radiotherapy." *Phys Med Biol* 52(7): 1831–43.

Almberg, S.S., Lindmo, T., and Frengen, J. 2011. "Superficial Doses in Breast Cancer Radiotherapy using Conventional and IMRT Techniques: A Film-Based Phantom Study." *Radiother Oncol* 100(2): 259–64.

Archambeau, J.O., Pezner, R., and Wasserman, T. 1995. "Pathophysiology of Irradiated Skin and Breast." *Int J Radiat Oncol Biol Phys* 31(5): 1171–85.

Avanzo, M., Stancanello, J., Trovo, M. et al. 2012. "Complication Probability Model for Subcutaneous Fibrosis Based on Published Data of Partial and Whole Breast Irradiation." *Phys Med* 28(4): 296–306.

Avanzo, M., Drigo, A., Ren Kaiser, S. et al. 2013. "Dose to the Skin in Helical Tomotherapy: Results of in vivo Measurements with Radiochromic Films." *Med Phys* 29(3): 304–11.

Balter, S., Hopewell, W., Miller, D.L. et al. 2010. "Fluoroscopically Guided Interventional Procedures: A Review of Radiation Effects on Patients' Skin and Hair." *Radiology* 254(2): 326–41.

Bentzen, S.M., Thames, H.D., and Overgaard, M. 1989. "Latent-Time Estimation for Late Cutaneous and Subcutaneous Radiation Reactions in a Single-Follow-Up Clinical Study." *Radiother Oncol* 15(3): 267–74.

Bentzen, S.M., and Overgaard, M. 1991. "Relationship between Early and Late Normal-Tissue Injury after Postmastectomy Radiotherapy." *Radiother Oncol* 20(3): 159–65.

Burman, C., Kutcher, G.J., Emami, B., and Goitein , M. 1991. "Fitting of Normal Tissue Tolerance Data to an Analytic Function." *Int J Radiat Oncol Biol Phys* 21(1): 123–35.

Chen, M.F., Chen, W.C., Lai, C.H., Hung, C.H., Liu, K.C., and Cheng, Y.H. 2010a. "Predictive Factors of Radiation-Induced Skin Toxicity in Breast Cancer Patients." *BMC Cancer* 10: 508.****

Chen, P.Y., Wallace, M., Mitchell, C. et al. 2010b. "Four-Year Efficacy, Cosmesis, and Toxicity Using Three-Dimensional Conformal External Beam Radiation Therapy to Deliver Accelerated Partial Breast Irradiation." *Int J Radiat Oncol Biol Phys* 76(4): 991–7.

Chin, M.S., Siegel-Reamer, L., Gordon, A. et al. 2017. "Association Between Cumulative Radiation Dose, Adverse Skin Reactions, and Changes in Surface Hemoglobin among Women Undergoing Breast Conserving Therapy." *Clin Transl Radiat Oncol* 4: 15–23.

Cox, J.D., Stetz, J., and Pajak, T.F. 1995. "Toxicity Criteria of the Radiation Therapy Oncology Group (RTOG) and the European Organization for Research and Treatment of Cancer (EORTC)." *Int J Radiat Oncol Biol Phys* 31(5): 1341–6.

De Langhe, S., Mulliez, T., Veldeman, L. et al. 2014. "Factors Modifying the Risk for Developing Acute Skin Toxicity after Whole-Breast Intensity Modulated Radiotherapy." *BMC Cancer* 14, (1): 711.

De Puysseleyr, A., Van De Velde, J., Speleers, B. et al. 2014. "Hair-Sparing Whole Brain Radiotherapy with Volumetric Arc Therapy in Patients Treated for Brain Metastases: Dosimetric and Clinical Results of a Phase II Trial." *Radiat Oncol* 9: 170.

De Rose, F., Franceschini, D., Reggiori, G. et al. 2017. "Organs at Risk in Lung SBRT." *Phys Med* 44: 131–8.

Emami, B., Lyman, J., Brown, A. et al. 1991. "Tolerance of Normal Tissue to Therapeutic Irradiation." *Int J Radiat Oncol Biol Phys* 21(1): 109–22.

Freedman, G.M., Li, T., Nicolaou, N., Chen, Y., Ma, C.C., and Anderson, P.R. 2009. "Breast Intensity-Modulated Radiation Therapy Reduces Time Spent with Acute Dermatitis for Women of all Breast Sizes during Radiation." *Int J Radiat Oncol Biol Phys* 74(3): 689–94.

Fu, H.J., Li, C.W., Tsai, W.T., Chang, C.C., and Tsang, Y.W. 2017. "Skin Dose for Head and Neck Cancer Patients Treated with Intensity-Modulated Radiation Therapy (IMRT)." *Radiat Phys Chem* 140: 435–41.

Gonzalez Sanchis, A., L. Brualla Gonzalez, J. L. Sanchez Carazo, J. C. Gordo Partearroyo, A. Esteve Martinez, A. Vicedo Gonzalez, and J. L. Lopez Torrecilla. 2017. Evaluation of Acute Skin Toxicity in Breast Radiotherapy with a New Quantitative Approach." *Radiother Oncol* 122 (1): 54–9.

Harsolia, A., Kestin L., Grills, I., et al. 2007. "Intensity-Modulated Radiotherapy Results in Significant Decrease in Clinical Toxicities Compared with Conventional Wedge-Based Breast Radiotherapy." *Int J Radiat Oncol Biol Phys* 68(5): 1375–8.

Haviland, J.S., Owen, J. R., J. A. Dewar, et al. 2013. "The UK Standardisation of Breast Radiotherapy (START) Trials of Radiotherapy Hypofractionation for Treatment of Early Breast Cancer: 10-Year Follow-Up Results of Two Randomised Controlled Trials." *Lancet Oncol* 14(11): 1086–94.

Hepel, J.T., Tokita, M., MacAusland, S.G., et al. 2009. "Toxicity of Three-Dimensional Conformal Radiotherapy for Accelerated Partial Breast Irradiation." *Int J Radiat Oncol Biol Phys* 75(5)1290–6.

Higgins, P.D., Han, E.Y., Yuan, J.L., Hui, DS and. 2007. "Evaluation of Surface and Superficial Dose for Head and Neck Treatments using Conventional Or Intensity-Modulated Techniques." *Phys Med Biol* 52(4): 1135–46.

Hijal, T., Al Hamad, A.A., Niazi, T., et al. 2010. "Hypofractionated Radiotherapy and Adjuvant Chemotherapy do Not Increase Radiation-Induced Dermatitis in Breast Cancer Patients." *Curr Oncol* 17(5): 22–7.

Hoppe, B.S., Laser, B., Kowalski, A.V., et al. 2008. "Acute Skin Toxicity Following Stereotactic Body Radiation Therapy for Stage I Non-Small-Cell Lung Cancer: Who's at Risk?" *Int J Radiat Oncol Biol Phys* 72(5): 1283–6.

Jagsi, R., Ben-David, M. A., Moran, J.M., et al. 2010. "Unacceptable Cosmesis in a Protocol Investigating Intensity-Modulated Radiotherapy with Active Breathing Control for Accelerated Partial-Breast Irradiation." *Int J Radiat Oncol Biol Phys* 76(1): 71–8.

Jagsi, R., Griffith, K. A., Boike T. P., et al. 2015 "Differences in the Acute Toxic Effects of Breast Radiotherapy by Fractionation Schedule: Comparative Analysis of Physician-Assessed and Patient-Reported Outcomes in a Large Multicenter Cohort." *JAMA Oncol* (7): 918–30.

Kawase, T., Takeda, A. Kunieda, E., et al. "Extrapulmonary Soft-Tissue Fibrosis Resulting from Hypofractionated Stereotactic Body Radiotherapy for Pulmonary Nodular Lesions." *Int J Radiat Oncol Biol Phys* 74(2):349–54.

Keller, L.M., Sopka, D.M., Li, T., et al. 2012. "Five-Year Results of Whole Breast Intensity Modulated Radiation Therapy for the Treatment of Early Stage Breast Cancer: The Fox Chase Cancer Center Experience." *Int J Radiat Oncol Biol Phys* 84(4): 881–7.

Lee, N., Chuang, C., Quivey, J.M., et al. 2002. "Skin Toxicity due to Intensity-Modulated Radiotherapy for Head-and-Neck Carcinoma." *Int J Radiat Oncol Biol Phys* 53: 630–7.

Lilla, C., Ambrosone, C.B., Kropp, S., et al. 2007. "Predictive Factors for Late Normal Tissue Complications Following Radiotherapy for Breast Cancer." *Breast Cancer Res Treat* 106(1): 143–50.

Marthinsen, A.B., Gisetstad, R., Danielsen, S., Frengen, J., Strickert, T. and Lundgren S. 2010. "Relative Biological Effectiveness of Photon Energies used in Brachytherapy and Intraoperative Radiotherapy Techniques for Two Breast Cancer Cell Lines." *Acta Oncol* 49(8): 1261–8.

Monten, C., Lievens Y., Olteanu, L.A.M., et al. 2017. "Highly Accelerated Irradiation in 5 Fractions (HAI-5): Feasibility in Elderly Women with Early Or Locally Advanced Breast Cancer." *Int J Radiat Oncol Biol Phys* 98(4): 922–30.

Moody, A.M., Mayles, W.P., Bliss, J.M., et al. 1994. "The Influence of Breast Size on Late Radiation Effects and Association with Radiotherapy Dose Inhomogeneity." *Radiother Oncol* 33(2): 106–12.

Mukesh, M.B., Barnett G.C., Wilkinson, J. S., et al. 2013a. "Randomized Controlled Trial of Intensity-Modulated Radiotherapy for Early Breast Cancer: 5-Year Results Confirm Superior overall Cosmesis." *J Clin Oncol.* 31(36): 4488–95.

Mukesh, M.B., Harris, E., Collette, S., et al. 2013b. "Normal tissue complication probability (NTCP) parameters for breast fibrosis: pooled results from two randomised trials." *Radiother Oncol* 108(2): 293–8.

Nagai, A., Shibamoto, Y., Yoshida, M., Inoda, K. and Kikuchi, Y.. 2017. "Intensity-Modulated Radiotherapy using Two Static Ports of Tomotherapy for Breast Cancer After Conservative Surgery: Dosimetric Comparison with Other Treatment Methods and 3-Year Clinical Results." *J Radiat Res* 58(4): 529–36.

Neal, A.J., Torr, M., Helyer, S. and Yarnold, J.R. 1995. Correlation of Breast Dose Heterogeneity with Breast Size using 3D CT Planning and Dose-Volume Histograms. *Radiother Oncol* 34(3): 210–8.

Pastore, F., M. Conson, V. D'Avino, et al. 2016. "Dose-Surface Analysis for Prediction of Severe Acute Radio-Induced Skin Toxicity in Breast Cancer Patients." *Acta Oncol* 55 (4): 466–73.

Pignol, J.P., Olivotto, I., Rakovitch, E., et al. 2008. "A Multicenter Randomized Trial of Breast Intensity-Modulated Radiation Therapy to Reduce Acute Radiation Dermatitis." *J Clin Oncol* 26(13): 2085–92.

Price, R.A., Koren, S., Veltchev, I., et al. 2014. "Planning Target Volume-to-Skin Proximity for Head-and-Neck Intensity Modulated Radiation Therapy Treatment Planning." *Pract Radiat Oncol* 4(1): e21–9

Withers, H.R., Taylor, J.M. and Maciejewski, B. 1998. "Treatment Volume and Tissue Tolerance." *Int J Radiat Oncol Biol Phys* 14(4): 751–9.

Rosenthal, D.I., Chambers, M. S., Fuller, C.D., et al. 2008. "Beam Path Toxicities to Non-Target Structures during Intensity-Modulated Radiation Therapy for Head and Neck Cancer." *Int J Radiat Oncol Biol Phys* 72(3): 747–55.

Rudat, V., Nour, A., Ghaida, S.A., and Alaradi, A. 2016. "Impact of Hypofractionation and Tangential Beam IMRT on the Acute Skin Reaction in Adjuvant Breast Cancer Radiotherapy." *Radiat Oncol* 11: 100.

Rudat, V., Nour, A., Alaradi, A.A., Mohamed, A., and Altuwaijri, S. 2014. "In Vivo Surface Dose Measurement using GafChromic Film Dosimetry in Breast Cancer Radiotherapy: Comparison of 7-Field IMRT, Tangential IMRT and Tangential 3D-CRT." *Radiat Oncol* 9: 156.

Saibishkumar, E.P., MacKenzie, M.A., Severin, D., et al. 2008. "Skin-Sparing Radiation using Intensity-Modulated Radiotherapy after Conservative Surgery in Early-Stage Breast Cancer: A Planning Study." *Int J Radiat Oncol Biol Phys* 70(2): 485–91.

Schefter, T.E., Kavanagh, B.D., Timmerman, R.D., Cardenes, H.R., Baron, A., and Gaspar, L.E. 2005. "A Phase I Trial of Stereotactic Body Radiation Therapy (SBRT) for Liver Metastases." *Int J Radiat Oncol Biol Phys* 62(5): 1371–8.

Schulte, K.W., Lippold, A., Auras, C., et al. "Soft x-Ray Therapy for Cutaneous Basal Cell and Squamous Cell Carcinomas." *J Am Acad Dermatol* 53(6): 993–1001.

Stephans, K.L., Djemil, T., . Tendulkar, R. D, Robinson, C.G., Reddy, C.A. and Videtic, G.M. 2012. "Prediction of Chest Wall Toxicity from Lung Stereotactic Body Radiotherapy (SBRT)." *Int J Radiat Oncol Biol Phys* 82(2): 974–80.

Studer, G., Brown, M., Salgueiro, E.B., et al. 2011. "Grade 3/4 Dermatitis in Head and Neck Cancer Patients Treated with Concurrent Cetuximab and IMRT." *Int J Radiat Oncol Biol Phys* 81(1): 110–7.

Swanick, C.W., Allen, P.K., Tao, R. et al. 2017. "Incidence and Predictors of Chest Wall Toxicity After High-Dose Radiation Therapy in 15 Fractions." *Pract Radiat Oncol* 7(1): 63–71.

Thomas, S.J., and Hoole, A.C. 2004. "The Effect of Optimisation on Surface Dose in Intensity Modulated Radiotherapy (IMRT)." *Phys Med Biol* 49(21): 4919–28.

Tommasino, Francesco, Marco Durante, Vittoria D'Avino, Raffaele Liuzzi, Manuel Conson, Paolo Farace, Giuseppe Palma, Marco Schwarz, Laura Cella, and Roberto Pacelli. 2017, Jun 3 "Model-Based Approach for Quantitative Estimates of Skin, Heart, and Lung Toxicity Risk for Left-Side Photon and Proton Irradiation After Breast-Conserving Surgery." *Acta Oncologica* 565: 730–6.

Turesson, I., and H.D. Thames. 1989, Jun. "Repair Capacity and Kinetics of Human Skin during Fractionated Radiotherapy: Erythema, Desquamation, and Telangiectasia After 3 and 5 Year's Follow-Up." *Radiother Oncol: Journal of the European Society for Therapeutic Radiology and Oncology* 152: 169–88.

von Essen, Carl F. 1963, Nov 1. "A Spatial Model of Time-Dose-Area Relationships in Radiation Therapy." *Radiology* 81(5): 881–3.

Welsh, J., J. Thomas, D. Shah, P.K. Allen, X. Wei, K. Mitchell, S. Gao, P. Balter, R. Komaki, and J. Y. Chang. 2011, Sep 1. "Obesity Increases the Risk of Chest Wall Pain from Thoracic Stereotactic Body Radiation Therapy." *Int J Radiat Oncol Biol Phys* 81(1): 91–96.

Werbrouck, J., K. De Ruyck, F. Duprez, L. Veldeman, K. Claes, M. Van Eijkeren, T. Boterberg et al. 2009. "Acute Normal Tissue Reactions in Head-and-Neck Cancer Patients Treated with Imrt: Influence of Dose and Association with Genetic Polymorphisms in DNA Dsb Repair Genes." *Int J Radiat Oncol Biol Phys* 73(4): 1187–95.

Yanagi, T., Kamada, T., Tsuji, H., Imai, R., Serizawa, I. and Tsujii, H. 2010. "Dose-Volume Histogram and Dose-Surface Histogram Analysis for Skin Reactions to Carbon Ion Radiotherapy for Bone and Soft Tissue Sarcoma." *Radiother Oncol* 95(1): 60–65.

CHAPTER 13

Bone Marrow and Hematological Toxicity

Elena S. Heide and Loren K. Mell

CONTENTS

Introduction 310
Historical Context 310
Anatomy 311
Physiology 313
Pathology.. 315
    Acute Effects of Bone Marrow Irradiation..316
    Late Effects of Bone Marrow Irradiation 317
Evaluating Bone Marrow Injury...318
    Peripheral Blood Counts 318
    Histopathology and Pre-Clinical Assays 319
    Stromal Tissue and Peripheral Stem Cells...319
Bone Marrow Imaging.. 320
    Computed Tomography.. 320
    Magnetic Resonance Imaging.. 320
    Single Photon Emission Computed Tomography 320
    Positron Emission Tomography 321
Clinical Effects of Bone Marrow Radiation 322
    Normal Tissue Complication Probability.. 322
    Effects of Chemotherapy with Radiotherapy 323
    Total Body Irradiation... 326
    Intensity-Modulated Total Marrow Irradiation.. 326
    Bone Marrow-Sparing Radiotherapy 327
Future Applications 328
    Proton Therapy 328
    Atlas-Based Treatment Planning 328
Conclusion 330
References.. 331

INTRODUCTION

The human bone and bone marrow (BM) constitute one of the body's largest and most important organ systems. Altogether, the bones (~15%), BM (~4%), and blood (~7%) account for approximately 25% of the entire human body weight (Paul 2013). The BM system is highly susceptible to harmful outside exposure, and is an important dose-limiting organ during cancer therapy, particularly radiation therapy (RT) and chemotherapy. Radiation is one of the most widely known and well-studied toxins impacting BM function. Physicians routinely obtain blood counts for diagnosing hematological disorders and when monitoring for toxicity. Recently, increased efforts have been focused on techniques to circumvent hematopoietic toxicities. This chapter will discuss the historical study of hematopoiesis and BM injury, the anatomic physiology of bone and the BM microenvironment, pathologic changes and clinical symptoms that occur when BM is exposed to differing doses of radiation, how BM damage is assessed, and modern clinical applications to both induce and mitigate BM injury.

HISTORICAL CONTEXT

In 1898, Marie and Pierre Curie announced their discovery of "Polonium" and "Radium" in July and December, respectively, eventually leading to their Nobel Prizes in Chemistry and Physics. Marie actively promoted the use of radium to treat neoplasms and alleviate suffering during World War II. Unfortunately, since the toxic effects of radiation were not known at this time, she often placed vials of radioactive substances in her pockets. As she reached the end of her life, the self-destructive consequences of her work began to present, first through cataracts leading to her near blindness, and later aplastic anemia that led to her passing in July of 1934 (Grandin et al. 2013). These devastating effects of radiation on bone and BM have been seen and recorded since the early 1900s.

Between 1917 and 1926, female factory workers at the United States Radium Corporation of New Jersey and Radiant Dial Company of Illinois painted radium onto clock dials to make the dials glow radiantly. Some placed paint brushes in their mouths as they painted to increase the speed of their work, and as a status symbol, even painted their nails, faces, and teeth. Not long after beginning work, many of these women began to fall ill with jaw necrosis, non-healing abscesses, fractures, aplastic anemia, and bone sarcomas leading to violent and agonizing deaths (Cohen and Kim 2017). Similar valency and properties between radium and calcium (in terms of weight, size, electronic shell structure, and orbital arrangements) allow radium to be actively taken up by bone on calcium transporters. Once inside the bone, radium becomes trapped causing aplasia of the bone and marrow. While many of these symptoms can now be attributed to BM damage, it was not until experiments in the early 1930s and radiation accidents worldwide that the causes of these deleterious effects were elucidated.

In 1931, Shouse, Warren, and Whipple published their findings after exposing dogs to Roentgen radiation, noting that fatal intoxication post-radiation was caused by BM aplasia. During autopsy, symptoms caused by decreased circulating cells included massive colonies of bacteria in the lungs, bleeding into tissues, and essentially all the megakaryocytes of the BM were destroyed. Each dog appeared to have a latent asymptomatic period, noted for six to seven days post-radiation, and it was not until day eight or nine that the dogs experienced loss of appetite, vomiting, severe intoxication, and death (Shouse et al. 1931).

Throughout history, there have been 440 major radiation accidents worldwide, with five having been escalated to International Nuclear Event Scale (INES) level 5 or higher, having serious consequences on the people and environment. In 1957, in Kyshtym, Soviet Union (now Russia), there was an event rated as a level 6 and in Windscale Piles, United Kingdom, there was an event rated as a level 5. In 1979, Three Mile Island in the United States had an incident that was rated as a level 5 and in 1986, Chernobyl, USSR, now Ukraine, had an incident rated as an INES level 7. The last incident, also rated as a level 7, occurred in Fukushima Daiichi Japan in 2011 (Hasegawa et al. 2015). Large scale epidemiology studies have tracked the long-term effects of these accidents. The Life Span Study, thought to be the most reliable source due to size and variability of the cohort, has shown those exposed to radiation to have amplified risk for myriad health disorders, including hematopoietic failure, secondary leukemias, cancers, and even kidney failure and cardiovascular diseases (Kamiya et al. 2015).

While the damaging effects of radium have been devastating, without the use of radiation, certain beneficial clinical procedures would not be possible. For example, total body irradiation (TBI) is widely used for therapeutic treatment of leukemias and lymphomas as part of BM transplant conditioning regimens. As the technology and treatment approaches have evolved, smaller and more precise fields are being used to target tumors, and the development of image-guided and highly modulated radiation techniques has enabled physicians to optimize dose to targets while drastically reducing normal tissue dose, also dramatically decreasing unwanted bone and BM irradiation. With decreasing impact on the marrow, the clinical importance of limiting BM injury diminished. However, more recently, more intensive chemotherapy regimens and the advent of immuno-oncology drugs, such as inhibitors of Programmed-Death 1 (PD-1) and Programmed Death-Ligand 1 (PD-L1), have been introduced to treat a wide variety of malignancies. Such work has renewed interest in examining the effects of radiation on the hematopoietic system.

ANATOMY

It is known from pathology and imaging studies that BM is comprised of both hematopoietically active, fat-poor, "red" regions and inactive, fat-rich, "yellow" regions (Basu et al. 2007; Duda et al. 1995; Vogler and Murphy 1988). At birth, all BM is hematopoietic marrow, whereas in adulthood half of the marrow is converted into fat-rich BM. Red BM or "active" BM contains approximately 40% fat, 40% water, and 20% protein while yellow BM contains approximately 80% fat, 15% water, and 5% protein (Vogler and Murphy 1988). The difference in appearance between these two types of marrow can be seen in Figure 13.1. In response to changing physiologic conditions, such as anemia due to blood loss, it is possible for yellow marrow to be converted back to red BM (Travlos 2006). The marrow responsible for the production of blood progenitors is found within flat bones such as the hip, sternum, skull, ribs, vertebrae, and shoulder blades as well as the metaphyseal and epiphyseal plates of long bones such as the femur, tibia, and humerus. Within the hollow diaphyseal shaft of bones, the marrow produces stromal cells.

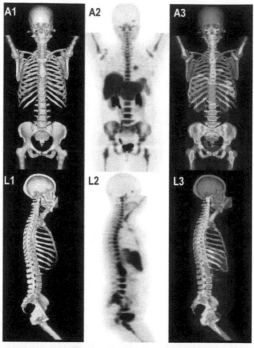

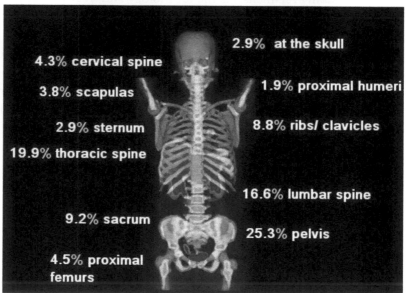

FIGURE 13.1   Location on active bone marrow within the adult human body as described by (Hayman et al. 2011). Permissions gained.

The first experiments to identify active BM used visual inspection of BM colour and the weight of bones before and after heating (Atkinson 1962; Custer and Hayhoe 1974; Woodard and Holodny 1960). In 1961, Ellis published estimates on the distribution of active BM. He estimated that for a 40-year-old man, approximately 40% of active marrow was localized to the pelvis, 25% in the lumbar and thoracic spine, 13% in the skull, 8% in the ribs, 3% in the cervical spine, 2% in the sternum, and 9% split between the humeri, clavicles, and scapula (Ellis 1961). Functional imaging studies, described later in this chapter, have further elucidated the distribution of active marrow (Figure 13.1).

The circular pattern of blood flow within bone marrow from center toward periphery and then back toward the center allows for extensive blood supply. In long and flat bones, a complex plexus of vessels make up the endosteal network which supplies blood to bone and bone marrow. Long bones have one or more nutrient canals, containing a nutrient artery and one to two nutrient veins, which pass through cortical bone and transverse the marrow's cavities. Flat bones have bone marrow that is served by a multitude of blood vessels of varying size. Arteries enter through large and small nutrient canals and split into ascending and descending branches that coil around the central longitudinal vein, the primary venous marrow channel. While much of the artery runs parallel to the long axis of the bone, branches of the artery including arterioles and capillaries extend outward toward the cortical bone into Haversian Canals, return to the inner marrow, and then drain into venous sinuses. The bone marrow system does not contain lymphatic vessels for draining. The drainage accumulates in the collecting venules which lead back to the central longitudinal vein (Munka and Gregor 1965).

PHYSIOLOGY

The main function of bone marrow is hematopoiesis, a compartmentalized process that allows for the generation of blood cells and platelets. An average human contains a total blood volume of approximately 5 liters, although this varies by gender, body size, chronic disease, and level of daily physical activity (Davy and Seals 1994). Plasma makes up 55% of blood volume while blood cells suspended in the plasma make up the other 45%. Plasma is an aqueous solution containing about 92% water by volume as well as proteins, glucose, mineral ions, hormones, and carbon dioxide. Blood cells, or formed elements, are generally divided into three categories: Leukocytes and thrombocytes constitute less than 1% of whole blood while erythrocytes constitute the other 45%. Assessment of the circulated levels of these elements through complete blood counts (CBCs) will be discussed later. To maintain steady state levels of circulating blood cells, a healthy adult produces approximately 10^{11}–10^{12} new blood cells daily which is achieved through the process of hematopoiesis (Parslow 2001).

Hematopoietic tissue is arranged in a highly specified pattern of organization, optimizing the bone marrow microenvironment for proliferation, differentiation, and maturation of stem cells along committed lineages (Figure 13.2). The microenvironment is made up of large variety of cell types including blood cells and their precursors, adventitial reticular or barrier cells, adipocytes, endothelial cells, macrophages, and elements of the extracellular matrix. Different hematopoietic processes take place within locations of varying specificity. Erythropoiesis, the production of erythrocytes, takes place within erythroblastic islands, which are distinct anatomical foci, while granulopoiesis, the production of granulocytes, occurs in less distinct locations. Megakaryopoiesis, the process of megakaryocyte maturation and differentiation, occurs adjacent to the sinus endothelium. These processes are regulated by humoral factors with varying sources and actions, some acting on primitive cells and others acting on specific cell lines. A list of some of these factors can be found

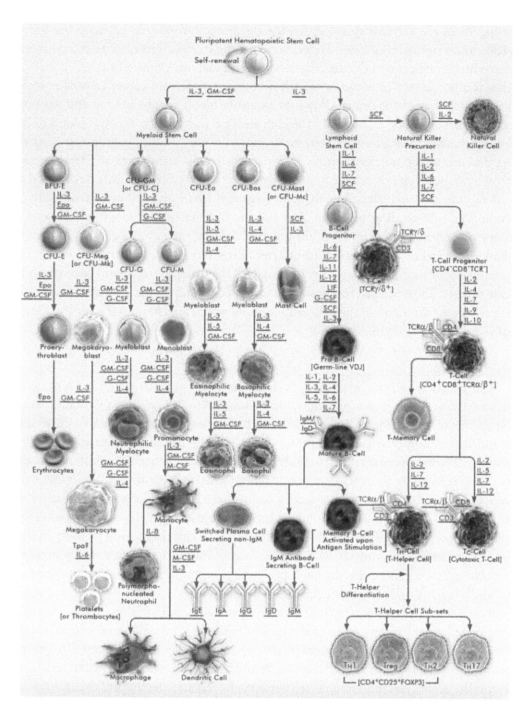

FIGURE 13.2 Myeloid and lymphoid cell lineages (https://www.thermofisher.com/us/en/home /life-science/antibodies/antibodies-learning-center/antibodies-resource-library/cell-signaling-pathways/hematopoiesis-pluripotent-stem-cells.html).

in Table 13.1. Other factors that affect the hematopoietic niche include hormones, dietary restriction, malnutrition, inflammation, toxicity, and proliferative or neoplastic disorders (Travlos 2006).

PATHOLOGY

The effects of ionizing radiation are commonly divided into acute and late effects, each characterized by distinct patterns of changes. Animal models have been the main mode of

TABLE 13.1 Examples of Humoral Factors of Varying Types with Different Sites of Action and Receptors, Some Acting on Primitive Cells and Others Acting on Specific Cell Lines

Humoral Factor	Type of Ligand	Site of Action	Receptor
Stem cell factor	Cytokine	Pluripotent Cells	c-KIT
Burst promoting activity (BPA)	Cytokine	Erythrocytes	BPA receptor
Erythropoietin	Glycoprotein cytokine	Erythrocytes	Erythropoeitin receptor (EpoR)
Granulocyte-macrophage colony stimulating factor (GM-CSF)	Glycoprotein	Erythrocytes Granulocytes Monoctyes	GM-CSF receptor
Thyroid hormone	Tyrosine-based hormone	Erythrocytes	Thyroid hormone receptor
Growth hormone	Peptide hormone	Erythrocytes	Growth hormone receptor
Testosterone	Steroid hormone	Erythrocytes	Androgen receptor
Granulocytopoeitin (G-CSF)	Glycoprotein cytokine	Granulocytes	G-CSF receptor
IL-1	Cytokine	Granulocytes Lymphocytes	IL-1 receptor
IL-2	Cytokine	Lymphocytes	IL-2 receptor
IL-3	Cytokine	Granulocytes Monoctyes Lymphocytes	IL-3 receptor
IL-4	Cytokine	Lymphocytes	IL-4 receptor
IL-5	Cytokine	Granulocytes	IL-5 receptor
IL-6	Cytokine	Pluripotent Cells	IL-6 receptor
Interferon (INF)	Protein	Granulocytes	Interferon-α receptor Interferon-β receptor
Tumor necrosis factor (TNF)	Cytokine	Granulocytes	TNF receptor superfamily
Macrophage colony stimulating factor (M-CSF)	Cytokine	Monoctyes Lymphocytes Megakaryocytes	CSF-1 receptor
Monocytopoietin	Protein	Monoctyes	Monocytopoietin receptor
Thymic Hormone	Peptide hormone	Lymphocytes	Multiple receptors
Lymphocyte mitogenic factor	Lymphokine	Lymphocytes	Lymphocyte mitogenic factor receptor
B-cell growth factor (BCGF)	Lymphokine	Lymphocytes	BCGF receptor
B-cell differentiation factor (BCDF)	Lymphokine	Lymphocytes	BCDF receptor
Thrombopoietin	Glycoprotein hormone	Megakaryocytes	Thrombopoietin receptor

understanding the mechanisms of toxicity and death due to marrow irradiation, although some data comes from previous radiation accidents.

Acute Effects of Bone Marrow Irradiation

Acute effects normally take place within hours, days, or weeks of radiation damage, and are due to the death of a large number of proliferating cells in a short period of time. These acute effects occur in rapidly proliferating tissues, such as the epidermal skin layers, gastrointestinal epithelium, and hematopoietic system (Fajardo et al. 2001). The response of a tissue to ionizing radiation is dependent on three key characteristics, the inherent sensitivity of the individual cells, the kinetics of the tissue as a whole and the way in which the cells are organized in the tissue (Hall and Giaccia 2006). The changes seen within bone marrow can be attributed to all three vulnerabilities. Casarett's classification and Michalowski's classification are commonly used to define a tissue's radiosensitivity using population kinetics and tissue architecture.

Michalowski's system divides tissues into H-type populations (hierarchal model) and F-type populations (flexible model). The H-type model is based upon a cell's ability to proliferate and differentiate accounting for acute damage to bone marrow, while the F-type model is used for tissues that are not mitotically active leading to the late effect symptoms. Casarett's system divides parenchymal tissues into four major classifications (I-IV) based on histologic observation of early cell death. Stem cells are a self-renewing tissue, continually dividing mitotically to provide a steady supply of primitive cells ready to differentiate and mature into a variety of functioning cells, determined by the body's need. Although a reservoir of these primitive stem cells is maintained, those cells that are dividing are inherently vulnerable to irradiation damage which causes a loss of functionality when attempting their next mitotic division. The time it takes for these effects to appear relates to the short-lifespan of these mature functional cells and can result in a shortage of peripheral red and white blood cells as mature cells degrade and no new functional cells take their place. Small lymphocytes, one of the most sensitive cells to radiation, defy systemic classification (Hall and Giaccia 2006). These cells disappear from circulating blood with radiation doses as low as 1–2 Gy (Mauch et al. 1995). While most cells die a mitotic death, lymphocytes rarely if ever divide mitotically and instead undergo apoptosis during interphase (Mauch et al. 1995).

Acute radiation syndrome is comprised of four separate syndromes (prodromal, cerebrovascular, gastrointestinal, and hematopoietic), with onset dependent on the extent of exposure. A summary these syndromes can be found in Table 13.2. Hematopoietic syndrome occurs at doses of 2.5–5 Gy, with peak effects typically observed weeks or months after onset of exposure (Hall and Giaccia 2006). In an experiment analyzing the mortality rate of rhesus monkeys, it was found that when exposed to a dose less than or equal to 2 Gy, almost no animals die. The whole-body dose required to kill 50% (LD_{50}) and 100% of animals were observed to be 5.3 Gy and 8 Gy, respectively (Henschke and Morton 1957).

Humans develop and recover from hematologic damage much more slowly than other mammals with various factors influencing the body's vulnerability to exposure. Females appear to be less radiosensitive than their male counterparts, and the young and old have

TABLE 13.2   Summary of Syndromes Comprising Acute Radiation Syndrome (ARS)

	Exposure (Gy)	Onset of Symptoms Post Exposure	Symptoms	Mechanism of Death
Prodromal Syndrome	>1	5–15 minutes	Anorexia Nausea Vomiting Fatigability	Not fatal unless symptoms merge into one of the other 3 syndromes
Cerebrovascular Syndrome	>100	24–48 hours	Nausea Vomiting Disorientation Loss of Coordination Respiratory Distress Diarrhoea Seizures Coma	Leakage of intracranial vessels, fluid content and pressure build up in the brain
Gastrointestinal Syndrome	5–12	Days	Nausea Vomiting Lethargy Diarrhoea	Apoptosis of cells within the GI tract without replacement
Hematopoietic Syndrome	2.5–5	Weeks to Months	Blood Cell Depression Hemorrhages Infection	Apoptosis of blood cells without replacement

less tolerance than adults. While death extends up to 60 days after exposure, the peak incidence of death occurs at 30 days. If a human is exposed to a dose close to the LD_{50}, he will experience a latent period followed by prodromal syndrome. Approximately three weeks following exposure, a depression of blood elements manifests itself as fever and infections from granulocyte depression, impairment of immune mechanisms, bleeding, and possibly anemia resulting from hemorrhages caused by platelet depression.

The best estimates of the LD_{50} for young, otherwise healthy humans without medical intervention are estimated to be between 3.25 to 4.5 Gy. These estimates arise from the experiences of those exposed to total body irradiation at Hiroshima and Nagasaki, patients undergoing TBI for bone marrow transplants, and *in vivo* and *in vitro* assays of hematopoietic stem cells (HSCs) (Hall and Giaccia 2006; Henschke and Morton 1957). It is possible that acute damage can be repaired and tissue function can be restored by the rapid proliferation of stem cells. However, if acute damage depletes stem-cell populations below the levels needed for reversal of damage, acute injuries may become chronic (i.e., a consequential late effect).

Late Effects of Bone Marrow Irradiation

In contrast, late radiation damage may improve but can never fully repair. Late effects tend to appear months or years after initial onset of damage and occur in tissues with slow proliferation cycles. Late effect symptoms are largely dependent on the organs that are irradiated. Long term, there is the possibility of secondary malignancies due to radiation exposure with increasing risks if exposed at a younger age and long-term residual damage to functioning BM (Kamiya et al. 2015). This damage is resistant to amelioration and can be caused by decreases in HSC reserves, defects within the HSC self-renewal process, and myeloid skewing.

Radiation not only affects the HSC populations by directly inducing HSC senescence, but also by interfering with the stem cell niche and therefore the entire regulatory network of processing that take place within the niche. Mounting evidence suggests that the bone marrow stem cell niche is comprised of at least two separate niches, defined primarily by local oxygen concentration (Schofield 1978; Spencer et al. 2014). The first niche is found in the periosteal region and provides a hypoxic environment, while the second is found in the perivascular location, both of which are vulnerable to the effects of radiation and chemotherapy. Parmar et al. found the hypoxic nature to be essential for the survival of the HSCs within the niche (Parmar et al. 2007). Spencer et al. reported an increase in local oxygen tension following both chemotherapy and radiation despite severely damaged vasculature causing the oxygen gradient between the two niches to essentially disappear (Spencer et al. 2014). In a later study, Wang et al. examined these changes in microvascular permeability post radiation and found them to be correlated with changes in fat content which has been known to be affected by stress including radiation-induced stress, suggesting a pathophysiological mechanism involving the fat-vasculature permeability (Wang et al. 2017).

EVALUATING BONE MARROW INJURY
Peripheral Blood Counts

The hematopoietic system is capable of repopulating entirely from HSCs such that they have retained the ability to differentiate into any lineage and have unlimited self-renewal capabilities. It is important to have a means of assessing bone marrow damage to determine if intervention is necessary. The most widely used test to determine toxicity and safety is a complete and differential blood count, which is used to analyze counts of specific cells within the peripheral blood. The Common Terminology Criteria for Adverse Events (CTCAE), published by the National Institutes of Health, sets forth a standard grading system for hematologic toxicities, which can be found in Table 13.3.

Several studies have analyzed the patterns of relative blood counts over the course of a patient's treatment. In 1976, Stutz and Slawson reported decreased peripheral white blood cell counts in a review of 203 treatment records of patients undergoing local radiotherapy for pituitary, gynecological, or head and neck malignancies (Stutz et al. 1976).

TABLE 13.3   Graded Hematologic Toxicity as Defined by the Common Terminology Criteria for Adverse Events (CTCAE) Version 4

	Grade 0	Grade 1	Grade 2	Grade 3	Grade 4
White Blood Cells (1000/mm³)	≥4.0	3.0–3.9	2.0–2.9	1.0–1.9	< 1.0
Platelets (1000/mm³)	WNL	75.0 – WNL	50.0–74.9	25.0–49.9	< 25.0
Hemoglobin (g/dL)	WNL	10.0 – WNL	8.0–10.0	6.5–7.9	< 6.5
Absolute Neutrophil Count (1000/mm³)	≥ 2.0	1.5–1.9	1.0–1.4	0.5–0.9	< 0.5
Lymphocytes (1000/mm³)	≥ 2.0	1.5–1.9	1.0–1.4	0.5–0.9	< 0.5

WNL, within normal limits.

Yang et al. and Zachariah et al. further described the patterns of decrease in patients with breast, prostate, lung, gynecological, or head and neck malignancies reporting the most dramatic decline occurring during the first week of treatment. The mean nadir of white blood cells is site dependent but occurs at the middle to end of radiation, with individuals with the lowest counts at the end of the first week of treatment likely to have the lowest nadir counts (Yang et al. 1995; Zachariah et al. 2001). Yang et al. found the greatest decrease to be within leukocytes and Zachariah et al. found lymphocytes to be the most radiosensitive leukocyte. These studies provide relevant information used to determine the necessity of weekly CBC monitoring during radiotherapy and chemotherapy. As previously described, the time it takes for blood cells counts to decrease following damage is dependent on each cell type's varying kinetics. The blood counts represent this delayed response which while important, may not accurately depict the current status of the bone marrow.

Histopathology and Pre-Clinical Assays

Histopathology and marrow smears are useful in assessing bone marrow architecture, cellularity, cell lineage, iron stores, and myeloid to erythroid (M:E) ratios. *In vitro* agars, methyl cellulose-based assays, and Colony Forming Cell assays (CFCs) can be used to determine human progenitor cell content of the blood and bone marrow by measuring late stem cells and multipotent cells. Cobblestone area-forming cells (CAFC) assays and long-term culture initiating cells (LTC-IC) assays measure the most primitive hematopoietic cell colonies by measuring long-term cultures seeded onto marrow-derived stromal layers. Flow cytometry, currently the fastest method of analysis, is only method that can prospectively identify and isolate HSCs but does not produce any functional data (Frisch and Calvi 2014). *In vivo* assays, serial transplantibility (ST and competitive repopulation (CR), involve transplantation of hematopoietic stem cells into lethally irradiated recipients (Frisch and Calvi 2014). These techniques, while accurate at measuring primitive hematopoietic stem cells and the ability of the marrow to regenerate, clearly cannot be used to test human BM cell's regenerative function.

Stromal Tissue and Peripheral Stem Cells

About 10% of the body's total population of human progenitor cells can be found circulating within the blood (Wintrobe and Greer 2009). These cells differ from those of the bone marrow in a few important ways. These cells are not cycling, do not have a tendency to differentiate, and express different cell-surface markers from that of the marrow (Ferrero et al. 1983). Multiple studies have found these cells to respond normally when exposed to colony stimulating factors (CSFs) and when transplanted, *in vivo*, these cells have been found to be capable of short-term marrow hematopoiesis (Mauch et al. 1995). However, the long-term hematopoietic potential of these cells *in vivo* and their relation to those of the bone marrow are still largely unknown. Mesenchymal stromal cells (MSCs) are multipotent cells with the capacity to repair tissue damaged making them vital for the HSC microenvironment. These repair mechanisms are thought to be related to MSC's differentiation capacity or paracrine effects and have been shown to enhance engraftment and improve bone marrow recover from radiation damage when transplanted with HSCs. Yang et al.

and Lange et al. found that a small number of MSCs could be found within a target organ after intravenous infusion of MSC leading to significant increases in hematopoietic recovery post irradiation damage (Yang et al. 2012; Lange et al. 2011). Wen et al. reported that vesicles derived from murine, human marrow MSCs, or murine whole bone marrow cells, have the able to reverse radiation injury to murine BM *in vivo* and *in vitro*, stimulate proliferation, and reverse radiation-induced damage and apoptosis in Factor Dependent Continuous-Paterson 1 (FDC-P1) cells (Wen et al. 2016).

BONE MARROW IMAGING

Computed Tomography

Computed tomography (CT) is useful for volumetric delineation of bone and bone marrow. Bone marrow can be visualized on a CT scan by assuming that the marrow exhibits lower CT values than water due to the adipose tissue. Fat-saturated, yellow BM within the long bones of adults is between −30 to −100 (Hounsfield Unit) HU and hematopoeitically active, red BM will measure variably higher at a value between subcutaneous fat and soft tissue (Nishida et al. 2015). When contouring bone marrow using CT, it is not possible to distinguish red and yellow marrow and the entire medullary canal must be contoured. This overestimates the amount of active bone marrow leading to unnecessary constraints during treatment planning.

Magnetic Resonance Imaging

Magnetic resonance imaging (MRI) provides more specific and sensitive images that can be used to measure bone and bone marrow through the assessment of molecular composition, allowing for increased ability to distinguish between soft tissues. Fat protons within fat-saturated BM have a relatively long T2 relaxation because of low efficiency spin-spin relaxation causing the signal intensity on T2-weighted MR images to be higher than muscle and slightly lower than subcutaneous fat. Red marrow has an SI slightly lower than that of yellow marrow. Hydrophobic carbon-hydrogen groups in adipose and adipose-rich BM have efficient spin-lattice relaxation resulting in short T1 relaxation times and hyperintense structures on the sequenced image. Red marrow appears darker than yellow marrow on the T1 sequence and can be distinguished using a comparison to intervertebral disks with 78% accuracy and adjacent muscle with 89% accuracy (Chan et al. 2016). Roeske and Mundt used T1-weighted MRI fused to traditional CT scans to aid in treatment planning by identifying active bone marrow using pixel values similar to muscle (Roeske and Mundt 2004). Contouring only active bone marrow decreases unnecessary constraints during the treatment planning process.

Single Photon Emission Computed Tomography

Single-photon computed emission tomography (SPECT) has been used in combination with radioactive nucleotides, including ^{59}Fe and ^{52}Fe, to localize erythroid cells while ^{99m}Tc-labeled nanocolloids have been used to image the reticuloendothelial system to create a three-dimensional map of active bone marrow (Datz and Taylor 1985). ^{99m}Tc-labeled sulfur colloids are sequestered by macrophages which are associated with active bone marrow. Boucek and Turner used an integrated SPECT/CT system to quantitatively measure the active spinal, pelvic, and femoral marrow before, during, and after administration of ^{131}I-anti-CD20 rituximab finding that BM activity was directly proportional to administered activity per unit weight, height, or body surface area (Boucek and Turner 2005). Roeske et al. reported that the use of SPECT and CT fusion during IMRT treatment planning decreased V_{20Gy} to bone marrow compared to CT alone without compromising target coverage (Roeske et al. 2005).

Positron Emission Tomography

Positron emission tomography (PET) using ^{18}F-flurodeoxyglucose (FDG) is taken up by metabolically active tissues, including many malignancies and bone marrow (Figure 13.3). Increased radiation to bone marrow sub-regions with higher ^{18}F-FDG-PET activity is associated with increased hematologic toxicity, supporting the hypothesis that reducing dose to active sub regions could mitigate this complication (Rose et al. 2012). However, sites in which bone marrow is known to be actively proliferating generally do not have high uptake of FDG, limiting the scans ability to elucidate the bone marrow distribution. In more recent years, the use of additional radiotracers has been examined including

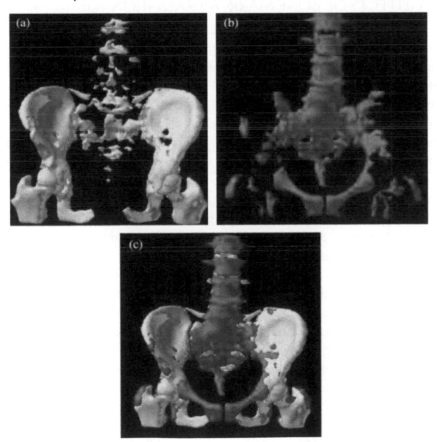

FIGURE 13.3 Representative distribution of active (red) and inactive (yellow) marrow from FDG-PET imaging. From (Rose et al. 2012), permissions gained.

3'-deoxy-3'-^{18}F-fluorothymidine (FLT) which is phosphorylated by the enzyme thymidine kinase 1, leading to intracellular trapping. FLT is taken up by tissues that are actively proliferating including malignancies and active bone marrow and can be used to develop BM-sparing radiation therapy plans (Everitt et al. 2009; McGuire et al. 2016).

Combining PET scans with computed tomography (CT) scans has allowed for the ability to link radiotracer uptake with specific anatomic structures to further clarify which structures contain active marrow. FLT PET/CT scans have been used to determine the mean percentage of bone marrow by atomic site (Hayman et al. 2011). Although not evenly distributed, in the anatomical locations where active marrow was present, the activity was relatively uniform. Several studies have noted decreased tracer uptake in bone marrow after irradiation with doses as low as 2 Gy, and complete absence of uptake after 10–20 Gy providing evidence that uptake only occurs in hematopoeitically active marrow (Everitt et al. 2009; Koizumi et al. 2011; Menda et al. 2010; Yue et al. 2010). Several studies indicate that ^{18}F-FLT is superior to ^{18}F-FDG for BM imaging due to the inability of ^{18}F-FDG to discriminate between proliferating cells and metabolically active non-proliferating cells (Hayman et al. 2011;Waarde et al. 2004; Waarde et al. 2006; Wyss et al. 2016).

CLINICAL EFFECTS OF BONE MARROW RADIATION
Normal Tissue Complication Probability

Historically, patients undergoing large (magna) field radiation treatment, including total TBI, hemibody irradiation (HBI), craniospinal irradiation, or extended nodal fields (e.g., total nodal or inverted Y irradiation courses) were at documented high risk for leukopenia and thrombocytopenia (Rubin et al. 1973). In addition, patients undergoing systemic radiation therapy with unsealed sources, such as strontium-89 or samarium-153, are at significant risk for high grade acute hematologic toxicity. Multiple studies have examined hematologic toxicity when smaller fields are used, finding decreasing rates of toxicity as field size decreases. For example, Yang et al. found that field sizes in which less than 40% of the total marrow was irradiated were unlikely to cause bone marrow depression significant enough to warrant clinical intervention (Yang et al. 1995). Hematologic effects of smaller radiation fields in general are less clinically apparent, largely due to compensatory hematopoiesis in unirradiated marrow.

As systemic therapies have evolved and the negative long-term consequences of large fields became realized, radiation field sizes have reduced, often in conjunction with an expanding role of systemic therapies. Multiple studies have analyzed the impact of bone marrow radiation dose to increased hematologic toxicity. Brixey et al. reported that gynecologic cancer patients treated with intensity modulated radiation therapy (IMRT) had lower probability of hematologic toxicity compared to those treated with conventional 4-field box techniques, ostensibly owing to lower doses delivered to the iliac crests. In particular, this effect was most apparent in patients undergoing treatment with concurrent chemotherapy (Brixey et al. 2002). Moreover, increased dose to pelvic bone marrow was associated with reduced peripheral cell counts (particularly neutropenia) in both cervical and anal cancer patients undergoing chemoradiotherapy (Mell et al. 2006, 2008). For example, increased pelvic BM receiving at least 10 Gy (BM-V_{10Gy}) was associated with an increased grade 2 or worse leukopenia and neutropenia (odds ratio [OR] = 2.09; 95% confidence interval [CI], 1.24–3.53; p = 0.006; and OR = 1.41; 95% CI, 1.02–1.94; p = 0.037, respectively) (Mell et al. 2006). Findings from others have reported similar associations (Klopp et al. 2013; Elicin et al. 2014).

Normal tissue complication probability (NTCP) models have been used to estimate the effect of increasing the dose to specific marrow regions on the probability of hematologic toxicity. For example, Mell et al. used log-linear models to correlate the volume of pelvic marrow receiving at least 20 Gy (V_{20Gy}) with acute white blood cell count (WBC) and absolute neutrophil count (ANC) nadirs in patients undergoing chemoradiotherapy (Mell et al. 2006; Mell et al. 2008). Patients with BM-V_{10Gy} > 90% had higher rates of grade 2 or worse leukopenia and neutropenia than did patients with BM-V_{10Gy} < 90% (11.1% vs. 73.7%, p < 0.01; and 5.6% vs. 31.6%, p = 0.09) and were more likely to have chemotherapy held on univariate (16.7% vs. 47.4%, p = 0.08) and multivariate (OR = 32.2; 95% CI, 1.67–622; p = 0.02) analysis. In contrast, Zhu et al. used linear mixed effects models to examine longitudinal changes in peripheral cell counts to correlate (V_{20Gy}, V_{30Gy}, and V_{40Gy}) with acute hematologic toxicity. They estimated that with every 1 Gy increase in mean pelvic bone marrow dose, the ln(ANC) was reduced by 9.6/µL per week (p = 0.015) (Zhu et al. 2015).

Because the pelvis and lower spine contain high concentrations of active marrow, much of the literature has focused on modulating dose to the marrow in this region. However, bone marrow irradiation can have an impact on cancers of other regions, including the lung, head/neck, and esophagus. For example, Deek et al. showed that increasing dose to the thoracic vertebrae was associated with increased hematologic toxicity in patients undergoing chemoradiotherapy for non-small cell lung cancer (Deek et al. 2016). Greater RT dose to the thoracic vertebrae (TV) was associated with higher risk of grade ≥3 leukopenia across multiple dose-volume histogram (DVH) parameters, including TV-V_{20Gy} (TV-V_{20Gy}) (OR 0 1.06; p = 0.025), TV-V_{30Gy} (OR = 1.07; p = 0.013), and mean vertebral dose (MVD) (OR = 1.13; p = 0.026). On multiple regression analysis, TV-V_{30Gy} (OR = 0.996 ; p = 0.018) and TV-V_{20Gy} (OR = 1.003; p = 0.048) were associated with white blood cell nadir. Sini et al. showed that there were significant differences (p < 0.005) in the DVH of BM volumes between IMRT, Volumetric Arc Modulated Therapy (VMAT), and tomotherapy. Tomotherapy associated with larger volumes receiving low doses (3–20 Gy) and smaller receiving 40–50 Gy and higher BM-V_{40Gy} were found to be associated with higher risk of acute grade 3 (OR = 1.018) or late grade 2 lymphopenia (OR = 1.005) (Sini et al. 2016). Recent planning studies have shown that proton beam radiotherapy can markedly reduce bone marrow dose and, hence, the risk for toxicity for esophageal cancer and other thoracic malignancies (Figure 13.4) (Warren et al. 2017).

Effects of Chemotherapy with Radiotherapy

Different chemotherapies are also known to be myelosuppressive to different extents, particularly those with alkylating abilities (Table 13.4). When used in conjunction, these toxicities compound complexly. Hematologic toxicity is a key barrier to intensifying chemoradiotherapy in patients with a wide array of malignancies, especially those with pelvic tumors such as anal, rectal, prostate, and gynecological malignancies.

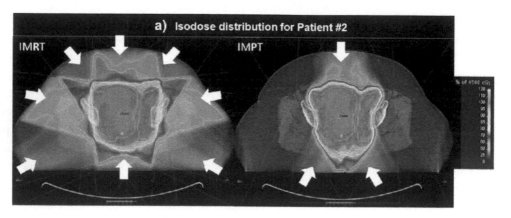

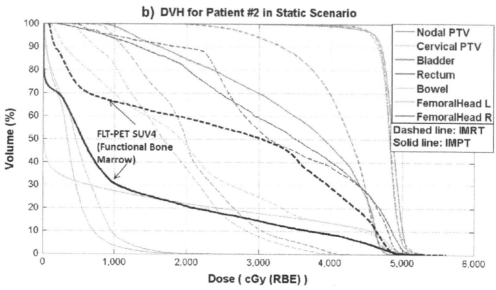

FIGURE 13.4 Intensity Modulated Radiation Therapy versus Intensity Modulated Proton Therapy isodose distribution and dose-volume histogram for model patient. From (Song et al. 2010).

Lyman et al. reported that 56% of early-breast cancer patients receive less than 85% of the targeted dose intensity because of dose reductions and delays due to neutropenia and other marrow suppression (Lyman, Dale, and Crawford 2003). The timing of treatment has been shown to affect efficacy of treatment and rates of toxicity. A multitude of trials have shown that concurrent chemoradiotherapy improves outcomes in a wide array of tumor types, but increases the rate of patients experiencing acute hematologic, gastrointestinal, and pulmonary toxicities (Aupérin et al. 2010; Gadducci et al. 2001; Li et al. 2016; Melcher and Sebag-Montefiore 2003). For instance, in a meta-analysis examining locally advanced non-small-cell lung cancer, concomitant radiochemotherapy increased acute esophageal toxicity (grade 3–4) from 4% to 18% with a relative risk of 4.9 (95% CI, 3.1 to 7.8; p = 0.001) (Aupérin et al. 2010). Currently, there are several single and multi-institutional studies looking at chemotherapy intensification in conjunction with concurrent or adjuvant radiotherapy.

TABLE 13.4   Timeline of Myelosuppressive Effects of Varying Chemotherapy Drugs

Drug or Drug Class	Degree of Suppression[a]	Myelosuppression	
		Nadir (Days)	Time to Marrow Recovery (Days)
Anthracycline	III	6–13	21–24
Vinca Alkaloids	I–II	4–9	7–21
Mustard Alkylator			
Nitrogen Mustard	III	7–14	28
Antifolates	III	7–14	14–21
Antipyrimidines	III	7–14	22–24
Antipurines	II	7–14	14–21
Podophyllotoxins	II	5–14	22–28
Alkylators	II	10–21	18–40
Nitrosoureas	III	26–60	35–85
Miscellaneous[b]			
Busulphan	III	11–30	24–54
Cisplatin	I	14	21
Dacarbazine	III	21–28	28–35
Hydroxyurea	II	7	14–21
Mithramycin	I	5–10	10–18
Mitomycin	II	28–42	42–56
Procarbazine	II	25–36	35–50
Razoxane (ICRF)	II	11–16	12–25

Source: From Mauch et al. (1995).

[a] I—mild, II—moderate. III—severe (based on common dose schedules).

[b] Agents differing from their class of compounds. Reprinted with permission from (62a).

One of the most promising intensified treatments for cervical malignancies is a doublet chemotherapy regimen with cisplatin and gemcitabine. Multiple trials have shown gemcitabine to have considerable activity against cervical cancer when given with cisplatin and radiotherapy leading to increased survival, although acute toxicity remains a significant barrier (Dueñas-González et al. 2005; Dueñas-González et al. 2011; Rose et al. 2007; Swisher et al. 2006). One of the most promising phase III randomized trials of cisplatin/RT with or without concurrent gemcitabine and adjuvant cisplatin/gemcitabine showed a significantly increased 3 year-progression-free survival of 74% versus 65% favouring the combined chemotherapy regimen. Overall survival and time to progression were also increased. However, acute toxicity was significant, with 87% of patients experiencing grade 3 or 4 toxicity, compared to 46% in the standard arm (Dueñas-González et al. 2011). Methods to reduce gastrointestinal and hematologic toxicity during chemoradiotherapy could mitigate this toxicity and take advantage of the therapeutic benefits chemotherapy intensifications.

For example, a phase I clinical trial has indicated that image-guided IMRT given concurrently with cisplatin and gemcitabine can permit escalated doses of chemotherapy beyond that obtained with conventional RT (Mell et al. 2017). The NRG GY006 phase 2 randomized trial (NCT02466971) is an ongoing multi-institutional study funded by the National Cancer Institute. It is currently testing the efficacy of RT and cisplatin with or without triapine (an inhibitor of ribonucleotide reductase) in gynecologic cancer patients. The trial uses 18-F-FDG PET to identify and spare active bone marrow in patients undergoing IMRT. The optimal constraints for total bone marrow (defined as the outer contour of L5, sacrum, os coxae, and proximal femora delineated on CT) for the GY006 trial are mean dose <27 Gy (29 Gy acceptable), $V_{10Gy} < 85.5\%$ (90% acceptable), and $V_{20Gy} < 66\%$ (75% acceptable). The optimal constraints for active bone marrow (defined as the subset of total bone marrow with standardized uptake value [SUV] greater than the mean) are mean dose < 28.5 Gy (30 Gy acceptable), $V_{10Gy} < 90\%$, and $V_{20Gy} < 70\%$ (75% acceptable). Similar trials are ongoing at Emory University, University of North Carolina, and other international institutions in other tumor sites with high rates of hematologic toxicity, such as colorectal and anal malignancies.

Total Body Irradiation

TBI is used mainly to treat hematopoietic diseases allowing for the successful transplant of hematopoietic stem cells including bone marrow stem cells (BMT), HLA-identical hematopoietic stem cells, or peripheral blood progenitor stem cells. These myeloablative doses of radiation combined with the immunoablative doses of chemotherapy eradicate abnormal cells that have eluded other treatments and suppress the patient's immune system to prevent rejection of donor bone marrow. The development of BMT therapy has promoted cures for leukemia and other blood cancers, earning Dr. E. Donnall Thomas the 1990 Nobel Prize in Physiology or Medicine. TBI followed by BMT has since then been used to treat widely disseminated malignant targets such as non-Hodgkin's lymphoma, chronic lymphocytic leukemia, chronic myeloid leukemia, acute myeloid and lymphoblastic leukemia, neuroblastoma, and Ewing's sarcoma (Aydogan et al. 2006). Modern TBI is typically delivered with conventional parallel-opposed fields in 6–10 total fractions given twice daily for 3–5 days. Fractionation has proved to be fairly successful in destroying malignant cells while allowing for adequate repair of normal tissue (Wilkie et al. 2008). However, several significant problems persist. As discussed previously, the effects of TBI can be extremely toxic, even deadly, with extensive bone marrow suppression, due predominantly to graft-versus-host disease.

Intensity-Modulated Total Marrow Irradiation

In the recent decades, multiple studies have reported that IMRT is highly effective at decreasing high dose to normal tissues and structures and increasing dose to targets during total marrow irradiation (TMI). Intensity-modulated total marrow irradiation (IM-TMI) is used to treat many of the same malignancies as TBI while conforming the dose to the target marrow using a multileaf collimator, allowing for a reduction in dose in the lungs, eyes, heart, liver, and kidneys. Hui et al. used the Tomotherapy Hi-Art system to deliver IMRT and reported a 35–70% reduction in dose to critical organs (Hui et al. 2005). In a separate study, Wong et al. used targeted tomotherapy and found a 1.7 to 7.5 fold reduction in median organ dose in comparison to conventional TBI (Wong et al. 2006). Aydogan et al. and Wilkie et al. have also shown the feasibility of linac-based IM-TMI (Figure 13.5), reporting dose reductions of 29–65% to critical organs while increasing heterogeneity of

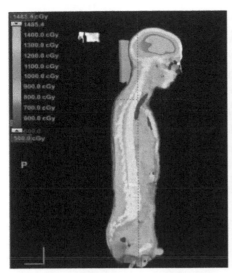
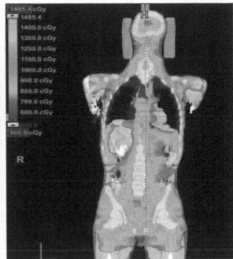

FIGURE 13.5   Intensity-modulated total marrow irradiation dose distribution in sagittal and frontal planes. From (Aydogan, Mundt, and Roeske 2006).

dose to bone and bone marrow (Aydogan et al. 2006; Wilkie et al. 2008). Liang et al. have also demonstrated the feasibility of dosimetric verification of volumetric-modulated arc therapy for TMI (Liang et al. 2013).

Decreasing dose to normal tissue allows for the potential of increasing total radiation dose. Dose escalation has been reported to decrease recurrence and relapse rates (Clift et al. 1990). While dose escalation can involve increasing dose to the entire skeletal system, it is also possible to selectively increase dose to hematopoeitically active bone marrow, perhaps, with the use of simultaneous boost technology and functional bone marrow imaging.

Bone Marrow-Sparing Radiotherapy

Subsequent investigations showed that IMRT plans can be optimized to decrease dose to bone marrow when BM is used as a constraint (Dueñas-González et al. 2005; Mell et al. 2017; Ahamad et al. 2005; Brixey et al. 2002; Roeske et al. 2000). As IMRT emerged as an effective treatment modality for many sites of malignancy, increased volumes of low dose to bone marrow and increased use of cytotoxic chemotherapy have reinvigorated interest in sparing bone marrow during treatment planning. Low blood counts due to hematologic toxicity are a significant clinical problem, due to delayed or missed cycles of chemotherapy, hospitalizations, and the need for growth factors decreasing the efficacy of treatment. Reducing hematologic toxicity is therefore an important goal, and has led to the development of bone marrow sparing techniques including IMRT and proton therapy.

As already pointed out, CT-based treatment planning can be used to identify marrow regions for sparing, but cannot distinguish active vs. inactive subregions of the marrow. Pressing treatment planning algorithms to avoid inactive subregions can significantly and unnecessarily constrain highly modulated radiotherapy plans. Thus, it has become increasingly important to elucidate the location of hematopoietically active, or functional bone marrow subregions. PET-guided bone marrow sparing techniques allow optimal customization of treatment based on an individual patient's active-bone marrow distribution. PET-based image-guided IMRT can be highly effective in reducing the incidence of acute neutropenia and hematologic toxicity (Mell et al. 2017) (Figure 13.6). For example, Mell et al. showed that using image guided bone marrow sparing IMRT (IG-BMS-IMRT) reduced the bone marrow V_{20Gy} from 71% to 56%, corresponding to a reduction in grade ≥ 3 neutropenia from 27% to 8% in cervical cancer patients undergoing chemoradiotherapy (Mell et al. 2017). The mean dose to active bone marrow in patients treated with IG-BMS-IMRT was 26.4 Gy.

FUTURE APPLICATIONS
Proton Therapy

Numerous studies have tested the use of proton beam therapy (PBT) in multiple cancer sites in efforts to decrease dose to normal critical organs. Intensity modulated proton therapy (IMPT) plans have the ability to decrease integral dose by decreasing the number of beams necessary to achieve target dose distributions. PBT and IMPT have also been shown to have the ability to decrease dose to bone marrow. As can be seen in Figure 13.4, Song et al. reported that passively scattered proton therapy is more effective than IMRT at decreasing bone marrow volume receiving low dose radiation (Song et al. 2010). Dinges et al. described increased bone marrow sparing with IMPT for cervical cancer patients, reporting the median dose reductions for functional bone marrow to be: 32% for V_{5Gy}, 47% for V_{10Gy}, 54% for V_{20Gy}, 57% for V_{40Gy}, all with $p < 0.01$ compared to IMRT (Dinges 2015). Even with range uncertainties and translational set up errors in the three principal dimensions, the IMPT plans were considered to produce a significant reduction to active bone marrow (ABM).

Atlas-Based Treatment Planning

As has been previously discussed, refining IMRT plans to focus on sparing ABM sub regions may be an effective strategy for individual optimisation during treatment planning but may be costly due to the use of functional imaging. Li et al. compared the efficacy of two treatment plans, one created using individual FDG-PET CT scan active bone marrow (ABM, ABM_{Custom}) and the other using CT scans and an ABMatlas (ABM_{Atlas}). 32 FDG-PET images were fused with a canonical template CT from an individual with an intermediate-sized pelvic bone marrow (PBM) volume chosen from the sample of 32 (Figure 13.7). The atlas was defined as the mean FDG-PET image using values above the mean SUV as a threshold. An atlas-based ABM plan was created for each patient by using a CT that had been deformed and resampled. The Dice coefficient indicated a high proportion of overlap and good agreement between the ABM_{Custom} and ABM_{Atlas}. Atlas-based plans were found to be as effective in reducing dose to active bone marrow as customized plans based of FDG-PET (Li et al. 2017). Therefore, atlas-based planning may be an effective method to decrease hematologic toxicity without the need for expensive functional imaging in every patient. In addition, this technique could improve patients' tolerance to chemotherapy and increase the therapeutic ratio of chemo-radiotherapy for pelvic malignancies in general. Such approaches are likely to be particularly important for underserved and resource-constrained communities.

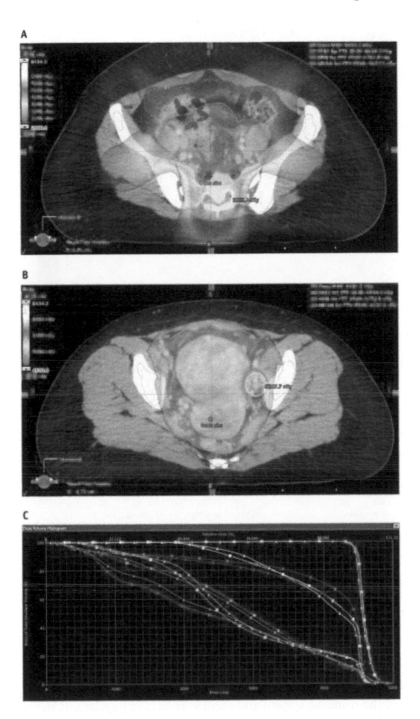

FIGURE 13.6 Intensity-Modulated Radiation Therapy dose distributions during treatment planning of cervical maliganancy. From (Mell et al. 2017).

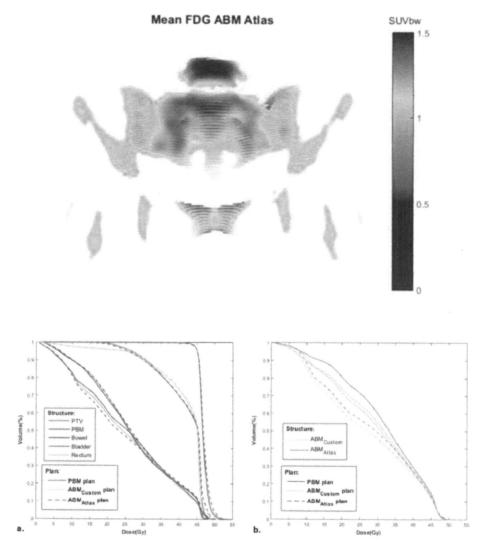

FIGURE 13.7 Heat map of the mean active bone marrow (ABM) generated using standardized FDG-PET images of 32 training patients and average dose-volume histogram comparisons for pelvic bone marrow plan ABM$_{Custom}$ and ABM$_{Atlas}$ plans. From (Li et al. 2017), permissions gained.

CONCLUSION

The competence of the bone marrow system is vital for the proper functioning of the body including the immune system and cardiac-respiratory system. It is, however, exceptionally vulnerable and many times the dose-limiting organ during cancer therapy, particularly radiation therapy and chemotherapy. Both acute and late bone marrow injury, reported throughout history, can be attributed to the bone marrow's unique pattern of pathology based on the multitude of factors present within the bone marrow niche and steady-state kinetics of lymphoid and myeloid progeny. This in part, makes bone marrow injury, whether the decreased ability to differentiate or decreased self-renewal capabilities, difficult to assess accurately and in a time-efficient manner with different *in vivo* and *in vitro* techniques providing different benefits and deterrents.

In turn, it has become increasingly important to find and assess the functionality of the bone marrow before and after radiation and chemotherapy, which is being done using various imaging modalities and NTCP models. There are some instances, such as with TBI or TMI followed by BMT, in which injuring diseased marrow is a vital portion of treatment. Most often, however, findings suggest it is important to limit radiation exposure to the marrow during treatment of malignancies due to the hematologic toxicity that ensues. This is being done in a number of ways including using image-guidance and atlas-guidance during treatment planning and set-up, making active bone marrow, a constraint during treatment planning, and through the use of novel techniques, such as proton beam therapy.

REFERENCES

Ahamad, A. et al. 2005. "Intensity-Modulated Radiation Therapy after Hysterectomy: Comparison with Conventional Treatment and Sensitivity of the Normal-Tissue–sparing Effect to Margin Size." *International Journal of Radiation Oncology Biology and Physics* 62: 1117–24.

Atkinson, H.R. 1962. "Bone Marrow Distribution as a Factor in Estimating Radiation to the Blood-Forming Organ." *Journal of the College of Radiologists of Australasia* 6: 149–54.

Aupérin, A. et al. 2010. "Meta-Analysis of Concomitant versus Sequential Radiochemotherapy in Locally Advanced Non-Small-Cell Lung Cancer." *Journal of Clinical Oncology* 28: 2181–90.

Aydogan, B. et al. 2006. "Linac-Based Intensity Modulated Total Marrow Irradiation (IM-TMI)." *Tech in Cancer Research and Treatment* 5: 513–9.

Basu, S. et al. 2007. "Magnetic Resonance Imaging Based Bone Marrow Segmentation for Quantitative Calculation of Pure Red Marrow Metabolism Using 2-Deoxy-2-[F-18]fluoro-D-Glucose- Positron Emission Tomography." *Molecular Imaging and Biology* 9: 361–5.

Boucek, J.A., and Turner, J.H. 2005. "Validation of Prospective Whole-Body Bone Marrow Dosimetry by SPECT/CT Multimodality Imaging in 131I-Anti-CD20 Rituximab Radioimmunotherapy of Non-Hodgkins Lymphoma." *European Journal of Nuclear Medicine and Molecular Imaging* 32: 458–69.

Brixey, C.J. et al. 2002. "Impact of Intensity-Modulated Radiotherapy on Acute Hematologic Toxicity in Women with Gynecologic Malignancies." *International Journal of Radiation Oncology Biology and Physics* 54: 1388–96.

Chan, B.Y. et al. 2016. "MR Imaging of Pediatric Bone Marrow." *RadioGraphics* 36(6), Radiological Society of North America: 1911–30.

Clift, R.A. et al. 1990. "Allogeneic Marrow Transplantation in Patients with Acute Myeloid Leukemia in First Remission: A Randomized Trial of Two Irradiation Regimens." *Blood* 76: 1867–71.

Cohen, D.E., and Kim, R.H. 2017. "The Legacy of the Radium Girls." *JAMA Dermatology.* 153: 801.

Custer, R.P., and Hayhoe, F.G.J., 1974. *An Atlas of the Blood and Bone Marrow.* Saunders.

Datz, F.L., and Taylor, A. 1985. "The Clinical Use of Radionuclide Bone Marrow Imaging." *Seminars in Nuclear Medicine* 15: 239–59.

Davy, K.P., and Seals, D.R. 1994. "Total Blood Volume in Healthy Young and Older Men." *Journal of Applied Physiology* 76: 2059–62.

Deek, M.P. et al. 2016. "Thoracic Vertebral Body Irradiation Contributes to Acute Hematologic Toxicity During Chemoradiation Therapy for Non-Small Cell Lung Cancer." *International Journal of Radiation Oncology Biology and Physics* 94: 147–54.

Dinges, E. 2015. "Bone Marrow Sparing in Intensity Modulated Proton Therapy for Cervical Cancer: Efficacy and Robustness under Range and Setup Uncertainties" *Radiotherapy and Oncology* 115: 373–9.

Duda, S.H. et al. 1995. "Normal Bone Marrow in the Sacrum of Young Adults: Differences between the Sexes Seen on Chemical-Shift MR Imaging." *American Journal of Roentgenology* 164: 935–40.

Dueñas-González, A. et al. 2005. "Pathologic Response and Toxicity Assessment of Chemoradiotherapy with Cisplatin versus Cisplatin plus Gemcitabine in Cervical Cancer: A Randomized Phase II Study." *International Journal of Radiation Oncology Biology and Physics* 61: 817–23.

Dueñas-González, A. et al. 2011. "Phase III, Open-Label, Randomized Study Comparing Concurrent Gemcitabine plus Cisplatin and Radiation Followed by Adjuvant Gemcitabine and Cisplatin versus Concurrent Cisplatin and Radiation in Patients with Stage IIB to IVA Carcinoma of the Cervix." *Journal of Clinical Oncology* 29: 1678–85.

Elicin, O. et al. 2014. "Clinical Investigation [18 F]FDG-PET Standard Uptake Value as a Metabolic Predictor of Bone Marrow Response to Radiation." *Radiation Oncology Biology* 90: 1099–107.

Ellis, R.E. 1961. "The Distribution of Active Bone Marrow in the Adult." *Physics in Medicine and Biology* 5: 255–8.

Everitt, S. et al. 2009. "Imaging Cellular Proliferation During Chemo-Radiotherapy: A Pilot Study of Serial 18F-FLT Positron Emission Tomography/Computed Tomography Imaging for Non–Small-Cell Lung Cancer." *International Journal of Radiation Oncology Biology and Physics* 75: 1098–104.

Fajardo, L.F. et al. 2001. *Radiation Pathology*. Oxford University Press.

Ferrero, D. et al. 1983. "Antigenically Distinct Subpopulations of Myeloid Progenitor Cells (CFU-GM) in Human Peripheral Blood and Marrow." *Proceedings of the National Academy of Sciences of the USA* 80: 4114–18.

Frisch, B.J., and Calvi, L.M. 2014. "Hematopoietic Stem Cell Cultures and Assays." *Methods in Molecular Biology* 1130: 315–24.

Gadducci, A. et al. 2001. "Neoadjuvant Chemotherapy and Concurrent Chemoradiation in the Treatment of Advanced Cervical Cancer." *Anticancer Research* 21: 3525–33.

Grandin, K.S. et al. 2013. *The Nobel Prizes 2012: Formerly Les Prix Nobel*. Science History Publications.

Hall, E.J., and Giaccia, A.J. 2006. *Radiobiology for the Radiologist*. 6th ed. Vol. 224. Philadelphia, PA: Lippincott Williams & Wilkins.

Hasegawa, A. et al. 2015. "Health Effects of Radiation and Other Health Problems in the Aftermath of Nuclear Accidents, with an Emphasis on Fukushima." *The Lancet* 386: 479–88.

Hayman, J.A. et al. 2011. "Distribution of Proliferating Bone Marrow in Adult Cancer Patients Determined Using FLT-PET Imaging." *International Journal of Radiation Oncology Biology and Physics* 79: 847–52.

Henschke, U.K., and Morton, J.L. 1957. "Mortality of Rhesus Monkeys after Single Total Body Irradiation." *The American Journal of Roentgenology Radium Therapy and Nuclear Medicine* 77: 899–909.

Hui, S. et al. 2005. "Feasibility Study of Helical Tomotherapy for Total Body or Total Marrow Irradiationa)." *Medical Physics* 32: 3214–24.

Kamiya, K. et al. 2015. "Long-Term Effects of Radiation Exposure on Health." *The Lancet* 386: 469–78.

Klopp, A.H. et al. 2013. "Hematologic Toxicity in RTOG 0418: A Phase 2 Study of Postoperative IMRT for Gynecologic Cancer." *International Journal of Radiation Oncology Biology and Physics* 86: 83–90.

Koizumi, M. et al. 2011. "Uptake Decrease of Proliferative PET Tracer 18FLT in Bone Marrow after Carbon Ion Therapy in Lung Cancer." *Molecular Imaging and Biology* 13: 577–82.

Lange, C. et al. 2011. "Radiation Rescue: Mesenchymal Stromal Cells Protect from Lethal Irradiation." *PLoS ONE* 6: e14486

Li, Y. et al. 2016. "A Review of Neoadjuvant Chemoradiotherapy for Locally Advanced Rectal Cancer." *International Journal of Biological Sciences* 12: 1022–31.

Li, N. et al. 2017. "Feasibility of Atlas-Based Active Bone Marrow Sparing Intensity Modulated Radiation Therapy for Cervical Cancer." *Radiotherapy and Oncology* 123: 325–30.

Liang, Y. et al. 2013. "Feasibility Study on Dosimetry Verification of Volumetric-Modulated Arc Therapy-Based Total Marrow Irradiation." *The Journal of Applied Clinical Medical Physics* 14: 3852.

Mauch, P. et al. 1995. "Hematopoietic Stem Cell Compartment: Acute and Late Effects of Radiation Therapy and Chemotherapy." *International Journal of Radiation Oncology Biology and Physics* 31: 1319–39.

McGuire, S.M. et al. 2016. "Using [18F]Fluorothymidine Imaged With Positron Emission Tomography to Quantify and Reduce Hematologic Toxicity Due to Chemoradiation Therapy for Pelvic Cancer Patients." *International Journal of Radiation Oncology Biology and Physics* 96: 228–39.

Melcher, A.A., and Sebag-Montefiore, D., 2003. "Concurrent Chemoradiotherapy for Squamous Cell Carcinoma of the Anus Using a Shrinking Field Radiotherapy Technique without a Boost." *British Journal of Cancer* 88: 1352–7.

Mell, L.K. et al. 2006. "Dosimetric Predictors of Acute Hematologic Toxicity in Cervical Cancer Patients Treated with Concurrent Cisplatin and Intensity-Modulated Pelvic Radiotherapy." *International Journal of Radiation Oncology Biology and Physics* 66: 1356–65.

Mell, L.K. et al. 2008. "Association Between Bone Marrow Dosimetric Parameters and Acute Hematologic Toxicity in Anal Cancer Patients Treated with Concurrent Chemotherapy and Intensity-Modulated Radiotherapy." *International Journal of Radiation Oncology Biology and Physics* 70: 1431–7.

Mell, L.K. et al. 2017. "Bone Marrow-Sparing Intensity Modulated Radiation Therapy with Concurrent Cisplatin For Stage IB-IVA Cervical Cancer: An International Multicenter Phase II Clinical Trial (INTERTECC-2)." *International Journal of Radiation Oncology Biology and Physics* 97: 536–45.

Menda, Y. et al. 2010. "Investigation of the Pharmacokinetics of FLT Uptake in the Bone Marrow before and Early after Initiation of Chemoradiation Therapy in Head and Neck Cancer." *Nuclear Medicine and Biology* 37: 433–8.

Munka, V., and Gregor, A. 1965. "Lymphatics and Bone Marrow." *Folia Morphologica* 13: 404–12.

Nishida, Y. et al. 2015. "Clinical and Prognostic Significance of Bone Marrow Abnormalities in the Appendicular Skeleton Detected by Low-Dose Whole-Body Multidetector Computed Tomography in Patients with Multiple Myeloma." *Blood Cancer Journal* 5: e329.

Parmar, K. et al. 2007. "Distribution of Hematopoietic Stem Cells in the Bone Marrow according to Regional Hypoxia." *Proceedings of the National Academy of Sciences of the USA* 104: 5431–6.

Parslow, T.G. 2001. *Medical Immunology.* Lange Medical Books/McGraw-Hill Medical Pub. Division.

Paul, W.E. 2013. *Fundamental Immunology.* Wolters Kluwer Health/Lippincott Williams & Wilkins.

Roeske, J.C., and Mundt, A.J. 2004. "Incorporation of Magnetic Resonance Imaging into Intensity Modulated Whole-Pelvic Radiation Therapy Treatment Planning to Reduce the Volume of Pelvic Bone Marrow Irradiated." *International Congress Series* 1268: 307–12.

Roeske, J.C. et al. 2005. "Incorporation of SPECT Bone Marrow Imaging into Intensity Modulated Whole-Pelvic Radiation Therapy Treatment Planning for Gynecologic Malignancies." *Radiotherapy and Oncology* 77: 11–17.

Rose, P.G. et al. 2007. "A Phase I Study of Gemcitabine Followed by Cisplatin Concurrent with Whole Pelvic Radiation Therapy in Locally Advanced Cervical Cancer: A Gynecologic Oncology Group Study." *Gynecologic Oncology* 107: 274–9.

Rose, B.S. et al. 2012. "Correlation Between Radiation Dose to 18F-FDG-PET Defined Active Bone Marrow Subregions and Acute Hematologic Toxicity in Cervical Cancer Patients Treated With Chemoradiotherapy." *International Journal of Radiation Oncology Biology and Physics* 83: 1185–91.

Roeske, J.C. et al. 2000. "Intensity-Modulated Whole Pelvic Radiation Therapy in Patients with Gynecologic Malignancies." *International Journal of Radiation Oncology Biology and Physics* 48: 1613–21.

Rubin, P. et al. 1973. "Bone Marrow Regeneration and Extension after Extended Field Irradiation in Hodgkin's Disease." *Cancer* 32: 699–711.

Schofield, R. 1978. "The Relationship between the Spleen Colony-Forming Cell and the Haemopoietic Stem Cell." *Blood Cells* 4: 7–25.

Shouse, S.S., Warren, S.L., and Whipple, G.H. 1931. "II. Aplasia of Marrow and Fatal Intoxication in Dogs Produced by Roentgen Radiation of All Bones." *Journal of Experimental Medicine* 53: 421–35.

Sini, C. et al. 2016. "Dose-Volume Effects for Pelvic Bone Marrow in Predicting Hematological Toxicity in Prostate Cancer Radiotherapy with Pelvic Node Irradiation." *Radiotherapy and Oncology* 118: 79–84.

Song, W.Y. et al. 2010. "Dosimetric Comparison Study between IMRT and Three-Dimensional Conformal Proton Therapy for Pelvic Bone Marrow Sparing in the Treatment of Cervical Cancer." *Journal of Applied Clinical Medical Physics* 11: 3255.

Spencer, J.A. et al. 2014. "Direct Measurement of Local Oxygen Concentration in the Bone Marrow of Live Animals." *Nature* 508: 269–73.

Stutz, L.T.C., Friedrich H., and Slowson, Robert G. 1976. "Cell Count: Review of 203 Treatment Records." *Military Medicine* 141: 390.

Swisher, E.M. et al. 2006. "Weekly Gemcitabine and Cisplatin in Combination with Pelvic Radiation in the Primary Therapy of Cervical Cancer: A Phase I Trial of the Puget Sound Oncology Consortium." *Gynecologic Oncology* 101: 429–35.

Travlos, G.S. 2006. "Normal Structure, Function, and Histology of the Bone Marrow." *Toxicologic Pathology* 34: 548–65.

Tubiana, M. et al. 1979. "Effects of Radiations on Bone Marrow." *Pathologie-Biologie* 27: 326–34.

Vogler, J.B., and Murphy, W.A. 1988. "Bone Marrow Imaging." *Radiology* 168: 679–93.

Waarde, A. et al. 2004. "Selectivity of 18F-FLT and 18F-FDG for Differentiating Tumor from Inflammation in a Rodent Model." *Journal of Nuclear Medicine* 45: 695–700.

Waarde, A. et al. 2006. "Comparison of Sigma-Ligands and Metabolic PET Tracers for Differentiating Tumor from Inflammation." *The Journal of Nuclear Medicine* 47: 150–54.

Wang, K. et al. 2017. "MRI Study on the Changes of Bone Marrow Microvascular Permeability and Fat Content after Total-Body X-Ray Irradiation." *Radiation Research* 189: 205–12.

Warren, S. et al. 2017. "Potential of Proton Therapy to Reduce Acute Hematologic Toxicity in Concurrent Chemoradiation Therapy for Esophageal Cancer." *International Journal of Radiation Oncology Biology and Physics* 99: 729–37.

Wen, S. et al. 2016. "Mesenchymal Stromal Cell-Derived Extracellular Vesicles Rescue Radiation Damage to Murine Marrow Hematopoietic Cells." *Leukemia* 30: 2221–31.

Wilkie, J.R. et al. 2008. "Feasibility Study for Linac-Based Intensity Modulated Total Marrow Irradiation." *Medical Physics* 35: 5609–18.

Wintrobe, M.M., and Greer, J.P. 2009. *Wintrobe's Clinical Hematology.* Wolters Kluwer Health/ Lippincott Williams & Wilkins.

Woodard, H.Q., and Holodny, E. 1960. "A Summary of the Data of Mechanik on the Distribution of Human Bone Marrow." *Physics in Medicine and Biology* 5: 57–59.

Wong, J.C. et al. 2006. "Targeted Total Marrow Irradiation Using Three-Dimensional Image-Guided Tomographic Intensity-Modulated Radiation Therapy: An Alternative to Standard Total Body Irradiation." *Biology of Blood and Marrow Transplantation* 12: 306–15.

Wyss, J.C. et al. 2016. "FDG versus FLT for Defining Hematopoietically Active Pelvic Bone Marrow in Gynecologic Patients." *Radiotherapy and Oncology* 118: 72–78.

Yang, F.E. et al. 1995. "Analysis of Weekly Complete Blood Counts in Patients Receiving Standard Fractionated Partial Body Radiation Therapy." *International Journal of Radiation Oncology Biology and Physics* 33: 607–17.

Yang, X. et al. 2012. "Marrow Stromal Cell Infusion Rescues Hematopoiesis in Lethally Irradiated Mice despite Rapid Clearance after Infusion." *Advances in Hematology* 8: 142530.

Yue, J. et al. 2010. "Measuring Tumor Cell Proliferation with 18 F-FLT PET During Radiotherapy of Esophageal Squamous Cell Carcinoma: A Pilot Clinical Study." *The Journal of Nuclear Medicine* 51: 528–34.

Zachariah, B. et al. 2001. "Effect of Fractionated Regional External Beam Radiotherapy on Peripheral Blood Cell Count." *Radiation Oncology* 50: 465–72.

Zhu, H. et al. 2015. "Longitudinal Study of Acute Haematologic Toxicity in Cervical Cancer Patients Treated with Chemoradiotherapy." *Journal of Medical Imaging and Radiation Oncology* 59: 389–93.

CHAPTER 14

Predicting Toxicity in External Radiotherapy

A Critical Summary

Tiziana Rancati and Claudio Fiorino

CONTENTS

Methodological Notes 338
Suggestions for SBRT 338
A Note of Caution 362
References 362

THE CENTRAL PART OF this book focused on a district/organ-based revision of the dose-volume relationships in external radiotherapy, mostly dealing with the main advancements since the publication of the QUANTEC Supplement in 2010. The improvement of our knowledge during the last decade was dramatic for several organs while it was limited for others. In few situations, most of the constraints suggested in QUANTEC were reinforced by more recent investigations, confirming their validity; however, in other cases, the suggestions have been overcome by new data that made them outdated.

A major merit of the QUANTEC is that it had the ability to summarize a large amount of knowledge in few, easy-to-use, tables ready to be practically applied in planning optimisation (Marks et al. 2010).

The availability of summary tables and of their periodic update is of paramount importance and planners are very reactive in looking for robust, quick and reliable tables of constraints to continuously improve the quality of their plans and keep them updated. Very importantly, this situation will not be changed by the rapid advent of more automatic optimisation systems using machine learning or artificial intelligence tools: Such systems need to be driven by constraints that the planners have to put in their optimisation boxes.

This is the main reason why we wrote the current chapter that was intentionally aimed to give a practical and updated instrument for external radiotherapy optimisation to radiation oncologists and planners.

METHODOLOGICAL NOTES

In short, we decided to merge the existing QUANTEC constraints (only in the case they are not outdated) with the suggestions coming from the new data and studies considered in Chapters 4 through 13. The suggestions collected in these chapters were translated into numbers, in part after an additional critical revision and referring, if necessary, to the original papers discussed in the same chapters and are summarized in Tables 14.1 through 14.9.

Conventional and moderate hypofractionation (<3 to 3.5 Gy/fraction) constraints were separated from severe hypofractionation/Stereotactic Body Radiation Therapy (SBRT) and suggested constraints were reported into different tables. In this second case, the milestone publications by Timmermann (2008) and Benedict et al. (2010) were considered as the starting point and consolidated new indications (reviewed in the book) were added, as explained later.

Regarding conventional and moderate hypofractionation, the tables grouped different organs by anatomical district (brain and central nervous system head and neck, thorax, abdomen, pelvis, and others including skin and hematological toxicity) and, where useful, different endpoints were considered for each organ. In general, the expected risk thresholds with respect to specific toxicity scales were reported for each constraint. In addition, we also reported if the constraint was derived from studies involving patients treated with three Dimensional Conformal Radiation Therapy (3DCRT) or Intensity Modulated Radiation Therapy (IMRT) or both. Major clinical predictors were added without adding their odds/hazard ratios in order to keep the tables reasonably light. Readers are invited to refer to the original chapters in this book for more detailed information.

Several organs were not considered in the chapters of this book (for instance, kidneys). The most relevant ones were included in the tables thanks to an additional effort, following the same criteria used for the other organs: Additional references referred to these "extra" organs were added at the bottom of the chapter; in a few cases, the suggestions reported in the chapters were integrated by additional references and indications (Weber et al. 2005; Sun et al. 2013; Hsiao et al. 2010; Gondi et al. 2013; Sanguineti et al. 2007, 2011; Palma et al. 2013; Darby et al. 2013; Cutter et al. 2015; Feng et al. 2012; Holyoake et al. 2017; Diavolitisis et al. 2011; Salerno et al. 2009; Landoni et al. 2016; Son et al. 2015; Lee et al. 2017; Sini et al. 2016).

SUGGESTIONS FOR SBRT

As mentioned, tables for SBRT constraints were provided (Tables 14.7 through 14.9). The constraints reported in the Task Group 101 (TG101) were first considered and updated/ integrated with constraints reported in the various chapters, as reviewed in this book. Then, few other consolidated suggestions as reported in recent reviews, guidelines and single papers (NRG Oncology 2011; Kimsey et al. 2016; Lawrence et al. 2010; Hanna et al. 2018) were integrated into a comprehensive picture, reporting constraints for one, three and five fractions SBRT.

TABLE 14.1 Central Nervous System

Organ	Endpoint	Rate	Volume Delineation	Constraint on Dose-Volume Parameter/EUD	RT Tech	Fraction	Source	Clinical Factors	Notes
Brainstem	Permanent cranial neuropathy or necrosis	<5%	Whole organ	$V_{60Gy} < 0.03$ cc	3DCRT	Conventional & moderate hypo	QUANTEC		
	Permanent cranial neuropathy or necrosis	<5%	Whole organ	Maximum dose <54 Gy	3DCRT	Conventional & moderate hypo	QUANTEC	• Skull-base surgeries • Diabetes	$\alpha/\beta = 3$ Gy
	Permanent cranial neuropathy or necrosis	<5%	Whole organ	$V_{64Gy} < 0.03$ cc peripheral volume $V_{54Gy} < 0.03$ cc central volume	3DCRT	Conventional & moderate hypo	Chapter 7, derived from single study (Weber et al. 2005)		
Brain (temporal lobe)	Temporal lobe necrosis	<5%	Whole organ	$D_{0.5cm3} < 70$ Gy	3DCRT & IMRT	Conventional & moderate hypo	Chapter 7, QUANTEC & two further studies		
	Temporal lobe necrosis	<5%	Whole organ	$V_{40Gy} < 10\%$ OR <5cc	IMRT	Conventional	Chapter 7, derived from a single study (Sun et al. 2013)		
	Temporal lobe necrosis	<20%	Whole organ	$V_{40Gy} < 15\%$ OR <10cc	IMRT	Conventional	Chapter 7, derived from a single study (Sun et al. 2013)	• Larger fraction doses • Shorter overall treatment time	
	Decrease in cognitive functioning	<50%	Whole organ	Mean dose < 36 Gy	IMRT	Conventional	Chapter 7, derived from a single study (Hsiao et al. 2010)		
	Decrease in cognitive functioning	<50%	Whole organ	$V_{50Gy} < 10\%$	IMRT	Conventional & moderate hypo	Chapter 7, derived from a single study (Hsiao et al. 2010)		
Brain (Hippocampus)	Neurocognitive function	<20%	Whole organ, bilateral hippocampi	Dose to 40% (EQD2Gy, $\alpha/\beta = 2$Gy) <7Gy	IMRT	Conventional & moderate hypo	Chapter 7, derived from a single study (Gondi et al. 2013)		
Spinal Cord	Myelopathy	<1%	Whole organ	Maximum dose <54Gy	3DCRT & IMRT	Conventional & moderate hypo	Chapter 7, QUANTEC and studies with proton RT		$\alpha/\beta = 0.9$ Gy
	Myelopathy	<10%	Whole organ	Maximum dose <60Gy	3DCRT & IMRT	Conventional & moderate hypo	Chapter 7, QUANTEC and studies with proton RT		

(Continued)

TABLE 14.1 (CONTINUED) Central Nervous System

Organ	Endpoint	Rate	Volume Delineation	Constraint on Dose-Volume Parameter/EUD	RT Tech	Fraction	Source	Clinical Factors	Notes
Cochlea	Sensory neural hearing loss	<15–20%	Whole organ	Mean dose < 45 Gy	3DCRT & IMRT	Conventional & moderate hypo	Chapter 7, QUANTEC and multiple studies	• Dose of cisplatin-based chemotherapy • Age • Pre-RT hearing level	
Optic nerves/ Chiasm	Grade ≥2 optic neuropathy	<5–10%	Whole organ	Maximum dose < 55 Gy	3DCRT & IMRT	Conventional & moderate hypo	Chapter 7, QUANTEC and multiple studies		
	Grade 4 optic neuropathy	<1%	Whole organ	Maximum dose < 55 Gy	3DCRT & IMRT	Conventional & moderate hypo	Chapter 7, QUANTEC and multiple studies		

EUD, Equivalent Uniform Dose; RT, radiotherapy; Tech, technique; Fraction, fractionation; 3DCRT, Three-Dimensional Conformal Radiation Therapy; IMRT, Intensity Modulated Radiation Therapy; V_{XGy}, organ volume receiving at least X Gy; D_{Xcm3}, minimum dose to X cm³; hypo, hypofractionation; QUANTEC, Quantitative Analyses of Normal Tissue Effects in the Clinic.

TABLE 14.2 Head and Neck

Organ	Endpoint	Rate	Volume delineation	Constraint on dose-volume parameter/EUD	RT Tech	Fraction	Source	Clinical Factors	Notes
Parotids	Xerostomia: long-term reduction to <25% of baseline function	<20%	Whole parotids	Mean dose < 25 Gy	3DCRT & IMRT	Conventional	QUANTEC		Suggested to reduce as much as possible the dose to the OAR volume not overlapped with PTVs
	Xerostomia: long-term reduction to <25% of baseline function	<20%	Contro-lateral parotid	Mean dose < 20 Gy	3DCRT & IMRT	Conventional	QUANTEC		
Larynx	Swallowing: Late PEG/TF dependence, RTOG grade ≥ 2	<20–40%	Whole organ/supra-glottic larynx	Mean dose < 50 Gy	3DCRT & IMRT	Conventional & moderate hypo	Chapter 9, merged from several studies	with CHT	
	Swallowing: EORTC H&N_35, MBS, Aspiration	<10–30%	Whole organ/supra-glottic larynx	Mean dose < 40 Gy	3DCRT & IMRT	Conventional & moderate hypo	Chapter 9, merged from several studies		
	Late vocal dysfunction	n.a.	True vocal cord	Mean dose < 20 Gy	3DCRT & IMRT	Conventional & moderate hypo	Chapter 9, merged from several studies		Rates depending on score and dose to other structures involved in swallowing
	Late vocal dysfunction	<20%	Whole organ	Maximum dose < 66Gy	3DCRT	Conventional	QUANTEC	With CHT	Suggested to reduce as much as possible the dose to the OAR volume not overlapped with PTVs
	Late dysphonia, FACT-HN4 voice changes	<25%	Whole organ	Mean dose < 50 Gy	IMRT	Conventional	Chapter 9, two studies	With CHT	
	Late dysphonia, FACT-HN4 voice changes	<25%	Whole organ	$V_{50Gy} < 33\%$	IMRT	Conventional	Chapter 9, two studies	With CHT	
	Late dysphonia, FACT-HN4 voice changes	<10%	Whole organ	Mean dose < 20 Gy	IMRT	Conventional	Chapter 9, two studies	With CHT	
	Edema, subacute/late Grade ≥ 2	<20%	Whole organ	Mean dose < 44 Gy	3DCRT	Conventional	QUANTEC	Without CHT	

(Continued)

TABLE 14.2 (CONTINUED) Head and Neck

Organ	Endpoint	Rate	Volume delineation	Constraint on dose-volume parameter/EUD	RT Tech	Fraction	Source	Clinical Factors	Notes
	Edema, subacute/late Grade ≥ 2	<20%	Whole organ	$V_{50Gy} < 27\%$	3DCRT	Conventional	Single study (Sanguineti et al. 2007)		
	Edema, Subacute/late Grade ≥ 2	<5%	Whole organ	Mean dose < 30 Gy	3DCRT	Conventional	Single study (Sanguineti et al. 2007)		Suggested to reduce as much as possible the dose to the OAR volume not overlapped with PTVs
Oral mucosa	Acute swallowing: need of PEG during RT	<30%	Oral cavity	$V_{9.5Gy/week} < 60$ cc	IMRT	Conventional	Chapter 9, derived from single study (Sanguineti et al. 2011)	Without CHT	
	Late swallowing problems	n.a.	Anterior oral cavity	Mean dose < 30 Gy	IMRT	Conventional & moderate hypo	Chapter 9		
Pharynx	Swallowing; Late PEG/TF dependence, RTOG grade ≥ 2	<20–40%	Sup (middle, inf)	Mean dose < 60 Gy	IMRT	Conventional & moderate hypo	Chapter 9, merged from several studies		Rates depending on score and dose to other structures involved in Swallowing
	Swallowing; EORTC H&N_35, MBS, Aspiration	<10–30%	Constrictors	Mean dose < 50 Gy	IMRT	Conventional & moderate hypo	Chapter 9, merged from several studies		
	Symptomatic dysphagia and aspiration	<20%	Constrictors	Mean dose < 50 Gy	3DCRT & IMRT	Conventional	QUANTEC		Suggested to reduce as much as possible the dose to OAR volume not overlapped with PTVs

(Continued)

TABLE 14.2 (CONTINUED)   Head and Neck

Organ	Endpoint	Rate	Volume delineation	Constraint on dose-volume parameter/EUD	RT Tech	Fraction	Source	Clinical Factors	Notes
Cervical Esophagous	Late PEG/TF dependence, RTOG grade ≥ 2	n.a.	Upper esophagous	Mean dose < 50 Gy	3DCRT & IMRT	Conventional & moderate hypo	Chapter 9, merged from several studies	With CHT	Rates depending on score and dose to other structures involved in Swallowing
	Aspiration: EORTC H&N_35, MBS	n.a.	Esophagous inlet	Mean dose < 40 Gy	3DCRT & IMRT	Conventional & moderate hypo	Chapter 9, merged from several studies		

EUD, Equivalent Uniform Dose; RT, radiotherapy; Tech, technique; Fraction, fractionation; 3DCRT, Three-Dimensional Conformal Radiation Therapy; IMRT, Intensity Modulated Radiation Therapy; V_{XGy}, organ volume receiving at least X Gy; $V_{XGy/week}$, organ volume receiving at least X Gy in a week; hypo, hypofractionation; CHT, chemotherapy; RTOG, Radiation Therapy Oncology Group; EORTC H&N_35, European Organization for Research and Treatment of Cancer Quality of Life Questionnaire Head-and-Neck module 35 questions; MBS, modified barium swallow grade; PEG, percutaneous endoscopic gastrostomy; TF, tube feeding; OAR, Organ at Risk; PTV, planning target volume; FACT-HN4, Functional Assessment of Chronic Illness Therapy-Head-and-Neck question 4; n.a., not available; QUANTEC, Quantitative Analyses of Normal Tissue Effects in the Clinic

TABLE 14.3 Thorax

Organ	Endpoint	Rate	Volume Delineation	Constraint on Dose-Volume Parameter/EUD	RT Tech	Fraction	Source	Clinical Factors	Notes
Lung	Grade ≥ 2 pneumonitis	<20%	Paired lungs-GTV	Mean dose <20 Gy	3DCRT	Conventional	Chapter 10, merged from several studies		For moderate hypo same dose-response relationship for mean dose can be used, provided that the value of the mean dose constraint is correctly scaled to the considered fractionation scheme. Details in Chapter 10
	Grade ≥ 2 pneumonitis	<10%	Paired lungs-GTV	Mean dose <13 Gy	3DCRT	Conventional	Chapter 10, merged from several studies		
	Grade ≥ 2 pneumonitis;	<20%	Paired lungs-GTV	$V_{20Gy} < 30\%$ $V_{13Gy} < 40\%$	3DCRT	Conventional	Chapter 10, merged from several studies	• Pulmonary comorbidities • Mid or inferior tumor location • Smoke	
	Grade ≥ 2 pneumonitis	<10%	Paired lungs-GTV	$V_{20Gy} < 20\%$ $V_{13Gy} < 30\%$	3DCRT	Conventional	Chapter 10, merged from several studies		
	Grade ≥ 3 pneumonitis	<20%	Paired lungs-GTV	Mean dose <25 Gy	3DCRT	Conventional	Chapter 10, merged from several studies		
	Grade ≥ 3 pneumonitis	<10%	Paired lungs-GTV	Mean dose <18 Gy	3DCRT	Conventional	Chapter 10, merged from several studies		
Esophagus	Grade ≥ 2 acute esophagitis	<40%	Whole organ	$V_{60Gy} < 17\%$	3DCRT & IMRT	Conventional & moderate hypo	Chapter 10, individual patient meta-analysis (Palma et al. 2013)		Comment from QUANTEC "the rapidity of dose accumulation might be more important than the final dose".
	Grade ≥ 2 acute esophagitis	<30%	Whole organ	$V_{60Gy} < 0.07\%$	3DCRT & IMRT	Conventional & moderate hypo	Chapter 10, individual patient meta-analysis (Palma et al. 2013)		Need for models accounting for the course of complication relative to the number of fractions delivered
	Grade ≥ 2 acute esophagitis	<30%	Whole organ	$V_{35Gy} < 50\%$ $V_{50Gy} < 40\%$ $V_{70Gy} < 20\%$	3DCRT	Conventional & moderate hypo	QUANTEC		
	Grade ≥ 3 acute esophagitis	<10%	Whole organ	$V_{60Gy} < 17\%$	3DCRT & IMRT	Conventional & moderate hypo	Chapter 10, individual patient meta-analysis (Palma et al. 2013)		

(Continued)

TABLE 14.3 (CONTINUED) Thorax

Organ	Endpoint	Rate	Volume Delineation	Constraint on Dose-Volume Parameter/EUD	RT Tech	Fraction	Source	Clinical Factors	Notes
	Grade ≥ 3 acute esophagitis	<5%	Whole organ	$V_{60Gy} < 0.07\%$	3DCRT & IMRT	Conventional & moderate hypo	Chapter 10, individual patient meta-analysis (Palma et al. 2013)		
	Grade ≥ 3 acute esophagitis	<5–20%	Whole organ	Mean dose <34Gy	3DCRT	Conventional & moderate hypo	QUANTEC		
Heart	Death from ischaemic heart disease (pts with no cardiac rf factors) at 30 yrs after RT	<1%	Whole organ	Mean dose <8Gy	2D-RT	Conventional	Chapter 11, derived from a single study (Darby et al. 2013)		
	Death from ischaemic heart disease (pts with no cardiac rf) at 30 yrs after RT	<1.5%	Whole organ	Mean dose <10Gy	2D-RT	Conventional	Chapter 11, derived from a single study (Darby et al. 2013)	• Age at RT • Presence of other factors for cardiac events • Smoking habits	Possible interaction with lung doses
	Death from ischaemic heart disease (pts with at least one cardiac rf) at 30 yrs after RT	<1%	Whole organ	Mean dose <4Gy	2D-RT	Conventional	Chapter 11, derived from a single study (Darby et al. 2013)		
	Death from ischaemic heart disease (pts with at least one cardiac rf) at 30 yrs after RT	<2%	Whole organ	Mean dose <8Gy	2D-RT	Conventional	Chapter 11, derived from a single study (Darby et al. 2013)		
	At least one acute coronary event (pts with no cardiac rf) at 30 yrs after RT	<2%	Whole organ	Mean dose <8Gy	2D-RT	Conventional	Chapter 11, derived from a single study (Darby et al. 2013)		

(Continued)

TABLE 14.3 (CONTINUED) Thorax

Organ	Endpoint	Rate	Volume Delineation	Constraint on Dose-Volume Parameter/EUD	RT Tech	Fraction	Source	Clinical Factors	Notes
	At least one acute coronary event (pts with no cardiac rf) at 30 yrs after RT	<3%	Whole organ	Mean dose <10Gy	2D-RT	Conventional	Chapter 11, derived from a single study (Darby et al. 2013)		
	At least one acute coronary event (pts with at least one cardiac rf) at 30 yrs after RT	<2%	Whole organ	Mean dose <4Gy	2D-RT	Conventional	Chapter 11, derived from a single study (Darby et al. 2013)		
	At least one acute coronary event (pts with at least one cardiac rf) at 30 yrs after RT	<4%	Whole organ	Mean dose <8Gy	2D-RT	Conventional	Chapter 11, derived from a single study (Darby et al. 2013)		
	At least one acute coronary event (pts with at least one cardiac rf) at 30 yrs after RT	<6%	Whole organ	Mean dose <10Gy	2D-RT	Conventional	Chapter 11, derived from a single study (Darby et al. 2013)		
	Long-term cardiac mortality	<1%	Whole organ	$V_{25Gy} < 10\%$	3DCRT	Conventional	QUANTEC		
	Valvular disease at 30 yrs after RT	<3%	Heart valves	Mean dose <30 Gy	3DCRT	Conventional	Chapter 11, derived from single study (Cutter et al. 2015)		
	Pericarditis	15%	Pericardium	Mean dose <26 Gy	3DCRT	Conventional	QUANTEC		
	Pericarditis	15%	Pericardium	$V_{30Gy} < 46\%$	3DCRT	Conventional	QUANTEC		

EUD, Equivalent Uniform Dose; RT, radiotherapy; Tech, technique; Fraction, fractionation; 3DCRT, Three-Dimensional Conformal Radiation Therapy; IMRT, Intensity Modulated Radiation Therapy; 2D-RT, two-dimensional radiotherapy; V_{XGy}, organ volume receiving at least X Gy; hypo, hypofractionation; yrs, years; pts, patients; GTV, Gross Tumor Volume; rf, risk factor; QUANTEC, Quantitative Analyses of Normal Tissue Effects in the Clinic

TABLE 14.4 Abdomen

Organ	Endpoint	Rate	Volume Delineation	Constraint on Dose-Volume Parameter/EUD	RT Tech	Fraction	Source	Clinical Factors	Notes
Stomach	Ulceration	<7%	Whole organ	$D_{100\%} < 45$ Gy	–	Conventional	QUANTEC	With CHT	
	Grade $\geq$ 2 CTCAE v3.0	n.a.	Whole organ	$V_{20Gy} < 31\%$	IMRT	44.25 Gy, 2.95Gy/fr @ PTV	Chapter 6	With CHT	
	Grade $\geq$ 2 CTCAE v4.0	<10%	Whole organ	$V_{50Gy} < 16$ cc	3DCRT	Conventional	Chapter 6	With CHT	
	Bleeding	<20%	Whole organ	EUD <46Gy (EQD2Gy, α/β=2.5 Gy, n=0.09)	3DCRT	Conventional, hyper & moderate hypo	Chapter 6 merged with Feng et al. (2012)	Normal baseline with/without CHT	
	Bleeding	<10%	Whole organ	EUD <41 Gy (EQD2Gy, α/β=2.5 Gy, n=0.09)	3DCRT	Conventional, hyper & moderate hypo	Chapter 6 merged with Feng et al. (2012)	Normal baseline with/without CHT	
	Bleeding	<50%	Whole organ	EUD <22 Gy (EQD2Gy, α/β=2.5 Gy, n=0.09)	3DCRT	Conventional, hyper & moderate hypo	Chapter 6 merged with Feng et al. (2012)	Cirrhotic with/without CHT	
Duodenum	Grade $\geq$ 2 CTCAE v.4.0	<10%	Duodenum	$V_{55Gy} < 1$ cc	3DCRT & IMRT	Conventional	Chapter 6	With CHT	
	Grade $\geq$3 CTCAE v.4.0	<5%	Gastro-duodenum	$V_{50Gy} < 30$ cc	3DCRT	Conventional	Chapter 6	With CHT	
	Grade $\geq$2 CTCAE v.3.0	<10%	Duodenum	$V_{25Gy} < 45\%$	3DCRT & IMRT	36 Gy, 2.4Gy/fr @PTV	Chapter 6	With CHT	
	Grade $\geq$ 3 CTCAE v.3.0	<5%	Duodenum	$V_{35Gy} < 20\%$	3DCRT & IMRT	36 Gy, 2.4Gy/fr @PTV	Chapter 6	With CHT	

(Continued)

TABLE 14.4 (CONTINUED) Abdomen

Organ	Endpoint	Rate	Volume Delineation	Constraint on Dose-Volume Parameter/EUD	RT Tech	Fraction	Source	Clinical Factors	Notes
	Grade ≥ 2 CTCAE v.3.0	<5%	Duodenum	V_{40Gy} <16%	IMRT	44.25 Gy, 2.95Gy/fr @ PTV	Chapter 6	With CHT	
	Grade ≥ 2 CTCAE v.4.0	<10%	Duodenum	V_{35Gy} <5%	3DCRT	35.5 Gy, 3.5 Gy/fr@PTV	Chapter 6		
	Grade ≥ 3 CTCAE v.4.0	<5%	Gastro-duodenum	V_{35Gy} <5%	3DCRT	36 Gy, 3Gy/fr @ PTV	Chapter 6	Cirrhotic	
	Grade ≥ 3 duodenal	<4%	Duodenum	EUD (n=0.07)<40 Gy	3DCRT & IMRT	Conventional & moderate hypo	Chapter 6, single study (Holyoake et al. 2017)	With CHT	
	Grade ≥ 3 duodenal	<7%	Duodenum	EUD (n=0.07)<60 Gy	3DCRT & IMRT	Conventional & moderate hypo	Chapter 6, single study (Holyoake et al. 2017)	With CHT	
Liver	RILD	<5%	Whole organ-GTV	Mean dose < 32 Gy	3DCRT & IMRT	Conventional	QUANTEC merged with Chapter 6	Good baseline functionality	
	RILD	<5%	Whole organ-GTV	Mean dose < 28 Gy	3DCRT & IMRT	Conventional	QUANTEC merged with Chapter 6	Poor baseline functionality	
	RILD	<50%	Whole organ-GTV	Mean dose < 42 Gy	3DCRT & IMRT	Conventional	QUANTEC merged with Chapter 6	Good baseline functionality	
	RILD	<50%	Whole organ-GTV	Mean dose < 36 Gy	3DCRT & IMRT	Conventional	QUANTEC merged with Chapter 6	Poor baseline functionality	

(Continued)

TABLE 14.4 (CONTINUED) Abdomen

Organ	Endpoint	Rate	Volume Delineation	Constraint on Dose-Volume Parameter/EUD	RT Tech	Fraction	Source	Clinical Factors	Notes
Kidney	Clinically relevant renal dysfunction	<5%	Single kidney	Mean dose <15 Gy	3DCRT	Conventional	QUANTEC		
	Clinically relevant renal dysfunction	<50%	Single kidney	Mean dose <28 Gy	3DCRT	Conventional	QUANTEC		
	Clinically relevant renal dysfunction	<5%	Combined kidneys	V_{12Gy} <55% V_{20Gy} <32% V_{23Gy} <30% V_{28Gy} <20%	3DCRT	Conventional	QUANTEC		
	Reduction of CrCl >10%	<5%	Single kidney (omolateral)	V_{5Gy} <50% V_{10Gy} <30% V_{20Gy} <10%	3DCRT		Single study (Diavolitisis et al. 2011)		
	Reduction of CrCl >10%	<5%	Single kidney (omolateral)	Mean dose <10 Gy	3DCRT		Single study (Diavolitisis et al. 2011)		
	Late Grade 1 RTOG/EORTC	<5%	Combined kidneys	Mean dose <10 Gy	3DCRT		Single study (Salerno May et al. 2009)	With CHT	

EUD, Equivalent Uniform Dose; RT, radiotherapy; Tech, technique; Fraction, fractionation; 3DCRT, Three-Dimensional Conformal Radiation Therapy; IMRT, Intensity Modulated Radiation Therapy; $D_{Y\%}$, threshold dose to Y% of the organ; V_{XGy}, organ volume receiving at least X Gy; hyper, hyperfractionation; hypo, hypofractionation; RILD, radiation-induced liver disease; CrCl, Creatinine Clearance; CTCAE, Common Terminology Criteria for Adverse Events; QUANTEC, Quantitative Analyses of Normal Tissue Effects in the Clinic; CHT, chemotherapy; PTV, Planning Target Volume; GTV, Gross Tumor Volume.

TABLE 14.5 Pelvis

Organ	Endpoint	Rate	Volume Delineation	Constraint on Dose-Volume Parameter/EUD	RT Tech	Fraction	Source	Clinical Factors	Notes
Rectum	Grade ≥ 2 late rectal bleeding	<10–15%	Anorectum whole organ	$V_{35Gy} < 80\%$ $V_{40Gy} < 60\%$ $V_{50Gy} < 50\%$ $V_{60Gy} < 40\%$ $V_{65Gy} < 30\%$ $V_{70Gy} < 10\%$ $V_{75Gy} < 5\%$	3DCRT & IMRT	Conventional	Chapter 4, merged from several studies	• Previous abdominal surgery • Cardiovascular comorbidities • Diabetes • Use of calcium channel blockers	Consequential effect with acute rectal toxicity both with conventionally fractionated and hypofractionated regimens
	Grade ≥ 2 late rectal bleeding	<10%	Anorectum whole organ	EUD (n = 0.1) < 65 Gy	3DCRT & IMRT	Conventional	Chapter 4, merged from several studies	• Use of anticoagulants • Age • Smoking	For moderate hypofractionation same dose-response relationship, convert doses and constraints using $\alpha/\beta = 5$ Gy
	Grade ≥ 2 late rectal bleeding	<5%	Anorectum whole organ	EUD (n = 0.1) < 60 Gy	3DCRT & IMRT	Conventional	Chapter 4, merged from several studies	• Presence of haemorrhoids • Whole pelvis/nodal irradiation • Reduce treatment time	
	Grade ≥ 2 fecal inco	<5%	Anorectum whole organ	Mean dose < 40 Gy	3DCRT & IMRT	Conventional & moderate hypo	Chapter 4, merged from several studies		
	Grade ≥ 2 fecal inco	<3%	Anorectum whole organ	Mean dose < 33 Gy	3DCRT & IMRT	Conventional & moderate hypo	Chapter 4, merged from several studies		
	Grade ≥ 2 fecal inco	<5%	Anal sphincter	Mean dose < 36 Gy	3DCRT & IMRT	Conventional & moderate hypo	Chapter 4, merged from several studies	• Previous abdominal surgery • Diabetes	
	Grade ≥ 2 fecal inco	<3%	Anal sphincter	Mean dose < 32 Gy	3DCRT & IMRT	Conventional & moderate hypo	Chapter 4, merged from several studies		

(Continued)

TABLE 14.5 (CONTINUED) Pelvis

Organ	Endpoint	Rate	Volume Delineation	Constraint on Dose-Volume Parameter/EUD	RT Tech	Fraction	Source	Clinical Factors	Notes
Bowel	Grade ≥ 2 bowel toxicity	<10–15%	Bowel loops	$V_{20Gy} < 350cc$ $V_{30Gy} < 200cc$ $V_{40Gy} < 150cc$ $V_{50Gy} < 100cc$ $V_{60Gy} < 70cc$ $V_{65Gy} < 20cc$	3DCRT & IMRT	Conventional & moderate hypo	Chapter 4, merged from several studies	• High dose cisplatin • Reduced treatment time • Age • Large tumor size • Smoking • Low BMI • Previous abdominal/pelvic surgery	Consequential effect with acute rectal toxicity For moderate hypofractionation same dose-response relationship provided the treatment time remains long, convert doses and constraints using $\alpha/\beta = 5$ Gy
Bladder	Grade ≥ 2 Late urinary toxicity OR Grade 3 Late urinary toxicity	<10–15% <5%	Whole organ	$V_{20Gy} < 74\%$ $V_{30Gy} < 60\%$ $V_{40Gy} < 50\%$ $V_{50Gy} < 35\%$ $V_{60Gy} < 20\%$ $V_{70Gy} < 10\%$ $V_{75Gy} < 1\%$	3DCRT & IMRT	Conventional	Chapter 5, merged from several studies	• Baseline urinary symptoms • Trans-urethral resection of the prostate • Smoking • Vascular comorbidities/use or cardiovascular drugs • Diabetes	for moderate hypofractionation same dose-response relationship, some studies indicate that doses and constraints should be converted using $\alpha/\beta = 1$ Gy for bladder (Landoni et al. 2016) (finding still controversial)
	Grade ≥ 2 Late urinary toxicity OR Grade 3 Late urinary toxicity	<10–15% <5%	Whole organ	$EUD (n=0.01) < 80Gy$	3DCRT & IMRT	Conventional	Chapter 5, merged from several studies		

(Continued)

TABLE 14.5 (CONTINUED) Pelvis

Organ	Endpoint	Rate	Volume Delineation	Constraint on Dose-Volume Parameter/EUD	RT Tech	Fraction	Source	Clinical Factors	Notes
Penile Bulb	Severe erectile dysfunction	<35%	Whole organ	$D_{90\%} < 50$ Gy	3DCRT	Conventional	QUANTEC	• Baseline functionality	
	IIEF 1–5 > 11	<10–20%	Whole organ	EQD2Gymax <70–74Gy	IMRT	Conventional & moderate hypo	Modified from Landoni et al. (2016)	• Age	
Vagina	Severe stenosis	<10%	Whole organ	Mean dose < 43 Gy	IMRT & 3DCRT	Conventional	Single study (Son et al. 2015)	• With CHT	From studies including rectal and anal cancer patients

EUD, Equivalent Uniform Dose; RT, radiotherapy; Tech, technique; Fraction, fractionation; 3DCRT, Three-Dimensional Conformal Radiation Therapy; IMRT, Intensity Modulated Radiation Therapy; $D_{Y\%}$, threshold dose to Y% of the organ; V_{XGy}, organ volume receiving at least X Gy; hypo, hypofractionation; QUANTEC, Quantitative Analyses of Normal Tissue Effects in the Clinic; CHT, chemotherapy; EQD2Gymax, maximum dose expressed as equivalent dose in 2Gy/fraction; IIEF 5, International Index of Erectile Function, questions 1 to 5; fecal inco, fecal incontinence

TABLE 14.6 Other

Organ	Endpoint	Rate	Volume Delineation	Constraint on Dose-Volume Parameter/EUD	RT Tech	Fraction	Source	Clinical Factors	Notes
Skin	Grade ≥ 2 acute dermatitits	<10%	Whole breast	Tolerance dose = 34Gy (α/β = 10Gy)	3DCRT & IMRT	Conventional & moderate hypo	Chapter 12		
	Grade ≥ 2 acute dermatitits	<20%	Whole breast	Tolerance dose = 40Gy (α/β = 10Gy)	3DCRT & IMRT	Conventional & moderate hypo	Chapter 12		
	Grade ≥ 2 acute dermatitits	<50%	Whole breast	Tolerance dose = 51Gy (α/β = 10Gy)	3DCRT & IMRT	Conventional & moderate hypo	Chapter 12		
	Acute moist desquamation	<10%	Whole breast	Tolerance dose = 38Gy (α/β = 10Gy)	3DCRT & IMRT	Conventional & moderate hypo	Chapter 12		Hypofractionation 2.5–6.7 Gy/fr Tolerance doses are expressed as prescribed doses to PTV in breast patients treated with high energy beams
	Acute moist desquamation	<20%	Whole breast	Tolerance dose = 44Gy (α/β = 10Gy)	3DCRT & IMRT	Conventional & moderate hypo	Chapter 12		
	Acute moist desquamation	<50%	Whole breast	Tolerance dose = 56Gy (α/β = 10Gy)	3DCRT & IMRT	Conventional & moderate hypo	Chapter 12		
	Telengectasia	<10%	Whole breast	Tolerance dose = 50Gy (α/β = 3Gy)	3DCRT & IMRT	Conventional & moderate hypo	Chapter 12		
	Telengectasia	<20%	Whole breast	Tolerance dose = 56Gy (α/β = 3Gy)	3DCRT & IMRT	Conventional & moderate hypo	Chapter 12		
	Telengectasia	<50%	Whole breast	Tolerance dose = 59Gy (α/β = 3Gy)	3DCRT & IMRT	Conventional & moderate hypo	Chapter 12		
Hematological Toxicity	Grade ≥ 2 acute neutropenia	<15%	Pelvic bones (definition according to Mell et al. 2006)	V_{10Gy} < 90%	3DCRT & IMRT		Chapter 13	With CHT	
	Grade ≥ 2 acute limphopenia	<10%							
	Grade ≥ 3 acute leukopenia	<10%	Lower pelvic bones (definition according to Mell et al. 2006)	V_{40Gy} < 23%	3DCRT & IMRT		Single study (Lee et al. 2017)	With CHT	Studies including anal cancer patients
	Grade ≥ 3 acute neutropenia	<10%							
	Grade ≥ 2 acute thrombocytopenia	<10%							

(Continued)

TABLE 14.6 (CONTINUED) Other

Organ	Endpoint	Rate	Volume Delineation	Constraint on Dose-Volume Parameter/EUD	RT Tech	Fraction	Source	Clinical Factors	Notes
	Grade ≥ 2 lymphopenia 1-year after RT	<10%	Iliac bones (definition according to Mell et al. 2006)	V_{40Gy} <95cc	IMRT		Single study (Sini et al. 2016)	Smoking, baseline lymphocyte count	Studies including prostate cancer patients without chemotherapy, with/without hormonal therapy

EUD, Equivalent Uniform Dose; RT, radiotherapy; Tech, technique; Fraction, fractionation; 3DCRT, Three-Dimensional Conformal Radiation Therapy; IMRT, Intensity Modulated Radiation Therapy; V_{XGy}, organ volume receiving at least X Gy; hypo, hypofractionation; CHT, chemotherapy; PTV, Planning Target Volume.

TABLE 14.7 Single Fraction

	Endpoint (Grade ≥ 3)[a]	Constraint	Dmax (Gy)[b]	Ref	Comments/Notes
Optic pathway	Neuritis	$V_{8Gy} < 0.2$ cm^3	10	Benedict et al. 2010	
Cochlea	Hearing loss		9	Benedict et al. 2010, Chapter 7	
Brain	Symptomatic necrosis	Brain – GTV $V_{12Gy} < 5$–10 cm^3		Marks et al. 2010	
	Cognitive deterioration	Brain – GTV $D_{10cm3} < 12$ Gy		Hanna et al. 2018	
		Brain – GTV $D_{50\%} < 5$ Gy		Hanna et al. 2018	
Lens	Cataract	$D_{0.1cm3} < 1.5$ Gy		Hanna et al. 2018	
Orbit	Retinopathy	$D_{0.1cm3} < 8$ Gy		Hanna et al. 2018	
Brainstem	Cranial neuropathy	$V_{10Gy} < 0.5$ cm^3	15	Benedict et al. 2010	
Spinal Cord	Myelitis	$V_{10Gy} < 0.35$ cm^3	14	Benedict et al. 2010	Sub-volume, 5–6mm above/below
		$V_{7Gy} < 1.2$ cm^3		Benedict et al. 2010	
		$V_{10Gy} < 10\%$		Benedict et al. 2010	
		$D_{1cm3} < 8$ Gy		Chapter 7	
Cauda equina	Neuritis	$V_{14Gy} < 5$ cm^3	16	Benedict et al. 2010	
Sacral plexus	Neuropathy	$V_{14.4Gy} < 5$ cm^3	16	Benedict et al. 2010	
Esophagous	Stenosis/fistula	$V_{11.9Gy} < 5$ cm^3	15.4	Benedict et al. 2010	
Brachial plexus	Neuropathy	$V_{14Gy} < 3$ cm^3	17.5	Benedict et al. 2010	
Heart/pericardium	Pericarditis	$V_{16Gy} < 15$ cm^3	22	Benedict et al. 2010	
Great vessels	Aneurysm	$V_{31Gy} < 10$ cm^3	37	Benedict et al. 2010	
Trachea & large bronchus	Stenosis/fistula	$V_{10.5Gy} < 4$ cm^3	20.2	Benedict et al. 2010	
Bronchus and smaller airways	Stenosis/fistula	$V_{12.4Gy} < 0.5$ cm^3	13.3	Benedict et al. 2010	
Skin	Ulceration	$V_{23Gy} < 10$ cm^3	26	Benedict et al. 2010	
Stomach	Ulceration/fistula	$V_{11.2Gy} < 10$ cm^3	12.4	Benedict et al. 2010	
Duodenum	Ulceration	$V_{11.2Gy} < 5$ cm^3	12.4	Benedict et al. 2010	
		$V_{9Gy} < 10$ cm^3			
	Stricture/Hemmorage	$V_{17Gy} < 1$ cm^3		Chapter 6	
		$V_{11.2Gy} < 5$ cm^3		Chapter 6	
Jejunum/ileum	Enteritis/obstruction	$V_{11.9Gy} < 5$ cm^3	15.4	Benedict et al. 2010	

(Continued)

TABLE 14.7 (CONTINUED) Single Fraction

	Endpoint (Grade ≥ 3)[a]	Constraint	Dmax (Gy)[b]	Ref	Comments/Notes
Colon	Colitis/fistula	$V_{14.3Gy} < 20$ cm^3	18.4	Benedict et al. 2010	
Rectum	Proctitis/fistula	$V_{14.3Gy} < 20$ cm^3	18.4	Benedict et al. 2010	
Bladder wall	Cystitis/fistula	$V_{11.4Gy} < 15$ cm^3	18.4	Benedict et al. 2010	
Penile bulb	Impotence	$V_{14Gy} < 3$ cm^3	34	Benedict et al. 2010	
Femoral heads (paired)	Necrosis	$V_{14Gy} < 10$ cm^3		Benedict et al. 2010	
Renal hilum/vascular trunk	Malignant hypertension	$V_{10.6Gy} < 66\%$	18.6	Benedict et al. 2010	
Renal cortex (paired)	Basic renal function	$D_{200cm3} < 8.4$ Gy		Benedict et al. 2010	
Lungs (paired)	Basic lung function	$D_{1500cm3} < 7$Gy		Benedict et al. 2010	
	Pneumonitis	$D_{1000cm3} < 7.4$Gy			
Chest wall/ribs	Pain or Fracture	$V_{22Gy} < 1$ cm^3	30	Benedict et al. 2010	
		$D_{2cm3} < 22.9$ Gy		Kimsey et al. 2016	Risk < 10%
		$D_{70cm3} < 9.3$ Gy		Kimsey et al. 2016	Risk < 10%
Liver	Basic liver function	$D_{700cm3} < 9.1$Gy		Benedict et al. 2010	

V_{XGy}, Threshold Volume (either %volume or absolute volume in cubic centimeters) that should receive less than X Gy; D_Y, threshold dose to indicated volume Y (either %volume or absolute volume in cubic centimeters).

[a] Excepting if differently specified.

[b] Dmax (maximum dose) expressed as the "point dose" referred to 0.035 cm^3 or less.

TABLE 14.8 Three Fractions

Organ	Endpoint (Grade $\geq$ 3)[a]	Constraint	Dmax (Gy)[b]	Ref	Comments/Notes
Optic pathway	Neuritis	$V_{15.3Gy} <0.2$ cm³	17.4	Benedict et al. 2010	
Cochlea	Hearing loss		17.1	Benedict et al. 2010	
Brainstem	Cranial neuropathy	$V_{18Gy} <0.5$ cm³	23.1	Benedict et al. 2010	
Spinal Cord	Myelitis	$V_{18Gy} <0.35$ cm³	21.9	Benedict et al. 2010	Sub-volume, 5–6mm above/below
		$V_{12.3Gy} <1.2$ cm³		Benedict et al. 2010	
		$V_{18Gy} <10\%$		Benedict et al. 2010	
		$D_{1cm3} <16$ Gy		Chapter 7	
Cauda equina	Neuritis	$V_{21.9Gy} <5$ cm³	24	Benedict et al. 2010	
Sacral plexus	Neuropathy	$V_{22.5Gy} <5$ cm³	24	Benedict et al. 2010	
Esophagous	Stenosis/fistula	$V_{17.7Gy} <5$ cm³	25.2	Benedict et al. 2010	
Brachial plexus	Neuropathy	$V_{20.4Gy} <3$ cm³	24	Benedict et al. 2010	
Heart/pericardium	Pericarditis	$V_{24Gy} <15$ cm³	30	Benedict et al. 2010	
Great vessels	Aneurysm	$V_{39Gy} <10$ cm³	45	Benedict et al. 2010	
Trachea & large bronchus	Stenosis/fistula	$V_{15Gy} <4$ cm³	30	Benedict et al. 2010	
Bronchus and smaller airways	Stenosis/fistula	$V_{18.9Gy} <0.5$ cm³	23.1	Benedict et al. 2010	
Skin	Ulceration	$V_{30Gy} <10$ cm³	33	Benedict et al. 2010	
Stomach	Ulceration/fistula	$V_{16.5Gy} <10$ cm³	22.2	Benedict et al. 2010	
Duodenum	Ulceration	$V_{16.5Gy} <5$ cm³	22.2	Benedict et al. 2010	
		$V_{11.4Gy} <10$ cm³		Benedict et al. 2010	
	Stricture/Hemmorage	$V_{25Gy} <1$ cm³		Chapter 6	
Jejunum/ileum	Enteritis/obstruction	$V_{17.7Gy} <5$ cm³	25.2	Benedict et al. 2010	
Colon	Colitis/fistula	$V_{24Gy} <20$ cm³	28.2	Benedict et al. 2010	
Rectum	Proctitis/fistula	$V_{24Gy} <20$ cm³	28.2	Benedict et al. 2010	
Bladder wall	Cystitis/fistula	$V_{16.8Gy} <15$ cm³	28.2	Benedict et al. 2010	
Penile bulb	Impotence	$V_{21.9Gy} <3$ cm³	42	Benedict et al. 2010	
Femoral heads (paired)	Necrosis	$V_{21.9Gy} <10$ cm³		Benedict et al. 2010	

(Continued)

TABLE 14.8 (CONTINUED)   Three Fractions

Organ	Endpoint (Grade ≥ 3)[a]	Constraint	Dmax (Gy)[b]	Ref	Comments/Notes
Renal hilum/vascular trunk	Malignant hypertension	$V_{18.6Gy} < 66\%$		Timmermann 2008	
Renal cortex (paired)	Basic renal function	$D_{200cm3} < 16$ Gy		Benedict et al. 2010	
Lungs (paired)	Basic lung function	$D_{1500cm3} < 11.6$ Gy		Benedict et al. 2010	
	Pneumonitis	$D_{1000cm3} < 12.4$ Gy			
Chest wall/ribs	Pain or Fracture	$V_{28.8Gy} < 1$ cm^3	36.9	Benedict et al. 2010	
		$D_{2cm3} < 37.8$ Gy		Kimsey et al. 2016	Risk < 10%
		$D_{70cm3} < 14.6$ Gy		Kimsey et al. 2016	Risk < 10%
Liver	Basic liver function	$D_{700cm3} < 19.2$ Gy		Benedict et al. 2010	Good baseline, 6 fractions
	RILD	Liver – GTV		Chapter 6	Poor baseline
		mean dose < 15 Gy		Chapter 6	
		$D_{800cm^3} < 18$ Gy			

V_{xGy}, Threshold Volume (either %volume or absolute volume in cubic centimeters) that should receive less than X Gy; D_Y, threshold dose to indicated volume Y (either %volume or absolute volume in cubic centimeters); RILD, radiation-induced liver disease.

[a] Excepting if differently specified.

[b] Dmax (maximum dose) expressed as the "point dose" referred to 0.035 cm^3 or less.

358 ■ Modelling Radiotherapy Side Effects

TABLE 14.9 Five Fractions

Organ	Endpoint (Grade ≥ 3)[a]	Constraint	Dmax (Gy)[b]	Ref	Comments/Notes
Optic pathway	Neuritis	$V_{23Gy} < 0.2$ cm^3	25	Benedict et al. 2010	
Cochlea	Hearing loss	$V_{23Gy} < 0.5$ cm^3	25	Benedict et al. 2010	
Brainstem	Cranial neuropathy		31	Benedict et al. 2010	
Spinal Cord	Myelitis	$V_{23Gy} < 0.35$ cm^3	30	Benedict et al. 2010	Sub-volume, 5–6mm above/below
		$V_{14.5Gy} < 1.2$ cm^3		Benedict et al. 2010	
		$V_{23Gy} < 10\%$		Benedict et al. 2010	
		$D_{1cm3} < 21$ Gy		Chapter 7	
Cauda equina	Neuritis	$V_{30Gy} < 5$ cm^3	32	Benedict et al. 2010	
Sacral plexus	Neuropathy	$V_{30Gy} < 5$ cm^3	32	Benedict et al. 2010	
Esophagous	Stenosis/fistula	$V_{19.5Gy} < 5$ cm^3	30.5	Benedict et al. 2010	
Brachial plexus	Neuropathy	$V_{27Gy} < 3$ cm^3	32	Benedict et al. 2010	
Heart/pericardium	Pericarditis	$V_{32Gy} < 15$ cm^3	38	Benedict et al. 2010	
Great vessels	Aneurysm	$V_{47Gy} < 10$ cm^3	53	Benedict et al. 2010	
Trachea & large bronchus	Stenosis/fistula	$V_{16.5Gy} < 4$ cm^3	40	Benedict et al. 2010	
Bronchus and smaller airways	Stenosis/fistula	$V_{21Gy} < 0.5$ cm^3	33	Benedict et al. 2010	
Skin	Ulceration	$V_{36.5Gy} < 10$ cm^3	39.5	Benedict et al. 2010	
Stomach	Ulceration/fistula	$V_{18Gy} < 10$ cm^3	32	Benedict et al. 2010	
Duodenum	Ulceration	$V_{18Gy} < 5$ cm^3	32	Benedict et al. 2010	
		$V_{12.5Gy} < 10$ cm^3		Benedict et al. 2010	
		$V_{28Gy} < 1$ cm^3		Chapter 6	
	Duodenal hystological damage	$V_{30Gy} < 5$ cm^3		Chapter 6	
		Mean dose < 20 Gy		Chapter 6	
Jejunum/ileum	Enteritis/obstruction	$V_{19.5Gy} < 5$ cm^3	35	Benedict et al. 2010	
Colon	Colitis/fistula	$V_{25Gy} < 20$ cm^3	38	Benedict et al. 2010	

(Continued)

TABLE 14.9 (CONTINUED)   Five Fractions

Organ	Endpoint (Grade $\geq$ 3)[a]	Constraint	Dmax (Gy)[b]	Ref	Comments/Notes
Rectum	Proctitis/fistula	$V_{25Gy} < 20$ cm^3	38	Benedict et al. 2010	
	CTCAE-grade 2	$V_{35Gy} < 1$ cm^3		Chapter 4	
		$V_{34.4Gy} < 3$ cm^3		NRG Oncology 2011	
		$V_{32.6Gy} < 10\%$		NRG Oncology 2011	
		$V_{29Gy} < 20\%$		NRG Oncology 2011	
		$V_{18Gy} < 50\%$		NRG Oncology 2011	
		$V_{24Gy} < 50\%$		Chapter 4	
Bladder wall	Cystitis/fistula	$V_{18.3Gy} < 15$ cm^3	38	Benedict et al. 2010	
	Urinary flare risk < 10%	$V_{37Gy} < 10$ cm^3		Chapter 5	Empty Bladder
		$V_{38Gy} < 1$ cm^3		Chapter 5	Empty Bladder
		$D_{90\%} < 32.6$ Gy		Chapter 5	Empty Bladder
		$D_{40\%} < 18$ Gy		Chapter 5	Empty Bladder
		$V_{33Gy} < 12\%$		Chapter 5	Empty Bladder
		EUD (n=0.13) < 30 Gy		Chapter 5	
Penile bulb	Impotence	$V_{30Gy} < 3$ cm^3	50	Benedict et al. 2010	
Femoral heads (paired)	Necrosis	$V_{30Gy} < 10$ cm^3		Benedict et al. 2010	
Renal hilum/vascular trunk	Malignant hypertension	$V_{23Gy} < 66\%$		Benedict et al. 2010	
Renal cortex (paired)	Basic renal function	$D_{200cm3} < 17.5$ Gy		Benedict et al. 2010	
Lungs (paired)	Basic lung function	$D_{1500cm^3} < 12.5$ Gy		Benedict et al. 2010	
	Pneumonitis	$D_{1000cm3} < 13.5$ Gy		Benedict et al. 2010	
Chest wall/ribs	Pain or Fracture	$V_{35Gy} < 1$ cm^3	43	Benedict et al. 2010	
		$D_{2cm3} < 50$ Gy		Kimsey et al. 2016	Risk < 10%
		$D_{70cm3} < 17.6$ Gy		Kimsey et al. 2016	Risk < 10%

(Continued)

TABLE 14.9 (CONTINUED) Five Fractions

Organ	Endpoint (Grade ≥ 3)[a]	Constraint	Dmax (Gy)[b]	Ref	Comments/Notes
Liver	Basic liver function			Benedict et al. 2010	
	RILD	$D_{700cm3} < 21$ Gy Liver – GTV		Chapter 6	Good baseline, 6 fractions
		mean dose < 20 Gy Liver – GTV		Chapter 6	Poor baseline (CPC-A)
		mean dose < 10 Gy $D_{33\%} < 10$ Gy Liver – GTV		Chapter 6	Poor baseline (CPC-B)
	Central liver (Hepatobiliary tox)	mean dose < 19 Gy Liver – GTV $V_{26Gy} < 37$ cm³ Liver – GTV $V_{21Gy} < 45$ cm³		Chapter 6 Chapter 6	

V_{xGy}, Threshold Volume (either %volume or absolute volume in cubic centimeters) that should receive less than X Gy; D_Y, threshold dose to indicated volume Y (either %volume or absolute volume in cubic centimeters); GTV, Gross Tumor Volume, CPC, Child-Pugh Class; tox, toxicity; EUD, Equivalent Uniform Dose; CTCAE, Common Terminology Criteria for Adverse Events; RILD, Radiation-Induced Liver Disease.

[a] Excepting if differently specified.

[b] Dmax (maximum dose) expressed as the "point dose" referred to 0.035 cm³ or less.

A NOTE OF CAUTION

The field of quantitative prediction of toxicity in external radiotherapy is continuously moving forward. In particular, as also depicted in this book, the availability of more comprehensive predictive models integrating individually assessed clinical and biological features with dose-volume information is still limited but is expected to grow in the next years. The tables reported here have to be considered as a "static" picture of the dynamic process and should not be used uncritically. This "criticism" should include the effort of the planners to keep updating their knowledge and of the radiation oncologists in carefully considering the individual situation where they apply current constraints, never forgetting that these numbers are not black and white threshold values and that the risk is a continuous function of dose-volume parameters, often modulated by non-dosimetry individual factors. On the other hand, the reported constraints showed to be effective in reducing toxicities, summarizing a long-term experience in a few numbers that may reasonably be used in the daily practice of treatment optimisation.

REFERENCES

Benedict SH et al. Stereotactic body radiation therapy: The report of AAPM Task Group 101. *Med Phys* 2010; 37:4078–101.

Cutter et al. Risk of valvular heart disease after treatment for Hodgkin lymphoma. *J Natl Cancer Inst* 2015; 107 (4).

Darby et al. Risk of ischemic heart disease in women after radiotherapy for breast cancer. *N Engl J Med* 2013; 368 :987–98.

Diavolitisis VM et al. Changes in creatinine clearance over time following upper abdominal irradiation: A dose-volume histogram multivariate analysis. *Am J Clin Oncol* 2011; 34:53–7.

Feng M et al. Dosimetric analysis of radiation-induced gastric bleeding. *Int J Radiat Oncol Biol Phys.* 2012; 84:e1–6.

Gondi et al. Hippocampal dosimetry predicts neurocognitive function impairment after fractionated stereotactic radiotherapy for benign or low-grade adult brain tumors. *Int J Radiat Oncol Biol Phys* 2013; 85 :345–54.

Hanna GG et al. UK consensus on normal tissue dose constraints for stereotactic radiotherapy. *Clin Oncol* 2018; 30:5–14

Holyoake DLP et al. Modeling duodenum radiotherapy toxicity using cohort dose-volume histogram data. *Radiother Oncol* 2017; 123:431–7.

Hsiao et al. Cognitive function before and after intensity-modulated radiation therapy in patients with nasopharyngeal carcinoma: A prospective study. *Int J Radiat Oncol Biol Phys* 2010; 77: 722–6.

Kimsey F et al. Dose response model for chest wall tolerance of stereotactic body radiation therapy. *Sem Rad Oncol* 2016; 26:129–38.

Landoni V et al. Predicting toxicity in radiotherapy for prostate cancer. *Phys Medica* 2016; 32:521–32.

Lawrence YR et al. Radiation dose-volume effects in the brain. *Int J Radiat Oncol Biol* 2010; 76:S20–S27

Lee AY et al. Hematologic nadirs during chemoradiation for anal cancer: Temporal characterization and dosimetric predictors. *Int J Radiat Oncol Biol Phys* 2017; 97:306–12.

Marks LB et al. Use of normal tissue complications models in the clinic. *Int J Radiat Oncol Biol* 2010; 76:S10–19.

Mell LK et al. Dosimetric predictors of acute hematologic toxicity in cervical cancer patients treated with concurrent cisplatin and intensity-modulated radiotherapy. *Int J Radiat Oncol Biol* 2006; 66:1356–65

NRG Oncology, RTOG-0938. A randomized phase II trial of hypofractionated radiotherapy for favorable risk prostate cancer. 2011. https://www.nrgoncology.org/Clinical-Trials/Protocol-Table

Palma et al. Predicting esophagitis after chemoradiation therapy for non-small cell lung cancer: An individual patient data meta-analysis. *Int J Radiat Oncol Biol Phys.* 2013; 87:690–6.

K Salerno May et al. Analysis of clinical and dosimetric factors associated with change in renal function in patients with gastrointestinal malignancies after chemoraiotherapy to the abdomen. *Int J Radiat Oncol Biol Phys* 2009; 76:1193–8.

Sanguineti G et al. Dosimetric predictors of laryngeal edema. *Int J Radiat Oncol Biol Phys.* 2007; 68:741–9.

Sanguineti G et al. Weekly dose-volume parameters of mucosa and constrictor muscles predict the use of percutaneous endoscopic gastrostomy during exclusive intensity-modulated radiotherapy for oropharyngeal cancer. *Int J Radiat Oncol Biol Phys* 2011; 79:52–9.

Sini C, et al. Dose-volume effects for pelvic bone marrow in predicting hematological toxicity in prostate cancer radiotherapy with pelvic node irradiation. *Radiother Oncol.* 2016; 118: 79–84.

Son CH et al. Dosimetric predictors of radiation-induced vaginal stenosis after pelvic radiotherapy for rectal and anal cancer. *Int J Radiat Oncol Biol Phys* 2015; 92:548–4.

Sun et al. Radiation-induced temporal lobe injury after intensity modulated radiotherapy in nasopharyngeal carcinoma patients: A dose-volume-outcome analysis. *BMC Cancer* 2013; 13:397.

Timmermann RD. An overview of hypofractionation and introduction to this issue of Seminars in Radiation Oncology. *Sem Rad Oncol* 2008; 18:215–22.

Weber et al. Results of spot-scanning proton radiation therapy for chordoma and chondrosarcoma of the skull base: The Paul Scherrer Institut Experience. *Int J Radiat Oncol Biol Phys* 2005; 63:401–9.

CHAPTER 15

Data Sharing and Toxicity Modelling

A Vision of the Near Future

Zhenwei Shi, Rianne Fijten, Zhen Zhou,
Andre Dekker, and Leonard Wee

CONTENTS

Introduction 366
General Data Requirements 366
  Available Data 366
    Structured Data Versus Unstructured Data.. 367
    Re-Use of Clinical Trials Data... 370
    Utilization of Real-World Clinical Data 370
  Generic Data Schema.. 372
    Patient Population Characteristics 372
    Intervention Details.. 373
    Outcomes Data.. 375
  Acute Versus Late Toxicity... 375
  Acute Versus Late Toxicity... 376
  Clinician-Graded Toxicities 337
  Patient-Reported Outcomes.. 337
  Timing of Toxicity Measurements... 337
  Linking Data from Multiple Sources 378
Data Sharing.. 379
  Advantages of Data Sharing... 379
  Barriers to Data Sharing 380
  Data-Sharing Architectures.. 382
    Centralized Data Sharing... 382
    Decentralized Data Sharing... 382
    Hybrid Architectures.. 382
  Distributed Machine Learning.. 383

Semantic Web Technologies..384
  Computer-Assisted Theragnostics (CAT) 384
Toxicity Modelling 385
  Current State of the Art 387
    Toxicity Modelling Process...387
    Handling Missing Values 387
    Model Fitting...389
    Bias-Variance Trade-Off 389
    Evaluating Prediction Performance..390
    Clinical Impact..390
  Toxicity Modelling in the Near Future 391
    Incorporating Diverse Data Types...391
    Exploiting Differences for Personalizing Treatment 392
    Use of Machine Learning and Artificial Intelligence 393
    Towards a Data-Sharing "Culture" in Oncology 394
Conclusion 394
References..394

INTRODUCTION

Optimal cancer treatment requires maximising the chance of tumour eradication while simultaneously minimising the risk of adverse treatment-induced side effects. The success of modern oncology implies that more and more patients (generally) live longer with the adverse outcomes of their treatment. Therefore, the ability to predict the likely trajectory of treatment-related toxicity is indispensable in value-based intervention against cancer. This can be achieved with predictive modelling of specific toxic events and their presumed severity. Estimation of probable toxicity touches multiple aspects of oncology decision-making, from modality selection down to individually adapted treatment plan optimisation.

In this chapter, a self-contained and compact overview of *data-driven toxicity modelling and prediction* is offered. The chapter opens with an examination of the data landscape and general data requirements to develop quantitative toxicity models. Next, data architectures that support data aggregation and data sharing for toxicity modelling are reviewed, along with the advantages and disadvantages of each respective architecture. Lastly, we consider the current condition of toxicity models and prediction model development, followed by an exposition of some of the current developments in toxicity modelling that may soon be realized in routine clinical practice.

GENERAL DATA REQUIREMENTS

Available Data

One of the major actions in the development of multi-variate prediction models of treatment-related toxicity is locating relevant data that exists (i) in a highly structured format, (ii) as a fully digital archive and (iii) with a sufficiently high probability of data completeness. One needs to understand the *data landscape* – where do we locate the data needed to answer the prediction question – before considering how to consolidate disparate elements together into a dataset on which a prediction model can developed, i.e., the data corpus.

Structured Data Versus Unstructured Data

Structured data refers to information that is organized into *key-value* pairs, such that the keys (or labels of *variables*) are uniformly used throughout the data corpus, and the values consist of a known range of results with a uniformly applied format through the corpus.

Typical keys include data fields such as a patient record number, their date of birth, clinical tumour staging and one or more outcomes of interest. Keys are generally defined in such a manner as to enable efficient processing by software – no spaces, no mixtures of upper and lower cases and only alphanumeric characters – but should nonetheless bear close resemblance to its human-readable label. Formats of values must be consistent and uniformly applied – for example, date formats are "*dd-mm-yyyy*" throughout the data corpus. Other permissible formats could be alphanumeric strings (e.g., "MRN012345"), Boolean labels (e.g., "y/n", "0/1" or "true/false"), categorical labels (e.g., "Grade 1", "Grade 2", "Grade 3", etc.) or floating-point numbers (e.g., "50.4"). A very common form of structured data is the table format, such that the keys are arranged as columns and the values are arranged in rows (example in Table 15.1). This is by no means the only manner of arranging the key value pairs; a different but equally valid format of structured data is shown in Figure 15.1.

Unstructured data consists of information that has not been organised into key-value pairs, or where the keys do not exist or are not uniformly applied, or where the values are neither of consistent range nor of uniform format. The two most common forms of unstructured data are free natural-language text (e.g., digitized speech, consultation notes, radiology reports, pathology findings) and images (e.g., medical imaging scans, facsimiled laboratory reports, photos of hand-written notes). Examples of unstructured data are shown in Figure 15.2. In comparison to structured data, it is very much evident that unstructured data requires sophisticated data pre-processing, either by human experts or by machine algorithms, in order to be appropriate for building mathematical toxicity models.

More recently, there has been increased effort to record clinical findings in a semi struc tured format by outlining consistent and well-defined sections within a textual report. However, the information within each of the sections remains free natural text, out of which semantically concrete values must be derived by pre-processing. Conversely, some efforts have been made to enforce labelled statements to be included in text reports such as "TUMOUR VOLUME: 3.5 cc" or "ESOPHAGITIS: GRADE 2". While this makes extraction of quantitative value from free text easier, such conventions are generally not feasible to uniformly enforce in the clinical setting, nor would there be any guarantee that any one clinic's labelling convention would be acceptable or used the same way by a different clinic.

TABLE 15.1   Example Presentation of Keys (Column Names) and Corresponding Values (Row Entries) as a Rectangular Table

Id	Name	Position	Location	Start_Date	Salary
1	Airi Satou	Accountant	Tokyo	2008/11/28	162700
2	Angelica Ramos	Chief Executive Officer	London	2009/10/09	1200000
3	Ashton Cox	Junior Technical Author	San Francisco	2009/01/12	86000

Source: Adapted from https://datatables.net/examples/ajax/custom_data_flat.html.

```
[
  {
  "id": 1,
  "name": "Airi Satou",
  "position": "Accountant",
  "salary": 162700,
  "start_date": "2008/11/28",
  "location": "Tokyo"
  },
  {
  "id": 2,
  "name": "Angelica Ramos",
  "position": "Chief Executive Officer",
  "salary": 1200000,
  "start_date": "2009/10/09",
  "location": "London"
  },
  {
  "id": 3,
  "name": "Ashton Cox",
  "position": "Junior Technical Author",
  "salary": 86000,
  "start_date": "2009/01/12",
  "location": "San Francisco"
  }
]
```

FIGURE 15.1   Example Ajax data, containing exactly the same information as in Table 15.1, as a flattened list of key-value pairs. (adapted from https://datatables.net/examples/ajax/custom_data_flat.html).

Efforts towards structured data collection are more advanced in certain domains, such as *structured reporting standards* for radiology notes, e.g., BI-RADS (Spak et al. 2017) and PI-RADS (Weinreb et al. 2016). However, such structured reporting standards are not yet universally adopted, and even in places where they are used, continuous quality assurance efforts are required to ensure compliance to the requirements.

Therefore, a major challenge before engaging in toxicity modelling is to impose a well-defined structure onto unstructured data so it can be used in quantitative analyses. Recent surveys of the types of data in medical care centres suggest that approximately 80% of healthcare data by volume is unstructured (see Figure 15.3). Advances in natural language

FIGURE 15.2   Example fragment of a pathology report from the National Cancer Institute training materials (https://training.seer.cancer.gov/abstracting/procedures/pathological/histologic/operative/example/ex1.html).

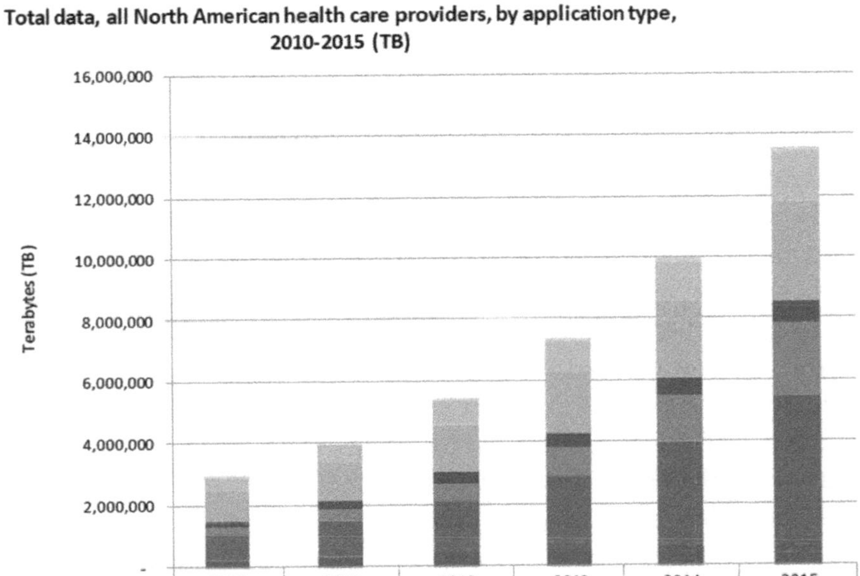

Total data, all North American health care providers, by application type, 2010-2015 (TB)

	2010	2011	2012	2013	2014	2015
Research Data	45,701	57,429	73,471	91,330	112,749	139,393
Non-Clinical Imaging	440,070	576,422	752,877	973,819	1,252,292	1,606,082
General Unstructured Data/File Services	950,980	1,219,132	1,562,994	1,998,615	2,551,631	3,262,842
E-Mail	242,711	300,632	372,166	457,259	559,598	684,522
Electronic Health Records	219,719	356,398	574,443	922,804	1,500,001	2,380,643
Clinical Imaging	824,493	1,179,802	1,695,289	2,397,323	3,371,689	4,730,891
Administrative Applications	272,337	340,466	423,563	523,403	643,292	789,160

FIGURE 15.3   Estimated volume of medical data in all North American hospitals from 2010–2015, showing proportions of different data types. The majority of the data is unstructured, and of these, medical imaging contributes the largest share. Reproduced with permission from Enterprise Strategy Group Research Publication "North American Health Care Provider Information Market Size & Forecast," January 2011.

processing (NLP) (Kreimeyer et al. 2017; Pons et al. 2016; Yim et al. 2016) may help to transform free text into structured data, but comprehensive clinical validation of such approaches are still lacking. The difficulty of obtaining clinician-expert labels on medical text, in order to provide machine algorithms with suitable data on which to "learn," remains the critical bottleneck in NLP development for medical text analysis. Better progress has been made in the area of deriving quantitative metrics from medical images (Kumar et al. 2012; Lambin et al. 2012; Larue et al. 2017; Yip and Aerts 2016), but once again the availability of expert clinician-annotated image sets is the limiting factor. An area that has seen significant development in recent years are digitization of medical records. Specifically, optical-character-recognition (OCR) tools are available to convert images of written or typed notes into digital text. Increasingly, speech-to-text may be used to transcribe dictations and spoken conversations directly into digital text. However the extant problems associated with extracting meaningful values from natural text, as described above, persist.

Re-Use of Clinical Trials Data

In light of difficulties in obtaining structured data for use in quantitative toxicity modelling, many researchers turn to secondary analysis, i.e., "re-use," of previously collected clinical trials data. In this respect, clinical trials generally excel, since vast amounts of trial financial resources would have been devoted to centralized data storage, rigorous data quality checks and systematic record keeping. Whenever this kind of data can be obtained, it is highly structured and has a high degree of data completeness.

However, as can be seen in Figure 15.3, structured research data (much of which can be attributed to the conduct of clinical trials) comprises only a small fraction of the entire healthcare data stack. In a rapidly evolving discipline such as radiotherapy, a randomized trial may be unethical (Sullivan et al. 2011) or unsuited to demonstrating clinical effectiveness (van Loon et al. 2012). Furthermore, clinical trials have been criticised for being unrepresentative of the wider patient population having the target disease (Geifman and Butte 2016; Jin et al. 2017) and for being excessively focussed on tumour control endpoints rather than treatment-induced toxicity (Bentzen and Trotti 2007; Secord et al. 2015). Clinician under-reporting of toxicity in trials has been well-documented (Gilbert et al. 2015).Sample size in most trials have been determined using clinical endpoints *other than toxicity*, therefore it is highly unlikely that reliable toxicity models might be obtained from secondary analysis of clinical trial data. A possible solution would be to perform meta-analysis for toxicity using individual patient-level data derived from multiple institutions and/or multiple trials. However, such efforts are generally impeded by concerns about patient privacy and legalities of material (data) transfer between institutions.

Utilization of Real-World Clinical Data

Real-world data refers to information about patients, medical interventions and clinical findings that have been derived from routine procedures in the standard-of-care care setting. The sheer *volume* and *variety* of real-world data available makes it attractive for developing multi-dimensional prediction models that can correctly stratify patients towards optimal interventions, according to their individual characteristics (Lambin et al. 2013) i.e., personalized medicine. Real-world data also accumulates at a significantly higher *velocity* than clinical trial data, thereby opening up the prospect of rapidly learning health-care systems (Abernethy et al. 2010).

However, the two major barriers in the way of real-world data utilization are (i) incompleteness of data collection in the routine clinical setting and (ii) high degree of fragmentation within the individual patient data landscape.

The first issue, data incompleteness or "missing data," refers to potentially useful explanatory variables that are either not measured during routine clinical practice or are not uniformly reported/recorded within the normal care setting. In the oncology setting (i.e., the empty circles in Figure 15.4), we suggest that data completeness rates could be routinely around 95% within clinical trial datasets and around 80% for specialized cancer registries, but one may reasonably expect upwards of 80% of real-world clinical data to be missing. An omnipresent risk of bias in real-world modelling studies is that the missing values may (or may not) be "missing purely at random," yet it would be difficult to prove that there is no systematic pattern in the data loss.

The second issue, data fragmentation, refers to the well-known fact that data elements of oncology patients are generally widely distributed across multiple hospital record systems (e.g., electronic patient journal, radiotherapy information system, treatment planning system and radiology image archive) and across multiple disciplinary divisions (e.g., surgery, medical oncology, radiology and radiotherapy). The problem of *horizontal partitioning* exists when one database contains all the variables, but only for a subset of patients. A clinical trial dataset is an idealized example of *horizontal* partitioning (see Figure 15.4a), where one records 100% of the variables of interest, but only for 3% of the target population. On the other hand, the problem of *vertical partitioning* exists when a database contains information about all of the patients, but only for a subset of variables of interest. The idealized example of *vertical* partitioning is a regional cancer registry, such that 100% of the patients

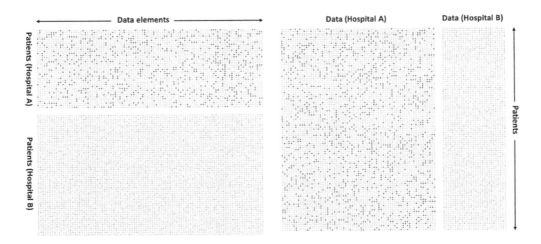

FIGURE 15.4 Illustration of two types of patient data partitioning; on the left (a) horizontal partitioning and, on the right (b) vertical partitioning. The unfilled circles illustrate the approximate prevalence of "missing values".

are included in the register, but only for 3% of the variables of interest. In practice, one would find that real-world data is subject to both horizontal and vertical partitioning; each of the departmental databases – surgery, oncology and radiology – only records some of the variables for some of the patients, but with incomplete overlap (see Figure 15.4b).

In general, a researcher needs to be able to: (i) use data derived from different settings (e.g., clinical trial, cancer registry and clinical routine), and (ii) connect variables and outcomes from multiple sources of data, that is also likely to involve data integration across multiple database architectures and non-uniform ways of encoding the data values.

Generic Data Schema

Having obtained a clear understanding of the possible complexities in the data landscape, we turn our attention to compiling a *data schema* comprising of all of the relevant clinical outcomes of interest, along with a wide range of potential explanatory variables that could be used to predict those outcomes. A comprehensive schema should not only identify the names of the data fields, but also the likely location of potential data, the possible range of values in the data and the existing (or desired) format of the values.

For this purpose, "standard sets" of fields, where published by international multi-disciplinary efforts will be particularly useful. The prime example of such interdisciplinary sets are the recommended outcomes tables produced by the International Consortium for Health Outcomes Measurement (ICHOM) (Kelley 2015). These standard sets are particularly strong in the domain of relevant outcomes per cancer type, and include those case-mix details, clinical baselines and diagnostic work-up elements that would be necessary for multivariate modelling studies.

In respect to intervention details, however, the ICHOM consensus sets are not comprehensive enough for modelling requirements in oncology. Cancer type-specific and intervention modality-specific expertise are required to understand the interplay between the clinical outcomes and the treatment variables. This process must be informed by an understanding of the data landscape, as some of the variables of interest are not feasible to access within a given data landscape.

Patient Population Characteristics

Modelling activities must always include certain general patient characteristics such as demographics, baseline clinical features, diagnostic findings (including any biopsy sampling performed as part of the diagnostic pathway) and any comorbidities.

Personal information is of importance in retrospective and/or secondary data analyses involving multiple cohorts, where basic personal characteristics modulate the outcome(s) of interest. This especially applies when incorporating quality of life or other patient-reported symptom where, for example, a person's marital status and living arrangements could affect the perception of outcome. Levels of educational attainment and ethnicity are known to correlate with quality of follow up, ability to respond to outcomes questionnaires (if used) and likely socio-economic disadvantages that may affect long-term general health outcomes (von dem Knesebeck et al. 2015). In some data fields, such as those pertaining to matters of cultural, political or social sensitivity, it is often preferable to allow "undis-closed" as a valid value, rather than to leave missing values in the data.

Baseline clinical features and comorbidities help to clarify the particular clinical case mix that the modelling study addresses and, where strong statistical heterogeneity may be encountered, multivariate correction for differences in case-mix may be attempted. Comorbidities should be recorded as potentially multiple-value categories using a universal co-morbidity instrument, such as the Charlson (Charlson et al. 1987) or the patient self-reported variant of Charlson (PRO-CCI) (Habbous et al. 2013), but without collapsing all the information into a single score.

Table 15.2 illustrates some of the general population variables that should be reasonably easily obtained for toxicity modelling studies, since these typically reside in the patient electronic medical record as a matter of routine registration for treatment.

The above fields must be supplemented by other baseline and diagnostic details deemed clinically relevant for the target disease to be addressed by modelling. For instance, menopausal status is likely to be highly relevant if studying breast cancer, and Gleason Score and prostate-specific antigen levels are important for prostate cancer models. Additional lifestyle information, such as the estimated number of pack-years of nicotine use, may be relevant for lung cancer. For toxicity modelling, baseline toxicity (i.e., complaints before treatment) is often crucial information and significantly predictive of post-therapy toxicity. Such information is likely to be located within the systematic medical departments e.g., urology, gynaecology or thoracic medicine, etc.

Since oncology becomes ever more reliant on multi-modality medical imaging, essential diagnostic information will also reside within the radiology department. Gene sequence analyses (i.e., of tumour specimen and of circulating tumour DNA) and proteomics analyses are also increasingly prominent in the oncology diagnosis pathway. If biopsies or tissue specimens were taken, it may be relevant to obtain the results of pathology findings such as the degree of tumour cell differentiation and surgical margin status. Wherever available and appropriate, diagnostic information from multiple imaging and pathologic modalities should be included.

Intervention Details

In many prediction studies, the clinical question of interest relates to how specific elements of the cancer treatment are likely to modulate the outcome(s). For instance, escalating a radiotherapy dose to the tumour may seem like an attractive option for a case with poor initial prognosis, but would this impact on the likelihood of severe adverse effects on this individual patient? Conversely, an oncologist may propose to skip chemotherapy altogether for a patient with poor overall health, but is the reduced risk for toxicity sufficiently appealing to the patient to outweigh their increased chance of earlier death?

The relevant treatment parameters needed to address these kinds of prediction questions are naturally highly specific to the target disease and must be significantly more detailed in the particular intervention that is supposed to be tailored towards an individual patient's outcome. However, over time, we may reasonably expect that a wider palette of surgical, pharmaceutical and radiotherapeutic options will become available to treating

TABLE 15.2   Representative Subset of Patient Population Characteristics that May Be Required for Toxicity Modelling, Including Possible Data Formats and Examples of Valid Values

Variable	Explanation	Data Type	Example Valid Values
Personal and Demographic Information			
Unique identifier	Denotes patients without using their name	Alphanumeric	MRN012345
Last name, first name	Patient's preferred name	Free text	Doe, John
Biological sex	Biological sex at birth, as opposed to gender which is a social/cultural construct	Categorical, single value	0 = female 1 = male 99 = undisclosed
Date of birth	Date of birth as defined on official documents such as passport or birth certificate	Date format	15-01-1970
Ethnicity	Generally defined in the context of the country where the study was performed	Categorical, single value	0 = Asian 1 = African 2 = Caucasian … 98 = Other 99 = undisclosed
Educational level	Highest educational level attained at time of enrolment in database	Categorical, single value	0 = No schooling 1 = Primary 2 = Secondary 3 = Tertiary 99 = undisclosed
Marital status	Relationship status at time of enrolment in database	Categorical, single value	0 = Never married or partnered 1 = Married or partnered 2 = Divorced or separated 3 = Widowed 99 = undisclosed
Living arrangement	Household living arrangements at time of enrolment in database	Categorical, single value	0 = Sole occupant 1 = Co-occupant, adult(s) 2 = Co-occupant - adult(s) and minor(s) 3 = Co-occupant, minor(s) 99 = undisclosed
Clinical Baseline and Diagnosis Information			
Comorbidities	Existing medical conditions, either clinically reported or patient self-reported, using an internationally recognized and validated schema such as the 19-item Charlson.	Categorical, one or more values	0 = no other conditions 1 = myocardial infarction … 19 = AIDS 99 = unknown
Height (in unit)	Height measured in clinic during enrolment, with pre-defined units e.g., cm or ft	Numeric	165

(Continued)

TABLE 15.2 (CONTINUED)   Representative Subset of Patient Population Characteristics that May Be
Required for Toxicity Modelling, Including Possible Data Formats and Examples of Valid Values

Variable	Explanation	Data Type	Example Valid Values
Weight (in unit)	Weight measured in clinic during enrolment, with pre-defined units e.g., kg or lbs	Numeric	78
Body mass index	Calculated from recorded height and weight	Numeric	28.7
Performance status	The ECOG/WHO performance status score	Categorical, single value	0 = WHO PS 0 … 4 = WHO PS 4 99 = unknown
Date of diagnosis	Date of diagnosis of the target condition	Date format	15-10-2015
Diagnosis code	International Classification of Disease (ICD)	Alphanumeric	C50.112
Diagnosis code version	ICD codebook version number	Numeric	10
Clinical T stage	Clinical tumour stage at time of diagnosis	Alphanumeric	IIa (or X if unknown)
Clinical N stage	Clinical nodal stage at time of diagnosis	Alphanumeric	0 (or X if unknown)
Clinical M stage	Clinical metastatic stage time of diagnosis	Alphanumeric	0 (or X if unknown)

physicians and their patients; this will lead to a high demand for the development of predictive models that handle variations in surgery, chemotherapy and radiotherapy treatment parameters. General data schema elements pertaining to each treatment modality are suggested in Table 15.3, along with caveats relating to data structure that may affect the extraction of such information.

Outcomes Data

The most crucial aspect defining the potential success of a predictive toxicity model is the availability and quality of outcomes data. In practice, the accessibility and reliability of outcomes data is the limiting factor on sample size. In comparison to treatment parameters and patient characteristics, the quantity of real-world outcomes data is often the single aspect in data-driven modelling studies that is most commonly overestimated by practicing clinicians. Simultaneously, the quality of real-world outcomes data is likely to be the least understood by practising clinicians. The quality aspect arises from a widespread assumption that the presence of a note or dictation in the patient record suffices for quantitative analysis, whereas such notes are often impractical to use as modelling data due to ambiguity and poorly structured reporting. It may be sometimes advantageous to consider pairing baseline observations and treatment parameters from real-world datasets to secondary analysis of clinical trials outcomes data, rather than to rely solely on systematic collection of outcomes data in routine clinical practice.

TABLE 15.3   An Example Subset of Interventional Data that May Be Required for Toxicity Modelling

Surgical notes (probably free text information archived in the local patient journal system in the treating hospital)	Pre-operative surgery
	Main cancer surgery date
	Main cancer surgical method
	Exploration during main surgery (and any relevant findings)
	Complications during any surgery
	Complications within 30 days of any surgery
	Wound complications (if any)
	   • Wound complication type
	   • Wound complication date

Chemotherapy notes (possibly structured fields in medical oncology information systems or otherwise free text; text mining of drug codes and billing codes may be feasible)	Chemotherapy agent(s) Dosage of chemotherapy agent(s) Number of chemotherapy cycles Timing of chemotherapy relative to other interventions Interval between chemotherapy cycles First cycle start date Last cycle end date

Radiotherapy (RT) details (most likely to be found as structured fields in RT information, treatment planning and verification systems; requires significant IT support when interacting with multiple computerized systems)	Prescribed dose
	Prescribed fraction size
	Number of fractions per day
	Elapsed time between fractions
	Treatment modality (x-rays, brachytherapy, protons, etc.)
	Timing of radiotherapy to surgery (pre-operative, post-operative, intra-operative, etc.)
	Timing of radiotherapy to chemotherapy (induction, concurrent, etc.)
	Start date of radiotherapy
	End date of radiotherapy
	   • Details of immobilization (e.g., stereotactic frame, abdominal compression, etc.)
	   • Details of localization imaging (e.g., cone-beam CT, port films, etc.)
	   • Organs at risk delineations archived in universal data interchange format such as DICOM-RT
	   • Organs at risk doses archived in universal data interchange format such as DICOM-RT
	   • Planning/simulation images (e.g., PET and CT) archived in universal data format such as DICOM

Acute Versus Late Toxicity

Short-term follow-up for treatment outcomes and induced toxicity are generally available for modelling studies. Surgical teams keep detailed notes of surgical complications, up to approximately 30 days following surgery. Likewise, one may expect observations during chemotherapy and radiotherapy to be reasonably systematic during the course of treatment, up to approximately 90 days after the end of treatment. Audits of follow-up in the routine oncology setting show that completeness of clinical observation data falls sharply after the initial period.

This implies that toxicity models are more likely to address short-term adverse events (acute toxicity) following oncological treatment, rather than long-term (late) toxicity. This naturally creates the potential for a conflict of interest between clinicians and patients in decision-making, since patients tend to care more about long-term toxicities that are perceived as debilitating or life-changing (Kunneman et al. 2015). It has been recognized that increasing numbers of patients are living with the long-term side effects of cancer treatment (Rowland et al. 2006), requiring oncology physicians to pay more attention to the dual impact of toxicity on quality *and* length of life after care.

Clinician-Graded Toxicities

The preferred method for clinician-assessed treatment-related toxicity is a structured data field, whereby the key can be unambiguously related to a specific toxicity in an internationally recognised toxicity schema such as the Common Terminology Criteria for Adverse Events, CTCAE (https://ctep.cancer.gov/protocolDevelopment/electronic_applications/ctc.htm), and the toxicity criteria of the Radiation Therapy Oncology Group (Cox et al. 1995). The outcome should be objectively reported as a numerical grade according to one of the abovementioned toxicity criteria. Free text descriptions of either the type of toxicity or its severity should be avoided, because this will require NLP pre-processing and/or human-expert re-interpretation before the data can be used for modelling – both of these are potentially time-consuming and/or error-prone.

Patient-Reported Outcomes

In light of systematic and logistic problems collecting long-term follow-up data in the routine clinical setting, some researchers have proposed either prospective (Basch et al. 2012; Riesen et al. 2017) or retrospective collection of patient-reported outcomes to supplement clinician assessments, though the latter is known to be particularly subject to recall bias (Lee et al. 2009). Some proponents argue that patient-reported outcomes (PROs) are potentially more useful for toxicity models informing patient-sensitive choices, since the side effects of treatment are couched in terms of the adverse experiences of the patient (Dueck et al. 2015; Kluetz 2016). However, opponents of PROs counter that a multitude of non-clinical factors may modulate the perception of toxicity, and therefore might not be accurate as an academic research instrument (Atkinson et al. 2016). However, clinically-validated PRO collection instruments now exist in multiple languages: such as the EORTC general quality of life (Fayers and Bottomley 2002) and its allied cancer-specific questionnaires, as well as the PRO variant of the CTCAE (PRO-CTCAE) (Basch et al. 2014).

Timing of Toxicity Measurements

Prediction models may address the probability of events within a given time interval (e.g., radiation pneumonitis within 6 months of completing chemo-radiotherapy for lung cancer), or most probable time-to-event. Repeated observations (or measurements) of toxicity over a long period of times may be instructive for the former class of model, since treatment-induced effects in normal tissue are known to grow and ebb away over time, but are mandatory for the latter class of model. For the reasons of data accessibility and quality stated previously, the required data to develop the latter class of model is significantly more challenging to obtain. Hence, the majority of toxicity prediction studies have focussed on the risk of events within a finite time interval.

Linking Data from Multiple Sources

The problem of populating a schema with variables and outcomes for individual patient data is one of *extraction, transformation and loading* to link the data elements from multiple different sources. This requires a detailed understanding of the data landscape, since it is highly unlikely (unless re-using previously curated datasets) that all of the required information exists in a structured form within a single data corpus.

The complexity of the linking challenge depends on the size of the dataset and the multiplicity of raw data sources. Smaller scale projects of several hundred cases, provided the data is already highly structured as tables, can be managed in office-desk applications such as STATA, SPSS and Excel. Less structured data may require sophisticated data reshaping functions available in R, Matlab and Mathematica. In R, the *dplyr* library includes functions for efficiently handling search, sub-setting and other manipulation of data shapes (including melting, casting and aggregation). The programming language python also supports an advanced data manipulation library (*pandas*). Data integration tools (such as Pentaho) permits linking of multiple data sources to compile a single database (PostGreSQL) without the need for database programming knowledge. However, large datasets (i.e., greater than the memory capacity of a single device) may be handled in a distributed computational framework such as Hadoop or Spark.

Extracting real-world data from live operational systems (such as electronic patient journals, hospital billing systems, etc.) requires transformation from the raw source to structured data. Some systems may have pre-defined APIs that allow users to dump structured data in plain text formats (e.g., character-separated tables). Other systems may publish their database architecture, therefore (assuming some programming skill) special queries (e.g., in SQL or Crystal Reports) may be written to extract data values directly from such systems. Patient demographics, procedural codes, diagnosis codes and billing numbers may be accessible in this manner.

However, other relevant information (such as radiology reports, pathology findings and surgical notes) will frequently be encountered as only free text. In some cases, a form of semi-structured reporting may have been used (see Section "*Structured Data Versus Unstructured Data*"). In any situation where values are embedded within free text, some form of natural language processing (Jovanović and Bagheri 2017) and semantic extraction approaches (Meystre et al. 2008) might be used to process these into structured data.

The combination of these approaches leads to the feasibility of data lakes, where large amounts of archived data are labelled with descriptive metadata (i.e., semantics) such that structured data can be efficiently retrieved using a semantic query language. A problem that quickly arises is semantically-labelled data in one such "lake" may not be inter-operable with other "lakes" unless there has been some uniformity in how the same concept shall be described. One has a choice of enforcing the same data structure over everyone or undertake a major challenge to re-map the metadata to a common scheme; but neither of these approaches are scalable in a practical manner.

The indefinitely scalable approach defined by Tim Berners-Lee in 2001 is the Semantic Web, based around formal, publicly open and infinitely extensible semantic ontologies.

Semantic Web defines the protocol by which any entity and any arbitrary relationship between entities can be assigned a persistently unique web address that is resolvable by widely-used web communication norms such as Hypertext Transfer Protocol (http). The indivisible data element is hence a triple, i.e., a statement in the format *subject-predicate-object*. Data that is stored as semantic triples can be easily rendered globally *Findable, Accessible, Interoperable and Re-useable* (FAIR) (Wilkinson et al. 2016).

DATA SHARING

Advantages of Data Sharing

Until recently, the majority of evidence-based healthcare practice has focussed on population-based treatment evaluation and randomized clinical trials.

The problem with this is three-fold. First, population-based findings only consider the mean effect size, and not the fact that some patients within the cohort perform much better than other patients within the same cohort; that is, there is always a range of probable effect sizes when looking at the outcome of any given patient. Personalized medicine asks – why would this be so, what are the other covariates that a population-based study glosses over that might help a doctor identify which patient will respond better (or which one will respond poorly)?

Secondly, addressing modern oncology treatment options by controlled trials is inefficient (van Loon et al. 2012), particularly in two instances:

- For technological innovations that either require large up-front capital investment or where the technological process evolves rapidly with respect to the time scale of clinical trials (as is usually the case in surgery and radiation oncology).

- When multiple factors are sought to stratify the variation in treatment outcome (thus leading to many study arms or multi-factorial randomization, both of which significantly increase the numbers needed to recruit).

Furthermore, as previously stated, most clinical trials have been designed to address clinical disease control or survival, and are hence underpowered to answer questions about toxicity.

A third problem that becomes evident in any systematic review of toxicity prediction models is that the input variables are divergent depending on the dataset used to develop the model. For example, there are multiple models of radiation-induced oesophagitis (Belderbos et al. 2005; Bradley et al. 2004; De Ruyck et al. 2011; Palma et al. 2013), but some models required volume-based dose metrics of the oesophagus, while others require the surface area of irradiated oesophagus as the input.

Into this arena, the combination of real-world big data and data-sharing approaches can be used to address the above concerns. In the Netherlands, several initiatives exist that encourage data sharing among Dutch radiotherapy clinics. For instance, the Translational Research IT (TraIT, http://www.ctmm-trait.nl) programme allows for data sharing for translational research projects within one clinic or among different (Dutch) clinics. Dutch national infrastructure initiatives that will exploit toxicity models based on this shared data include *PRODECIS* (Cheng et al. 2016) and *proTRAIT*.

PRODECIS is a prediction platform that evaluates the benefit a patient might experience from proton therapy compared to conventional x-ray therapy. This benefit is calculated on three levels: (i) the dosimetric level, where it is evaluated whether a radiotherapy plan meets a pre-defined dosimetric threshold for the organs at risk; (ii) the toxicity level, where any differences in toxicity of normal tissue are expected; and (iii) the cost-effectiveness level, where it is evaluated whether the benefits for the patient outweigh the extra costs for proton therapy. *ProTRAIT* (http://www.rug.nl/news/2017/08/1_5-miljoen-euro-voor-onderzoek-protonentherapie?lang=en) is a new infrastructure project, funded by the Dutch Cancer Society, to enable data exchange for proton therapy between institutes serving both clinical and research purposes.

Data-sharing approaches combining multiple datasets from different clinics have the advantage of reducing the risk of cohort-dependent biases when developing a prediction model, resulting in a more globally-applicable consensus model across multiple centres. Data sharing enables either a larger combined data corpus for training and cross-validation, or the option for a model developed on one corpus to be independently validated against the other corpus.

An additional advantage of data sharing is the ability to link data from multiple sources and/or overcome the data partitioning problems discussed in the earlier section of this chapter. Data from cancer registries, population databases, clinical trials and electronic medical records could be integrated, particularly by using a semantic data lakes approach with a universal domain ontology. A much richer (i.e., more potential explanatory variables) and possibly larger sample size may be feasible with data sharing versus an institutionalized approach. This is particularly advantageous for rare cancers (e.g., anal cancers and sarcomas) where it is exceedingly unlikely that a single centre receives enough cases to develop a broadly externally valid model using solely its own data.

One example of this approach is the study described by Dekker et al. (2014) who built a model for overall survival in Stage IV lung cancer patients at the MAASTRO clinic in The Netherlands, then validated this model in a similar set of patients treated at two centres in Australia. The model successfully differentiated between a good prognosis group and a medium/poor prognosis group, suggesting that inter-operability between different clinics is possible. Several initiatives apply this approach to combine data from different clinical centres for outcome and toxicity modelling: euroCAT (Deist et al. 2017), ozCAT (Lustberg et al. 2016) and meerCAT (Jochems et al. 2017). This approach will likely also be successful in the case of toxicity modelling, especially in rare cancer types. Data sharing further allows researchers to exploit heterogeneity in the data in order to learn something of immense clinical importance, such as a dose-dependent response to treatment.

Barriers to Data Sharing

Despite the advantages of data sharing, many barriers exist that hinder data sharing. These can be divided in administrative, ethical and technical barriers.

Two main administrative barriers hinder data sharing among multiple institutes are data completeness and coding inconsistency. Data completeness is difficult, if not impossible, to accomplish. It is generally not practical to collect every available data element of an individual patient, which severely hampers analysis. A pragmatic compromise always results in an incomplete dataset. The second barrier, coding inconsistency, occurs when system-dependent settings (such as standard operating procedure, treatment protocols, equipment settings and the way different observations are recorded) vary widely among sites. These variations impede data integration and, if not correctly reconciled under a consistent data schema, may lead to incompatibility or clinically biased results between datasets across different sites.

Ethical barriers refer to data privacy regulations and other restrictions pertaining to re-use of patient data for some other purpose than originally collected. First, data privacy is very different in its definition, legislation and implementation across countries (Skripcak et al. 2014). This privacy barrier needs to be overcome for multi-centre infrastructure in order to achieve data sharing between institutions, implemented in either a centralized or distributed infrastructure.

A distributed solution may be more secure than a centralized solution, since very few (rather than all) of the potentially sensitive data parameters are communicated between multiple centres, thus overcoming the privacy barrier partially even if not perfectly entirely. A security breach in the distributed sharing network compromises hardly any individual patient-level data, whereas the same failure in a centralized system necessarily compromises *all* of the data. However, the specific methods to overcome this privacy barrier needs to be tailored to the project at hand, as there is no global solution that works in all cases.

Furthermore, re-use of research data may be a potential barrier since data sharing is not yet fully embedded in the scientific culture, although this is improving in recent times. Nevertheless, if data is shared among scientists, more data will be thus become available for modelling, resulting in better models for toxicity predictions. To promote data sharing in the scientific community, more scientific evidence of the benefits of data exchange needs to be published, thus persuading data holders to participate in the collaborative research community and subsequently share their data.

Even if these administrative and ethical barriers can be overcome, technical barriers such as interoperability and standards remain, which may still impede open data sharing between institutes. First, achieving interoperability between IT systems within a clinic is important for the generation of anonymized data, which should ideally be managed by the institutional data warehouse and be accessible through semantic web technologies. However, it is often difficult to achieve this ideal situation due to (oftentimes) incomplete support of internationally standardized protocols, formats and semantic web technologies. In order to improve the support for these technologies, investments are needed to implement these resources if not yet present at each contributing site.

Second, standardization of radiotherapy-specific terminologies and ontologies is still under development (Skripcak et al. 2014), resulting in a situation where various hospitals may use different representations to describe patient-specific information, such as the biological sex of a patient, which is denoted as "female" or "male" in one clinic and "f" or "m" in another. When combining data from two clinics, this difference results in the presence of both representations in this combined dataset. This problem can be solved by linking each clinical representation to a semantic ontology that can provide the standard definition of the variable. For instance, the biological sex of a patient is represented by NCI Thesaurus codes *C16576* for female and *C20197* for male. Thus, the meaning of a clinical variable is only related to its ontological label rather than through its literal representation as a text string.

Data-Sharing Architectures

Centralized Data Sharing

When data is shared within a centralized architecture, the infrastructure itself has complete control of all of the data. This data is not stored in each of the individual clinics, but must be pooled in a centralized repository. In this situation, all operations occur at a central location and no real-time communication occurs between participating institutions. Even though this architecture type is conceptually simple, the ethical barriers mentioned above have to be managed at the central meta-institutional level. Furthermore, as new data fields are added, or the existing data elements are periodically re-organized in some fashion, the effort of centralized data management increases much more rapidly than either size or complexity. Additionally, there is redundancy related to duplication of data, transformation of the local data to the central data model (usually resulting in manual data entry and/or manual copying) and negotiation of intellectual property (IP) rights in the form of data ownership agreements. National cancer registries and health service quality monitoring databases commonly adopt a centralized data architecture.

Decentralized Data Sharing

In a decentralized sharing architecture, each local source of data retains full ownership (in terms of IP) over its own data. Unlike a centralized model, peer-to-peer data sharing is now possible without the need for a centralized governance structure to pool data from all of the sources. A decentralized data-sharing architecture allows for multiple levels of granularity for data access control (e.g., project-by-project based or network-wide protocols) between multiple institutes (Skripcak et al. 2014). The downside of the arrangement is, every data station need to be established and exposed to the other partners in an interoperable manner at each site to enable effective data exchange, and each clinic needs to comply with a standard data communication protocol, such as *http*. Governance of the network is almost non-existent, involving only the self-imposed adoption of communication protocols and inter-operability standards. This architecture scales readily with the size of network and dimensional complexity of data, since there is no overarching central coordination whatsoever. The most commonly quoted example of a decentralized architecture is the World Wide Web.

Hybrid Architectures

The final option is to use a hybrid data-sharing architecture, in which elements of both centralized and decentralized architectures are combined. In this situation, direct peer-to-peer communication and data sharing occurs between sites, but the information about the infrastructure, data representation format, controlled terminologies and other required metadata are maintained at a central location, which facilitates the maintenance and modification of data exchange. One obvious advantage of a hybrid architecture is that the data is stored locally at each site, but it is only conceptually centralized by means of the controlled protocols, thus addressing some of the technical barriers such as site-specific terminology at the level of multi-centric sharing (Skripcak et al. 2014).

Distributed Machine Learning

An increasing amount of data is being generated in different fields of modern medicine including medical imaging, transcriptomics, metabolomics and proteomics. This highly dimensional data is potentially very valuable for machine learning to build a reliable predictive models for diagnosis and treatment outcomes. Generally speaking, the more data that is used from diverse sources, the more robust and externally valid a predictive model becomes. Each institution generally possesses data for a limited number of patients, that may be insufficient for reliable predictive modelling. Therefore, multi-centric collaboration is key to ensuring that a sufficient amount of data is collected for predictive modelling. This can be practically implemented in two archetypal ways, as described in Section "*Data-Sharing Architectures*".

A general overview for the centralized multi-centre architecture for machine learning purposes is shown in Figure 15.5a. In this configuration, a central collection point is responsible for storing all of the data transmitted by each contributing institution and then processing the site-specific data into a uniform format for analysis. Each institute therefore provides one communication point, which communicates in one direction only with the

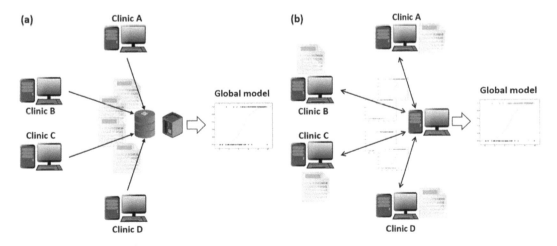

FIGURE 15.5 Centralized and decentralized approaches to sharing data for model-building. (a) Centralized architecture where data is sent to a central repository, where all of the processing and analysis takes place to build a global multi-institutional model. (b) Decentralized architecture where local data is retained within each institution and only fitting coefficients are transmitted to a central processor, which combines the local models (possibly over multiple iterations) to produce a global model.

central data server. Statistical analysis and model building thus occurs only at the centre (which typically hosts high-performance computing hardware), which results in a global model that is internally valid across all of the contributing centres.

An alternative method for machine learning across multiple institutions is the distributed learning architecture (see Figure 15.5b). It differs from the centralized architecture in regards to the location at which the numerically intensive computations are performed (Damiani et al. 2015; Soest et al. 2015). According to this paradigm, an institution with a research question (which may also be any one of the centres intending to contribute data) splits the problem into multiple sub-computations which will be deployed to each participating site. The data-intensive computations are performed within each site on its own respective database, and the coefficients of the local model are shared with the researcher. Crucially, none of the individual patient-level data needs to leave the contributor's database. The multiple local models must then be compared at the original source of the research question (Soest et al. 2015). Depending on the algorithm used, such as Boyd's Alternating Direction Method of Multipliers (Boyd et al. 2011), modified constraints on the coefficients may be sent back out to the contributing sites for re-calculation, and the above cycle repeated many times until a globally convergent model emerges. For specific machine learning algorithms (such as logistic regression), it can be mathematically proven that the final convergent model obtained by decentralized learning must be the same as a centrally learned model (Boyd et al. 2011). In-depth explanations of distributed machine learning algorithms may be found in literature reviews (Boyd et al. 2011; Wu et al. 2012).

Semantic Web Technologies

For the abovementioned distributed learning methodology to work, the local data needs to be parsed in a format that is fully machine-readable and machine-understandable (i.e., objectively semanticized). This is achieved by defining domain ontologies conforming to the standard of the Semantic Web. The Semantic Web is an extension of the internet in which data is intentionally designed to be interpreted by machines rather than humans. In order to achieve machine-readability of the data, structured data is labelled with a publicly accessible semantic ontology. Machine learning algorithms can thus access this via a universal Resource Description Framework (RDF) and the SPARQL Protocol and RDF Query Language (SPARQL). Both sources use Uniform Resource Identifier (URIs) as links between expert semantic meaning and actual physical resources, so that data published through a web "endpoint" is amenable to queries by both machines and people.

Computer-Assisted Theragnostics (CAT)

A real-world decentralized machine-learning network where the concepts described above have been realized in practice is known as Computer Assisted Theragnostics (CAT). Within CAT and its related projects, data is curated from a variety of sources at each collaborating institute and stored locally in standard (PostgreSQL) data warehouses. The "warehoused" data is then converted into a semantic ontology-labelled RDF store and published to other collaboration partners as a SPARQL-compatible endpoint. A set of site-based connector applications and webpage access portals are then used to distribute machine learning algorithms to address prediction modelling questions. The use of CAT methodology allows research networks to address the barriers to data sharing and model-building as discussed earlier in Section 3.2, such as patient privacy, data ownership and lack of inter-operability between data warehouses.

A number of real-world projects have demonstrated the practical feasibility of distributed multi-centric machine learning including: I2B2 (https://www.i2b2.org), EuroCAT (Deist et al. 2017), VATE (Meldolesi et al. 2016), meerCAT (Jochems et al. 2017). Additionally, improving the robustness of the models by external validation will benefit the reproducibility and validity of prediction models (Dekker et al. 2014). This has also been demonstrated within the EuroCAT project (Deist et al. 2017).

TOXICITY MODELLING

The efficacy of radiotherapy against malignant neoplasms is based on the deterministic effect of ionising radiation-induced damage on living cells. Increasing the absorbed dose of radiation always leads to greater detrimental effect on cellular function. The relationship between the physics behind the interaction between ionising radiation and matter and the macroscopic effects observable at the clinical level is a complex process (West and Barnett 2011). Energy transfer from radiation to matter via ionization and collision events induces reactive oxygen species (ROS; free radicals) that attack deoxyribonucleic acid (DNA), resulting in sub-lethal damage (eventually repaired by the cell) and lethal damage (triggers apoptosis, i.e., cell death). Chemical reactions also trigger a cascade of stress-related responses, immune system responses, release of tumour growth and tumour necrotic factors and release of DNA repair molecules. Understanding radiation toxicity through mechanistic arguments remains a large and active area of research. Furthermore, certain genetic mutations may predispose or protect against some form of radiation-induced toxicity (West and Barnett 2011).

What is lacking from mechanistic explanations of the interaction between radiation and cells is an understanding of how highly localized, compartmentalised functional damage accumulates over spatial and temporal scales to give rise to clinically observable toxicity (see Table 15.4). Radiation-induced toxicity is known to have dependencies on dose magnitude, spatial dose distribution and dose-time (fractionation) patterns.

The outstanding technical success of modern-era radiotherapy, specifically ultra-conformal planning techniques (Bauman et al. 2012; Co et al. 2016; Marta et al. 2014; Zhang et al. 2015) and use of image-guidance during treatment (Gwynne et al. 2012; Jadon et al. 2014) have been based on one relatively simple paradigm: increasing the absorbed dose to the tumour while simultaneously reducing, as much as possible, the incidental dose to the healthy tissues in the vicinity of the tumour. Constraints imposed by the physics of radiation transport in matter make it impossible to completely eliminate radiotherapy related toxicity. Therefore, the ability to predict toxicities associated with radiotherapy is essential.

One approach to reduce toxicity to healthy tissues is phenomenologically based toxicity modelling, where a deeply mechanistic description of the underlying process is temporarily set aside in favour of using only a few explanatory variables to predict clinically measurable toxicity.

TABLE 15.4   Some Examples of Commonly Observed Toxicities, Listed by Radiotherapy Treatment Site

Radiotherapy Site	Acute Toxicities	Late Toxicities
Any/unspecific location	Skin erythema Skin desquamation Dry and/or painful skin Tiredness Nausea Anorexia Oedema (swelling) Paresthesia	Telangiectasia Fibrosis Ulceration Obstruction Stenosis Radiation osteonecrosis
Head and Thoracic	Oral mucositis Oesophagitis Odynophagia Dysphagia Radiation pneumonitis Dyspnea	Lymphoedema Breast atrophy Xerostomia Dyspnea Myelopathy Cognitive decline Hearing loss Hypopituitarism
Abdominal	Gastritis Stomach pain Vomiting Diarrhoea Malabsorption Mucus colitis Bleeding (melena or haematochezia)	Strictures Adhesions Fistulae
Pelvic	Dysuria Frequent micturition Bladder cystitis Radiation nephritis Mucosal oedema Leukopenia Thrombocytopenia	Proteinuria Incontinence Nocturia Fistulae Proctitis Bleeding
Reproductive organs	Vaginal mucositis Cessation of menstruation Dyspareunia	Infertility or sterility Induced menopause Erectile dysfunction Vaginal stenosis Vaginal obstruction Vaginal dryness Vaginal fistulae

The most commonly quoted Normal Tissue Complication Probability (NTCP) models such as QUANTEC (Bentzen et al. 2010; Marks et al. 2010) are based on the mathematical formalism developed by Lyman (1985), Kutcher 1991 and Burman 1991 to equate a non-uniform dose distribution on a healthy organ to a generalized uniform dose. Given a sparse collection of toxicity data, such as Emami et al.,1991 the coefficients of the Lyman-Kutcher-Burman (LKB) model allows the probability of a given toxicity to be estimated for any arbitrary dose distribution in the healthy organ of concern.

Of paramount importance in the phenomenological approach is the availability of large volumes of individual level patient data on treatment parameters as well as systematic longitudinal (i.e., repeated over time) observations of treatment-related side effects. Richly multidimensional data is required to characterize an individual patient's phenotype as completely as possible, along the lines discussed in the preceding sections of this chapter, so as to accurately predict the expected toxicity.

A data-sharing approach to toxicity modelling (Deasy et al. 2010) is important due to the relative sparsity of toxicity observations in post-treatment follow-up (compared to clinical outcomes such as survival or local control) as well as strong heterogeneity in individual radio-sensitivity within populations. The data-sharing architectures previously described are of significance here, as suitable datasets for toxicity modelling are more likely if individual patient-level observations from multiple institutions have been pooled together. Pooled individual-level toxicity data from multiple institutions has the added advantage that the distribution of radiation doses to organs-at-risk will naturally extend over a wider range, leading to a more robust statistical model. Some heterogeneity in the model covariates is an absolute requirement to identify predictive features and hence estimate the change in toxicity outcome with respect to changing one or more of these features.

Current State of the Art
Toxicity Modelling Process
The procedure for modelling treatment-related toxicity is generally similar and broadly analogous to the procedure for developing diagnostic tests, genetic markers, blood-borne biomarkers and predictive models of tumour control (e.g., overall survival or time to recurrence). The general procedure begins with a clear formulation of the clinical question at hand, and proceeds onwards to identifying the relevant covariates and assembling the data into a structured format. Thereafter, technicalities of the methods used may diverge somewhat, depending on (i) the clinical question, (ii) the data landscape and (iii) the data-sharing architecture. To conclude the process, there must be some evaluation of the predictive performance of the model. The model should be published according to a standard quality of reporting checklist such as TRIPOD (Collins et al. 2015) and, ideally, an anonymized data repository is made openly accessible to other researchers. A methodological framework (Lambin et al. 2012) is shown in Figure 15.6.

As previously mentioned, re-use of clinical trial data and observations collected from routine clinical practice is common practice for toxicity model development; one must remain cognizant of potential problems that could impact the model. For example, clinical trial data tends to be relatively complete for a selective sample of patients with a given condition. In contrast, real-world data might be unselectively sampled and available in huge quantity, but is likely to be missing many of the covariates that would be relevant for modelling.

Handling Missing Values
The question of imputation of missing data (either covariates that are entirely missing or only missing values) is crucially important and is described in detail elsewhere (Schafer and Olsen 1998; Sterne et al. 2009). Generally, a small proportion of values that are *missing purely at random* may be imputed. However, it is not advisable to impute values that

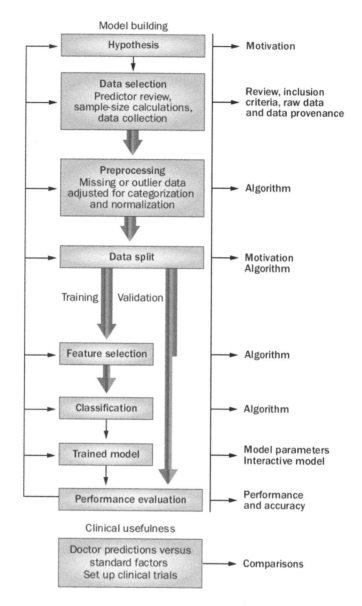

FIGURE 15.6 Schematic overview of methodological processes in clinical decision-support system development, describing model development, assessment of clinical usefulness and what ideally to publish. The differently-shaded, parallel lines represent heterogeneous data, which have been split early for independent validation (but without internal cross-validation). Reproduced with permission from Lambin et al.2012.

are missing in some systematic (but potentially unexplained) way. In particular regard to toxicity observations, one must be wary that moderate toxicities (≤ Grade 2) might be either significantly under-reported by clinicians or not recorded at all; the potential bias introduced in the model would be to assume such missing values as "no toxicity".

The difficulty that arises in routine treatment data, where a large proportion of values (e.g., greater than 80%) may be missing, is that a systematic pattern of missing values may be impossible to detect. Missing covariates may be imputed, but hinges on strong correlation with other (non-missing) covariates. If so, the benefit of imputing the missing covariate is somewhat questionable, since it is now some functional combination (and therefore no longer statistically independent) of the other surrogate variables.

Model Fitting

In the next phase of the process, a carefully curated dataset is used to determine the coefficients of the model such that it describes the observed outcomes with the least bias in residual error. Suitable models for categorical or dichotomous (i.e., binary) outcomes include logistic regression, support vector machines (Noble 2006), Bayesian network classifiers (Friedman et al. 1997) and classification trees (Loh 2011). Treatment-related toxicities are almost always recorded as binary outcomes (patient either does or does not have a given severity of toxicity) with asymptotic upper and lower bounds on probability, therefore one of the above models would be appropriate for toxicity modelling.

Toxicity outcomes may sometimes be framed as a time-to-event problem, such as an expected time interval when a certain severity of late toxicity is observed. For these, Cox proportional hazards or competing risks models are commonly used.

Unless there is a prospective plan for an independent dataset to validate the model, the available data should first be partitioned into a "training" set and a "testing" set. Splitting the overall sample 80/20 between training and testing, respectively, is a commonly applied rule-of-thumb. The splitting may be determined entirely at random, or one may retain the same relative proportion of outcome events on each side of the split (i.e., a stratified split). The latter may be advisable if toxicity events are rare, since a purely random split might result in no events falling within the training subset. Splitting based on the outcome has an attendant risk of unknowingly contaminating the testing subset with features that are surrogates of the outcome, and would therefore not be truly independent from the training subset.

Bias-Variance Trade-Off

An over-fitted model is one that so specifically and so exclusively describes the training subset (i.e., low variance in the model's residual errors) that it performs poorly when presented with a previously unseen testing subset (i.e., highly biased). Conversely, an under-fitted model is poor at describing the observed variation of outcomes in the training subset (high variance in the residual) but has the same performance in a testing subset (low bias). Neither of the above is desirable in prediction modelling; an optimally fitted model needs to be a trade-off between in-training set variance versus in-validation set bias.

Various strategies can be employed, either singly or in combination, to reduce the risk of over-fitting. First, the number of covariates relative to the number of events should be as low as possible. Suitable candidates for elimination are covariates that are either strongly correlated to others or where its values hardly change.

Secondly, data-driven covariate selection methods, such as stepwise-forward or stepwise-backward regression, may be used. Maximum likelihood estimation or information criteria metrics, such as Akaike's and Bayes' Information Criteria (Akaike 2011;

Posada et al. 2004), can be used to iteratively build up an optimal set of explanatory covariates (or drop them from further consideration). However, the implicit risk is that potentially predictive variables may be eliminated because they do not appear to be sufficiently predictive in the training cohort. Another recommended method is ranked univariate selection, where each candidate covariate is ordered according to the strength of its correlation with the outcome. A *maximally parsimonious model* can then be designed by progressively dropping covariates from the bottom of the ranking, until the best predictive performance is obtained using the smallest number of covariates. In a similar way, regularization approaches (Friedman et al. 2010) may be used to progressively and automatically filter out covariates.

Thirdly, the training subset may be further split randomly into multiple internal subgroups (i.e., "folds"). Each fold is tested against a model trained using all of the other folds combined, and the process iterates over all folds. This is known as *K-fold cross-validation* (Stone 1977; Stone 1974), where *K* represents the number of folds. The value of the model coefficient is averaged over all folds. If there is a concern about selective sampling among the folds, the above cycle may be repeated several times with a new pseudo-random number generator seed in each cycle.

Evaluating Prediction Performance

Minimally, the predictive performance of the model needs to be evaluated in the testing cohort. The typical measure of discrimination/classification performance is the area under the receiver-operator characteristic curve (AUC). Overall accuracy, sensitivity and specificity and the calibration index are also commonly reported for categorical and/or binary toxicity outcomes. However, for time-to-adverse event predictions, the calibration index and the hazard ratio between different patient sub-populations are commonly reported.

The number of published multifactorial prediction models has grown rapidly in recent years. Attention to methodological transparency and adherence to reporting guidelines supports the evaluation of models, and helps identify potential sources of bias that are to be avoided. Objectiveness in model performance evaluations ought to be based on repeatability and reproducibility of the published results. Hence, it is important that journals encourage open access to datasets and the software or computer code used to generate the prediction models.

Clinical Impact

At present, toxicity prediction models can only distinguish between broad categories of interventions (such as radical surgery versus active surveillance, or sequential versus concurrent chemo-radiotherapy for lung cancer). In the radiotherapy domain, physicians and patients are likely more interested in tailoring each specific radiotherapy intervention, at an individual patient level, towards avoidance of radiation-induced injury. The vast majority of present day toxicity models are neither sufficiently detailed nor independently validated to be used to select interventions and included in the radiotherapy treatment plan optimisation (i.e., inverse planning). The present generation of toxicity models are largely based on dose-volume metrics alone.

Ultimately, the utility of all multifactorial toxicity prediction models must be established within a clinical context (Lambin et al. 2012). The predictive model needs to demonstrate added clinical value when used in the treatment decision-making process, which is over and above the utility of existing prognostic indicators. A positive clinical impact may be framed in terms of reduction in frequency (and/or severity) of adverse events due to treatment, or better quality of life of patients in after-treatment care. After a prediction model has been shown to have significant clinical impact and be broadly valid over a range of clinical settings, one may then expect rapid adoption and wide usage of the model in routine clinical decision-making.

Toxicity Modelling in the Near Future
Incorporating Diverse Data Types
Though present day NTCP models in radiotherapy are still largely based on only dose and/or dose-volume metrics, efforts are under way to include a wider diversity of clinical data types to improve the predictive performance of such models. In the domain of "big data" relevant to oncology, there are major research efforts under way to develop more holistic models in which covariates from many different disciplines (such as surgery, radiology, chemo-radiotherapy, genetic testing, immunotherapy and blood-borne biomarkers) can be combined.

The first among these can be seen in efforts to incorporate clinical risk factors into the QUANTEC dose-based NTCP models (Appelt et al. 2014).

Another deep vein of richly multi-dimensional data that can be mined for potential predictors is routine medical imaging. Medical images constitute a large proportion of real-world healthcare data on individual patients. Extraction of data from images requires high-throughput and (ideally) fully automated image processing pipelines to reduce the qualitative aspects of the images into a finite number of structured features. Such "radiomic" features have been shown in diagnostic and prognostic studies to be potentially correlated with tumour histology and overall survival (Kumar et al. 2012; Lambin et al. 2012; Larue et al. 2017; Aerts et al. 2014; Miles et al. 2013). Forthcoming developments in radiomics-boosted clinical toxicity prediction models will address radiation responses in tumour and healthy tissues over time (i.e., "delta radiomics"), including the individual patient's susceptibility for treatment-induced toxicity.

A further active area of research involves the study of genetic susceptibilities, including innate radio-sensitivity based on single-nucleotide polymorphism (Kerns et al. 2015), to inform toxicity prediction. Drug-gene interaction studies are being performed in the chemotherapeutic domain. In radiotherapy, the rapidly advancing field of "radiogenomics" utilizes genome-wide association studies (GWAS) to correlate genetic markers with individual predisposition to radiation-induced toxicity (Andreassen et al. 2016; Herskind et al. 2016; Kerns et al. 2016; Kerns et al. 2014). Therefore, one may also expect to see trials and validation studies of gene-boosted toxicity prediction models in the near future.

It has been hypothesised that inherent differences in radio-sensitivity at the cellular level is the dominant contributing factor to clinically observable normal-tissue reactions (Lambin et al. 2012). This suggests that the effect estimate in a cohort study of toxicity may be unduly biased by the potential inclusion of a small number of highly radiosensitive patients. Rapid-turnaround blood assays, potentially exploiting lab-on-a-chip technology, may be used to measure inflammatory responses and other molecular expressions of cellular damage (such as tumour growth factors). This could be used to stratify patient subgroups by innate cellular radiosensitivity, and the potentially predictive molecular markers may themselves be incorporated into future predictive models of toxicity.

Exploiting Differences for Personalizing Treatment
Tailoring interventions to the specific characteristics of the individual patient has become an important focus of modern cancer research. This necessarily requires that individual variability, that is, the comprehensive "human phenotype" must be taken into account when preparing an intervention (Collins and Varmus 2015). However, the observable phenotype extends beyond purely genetic and biological aspects – the patient must be also be understood as a holistic combination of cognitive abilities, experiences and preferences, as well as the accumulated effects of lifestyle and environmental exposures (Wild 2005).

Pre-existing illnesses and co-morbidities are known to be correlated with radiation-induced injuries. A study by Nalbantov et al. (2013) proposed that cardiac co-morbidity and baseline dyspnea were significantly correlated with dyspnea at 6 months after radiotherapy. However, few toxicity models presently take account of pre-existing medical conditions and co-morbidities as potential predictive features of post-treatment toxicity. Chen et al. (2012) suggests that a subset of elderly cancer patients might tolerate and benefit from more aggressive radiotherapy, but more precise stratification of such patients based on age and presence of comorbid illnesses would be required. This hypothesis was supported by Dekker et al. (2014), who reported that a subset of palliatively treated lung cancer patients with promising treatment response (according to a multifactorial prediction model) may have had an additional 18 month benefit in median survival time, if they had been treated aggressively.

The existence of two or more treatment options within a given range of clinical equipoise satisfies a necessary condition for preference-sensitive decision-making. Consensus guidelines now recommend that post-mastectomy radiotherapy for women with T1–2 breast cancer (with one to three positive axillary nodes) should be discussed with patients in connection to their individual situation and personal preferences (Recht et al. 2016). Similarly, paucity of toxicity data from large international trials in anal cancer chemo-radiotherapy has led to equivocation about the optimal radiotherapy prescription dose (Fakhrian et al. 2013; Han et al. 2014; Kachnic et al. 2013; Northover et al. 2010), even though radiation-related side effects occur frequently. Although there is no evidence pointing to any single dose prescription being clinically superior, the risks of normal tissue complication are thought to be greater at the highest dose levels. Rønde et al. (2017) examined the feasibility of including physician and patient preferences for treatment outcome into treatment plan optimisation, but were unable to pose the optimisation problem in regards to probabilities of genito-urinary and gastro-intestinal toxicity due to lack of data.

Accumulation and sharing of multi-dimensional toxicity data, along the lines discussed in this chapter, would enable the development of detailed toxicity models that support physician decision-making, patient counselling and shared decision making. Shared decision making is a structured collaborative consultation between a patient and their treating physician, such that each brings their own perspective into a conversation about the *trade-off between risks and benefits* among a set of preference-sensitive treatment options (Barry and Edgman-Levitan 2012; Stiggelbout et al. 2015), and in so doing arrive at a mutually agreed choice of treatment.

Use of Machine Learning and Artificial Intelligence

Up until recently, the vast majority of the data used for radiotherapy modelling has been processed by human hands. That is, the data is collected, cleaned, linked and structured by human operators. Thereafter, the data is fed to computers to compute statistical models and machine classifiers. While this approach may be suited to small amounts of information, it is hopelessly inappropriate for scaling to the volume, variety and velocity of data required to address the clinical challenges addressed in this chapter.

The only possible option for the future is to generate data handling workflows where machines perform almost the entirety of the work of data pre-processing and data linkage from a large number of sources. Such workflows can be adapted to any one of the data architectures previously discussed, and thus efficiently manage the modelling workflow for human experts to afterwards validate and use.

With increasing size and dimensionality of data, a form of machine-based intelligence is required to rapidly derive clinical insight from the raw data with minimal need for human guidance (Bibault et al. 2016). Artificial neural networks (i.e., neural "nets" or ANNs) are increasingly being set this task in oncology, including prediction of treatment toxicity.

Such neural nets consist of multiple layers of individual "neurons," that are each univariate logistic regression classifiers. The input to every neuron (other than the lowest input layer) is a linear combination of the output of every neuron in the layer below. Therefore, a "stack" of such layers are capable of encoding astoundingly complex and nuanced "responses" to a given input. The drawback of such an approach is, a vast volume of data (in the magnitude of thousands or possibly millions of independent cases) are required in the training set.

The earliest examples of application already show that ANNs can feasibly predict nocturia and rectal bleeding following prostate radiotherapy (Tomatis et al. 2012; Gulliford et al. 2004), as well as radiation pneumonitis in lung radiotherapy (Su et al. 2005; Chen et al. 2007). The current performance of ANNs remains sub-optimal, and this may be a combined effect of poor data quality, paucity of individual-patient level data and low "learning" efficiency of ANNs (for the typical number of training cases that can be feasibly mustered in radiation oncology).

"Deep learning," as a variant of simpler ANNs, consists of multiple stacks of neurons in the intermediate zone between the input layer and the output layer. In non-medical applications, deep learning networks have been shown to be capable of remarkable feats of object recognition, target object segmentation, natural language processing, competitive game-playing and spontaneous generation of simple narratives (i.e., auto-generating captions when given a previously unseen photograph). The anticipation is that in the near future, deep learning systems will also have immense impact in the field of radiotherapy in the form of diagnosis support systems, region-of-interest autosegmentation, conversion of natural text (e.g., pathology reports and follow-up notes) into structured data and sophisticated prediction models utilising a wide range of "-omics" inputs.

Towards a Data-Sharing "Culture" in Oncology

The future development of reliable and consistent toxicity model for clinical use will be critically dependent on the evolution of a data-sharing culture across multiple disciplines that intersect in oncology. As Deasy et al. (2010) have pointed out, NTCP models based only on summary reports of dose-volume effects can be inconsistent. These authors make a compelling case for data pooling (rather than the so-called "data-to-trash can" approach). Given the natural heterogeneity and variability inherent in these once-off cohort studies, secondary analysis of pooled individual patient-level data for toxicity modelling is a valuable and under-exploited opportunity for clinical learning, given that data architectures now exist to support large-scale learning and modelling efforts that do not divulge personal patient information (see Section 3 above). Fortunately, a number of data linkage efforts are under way to render institutional data stores FAIR and discoverable for the purpose of collaborative learning, including the development of multi-institutional toxicity models (e.g., OncoSpace - https://oncospace.radonc.jhmi.edu/about/default.aspx, and CORAL - http://www.eurocat.info/community.html).

CONCLUSION

Development of reliable, accurate and clinically validated prediction models of toxicity requires a concerted effort to address systemic healthcare challenges pertaining to (i) structured data collection and data linkage, (ii) multiple barriers against sharing data for cooperative clinical learning of toxicity models and (iii) methodological approaches for clinical learning on vast volumes of richly dimensional healthcare data. We conclude on a note of optimism, that present-day research and development projects are attempting to solve these systemic challenges, and hence to usher in an era where improved toxicity prediction models will support precision personalized medicine and shared decision-making in the very near future.

REFERENCES

Abernethy, A. P. et al. Rapid-learning system for cancer care. *J. Clin. Oncol.* **28(27)**, 4268–4274 (2010).

Aerts, H. J. W. L. et al. Decoding tumour phenotype by noninvasive imaging using a quantitative radiomics approach. *Nat. Commun.* **5**, 4006 (2014).

Akaike, H. Akaike's Information criterion. in *International Encyclopedia of Statistical Science* (ed. Lovric, M.) 25–25 (Springer Berlin Heidelberg, 2011).

Andreassen, C. N., Schack, L. M. H., Laursen, L. V. & Alsner, J. Radiogenomics - current status, challenges and future directions. *Cancer Lett.* **382(1)**, 127–136 (2016).

Appelt, A. L., Vogelius, I. R., Farr, K. P., Khalil, A. A. & Bentzen, S. M. Towards individualized dose constraints: adjusting the QUANTEC radiation pneumonitis model for clinical risk factors. *Acta Oncol. Stockh. Swed.* **53(5)**, 605–612 (2014).

Atkinson, T. M. et al. The association between clinician-based common terminology criteria for adverse events (CTCAE) and patient-reported outcomes (PRO): a systematic review. *Support. Care Cancer.* **24(8)**, 3669–3676 (2016).

Barry, M. J. & Edgman-Levitan, S. Shared decision making--pinnacle of patient-centered care. *N. Engl. J. Med.* **366(9)**, 780–781 (2012).

Basch, E. et al. Recommendations for incorporating patient-reported outcomes into clinical comparative effectiveness research in adult oncology. *J. Clin. Oncol.* **30(34)**, 4249–4255 (2012).

Basch, E. et al. Development of the National Cancer Institute's Patient-Reported Outcomes Version of the Common Terminology Criteria for Adverse Events (PRO-CTCAE). *JNCI J. Natl. Cancer Inst.* **106(9)**, (2014).

Bauman, G. et al. Intensity-modulated radiotherapy in the treatment of prostate cancer. *Clin. Oncol.* **24(7)**, 461–473 (2012).

Bentzen, S. M. & Trotti, A. Evaluation of early and late toxicities in chemoradiation trials. *J. Clin. Oncol. Off. J. Am. Soc. Clin. Oncol.* **25(26)**, 4096–4103 (2007).

Belderbos, J. et al. Acute esophageal toxicity in non-small cell lung cancer patients after high dose conformal radiotherapy. *Radiother. Oncol.* **75(2)**, 157–164 (2005).

Bentzen, S. M. et al. Quantitative Analyses of Normal Tissue Effects in the Clinic (QUANTEC): An Introduction to the Scientific Issues. *Int. J. Radiat. Oncol. Biol. Phys.* **76(S3)**, S3–S9 (2010).

Bibault, J.-E., Giraud, P. & Burgun, A. Big Data and machine learning in radiation oncology: state of the art and future prospects. *Cancer Lett.* **382(1)**, 110–117 (2016).

Boyd, S., Parikh, N., Chu, E., Peleato, B. & Eckstein, J. Distributed optimisation and statistical learning via the alternating direction method of multipliers. *Found Trends Mach Learn* **3(1)**, 1–122 (2011).

Bradley, J., Deasy, J. O., Bentzen, S. & El-Naqa, I. Dosimetric correlates for acute esophagitis in patients treated with radiotherapy for lung carcinoma. *Int. J. Radiat. Oncol. Biol. Phys.* **58(4)**, 1106–1113 (2004).

Burman, C., Kutcher, G. J., Emami, B. & Goitein, M. Fitting of normal tissue tolerance data to an analytic function. *Int. J. Radiat. Oncol. Biol. Phys.* **21(1)**, 123–135 (1991).

Charlson, M. E., Pompei, P., Ales, K. L. & MacKenzie, C. R. A new method of classifying prognostic comorbidity in longitudinal studies: development and validation. *J. Chronic Dis.* **40(5)**, 373–383 (1987).

Chen, S. et al. A neural network model to predict lung radiation-induced pneumonitis. *Med. Phys.* **34(9)**, 3420–3427 (2007).

Cheng, Q. et al. Development and evaluation of an online three-level proton vs photon decision support prototype for head and neck cancer – comparison of dose, toxicity and cost-effectiveness. *Radiother. Oncol.* **118(2)**, 281–285 (2016).

Chen, R. C., Royce, T. J., Extermann, M. & Reeve, B. B. Impact of age and comorbidity on treatment and outcomes in elderly cancer patients. *Semin. Radiat. Oncol.* **22(4)**, 265–271 (2012).

Co, J., Mejia, M. B. & Dizon, J. M. Evidence on effectiveness of intensity-modulated radiotherapy versus 2-dimensional radiotherapy in the treatment of nasopharyngeal carcinoma: meta-analysis and a systematic review of the literature. *Head Neck* **38** (S1), E2130–2142 (2016).

Collins, G. S., Reitsma, J. B., Altman, D. G. & Moons, K. G. M. Transparent reporting of a multivariable prediction model for individual prognosis or diagnosis (TRIPOD): the TRIPOD statement. *BMJ* **350**, g7594 (2015).

Collins, F. S. & Varmus, H. A new initiative on precision medicine. *N. Engl. J. Med.* **372(9)**, 793–795 (2015).

Cox, J. D., Stetz, J. & Pajak, T. F. Toxicity criteria of the Radiation Therapy Oncology Group (RTOG) and the European Organization for Research and Treatment of Cancer (EORTC). *Int. J. Radiat. Oncol. Biol. Phys.* **31(5)**, 1341–1346 (1995).

Damiani, A. et al. Distributed learning to protect privacy in multi-centric clinical studies. In: Holmes J., Bellazzi R., Sacchi L., Peek N. (eds) Artificial Intelligence in Medicine. AIME 2015. Lecture Notes in Computer Science, vol 9105, 66–75. Springer, Cham Springer Cham Heidelberg New York Dordrecht London.

Deasy, J. O. et al. Improving normal tissue complication probability models: the need to adopt a "data-pooling" culture. *Int. J. Radiat. Oncol. Biol. Phys.* **76(S3)**, S151–S154 (2010).

De Ruyck, K. et al. Development of a multicomponent prediction model for acute esophagitis in lung cancer patients receiving chemoradiotherapy. *Int. J. Radiat. Oncol. Biol. Phys.* **81(2)**, 537–544 (2011).

Deist, T. M. et al. Infrastructure and distributed learning methodology for privacy-preserving multi-centric rapid learning health care: euroCAT. *Clin. Transl. Radiat. Oncol.* **4**, 24–31 (2017).

Dekker, A. et al. Rapid learning in practice: A lung cancer survival decision support system in routine patient care data. *Radiother. Oncol.* **113(1)**, 47–53 (2014).

Dueck, A. C. et al. Validity and reliability of the US National Cancer Institute's Patient-Reported Outcomes Version of the Common Terminology Criteria for Adverse Events (PRO-CTCAE). *JAMA Oncol.* **1(8)**, 1051–1059 (2015).

Emami, B. et al. Tolerance of normal tissue to therapeutic irradiation. *Int. J. Radiat. Oncol. Biol. Phys.* **21(1)**, 109–122 (1991).

Fakhrian, K. et al. Chronic adverse events and quality of life after radiochemotherapy in anal cancer patients. A single institution experience and review of the literature. *Strahlenther. Onkol.* **189(6)**, 486–494 (2013).

Fayers, P., Bottomley, A., EORTC Quality of Life Group & Quality of Life Unit. Quality of life research within the EORTC-the EORTC QLQ-C30. European Organisation for Research and Treatment of Cancer. *Eur. J. Cancer* **38(S4)**, S125–133 (2002).

Friedman, N., Geiger, D. & Goldszmidt, M. Bayesian network classifiers. *Mach. Learn.* **29(2-3)**, 131–163 (1997).

Friedman, J., Hastie, T. & Tibshirani, R. Regularization paths for generalized linear models via coordinate descent. *J. Stat. Softw.* **33(1)**, 1–22 (2010).

Geifman, N. & Butte, A. J. Do cancer clinical trial populations truly represent cancer patients? A comparison of open clinical trials to the cancer genome atlas. *Pac. Symp. Biocomput. Pac. Symp. Biocomput.* **21**, 309–320 (2016).

Gilbert, A. et al. Systematic review of radiation therapy toxicity reporting in randomized controlled trials of rectal cancer: a comparison of patient-reported outcomes and clinician toxicity reporting. *Int. J. Radiat. Oncol. Biol. Phys.* **92(3)**, 555–567 (2015).

Gulliford, S. L., Webb, S., Rowbottom, C. G., Corne, D. W. & Dearnaley, D. P. Use of artificial neural networks to predict biological outcomes for patients receiving radical radiotherapy of the prostate. *Radiother. Oncol.* **71(1)**, 3–12 (2004).

Gwynne, S. et al. Image-guided radiotherapy for rectal cancer: a systematic review. *Clin. Oncol.* **24(4)**, 250–260 (2012).

Jovanović, J. & Bagheri, E. Semantic annotation in biomedicine: the current landscape. *J. Biomed. Semant.* **8(1)**, 44 (2017).

Habbous, S. et al. Validation of a one-page patient-reported Charlson comorbidity index questionnaire for upper aerodigestive tract cancer patients. *Oral Oncol.* **49(5)**, 407–412 (2013).

Han, K. et al. Prospective evaluation of acute toxicity and quality of life after IMRT and concurrent chemotherapy for anal canal and perianal cancer. *Int. J. Radiat. Oncol. Biol. Phys.* **90(3)**, 587–594 (2014).

Herskind, C. et al. Radiogenomics: A systems biology approach to understanding genetic risk factors for radiotherapy toxicity? *Cancer Lett.* **382(1)**, 95–109 (2016).

Jadon, R. et al. A systematic review of organ motion and image-guided strategies in external beam radiotherapy for cervical cancer. *Clin. Oncol.* **26(4)**, 185–196 (2014).

Jin, S., Pazdur, R. & Sridhara, R. Re-evaluating eligibility criteria for oncology clinical trials: analysis of investigational new drug applications in 2015. *J. Clin. Oncol.* **35(33)**, 3745–3752 (2017).

Jochems, A. et al. Developing and validating a survival prediction model for NSCLC patients through distributed learning across 3 countries. *Int. J. Radiat. Oncol. Biol. Phys.* **99(2)**, 344–352 (2017).

Kachnic, L. A. et al. RTOG 0529: a phase 2 evaluation of dose-painted intensity modulated radiation therapy in combination with 5-fluorouracil and mitomycin-C for the reduction of acute morbidity in carcinoma of the anal canal. *Int. J. Radiat. Oncol. Biol. Phys.* **86(1)**, 27–33 (2013).

Kelley, T. A. International Consortium for Health Outcomes Measurement (ICHOM). *Trials* **16**, O4 (2015).

Kerns, S. L., Ostrer, H. & Rosenstein, B. S. Radiogenomics: using genetics to identify cancer patients at risk for development of adverse effects following radiotherapy. *Cancer Discov.* **4(2)**, 155–165 (2014).

Kerns, S. L. et al. The prediction of radiotherapy toxicity using single nucleotide polymorphism–based models: a step toward prevention. *Semin. Radiat. Oncol.* **25(4)**, 281–291 (2015).

Kerns, S. L. et al. Meta-analysis of genome wide association studies identifies genetic markers of late toxicity following radiotherapy for prostate cancer. *EBioMedicine* **10**, 150–163 (2016).

Kluetz, P. G., Chingos, D. T., Basch, E. M. & Mitchell, S. A. Patient-reported outcomes in cancer clinical trials: measuring symptomatic adverse events with the national cancer institute's patient-reported outcomes version of the common terminology criteria for adverse events (PRO-CTCAE). *Am. Soc. Clin. Oncol. Educ. Book.* **35**, 67–73 (2016).

Kreimeyer, K. et al. Natural language processing systems for capturing and standardizing unstructured clinical information: A systematic review. *J. Biomed. Inform.* **73**, 14–29 (2017).

Kutcher, G. J., Burman, C., Brewster, L., Goitein, M. & Mohan, R. Histogram reduction method for calculating complication probabilities for three-dimensional treatment planning evaluations. *Int. J. Radiat. Oncol. Biol. Phys.* **21(1)**, 137–146 (1991).

Kumar, V. et al. Radiomics: the process and the challenges. *Magn. Reson. Imaging* **30(9)**, 1234–1248 (2012).

Kunneman, M., Pieterse, A. H., Stiggelbout, A. M. & Marijnen, C. A. M. Which benefits and harms of preoperative radiotherapy should be addressed? A Delphi consensus study among rectal cancer patients and radiation oncologists. *Radiother. Oncol.* **114(2)**, 212–217 (2015).

Lambin, P. et al. Predicting outcomes in radiation oncology—multifactorial decision support systems. *Nat. Rev. Clin. Oncol.* **10(1)**, 27–40 (2012).

Lambin, P. et al. Radiomics: extracting more information from medical images using advanced feature analysis. *Eur. J. Cancer* **48(4)**, 441–446 (2012).

Lambin, P. et al. 'Rapid Learning health care in oncology' - an approach towards decision support systems enabling customised radiotherapy. *Radiother. Oncol.* **109(1)**, 159–164 (2013).

Larue, R. T. H. M., Defraene, G., De Ruysscher, D., Lambin, P. & van Elmpt, W. Quantitative radiomics studies for tissue characterization: a review of technology and methodological procedures. *Br. J. Radiol.* **90(1070)**, 20160665 (2017).

Lee, C., Sunu, C. & Pignone, M. Patient-reported outcomes of breast reconstruction after mastectomy: a systematic review. *J. Am. Coll. Surg.* **209(1)**, 123–133 (2009).

Loh, W.-Y. Classification and regression trees. *Wiley Interdiscip. Rev. Data Min. Knowl. Discov.* **1**, 14–23 (2011).

Lustberg, T. et al. Implementation of a rapid learning platform: predicting 2-year survival in laryngeal carcinoma patients in a clinical setting. *Oncotarget* **7(24)**, 37288–37296 (2016).

Lyman, J. T. Complication probability as assessed from dose-volume histograms. *Radiat. Res. Suppl.* **8**, S13–19 (1985).

Marta, G. N. et al. Intensity-modulated radiation therapy for head and neck cancer: systematic review and meta-analysis. *Radiother. Oncol.* **110(1)**, 9–15 (2014).

Marks, L. B. et al. Use of normal tissue complication probability models in the clinic. *Int. J. Radiat. Oncol. Biol. Phys.* **76(S3)**, S10–19 (2010).

Meldolesi, E. et al. Standardized data collection to build prediction models in oncology: a prototype for rectal cancer. *Future Oncol.* **12(1)**, 119–136 (2016).

Meystre, S. M., Savova, G. K., Kipper-Schuler, K. C. & Hurdle, J. F. Extracting information from textual documents in the electronic health record: a review of recent research. *Yearb. Med. Inform.* 128–144 (2008).

Miles, K. A., Ganeshan, B. & Hayball, M. P. CT texture analysis using the filtration-histogram method: what do the measurements mean? *Cancer Imaging Off. Publ. Int. Cancer Imaging Soc.* **13(3)**, 400–406 (2013).

Nalbantov, G. et al. Cardiac comorbidity is an independent risk factor for radiation-induced lung toxicity in lung cancer patients. *Radiother. Oncol.* **109(1)**, 100–106 (2013).

Noble, W. S. What is a support vector machine? *Nat. Biotechnol.* **24(12)**, 1565–1567 (2006).

Northover, J. et al. Chemoradiation for the treatment of epidermoid anal cancer: 13-year follow-up of the first randomised UKCCCR Anal Cancer Trial (ACT I). *Br. J. Cancer* **102(7)**, 1123–1128 (2010).

Palma, D. A. et al. Predicting esophagitis after chemoradiation therapy for non-small cell lung cancer: an individual patient data meta-analysis. *Int. J. Radiat. Oncol. Biol. Phys.* **87(4)**, 690–696 (2013).

Pons, E., Braun, L. M. M., Hunink, M. G. M. & Kors, J. A. Natural language processing in radiology: a systematic review. *Radiology* **279(2)**, 329–343 (2016).

Posada, D., Buckley, T. R. & Thorne, J. Model selection and model averaging in phylogenetics: advantages of Akaike information criterion and bayesian approaches over likelihood ratio tests. *Syst. Biol.* **53(5)**, 793–808 (2004).

Recht, A. et al. Postmastectomy radiotherapy: an American society of clinical oncology, American society for radiation oncology, and society of surgical oncology focused guideline update. *Pract. Radiat. Oncol.* **6(6)**, e219–e234 (2016).

Riesen, I. N., Boersma, L., Brouns, M., Dekker, A. & Smits, K. PO-0755: Implementation of structural patient reported outcome registration in clinical practice. *Radiother. Oncol.* **123(S1)**, S398 (2017).

Rønde, H. S., Wee, L., Pløen, J. & Appelt, A. L. Feasibility of preference-driven radiotherapy dose treatment planning to support shared decision making in anal cancer. *Acta Oncol.* **56(10)**, 1277–1285 (2017).

Rowland, J. H., Hewitt, M. & Ganz, P. A. Cancer survivorship: a new challenge in delivering quality cancer care. *J. Clin. Oncol.* **24(32)**, 5101–5104 (2006).

Schafer, J. L. & Olsen, M. K. Multiple imputation for multivariate missing-data problems: a data analyst's perspective. *Multivar. Behav. Res.* **33(4)**, 545–571 (1998).

Secord, A. A. et al. Patient-reported outcomes as end points and outcome indicators in solid tumours. *Nat. Rev. Clin. Oncol.* **12(6)**, 358–370 (2015).

Skripcak, T. et al. Creating a data exchange strategy for radiotherapy research: Towards federated databases and anonymised public datasets. *Radiother. Oncol.* **113(3)**, 303–309 (2014).

Soest, J. P. A. van, Dekker, A. L. A. J., Roelofs, E. & Nalbantov, G. Application of machine learning for multicenter learning. in El Naqa I., Li R., Murphy M.J. (eds) *Machine Learning in Radiation Oncology* 71–97 (Springer, Cham Springer Cham Heidelberg New York Dordrecht London 2015).

Spak, D. A., Plaxco, J. S., Santiago, L., Dryden, M. J. & Dogan, B. E. BI-RADS® fifth edition: A summary of changes. *Diagn. Interv. Imaging* **98(3)**, 179–190 (2017).

Sterne, J. A. C. et al. Multiple imputation for missing data in epidemiological and clinical research: potential and pitfalls. *BMJ* **338**, b2393 (2009).

Stiggelbout, A. M., Pieterse, A. H. & De Haes, J. C. J. M. Shared decision making: concepts, evidence, and practice. *Patient Educ. Couns.* **98(10)**, 1172–1179 (2015).

Stone, M. Cross-validatory choice and assessment of statistical predictions. *J. R. Stat. Soc. Ser. B Methodol.* **3682)**, 111–147 (1974).

Stone, M. An asymptotic equivalence of choice of model by cross-validation and Akaike's criterion. *J. R. Stat. Soc. Ser. B Methodol.* **3981)**, 44–47 (1977).

Su, M. et al. An artificial neural network for predicting the incidence of radiation pneumonitis. *Med. Phys.* **32(2)**, 318–325 (2005).

Sullivan, R. et al. Delivering affordable cancer care in high-income countries. *Lancet Oncol.* **12(10)**, 933–980 (2011).

Tomatis, S. et al. Late rectal bleeding after 3D-CRT for prostate cancer: development of a neural-network-based predictive model. *Phys. Med. Biol.* **57(5)**, 1399–1412 (2012).

von dem Knesebeck, O. et al. Social inequalities in patient-reported outcomes among older multimorbid patients – results of the MultiCare cohort study. *Int. J. Equity Health* **14**, 17 (2015).

van Loon, J., Grutters, J. & Macbeth, F. Evaluation of novel radiotherapy technologies: what evidence is needed to assess their clinical and cost effectiveness, and how should we get it? *Lancet Oncol.* **13(4)**, e169–177 (2012).

Weinreb, J. C. et al. PI-RADS prostate imaging - reporting and data system: 2015, version 2. *Eur. Urol.* **69(1)**, 16–40 (2016).

West, C. M. & Barnett, G. C. Genetics and genomics of radiotherapy toxicity: towards prediction. *Genome Med.* **3(8)**, 52 (2011).

Wilkinson, M. D. et al. The FAIR guiding principles for scientific data management and stewardship. *Sci. Data* **3**, 201618 (2016).

Wild, C. P. Complementing the genome with an 'exposome': the outstanding challenge of environmental exposure measurement in molecular epidemiology. *Cancer Epidemiol. Biomark. Prev.* **14(8)**, 1847–1850 (2005).

Wu, W.-H. et al. Review of trends from mobile learning studies: A meta-analysis. *Comput. Educ.* **59(2)**, 817–827 (2012).

Yim, W.-W., Yetisgen, M., Harris, W. P. & Kwan, S. W. Natural language processing in oncology: a review. *JAMA Oncol.* **2(6)**, 797–804 (2016).

Yip, S. S. F. & Aerts, H. J. W. L. Applications and limitations of radiomics. *Phys. Med. Biol.* **61(13)**, R150 (2016).

Zhang, B. et al. Intensity-modulated radiation therapy versus 2D-RT or 3D-CRT for the treatment of nasopharyngeal carcinoma: A systematic review and meta-analysis. *Oral Oncol.* **51(11)**, 1041–1046 (2015).

Quantitative Imaging for Assessing and Predicting Toxicity

Maria Thor and Joseph O. Deasy

CONTENTS

Introduction 402
Quantitative Imaging to Assess Radiation-Induced Normal Tissue Toxicity 402
  Central Nervous System Cancer (CNS) 402
    Brain Injuries (Brainstem, Cerebral Cortex [Gray Matter], and White Matter) 402
  Head and Neck Cancer (HNC) 403
    Xerostomia (Parotid and Submandibular Glands) 403
    Laryngeal and Pharyngeal Toxicity (Primarily Larynx and Pharynx) 404
    Trismus (Masticatory Muscles) 404
  Thoracic Cancer 404
    Cardiac Toxicity (Heart) 404
    Esophageal Toxicity (Esophagus) 404
    Lung Toxicity (Tumor-Subtracted Lung) 405
  Abdominal Cancer 406
    Kidney Toxicity (Kidneys) 406
    Liver Toxicity (Liver) 406
    Small Bowel and Stomach Toxicity (Duodenum, Ileum, Jejunum, and Stomach) 406
  Pelvic Cancer 407
    Gastrointestinal Toxicity (Anal Sphincter, Rectum, and Sigmoid Colon) 407
    Genitourinary Toxicity 407
    Sexual Dysfunction (Penis, Vagina) 407
  Summary and Perspectives 409
Radiomics to Assess Radiation-Induced Normal Tissue Toxicity 409
  Head-and-Neck Cancer RT 409
    Xerostomia (Parotid Glands) 409
    Trismus (Masticatory Muscles) 409

Thoracic RT 409
    Radiation Pneumonitis and Miscellaneous Lung Injury (Tumor-Subtracted Lung) 409
    Summary and Perspectives...410
References..411

INTRODUCTION

The utilization of imaging to quantify radiotherapy (RT) response has recently emerged in parallel to the introduction of radiomics, i.e., high-throughput mining of quantitative features from standard-of-care medical imaging (Lambin et al. 2012). Established quantitative imaging approaches to assess RT response have been explored for both tumors and normal tissues, but radiomics has primarily been exploited for tumors only (Scalco et al. 2017). An interesting question is whether quantification of normal tissue toxicity via imaging could augment, or in some cases even replace, traditional toxicity scoring methods that are often more subjective.

- In this chapter, methods to identify persistent radiation-induced normal tissue toxicity in dose-limiting organs, as quantified by commonly used 3D-imaging modalities in oncology (Computed Tomography [CT], Magnetic Resonance Imaging [MRI], Nuclear Medicine [NM], and Ultrasound [US]) are summarized. In the first section, established quantitative imaging approaches are outlined, giving an overview of the most common modalities with examples, while the second section covers radiomics studies. Similar to the structure in the Quantitative Analyses of Normal Tissue Effects in the Clinic (QUANTEC) papers (Bentzen et al. 2010), the chapter is divided according to tumor site (central nervous system to pelvis), and further divided according to organ and etiology, but not limited to the toxicities conveyed by QUANTEC. In general, studies exploiting state of the art RT regimens (published post-QUANTEC) with positive findings were considered, and a study was included if:

  - available in full-text in English

  - using the same imaging modality pre- and post-RT suggesting quantification of radiation-induced toxicity (exception: post-RT imaging only but no patient presenting with normal tissue toxicity pre-RT)

  - using metrics beyond structure volume (not considered an imaging metric)

  - anchoring imaging findings with clinically observed normal tissue toxicity assessed by physicians and/or patients (exception: toxicities where there is no established clinical method of assessment).

QUANTITATIVE IMAGING TO ASSESS RADIATION-INDUCED NORMAL TISSUE TOXICITY

Central Nervous System Cancer (CNS)

Brain Injuries (Brainstem, Cerebral Cortex [Gray Matter], and White Matter)
Radiation-induced toxicities following CNS RT are infrequent, and thus late and non-transient toxicities such as radionecrosis and cognitive deterioration are of primary

interest (Lawrence et al. 2010). Quantifying these toxicities using imaging requires high sensitivity and specificity as provided by, e.g., MRI in order to differentiate them from tumor recurrence (Walker et al. 2014). In one of the largest descriptive cohort studies on this topic (N = 148) and based on T1-weighted contrast-enhanced MRIs after RT (≤2 years post-RT) compared to before RT, Kumar et al. (2000) found that 20 patients with histo-pathologically confirmed radionecrosis without evidence of tumor progression presented to a larger extent with contrast-enhanced areas with a non-enhanced necrotic core, and scattered necrotic areas of different size within the white matter (referred to as 'Soap bubbles' and 'Swiss cheese', respectively). In a study by Karunamuni et al. (2016), cognitive deterioration as surrogated by the degree of atrophy of the cerebral cortex on MRIs (1 year post-RT subtracted pre-RT scans) significantly correlated with cortical dose (p < 0.001). In a proof-of-concept study of diffusion tensor imaging for evaluation of brainstem injuries in 42 pediatric medulloblastoma patients by Uh et al. (2013), a significant decrease in fractional anisotropy (median 5 years post-RT subtracted pre-RT of diffusion tensor imaging scans) was detected in the transverse pontine fiber (p < 0.001), but not in other regions of the brainstem.

Head and Neck Cancer (HNC)
Xerostomia (Parotid and Submandibular Glands)
Insufficient salivary function (xerostomia) is one of the most prevalent and reported toxicities following HNC RT, and numerous studies have been conducted to quantify salivary production using imaging (Deasy et al. 2010). Imaging with and without gustatory stimulation is used to measure the salivary function within the parotid glands, whereas submandibular salivary gland function is primarily measured with a non-stimulated approach. Recent examples with relatively large cohorts include the study by Cannon et al. (2012) in which the voxel-based pre-RT normalized 18-fluorodeoxyglucose positron emission tomography (18F-FDG PET) signal in the parotid glands ~4 months post-RT in their training cohort (N = 98) accurately predicted the ranking of observed xerostomia rates (±stimulation) in their validation cohort (N = 14; Spearman's rank correlation coefficient, [Rs]: 0.94; p = 0.005). In another HNC cohort (N = 31), Lee et al. (2012) found that the dose-response relationship for xerostomia using salivary excretion function (SEF) from 99mTc scintigraphy with stimulation (1–2 year normalized pre-RT values) as endpoint had a similar performance to that of using clinically assessed xerostomia (area under the Receiver Operating Characteristics curve [AUC]: 0.75). In a subsequent study by this group (Chen et al. 2013), the mean parotid gland dose significantly predicted SEF 1 year post-RT (Rs: 0.65; p < 0.001), and the only correlation established between SEF and clinically assessed xerostomia was between SEF recovery in the submandibular glands and sticky saliva (Rs: 0.43; p = 0.02). In an MRI-directed study, Zhou et al. (2016) found that the four investigated parameters from intravoxel incoherent motion MRI (IVIM), and the two dynamic contrast-enhanced MRI (DCE-MRI) parameters significantly increased one month after, compared to before, RT (p < 0.001–0.04). Also, the change in one diffusion parameter (IVIM), and two perfusion parameters (IVIM, DCE-MRI), was significantly correlated with the change in volume, i.e., atrophy (Rs: −0.45, p = 0.006, and Rs: −0.37; p = 0.01, respectively), but not

with clinically assessed grade $\geq$ 2 xerostomia). It should be pointed out that no gustatory stimulation was performed, and xerostomia rates or xerostomia symptoms at baseline were not specified.

Laryngeal and Pharyngeal Toxicity (Primarily Larynx and Pharynx)
Clinically significant radiation-induced toxicity within the larynx and pharynx typically presents as laryngeal edema, laryngeal fibrosis and/or dysphagia (Rancati et al. 2010). Videofluoroscopy has traditionally been the principal imaging modality of relevance. As in the case of CNS cancer, the imaging method needs to be specific and sensitive to avoid misclassifying tumor progression as toxicity. In a hypothesis-generating study (with no clinically assessed toxicity), based on T1- and T2-weighted MRIs of 72 nasopharyngeal patients, Messer et al. (2016) found that a decrease in late (median of 3.4 years post-RT) T1-weighted signal relative to that of pre-RT is associated with higher superior pharyngeal constrictor mean doses (p = 0.007).

Trismus (Masticatory Muscles)
Radiation-induced mouth-opening inability (trismus) was not covered by the QUANTEC reports. Only Hsieh et al. (2014) have, thus far, conducted an image-based study to detect trismus. In all 22 investigated HNC patients, muscle abnormalities, muscle atrophy and obvious enhancement within T1-weighted contrast-enhanced MRIs or high-intensity on T2-weighted MRIs, were identified and these were further correlated with the clinical staging of trismus within two years post-RT (Rs = 0.52).

Thoracic Cancer
Cardiac Toxicity (Heart)
Cardiac toxicity after RT for thoracic cancer typically presents as pericardial (acute pericarditis and pericardial effusion, and constrictive pericarditis) or valvular (within the left ventricular, and chronic aortic and mitral stenosis) (Filopei et al. 2012). While echocardiography (US of the heart) in which geometric and systolic/diastolic information of the left ventricle and the valve of the pericardium is quantified has been considered the standard imaging modality to identify radiation-induced cardiac toxicity (Moonen et al. 2017), cardiovascular magnetic resonance imaging (CMR) has proven superior in terms of detection reproducibility in non-cancer subjects with cardiac toxicity (Grothues et al. 2002), and should, thus, be further explored as an alternative modality also within oncology (Moonen et al. 2017). Large-scale data on imaging to assess radiation-induced cardiac toxicity is underway (Jacob et al. 2016), however, in a small cohort study, Cella et al (2011) found indications that valvular defect, i.e., regurgitation/stenosis in any cardiac valve as measured with echocardiography, in 56 Hodgkin's lymphoma patients is explained by intermediate doses to both ventricles and to the left atrium (Odds Ratios: 4.4 and 7.2, respectively).

Esophageal Toxicity (Esophagus)
Esophageal toxicity following RT is classified as acute and/or late esophagitis, including stricture and dysphagia (Werner-Wasik et al. 2010). In two studies that included non small

cell lung cancer (NSCLC) patients (N = 27 in (Mehmood et al., 2016; N = 50 in Yuan et al., 2014), pre-RT normalized 18F-FDG PET scans during RT were found to predict esophagitis during RT (grade ≥2 rates: 74% and 32%, respectively). In particular, Mehmood et al. (2016) observed that patients with acute esophagitis at a median of three weeks post-RT had significantly higher esophagus peak standardized uptake values (SUVpeak) at the last week of RT (AUC: 0.69; p = 0.03) in addition to larger percent esophageal volumes receiving ≥50 Gy (V_{50Gy}) and and ≥ 60 Gy (V_{60Gy}) (AUC: 0.67 and 0.66, respectively). Patients with esophagitis (minimum follow-up: two years) in the study by Yuan et al. (2014) had significantly higher esophagus maximum SUVmax about midway through the RT course, higher maximum esophageal dose, and received concurrent chemotherapy to a larger extent (AUC range: 0.70–0.78; p 0.01–0.03). In a geometrically driven study based on weekly CT scans of 85 NSCLC patients, Niedzielski et al. (2016) found that in particular the maximum esophageal expansion and the esophageal length with the axial expansion ≥30% during RT relative to pre-RT predicted acute esophagitis observed on average at the same treatment fraction (AUC: 0.93 and 0.91, respectively; p < 0.0001 in both cases).

Lung Toxicity (Tumor-Subtracted Lung)

Radiation pneumonitis (RP) has been by far the most studied radiation-induced lung toxicity within a dose-response setting, and typically manifests as ground-glass opacity within the lung, although it is also associated with an episodic inflammatory phase (Marks et al. 2010). Lung toxicity analyses have typically been performed on the tumor-subtracted lung. In a recent CT-based study including 220 NSCLC patients, a multivariate model for RP (RP: four-grade categorization based on appearance on CT scans six months post-RT) included longer time to RP (30 days) and larger relative lung volumes irradiated to ≥40 Gy (V_{40Gy}, p < 0.001) (Bernchou et al. 2017). Furthermore, the final multivariate models for the two additionally investigated lung injuries, i.e., interstitial and consolidation, included the same two variables as well as increased age and current smoking status (predisposing), respectively. Based on the change in Hounsfield Units (ΔHU, three months post-RT subtracted pre-RT) from CT images (for 95 primarily NSCLC patients (75% of the total cohort), De Ruysscher et al. (2013) generated a linear CT and dose-response relationship (ΔHU/Gy) within the lung. A significantly larger fraction of patients experienced dyspnea if ΔHU/Gy > 0.5/Gy: The rate of grade ≥2 dyspnea within four months post-RT in the whole was 35%, while the rate was 8/26 (30.8%) if ΔHU/Gy ≤ 0.5/Gy vs. 25/69 (36.3%) if ΔHU/Gy > 0.5/Gy (p < 0.001). ΔHU/Gy > 0.5/Gy was suggested as a potentially clinical significant cutoff. From the same group and using a similar path of analysis (ΔHU: within six months post-RT subtracted pre-RT) but within a smaller NSCLC cohort (N = 58), mean ΔHU significantly correlated with physician-scored radiological pulmonary fibrosis (grade ≥1 within six months post-RT; rate in the whole cohort 53%; X^2 test: p = 0.01) (Sharifi et al. 2016). In a longitudinal study comprising 136 CT scans (median: 26 months post-RT) from 22 selected NSCLC patients, Mattonen et al. (2013) successfully separated fibrosis from recurrence (rates: 59% and 41%, respectively), and in particular uncovered that, from 9 months after RT and onwards, patients with recurrence had denser lungs than patients with fibrosis (ΔHUmean-: −96 vs. −143;

p = 0.05). A comparable separation could not be made using pre-RT CT scans only. Using instead cone beam CT (CBCT) scans of 135 NSCLC patients, Bernchou et al. (2015) established a multivariate model for lung radioresponsivness based on changes in lung density (as measured by HU on CBCT scans) between CBCT at the first treatment fraction and CBCT scans acquired later during RT or during follow-up (within 6 months post-RT). These models included advanced age and changes in lung density at either the 10th, 20th, or 30th treatment fraction. All models presented with comparable discriminative ability (AUC 0.81–0.82; p < 0.001 in all cases).

Abdominal Cancer

Kidney Toxicity (Kidneys)

Chronic kidney injuries typically involve malignant hypertension, elevated creatine levels, anemia and renal failure, and have commonly been quantified by imaging using single-photon emission computed tomography (SPECT) with 99mTc (Dawson et al. 2010). In a prospective renal cell carcinoma trial (N = 21), the net change in glomerular filtration rate, assessed from Chromium-51 ethylene diamine tetracetic (51Cr-EDTA) test and 99mTc-dimercaptosuccinin (99mTc-DMSA) SPECT, was evaluated during follow-up and found to be significantly decreased one year post- compared to pre-RT. In addition, for every 10 Gy of physical dose delivered, an exponential decline in glomerular filtration rate was observed (p range: < 0.01–0.03) (Siva et al. 2016). Furthermore, a tolerance dose level at 10 Gy was further identified, below which damage was negligible.

Liver Toxicity (Liver)

Radiation-induced liver toxicity can lead to liver failure and death (Schuffenegger et al. 2017). The incomplete etiological pattern involves blockage and destruction of veins within the hepatic lobules, retrograde congestion, and hepatocyte necrosis. Caution interpreting imaging findings such as decreased signals on CT or MRI scans should be made in order to avoid misclassifying tumor progression as liver toxicity. In a cohort of 23 intrahepatic patients, Wang et al. (2016) derived a local-and-global liver function model based on normalized DCE-MRI measures obtained before, during and one month after RT, and on simultaneously assessed indocyanine green clearance. A relatively steep perfusion probability function of indocyanine green clearance was found; this finding was in addition consistent across fractionation differences and between the hepatocellular and non-hepatocellular groups of patients.

Small Bowel and Stomach Toxicity (Duodenum, Ileum, Jejunum, and Stomach)

The most common radiation-induced persistent toxicities within the small bowel and the stomach include fibrosis of the bowel walls and mucositis of the stomach leading to inflammation, i.e., enteritis (small bowel) and gastritis (stomach), and is expressed as e.g., bleeding, diarrhoea, fistula, obstruction, perforation, and ulceration (Kavanagh et al. 2010). Enteritis could hypothetically be detected with anatomical imaging such as CT and MRI, but so far only one study has exploited MRI post-RT, including just four cases (Algin et al. 2011), in which indications of abnormalities (thickening, accumulated contrast in long segments, mesenteric fat with stranding, and luminal narrowing) were observed primarily in the distal ileum. Imaging has not yet been exploited to quantify radiation-induced gastritis.

Pelvic Cancer

Gastrointestinal Toxicity (Anal Sphincter, Rectum, and Sigmoid Colon)

RT-induced gastrointestinal toxicity primarily presents as colitis and proctitis (e.g., bleeding, diarrhoea, obstruction from stricture/fistula, pain, and urgency). This is attributed to damage of the anal sphincter, rectum, or sigmoid colon, and appears distinctly on CT scans as thickening of the colonic and rectal wall, thickening of the perirectal fascia, as well as increased perirectal fat (Viswanathan et al. 2012). In an exploratory analysis of 34 prostate cancer patients, based on anal sphincter morphology from US, Yeoh et al. (2012) observed inner anal sphincter wall thinning ~2–3 years compared to pre-RT (2.1 mm vs. 2.4 mm). However, this was not a statistically significant finding, and the authors did not investigate whether it was related to any of the six physician-assessed gastrointestinal symptoms (stool consistency, frequency; rectal bleeding, mucous discharge, pain; urgency of defecation).

Genitourinary Toxicity

The onset of chronic genitourinary toxicity has a wide range of latency times, and can take up to 20 years to manifest after RT (Smit et al. 2010). While the radiation-induced etiology is known as 'unfinished business' (Viswanathan et al. 2010), most symptoms such as dysuria, frequency, hematuria, and urgency are thought to arise from inflammation of the urinary tract (cystitis). In the review by Smit et al. (2010), US of both the lower and upper urinary tract was recommended to accurately identify cystitis, but, unfortunately, such an approach hasn't been carried out following state-of-the-art RT regimens.

Sexual Dysfunction (Penis, Vagina)

No image-driven study has elaborated further on radiation-induced sexual dysfunction in men since the publication of the related QUANTEC report in which it was concluded that the penile bulb is only a surrogate of the multi organ apparatus behind sexual dysfunctional (Roach et al. 2010). In a pre-QUANTEC study by Zelefsky et al. (1998) that comprised 38 prostate cancer patients without erectile dysfunction prior to RT, duplex US (at a median time of 14 months post-RT), measuring the penile blood flow, suggested that erectile dysfunction is primarily caused by insufficient arterial blood flow (arteriogenic) followed by abnormal distensibility of the corpora cavernosa (cavernosal). Functional MRI is also promising to assess early and chronic changes of the structures involved in erection: An example is shown in Figure 16.1, focusing on diffusion-weighted MRI for patients treated for prostate cancer.

Radiation-induced sexual dysfunction in women was not covered by QUANTEC, though it also involves multiple organs and symptoms (e.g., vaginal decreased elasticity, dryness, and pain) (Agarwal et al. 2017). In a case-control study (N = 24), Yang et al. (2013) observed that all four investigated transvaginal US parameters (vaginal wall thickness, intensity,

Nakamagi shape, and probability density function) significantly increased in the cases with mild-moderate vaginal fibrosis when compared to controls (p range: < 0.001–0.03). In a recently published study by Agarwal et al. (2017), various T1- and T2-weighted MRI sequences together with diffusion-weighted MRI were suggested to improve visualization of the anatomy of the female perineum. The foregoing imaging techniques could also have implications in the setting of quantifying sexual dysfunction following RT.

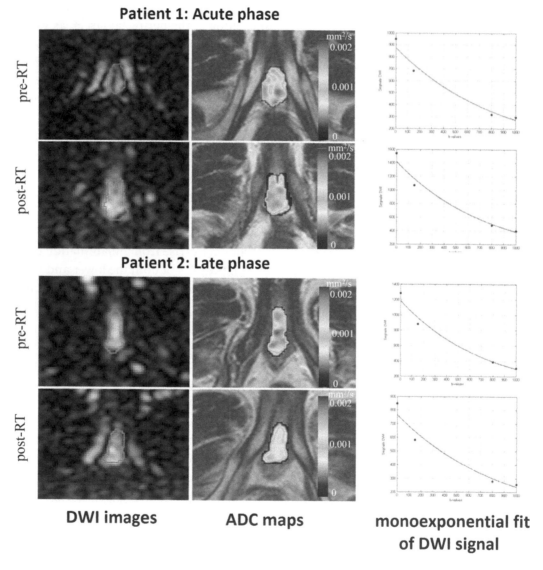

FIGURE 16.1 Pre and post-RT DWI images (left column), ADC maps (middle column) and mono-exponential fit of DWI signal (right column) for two example patients sorted from a cohort of patients with MRI acquired in an acute phase (patient 1, post-RT MRI at 1 month after RT completion) and from a cohort of patients with MRI acquired in a late phase (patient 2, post-RT MRI at 12 months after RT completion). Courtesy of Elisa Scalco and Ileana Pirovano, unpublished.

Summary and Perspectives

Quantitative imaging methods established for tumor response assessment have been explored to a much lesser degree for normal tissue toxicity. Some of the considered toxicities were covered by a couple of image-based studies (e.g., for esophagitis and xerostomia), but overall only one or even no study was identified. There is, therefore, an opportunity for future studies to explore quantitative imaging in this regard, and to seek for objective assessments of normal tissue toxicity.

RADIOMICS TO ASSESS RADIATION-INDUCED NORMAL TISSUE TOXICITY

Head-and-Neck Cancer RT

Xerostomia (Parotid Glands)

Using US images, Yang et al. (2012) found that seven of the eight investigated second order radiomic features across the four investigated angles (0°, 45°, 90°, 135°) were significantly different (p < 0.001) between the 24 parotid glands from 12 HNC patients after RT (median time: 17 months post-RT), and the 14 parotid glands from seven healthy volunteers . The post-RT parotid glands were more heterogeneous (lower angular second movement and correlation; higher contrast, entropy, and variance), and more asymmetric (higher cluster prominence and cluster shade) than those of the healthy volunteers. In a cohort of 107 HNC patients, van Dijk et al. (2017) observed that CT-based features related to a reduction of the parotid gland surface (six weeks post- non-contrast enhanced CT scans normalized pre-RT contrast-enhanced CT scans) together with baseline and acute xerostomia (six weeks post-RT) improved the ability to predict moderate-severe xerostomia one year after RT (rate in the cohort: 30%) compared to using baseline and acute xerostomia alone (bootstrapped AUC including imaging features 0.90 vs. without imaging features 0.85). Although replacing the parotid gland surface with volume (the latter initially excluded due to Pearson's correlation = 0.91) did not change their model performance (non-corrected bootstrapped AUC: 0.90 vs. 0.89), the authors hypothesized that the surface contains more shape information than that of the volume. Based on 48 parotid glands from 24 patients, Belli et al. (2014) found that CT changes within the first two weeks of RT significantly predicted the average xerostomia score during RT (dichotomized at mean grade of xerostomia during treatment = 1.57, OR = 0.15; AUC = 0.74; p = 0.02). The best cut-off value for CT changes was −0.50 HU/day.

Trismus (Masticatory Muscles)

In an exploratory case-control study of mild-severe trismus within one year post-RT (N = 20) and based on 24 radiomic features extracted from T1-weighted MR scans (three months post-RT), Thor et al. (2017) observed a trend of radiomics features within the masseter and the medial pterygoid muscles being more heterogeneous (higher Haralick correlation; p range: 0.12–0.14) for cases compared to controls.

Thoracic RT

Radiation Pneumonitis and Miscellaneous Lung Injury (Tumor-Subtracted Lung)

In an esophageal cancer cohort (N = 106), Cunliffe et al. (2015) found that all ΔHU (ΔHU: within four months post-RT subtracted pre-RT) representations, i.e., the 20 selected radiomic features, better separated grade ≥ 2 RP (onset within one year post-RT; rate in the whole cohort: 19%) within higher dose regions as opposed to within lower dose regions (mean AUC across 20 radiomic features: 0.71 vs. 0.64–0.68). Twelve radiomic features significantly predicted RP ($p < 0.0025$), and an overall linear relationship between monotonously increasing dose and ΔHU was found. In the study by Palma et al. (2011), the mean ΔHU (ΔHU: 3 months post-RT subtracted pre-RT) within 3 cm outside the tumor significantly correlated with RP (grade 0–3, $p < 0.001$) in a cohort of 18 NSCLC patients (distributed uniformly over the four RP categories), while the mean ΔHU of the ipsilateral lung did not ($p = 0.22$). Within a preceding lung cancer study by Cunliffe et al. (2014) (N = 26; NSCLC: 69%; ΔHU: within 3 months post-RT subtracted pre-RT), a linear relationship between 19/20 investigated radiomic features and physician-scored radiological lung abnormality was established (1–4 scale for lung abnormalities, Mild-severe abnormality identified on three month post-RT scans), and the highest AUCs were obtained when distinguishing severe from no-moderate lung abnormality (mean AUC across 20 radiomics: 0.84 vs. 0.68-0.81). In a longitudinal analysis of lung injury for 14 NSLC patients (0–3 scale for lung injury, No-severe injury identified on 3, 6, and 9 months post-RT scans), Moran et al. (2017) found that 8/9 ΔHU radiomics resulted in robust and significant dose-response relationships at all three investigated time points ($p < 0.001$–0.009). In particular, three second order radiomic features significantly distinguished moderate-severe from no-mild lung injury (correlation, dissimilarity, contrast, p 0.01–0.04).

Summary and Perspectives

Radiomics from conventional imaging could have potential for poorly understood RT-induced injuries (cf. first section of this chapter). However, to date, accumulated results from studies in which radiomics has been used to quantify radiation-induced normal tissue toxicity are limited, in particular beyond CT quantifications of radiation-induced toxicities within the lung. The six lung toxicity studies together pointed out that radiomics assessed from CT changes post- and pre-RT conveys information related to dyspnea, fibrosis, RP, and lung abnormality/injury. Nevertheless, to validate radiomic features as surrogates for normal tissue toxicity, and for a broader use in general within the setting of patient-specific RT, we agree with key points recently suggested in the review by Lambin et al. (2017) summarized here:

- Input: use standardized imaging protocols (and uncover complete protocol specifics), including radiomics definitions

- Robustness: investigate robustness of radiomic features with respect to scanning-specific parameters (phantom studies), and to organ segmentation and motion (multiple segmentations and repeat imaging)

- Analysis: define a composite cohort where radiomic features are added to routinely acquired characteristics, extract radiomic features in an unbiased manner (remove highly correlated features), exploit the use of multiple modeling methodologies

- Validate results internally/externally

- Provide open-access to data

The use of quantitative and standardized imaging features to objectively measure normal tissue toxicity due to radiation therapy is likely to be a rewarding endeavour. However, we are surely still nearer the base of the mountain than the summit.

REFERENCES

Agarwal, M.D., Resnick, E.L., Mhuircheartaigh JN, and Mortele KJ. MR imaging of the female perineum: Clitoris, labia, and introitus. *Magn Reson Imaging Clin N Am* 2017; 25(3):435–55.

Algin O, Turkbey B, Ozmen E, and Algin E. Magnetic resonance enterography findings of chronic radiation enteritis. *Cancer Imaging* 2011; 11:189–94.

Belli ML, Scalco E, Sanguineti G, Fiorino C, Broggi S, Dinapoli N, Ricchetti F, Valentini V, Rizzo G, and Cattaneo GM. Early changes of parotid density and volume predict modifications at the end of therapy and intensity of acute xerostomia. *Strahlenther Onkol* 2014; 190(11):1001–7.

Bentzen SM, Constine LS, Deasy JO, Eisbruch A, Jackson A, Marks LB, Ten Haken RK, and Yorke ED. Quantitative Analyses of Normal Tissue Effects in the Clinic (QUANTEC): an introduction to the scientific issues. *Int J Radiat Oncol Biol Phys* 2010; 76(3 suppl):3–9.

Bernchou U, Hansen O, Schytte T, Bertelsen A, Hope A, Moseley D, and Brink C. Prediction of lung density changes after radiotherapy by cone beam computed tomography response markers and pre-treatment factor for non-small cell lung cancer patients. *Radiother Oncol* 2015; 117(1):17–22.

Bernchou U, Christiansen RL, Asmussen JT, Schytte T, Hansen O, and Brink C. Extent and computed tomography appearance of early radiation induced lung injury for non-small cell lung cancer. *Radiother Oncol* 2017; 123(1):93–8.

Cannon B, Schwartz DL, and Dong L. Metabolic imaging biomarkers of postradiotherapy xerostomia. *Int J Radiat Oncol Biol Phys* 2012; 83(5):1609–16.

Cella L, Liuzzi R, Conson M, Torre G, Caterino M, De Rosa N, Picardi M, Camera L, Solla R, Farella A, Salvatore M, and Pacelli R. Dosimetric predictors of asymptomatic heart valvular dysfunction following mediastinal irradiation for Hodgkin's lymphoma. *Radiother Oncol* 2011; 101(2):316–21.

Chen WC, Lai CH, Lee TF, Hung CH, Liu KC, Tsai MF, Wang WH, Chen H, Fang FM, and Chen MF. Scintigraphic assessment of salivary function after intensity-modulated radiotherapy for head and neck cancer: Correlations with parotid dose and quality of life. *Oral Oncol* 2013; 49(1):42–8.

Cunliffe AR, Armato SG, Straus C, Mailk R, and Al-Hallaq HA. Lung texture in serial thoracic CT scans: Correlation with radiologist-defined severity of acute changes following radiation therapy. *Phys Med Biol* 2014; 59(18):5387–98.

Cunliffe AR, Armato SG, Castillo R, Pham N, Guerrero T, and Al-Hallaq HA. Lung texture in serial thoracic computed tomography scans: Correlation of radiomics-based features with radiation therapy dose and radiation pneumonitis development. *Int J Radiat Oncol Biol Phys* 2015; 91(5):1048–56.

Dawson LA, Kavanagh BD, Paulino AC, Das SK, Miften M, Li XA, Pan C, Ten Haken RK, and Schultheiss TE. Radiation-associated kidney injury. *Int J Radiat Oncol Biol Phys* 2010; 76(3 suppl):108–15.

De Ruysscher D, Sharifi H, Defraene G, Kerns SL, Christiaens M, De Ruyck K, Peeters S, Vansteenkiste J, Jeraj R, Van Den Heuvel D, and van Elmpt W. Quantification of radiation-induced lung damage with CT scans: The possible benefit for radiogenomics. *Acta Oncol* 2013; 52(7):1405–10.

Deasy JO, Moiseenko V, Marks L, Chao C, Nam J, and Eisbruch A. Radiotherapy dose-volume effects on salivary gland function. *Int J Radiat Oncol Biol Phys* 2010; 76(3 suppl): 58–63.

Filopei J and Frishman W. Radiation-induced heart disease. *Cardiol Rev* 2012; 20:184–8.

Grothues F, Smith GC, Moon JC, Bellenger NG, Collins P, Klein HU, and Pennel DJ. Comparison of interstudy reproducibility of cardiovascular magnetic resonance with two-dimensional echocardiography in normal subjects and in patients with heart failure or left ventricular hypertrophy. *Am J Cardiol* 2002; 90(1):29–34.

Hsieh LC, Chen JW, Wang LY, Tsang YM, Shueng PW, Liao LJ, Lin YC, Tseng CF, Kuo YS, Jhuang JY, Tien HJ, Juan HF, and Hsieh CH. Predicting the severity and prognosis of trismus after intensity-modulated radiation therapy for oral cancer patients by magnetic resonance imaging. *PLoS One* 2014; 9(3):e92561.

Jacob S, Pathak A, Franck D, Latzoreff I, Jiminez G, Fondard O, Lapeyre M, Colombier D, Brugiere E, Lairez O, Fontenel B, Milliat F, Tamarat R, Broggio D, Derremumaux S, Ducassou M, Ferrieres J, Laurier D, Benderitter M, and Berner MO. Early detection and prediction of cardiotoxicity after radiation therapy for breast cancer: the BACCARAT prospective cohort study. *Radiat Oncol* 2016; 11:1–10.

Karunamuni RA, Moore KL, Seibert TM, Li N, White N, Bartsch H, Carmona R, Marshall D, McDonald CR, Farid N, Krishnan A, Kuperman J, Mell LK, Brewer J, Dale AM, Moiseenko V, and Hattangadi-Gluth JA. Radiation sparing of cerebral cortex in brain tumor patients using quantitative neuroimaging. *Radiother Oncol* 2016; 118(1):29–34.

Kavanagh BD, Pan CC, Dawson LA, Das SK, Li XA, Ten Haken RK, and Miften M. Radiation dose-volume effects in the stomach and bowel. *Int J Radiat Oncol Biol Phys* 2010; 76(3 suppl):101–7.

Kumar AJ, Leeds NE, Fuller GN, van Tassel P, Maor MH, Sawaya RE, and Levin VA. Malignant gliomas: MR imaging spectrum of radiation therapy- and chemotherapy-induced necrosis of the brain after treatment. *Radiology* 2000; 217(2):377–84.

Lambin P, Rios-Velasquez E, Leijenaar Rm Carvalho S, van Stiphout RG, Granton P, Zegers CM, Gillies R, Boellard R, Dekker Am abd Aerts HJ. Radiomics: Extracting more information from medical images using advanced feature analysis. *Eur J Cancer* 2012; 48(4):441–6.

Lambin P, Leijnaar RTH, Deist TM, Peerlings J, de Jong EEC, van Timmeren J, Sandauleanue S, Larue RTHM, Evan AJG, Jochems A, van Wijk Y, Woodruff H, van Soest J, Lustberg T, Roelofs E, van Elmpt W, Dekker A, Mottaghy FM, Wildberger JE, and Walsh S. Radiomics: The bridge between medical imaging and personalized medicine. Nat Rev Clin Oncol. 2017 14(12):749-762.

Lawrence YR, Li XA, El Naqa I, Hahn CA, Marks LB, Merchant TE, and Dicker AP. Radiation dose-volume effects in the brain. *Int J Radiat Oncol Biol Phys* 2010; 76(3 suppl): 20–7.

Lee TF, Chao PJ, Wang HY, Hsu HC, Chang P, and Chen WC. Normal tissue complication probability model parameter estimation for xerostomia in head and neck cancer patients based on scintigraphy and quality of life assessments. *BMC Cancer* 2012; 12:1–9.

Marks LB, Bentzen SM, Deasy JO, Kong FM, Bradley JD, Vogelius IS, El Naqa I, Hubbs JL, Lebesque JV, Timmerman RD, Martel MK, and Jackson A. Radiation dose-volume effects in the lung. *Int J Radiat Oncol Biol Phys* 2010; 76(3 suppl): 70–6.

Mattonen SA, Palma DA, Haasbeek JA, Senan S, and Ward AD. Distinguishing radiation fibrosis from tumour recurrence after stereotactic ablative radiotherapy (SABR) for lung cancer: A quantitative analysis of CT density changes. *Acta Oncol* 2013; 52(5):910–8.

Mehmood Q, Sun A, Becker N, Higgins J, Marshall A, Le LW, Vines DC, McCloskey P, Ford V, Clarke K, Yap M, Bezjak A, and Bissonnette JP. Predicting radiation esophagitis using 18F-FDG PET during chemoradiotherapy for local advanced non-small cell lung cancer. *J Thorac Oncol* 2016; 11(2):213–21.

Messer JA, Mohamed AS, Hutcheson KA, Ding Y, Lewin JS, Wang J, Lai SY, Frank SJ, Garden AS, Sandulache V, Eichelberger H, French CC, Colen RR, Kalpathy-Cramer J, Hazle JD, Rosenthal DI, Gunn GB, and Fuller CD. Magnetic resonance imaging of swallowing-related structures in nasopharyngeal carcinoma patients receiving IMRT: Longitudinal dose-response characterization of quantitative signal kinetics. *Radiother Oncol* 2016; 118(2):315–22.

Moonen M, Oury C, and Lancelotti P. Cardiac Imaging: Multimodality and surveillance strategies in detection of cardiotoxicity. *Curr Oncol Rep* 2017; 19(10): 63.

Niedzielski JS, Yang J, Stingo F, Martel MK, Mohan R, Gomez DR, Briere TM, Lia Z, and Court LE. Objectively quantifying radiation esophagitis with novel computed tomography-based metrics. *Int J Radiat Oncol Biol Phys* 2016; 94(2):385–93.

Palma DA, van Sörnsen de Koste JR, Verbakel WF, and Senan S. A new approach to quantifying lung damage after stereotactic body radiation therapy. *Acta Oncol* 2011; 50(4):509–17.

Rancati T, Schwarz M, Allen AM, Fenf F, Popovtzer A, Mittal B, and Eisbruch A. Radiation dose-volume effects in the larynx and pharynx. *Int J Radiat Oncol Biol Phys* 2010; 76(3 suppl):64–9.

Roach M, Nam J, Gagliardi G, El Naqa I, Deasy JO, and Marks LB. Radiation dose-volume effects and the penile bulb. *Int J Radiat Oncol Biol Phys* 2010;76(3 suppl):130–4.

Scalco E, and Rizzo G. Texture analysis of medical images for radiotherapy applications. *Br J Radiol* 2017; 90(1070):1–15.

Schuffenegger PM, Ng S, and Dawson LA. Radiation-induced liver toxicity. *Semin Radiat Oncol* 2017; 2784):350–7.

Sharifi H, van Elmpt W, Oberije C, Nalbantov G, Das M, Öllers M, Lambin P, Dingmans AMC, and De Ruysscher D. Quantification of CT-assessed radiation-induced lung damage in lung cancer patients treated with or without chemotherapy and cetuximab. *Acta Oncol* 2016; 55(2):156–62.

Siva S, Jackson P, Kron T, Bressel M, Lau E, Hofman M, Shaw M, Chander S, Pham D, Lawrentschuk N, Wong LM, Goad J, and Foroudi F. Impact of stereotactic radiotherapy on kidney function in primary renal cell carcinoma: Establishing a dose-response relationship. *Radiother Oncol* 2016; 118(3):540–6.

Smit SG and Heyns CF. Management of radiation cystitis. *Nat Rev Urol* 2010; 7(4):206–14.

Thor M, Tyagi N, Hatzoglou V, Apte A, Saleh Z, Riaz N, Lee NY, and Deasy JO. A Magnetic Resonance Imaging-based approach to quantify radiation-induced normal tissue injuries applied to trismus in head and neck cancer. *Phys Imaging Radiat Oncol* 2017; 34–40.

Uh J, Merchant TE, Li Y, Feng T, Gajjar A, Ogg RJ, and Hua C. Differences in brainstem fiber tract response to radiation: A longitudinal diffusion tensor imaging study. *Int J Radiat Oncol Biol Phys* 2013; 86(2):292–97.

Van Dijk LV, Brouwer CL, van der Laan HP, Burgerhof JGM, Langendijk JA, Steenbakkers RJHM, and Sijtsema NM. Geometric image biomarker changes of the parotid gland are associated with late xerostomia. *Int J Radiat Oncol Biol Phys* 2017; 99(5):1101-1110.

Viswanathan AN, Yorke ED, Marks LB, Eifel PJ, and Shipley WU. Radiation dose-volume effects of the urinary bladder. *Int J Radiat Oncol Biol Phys* 2010; 76(3 suppl):116–22.

Viswanathan C, Bhosale P, Ganeshan DM, Truong MT, Silverman P, and Balachandran A. Imaging of complications of oncological therapy in the gastrointestinal system. *Cancer Imaging* 2012; 12:163–72.

Walker AJ, Ruzevick J, Malayeri AA, Rigamonti D, Lim M, Redmond KJ, and Kleinberg L. Postradiation imaging changes in the CNS: How can we differentiate between treatment effect and disease progression? *Future Oncol* 2014; 10(7):1277–97.

Wang H, Feng MF, Jackson A, Ten Haken RK, Lawrence TS, and Cao Y. Local and global function model of the liver. *Int J Radiat Oncol Biol Phys* 2016; 94(1):181–8.

Werner-Wasik M, Yorke E, Deasy JO. Nam J, and Marks LB. Radiation dose-volume effects in the esophagus. *Int J Radiat Oncol Biol Phys* 2010; 76(3 suppl): 86–93.

Yang X, Tridandapani S, Beitler JJ, Yu DS, Yoshida EJ, Curran WJ, and Liu T. Ultrasound GLCM texture analysis of radiation-induced parotid-gland injury in head-and-neck cancer radiotherapy: An *in vivo* study of late toxicity. *Med Phys* 2012; 39(9):5732–9.

Yang X, Rossi P, Bruner DW, Tridandapani S, Shelton J, and Liu T. Noninvasive evaluation of vaginal fibrosis following radiotherapy for gynecological malignancies: A feasibility study with ultrasound B-mode and Nakagami parameter imaging. *Med Phys* 2013; 40(2):022901.

Yeoh EK, Holloway RH, Fraser RJ, Botten RJ, Di Matteo AC, and Butters J. Pathophysiology and natural history of anorectal sequelae following radiation therapy for carcinoma of the prostate. *Int J Radiat Oncol Biol Phys* 2012; 84(5):593–9.

Yuan ST, Brown RKJ, Zhao L, Ten Haken RK, Gross M, Cease KB, Schipper M, Stanton P, Yo U, and Kong FM. Timing and intensity of changes in FDG uptake with symptomatic esophagitis during radiotherapy or chemo-radiotherapy. *Radiat Oncol* 2014; 9(1):37.

Zelefsky MJ and Eid JF. Elucidating the etiology of erectile dysfunction after definitive therapy for prostatic cancer. *Int J Radiat Oncol Biol Phys* 1998; 40(1):129–33.

Zhou N, Chu C, Dou X, Li M, Liu S, Zhu L, Liu B, Guo T, Chen W, He J, Yan J, Zhou Z, Yang X, and Liu T. Early evaluation of irradiated parotid glands with intravoxel incoherent motion MR imaging: Correlation with dynamic contrast-enhanced MR imaging. *BMC Cancer* 2016; 16(1):865.

CHAPTER 17

Beyond DVH

2D/3D-Based Dose Comparison to Assess Predictors of Toxicity

Oscar Acosta and Renaud de Crevoisier

CONTENTS

Introduction 415
   Beyond the Individual Organ Dose-Volume Relationship: Adding Spatial
   Considerations to the Analysis 417
   Towards a Local Analysis of Dose-Effect Relationships 422
      Dose-Surface Maps 422
      3D Dose-Volume Maps (DVMs) 422
      DSM and 3D DVM Methodological Challenges 422
   Spatial Normalization for Dose Mapping 423
Inter-Individual Registration and Validation 425
   Template Selection 426
Pixel/Voxel-Wise Analysis in a Common Coordinate System 429
Towards a Patient-Specific Planning 430
   Sub-Region(s) 432
Discussion and Conclusion 433
Acknowledgements 436
References 436

INTRODUCTION

To estimate the risk of radio-induced toxicities, predictive models have exploited dosimetric data at different spatial scales. Scalar values such as the Effective Dose or the Equivalent Uniform Dose, representing with a single parameter the dose to an organ, have been proved to be correlated with toxicity (Boulé et al. 2009). Most of the Normal Tissue Complication Probability (NTCP) models are based on the dose-volume histograms (DVH) (Fiorino et al. 2009). Lyman's model (Lyman 1985) appears among the most frequently cited DVH-models. However, the DVHs reduce the 3-dimensional (3D) dose distribution within an organ to a uni-dimensional and discrete representation of the dose-volume relationship.

- Several limitations arise when only scalar values or DVHs are used within predictive models:

  - A DVH is limited to a single organ.

  - By construction, different 3D dose distributions may lead to the same DVH.

  - The information on the spatial distribution of dose is lost by merely considering the organ volume, thus ignoring the local 3D variations and thus the likely heterogeneous intra-organ radiosensitivity.

  - Correlation may exist between adjacent DVH bins.

  - More broadly, toxicity is a complex multiparametric phenomenon that may involve structures at different scales, from sub-organ parcels to large structures or regions whose response may additionally depend on individual radiosensitivity.

These inconvenient factors may explain the limited prediction capability of DVH-based models.

Dimensionality reduction, feature extraction strategies together with machine learning methodologies aimed at exploiting more available multimodal data have emerged to overcome some of these issues in the last few years, exhibiting promising prediction capabilities. Among them, Principal Component Analysis (PCA) was proposed to reduce the dimensionality of the DVH data and quantify the variability of DVH shapes (Söhn et al. 2007); functional data analysis (Dean et al. 2016) enabled a representation of the DVH as a curve rather than discrete measurements through the use of basic functions. This allows for overcoming of the correlation issues between adjacent bins. Functional PCA was also proposed for simultaneous functional data analysis and PCA dimensionality reduction (Benadjaoud et al. 2014); Independent Component Analysis (ICA) was proposed for feature extraction exploiting more statistical information, such as mutual independence (Fargeas et al. 2018). By minimizing statistical dependence between components extracted from the DVH a better separation between patients is achieved. Other machine learning methods such as artificial neural networks (Gulliford et al. 2004; Tomatis et al. 2012), together with genetic algorithms and comparisons with support vector machines (Pella et al. 2011) or random forest (Ospina et al. 2014), have also been investigated, reporting competitive, predictive results. A thorough review and comparison of different machine learning methods was recently undertaken by Yahya et al. (2016) for the prediction of urinary symptoms following prostate cancer radiotherapy. Figure 17.1 offers a global overview of some of the existing predictive models with respect to the origin and scale of gathered data. For example, at the left part of Figure 17.1 we consider the DVH as a source of dosimetric information, where the two-step workflow of dimensionality reduction-feature extraction and modelling

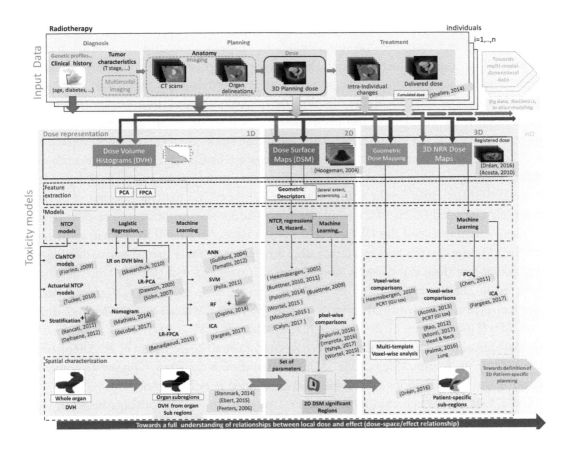

FIGURE 17.1 Overview of the state of the art of predictive models of toxicity. They are all based on population data and appear divided in three different categories depending on how the 3-dimensional (3D) dose and patients' characteristics are considered. Either through the dose-volume histogram (DVH), dose-surface maps, or 3D Non-Rigidly registered dose maps.

is shown. This workflow has been followed by some of the aforementioned DVH-based models (NTCP, regressions, machine learning).

Beyond the Individual Organ Dose-Volume Relationship: Adding Spatial Considerations to the Analysis

Although these models are continuously being improved and are bringing new insights into the understanding of dose-volume effects from DVHs, the data collected through the whole radiotherapy treatment (clinical history, tumor stage, multimodal imaging, organ delineations, 3D dose distribution, intra-individuals changes, etc.) are very rich and on most occasions not thoroughly exploited. The planning dose for instance is a 3D function presenting a high gradient within the organs at risk, a large variability across the treated populations and it is strongly linked to the individual's anatomy, which in addition may change between-fractions sometimes at the expense of the prediction (Shelley et al. 2017; Palorini et al. 2016a). With the steadily increasing computational capabilities, exploiting information from more available data within integrative approaches becomes nowadays more feasible.

To address some of the predictive limitations of organ DVHs, recent approaches aim at investigating more localized dose-toxicity relationships by analyzing the dose at lower spatial scales. This trend is illustrated at the bottom part in Figure 17.1 (spatial characterization) whereby identification of sensitive sub-regions is improving the prediction capabilities. Thus, through the separate analysis of dosimetric parameters of the lower gastrointestinal (GI) anatomy, for example, it has been shown that dose on some sub-regions correlated better with specific toxicities. This is the case of the works undertaken by Stenmark et al. (2014) dividing the rectum in three different regions, or Ebert et al. (2015), Peeters et al. (2006), and Heemsbergen et al.(2005), who have demonstrated an anatomical dependence of specific GI toxicities. In the study of genitor-urinary (GU) toxicity Ghadjar et al. (2014) separately analyzed DVHs of the whole bladder, bladder wall, urethra, and bladder trigone.

Going further beyond the DVH, recent models have also sought to geometrically represent the 3D dose distribution in a single coordinate system via a spatial normalization for a joint analysis of dose at the lowest sampling scale (pixel and voxel levels). Different studies appeared in the literature, either by building a parametric mapping via an intermediate spherical/cylindrical coordinate system such as the Dose-Surface Maps (DSM) in 2-dimensions (2D) (Hoogeman et al. 2004), in 3D (Heemsbergen et al. 2010; Witte et al. 2010), or via 3D anatomical non-rigid registration (NRR) (Acosta et al. 2013; Drean et al. 2016a), herein called 3D Dose-Volume Maps (3D DVM). This workflow is depicted in Figure 17.2 where 3D planning dose distributions and anatomy are exploited in either 2D or 3D.

The DSMs have been exploited in several ways to show relationships between toxicity and 2D local dose distributions. Either via the extraction of geometric features from iso-dose curves (Buettner et al. 2009, 2011, 2012; Calyn et al. 2017) or through pixel-wise comparisons (Palorini et al. 2016b; Improta et al. 2016; Yahya et al. 2017; Wortel et al. 2015) for identifying the regions that better discriminate the patients with and without toxicity. More recently the emerging deep learning field has also been applied to the study of dosimetric effects based on a convolutional neural network model to analyse rectum dose distribution on DSMs (Xin et al. 2017).

With regards to the 3D DVM, voxel-wise comparisons in a common coordinate system represent a suitable strategy to reveal local differences across individuals within a whole volume at low spatial scales. After a pre-processing step of inter-individual spatial alignment, voxel-wise statistical tests are then used to produce 3D maps depicting the localization of regions where statistically meaningful differences between or within groups may exist. These methods are inspired from the voxel-based morphometry (Ashburner et al. 2000). Applied to toxicity studies, the works undertaken in this field have allowed the identification of more predictive sub-regions within the organs in several locations (Heemsbergen et al. 2010; Acosta et al. 2013; Drean et al. 2016b; Ziad et al. 2012; Rao et al. 2012; Monti et al. 2017; Palma et al. 2016).

One of the major advantages of the voxel-wise analysis is that no prior assumptions are made regarding the location of regions correlating with toxicity: The whole planned 3D dose distribution is considered and compared in the subgroups of patients with/without toxicity. As depicted in Figure 17.1 (right side part) these low spatial scale methodologies have allowed the unravelling of the local dose-effect relationship across a population

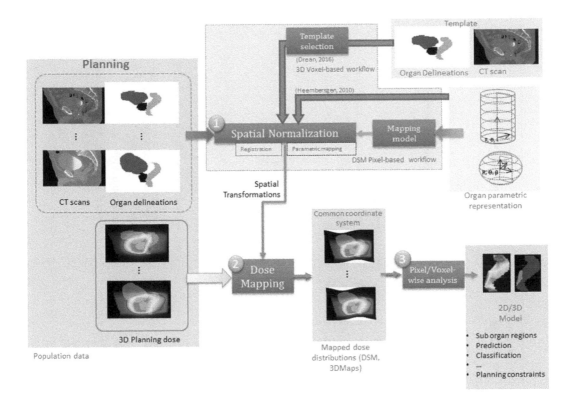

FIGURE 17.2 Methodological aspects of voxel/pixel-based models. These models require the dose to be mapped to a single coordinate system and thus different pre-processing steps are performed: (1) the spatial normalization of a population of individuals to a common coordinate system (A template or an organ parametric representation); (2) the mapping of the 3D dose distributions according to the transformation beforehand obtained; (3) a reliable methodology to perform statistical analysis of local dose-effect relationships.

at each single voxel in a common coordinate system in different organs. The literature reports results for the bladder (Heemsbergen et al. 2010), the rectum (Acosta et al. 2013; Drean et al. 2016b; Ospina et al. 2013), and the lung (Palma et al. 2016). With this kind of analysis, the implication of multiple structures was identified, such as in the head and neck area (Monti et al. 2017), or in the heart in patients treated for lung cancer (McWilliam et al. 2016; 2017). Improved classification performances were achieved (Fargeas et al. 2015; Coloigner et al. 2015; Liu et al. 2015), that reveal significant between-group differences on some sub-regions or combined with other voxel-based variables such as density changes from computed tomography (CT) scans for explaining radiation-induced lung injury (Avanzo et al. 2017).

Table 17.1 summarizes some representative works in this field where both DSMs and DVMs have been used in several clinical locations. As mentioned before, once the dose distribution is normalized to a single coordinate system, which is central to this methodology, a comparison of extracted features or a pixel- or voxel-wise analysis can be performed.

These methods are impacting not only the prediction capabilities but also helping to improve the patient-specific therapy planning as raised in Drean et al. (2016b) and

TABLE 17.1 Summary of Main Works Exploiting 2-Dimensional (2D) Dose-Surface Maps (DSMs) and 3-Dimensional (3D) Dose-Volume Maps (DVM) for Prediction of Toxicity

Reference	Model	Cancer Location	Studied Organ/Toxicity	Spatial Normalization/Dose Mapping	Pixel/Voxel-wise Analysis
Munbodh et al. (2008)	DSM	Prostate	GI	Conformal mapping	Geometric features
Buettner et al. (2011); Buettner et al. (2012)	DSM	Prostate	GI	Geometric dose-surface map	Geometric features
Buettner et al. (2009)	DSM	Prostate	GI	Geometric dose-surface map	Geometric features/neural networks
Palorini et al. (2014)	DSM	Prostate	GU	Geometric dose-surface map	Geometric features
Wortel et al. (2015)	DSM	Prostate	GI	Geometric dose-surface map	Dose-surface features
Palorini et al. (2016b)	DSM	Prostate	GU	Geometric dose-surface map	Pixel-wise comparison
Improta et al. (2016)	DSM	Prostate	GU	Geometric dose-surface map	Pixel-wise comparison/spatial descriptors
Calyn et al. (2017)	DSM	Prostate	GI	Geometric dose-surface map	Spatial features
Yahya et al. (2017)	DSM	Prostate	GU	Geometric dose-surface map	Pixel-wise comparison/spatial descriptors
Xin et al. (2017)	DSM	Cervix	GI	Geometric dose-surface map	Deep learning
Heemsbergen et al. (2010)	3D DVM	Prostate	GU	Geometric 3D mapping	Voxel-wise comparisons
Ziad et al. (2012); Rao et al. (2012)	3D DVM	H&N	Trismus	NRR	Voxel-wise comparisons
Cologuer et al. (2015)	3D DVM	Prostate	GI	NRR	ICA for classification
Chen et al. (2011); Fargeas et al. (2013)	3D DVM	Prostate	GI	NRR	PCA for feature extraction and classification
Fargeas et al. (2015)	3D DVM	Prostate	GI	NRR	Tensor decompositions
Ospina et al. (2013)	3D DVM	Prostate	GI	NRR	Tensor value decomposition for sub-region identification
Liu et al. (2015)	3D DVM	Prostate	GI	NRR	Non-negative matrix factorization for classification
Acosta et al. (2013)	3D DVM	Prostate	GI	NRR	Voxel-wise comparisons
Drean et al. (2016b)	3D DVM	Prostate	GI	NRR on different templates	Repeated voxel-wise comparisons definition of a generic 3D patient-specific region for prediction
McWilliam et al. (2016; 2017)	3D DVM	Lung	Heart	NRR	Voxel-wise

(Continued)

TABLE 17.1 (CONTINUED) Summary of Main Works Exploiting 2-Dimensional (2D) Dose-Surface Maps (DSMs) and 3-Dimensional (3D) Dose-Volume Maps (DVM) for Prediction of Toxicity

Reference	Model	Cancer Location	Studied Organ/ Toxicity	Spatial Normalization/ Dose Mapping	Pixel/Voxel-wise Analysis
Palma et al. (2016)	3D DVM	Lung	Lung	NRR	Voxel-wise differences
Monti et al. (2017)	3D DVM	H&N	Acute dysphagia	NRR	Voxel-wise comparisons
Avanzo et al. (2017)	3D VVM	Lung	RRLI	NRR	Voxel-based longitudinal comparison of CT density & dose

H&N= Head and Neck; NRR= Non-Rigid Registration; GI= Gastrointestinal Toxicity; GU=Genitourinary Toxicity; RRLI=Radiation-Induced Lung Injury. PCA= principal component analysis; ICA=independent component analysis; VVM=voxel by voxel map; DVM=dose-volume map; DSM=dose-surface map; 3D=3-dimensional; 2D=2-dimensional.

Lafond et al. (2017a). New planning systems are steadily moving from the era of DVH constraints applied as suggested by international recommendations and available on the commercial treatments planning systems (TPS) towards the definition of 3D patient-specific constraints as part of the TPS optimisation.

In the following sections we describe the methodological details for the construction of DSM and 3D DVM. We discuss the challenges that these methods are facing to finally open some perspectives for these models as the inclusion of new available data and the consequences for improved patient-specific planning.

Towards a Local Analysis of Dose-Effect Relationships

Dose-Surface Maps

A DSM is a mapping of the 3D dose to a 2D virtual unfolded representation of an organ wall. Different algorithms exist for generating DSMs from the 3D dose distribution (Sanchez-Nieto et al. 2001; Munbodh et al. 2008; Tucker et al. 2006; Hoogeman et al. 2004). To build the map, a 2D image is constructed and a parametric mapping is established between the 3D coordinate system of the organ wall and the 2D image. Thus, each pixel in the 2D image corresponds to a portion of the organ wall with the local dose computed, for instance, by interpolation at that 3D point. By construction, these DSMs reflect the dose of the organ wall. A dedicated module of VODCA (MSS Medical Software Solutions, Hagendorn, Switzerland) allows the generation of DSMs from organ contours and the calculated dose distributions. As mentioned before, the DSMs can be exploited via the extraction of geometric features from iso-dose curves (Buettner et al. 2009, 2011, 2012; Calyn et al. 2017) or through direct pixel-wise comparisons (Palorini et al. 2016b; Improta et al. 2016; Yahya et al. 2017; Wortel et al. 2015; Xin et al. 2017).

3D Dose-Volume Maps (DVMs)

The 3D DVM stands upon the 3D inter-individual spatial alignment, allowing further voxel-wise comparisons. The dose mapping to a common coordinate system remains a central question and may be obtained via a parametric representation of the anatomy in a spherical or cylindrical coordinate system as in Heemsbergen et al. (2010) or Witte et al. (2010), or can be more precisely computed through existing NRR methods as described in Monti et al. (2018) or tailored to a particular anatomy as proposed in Drean et al. (2016a).

DSM and 3D DVM Methodological Challenges

Pixel/voxel-based methods share several methodological aspects as they require the dose to be mapped to a single coordinate system and thus different steps must be performed for the comparisons to be anatomically meaningful. These steps are summarized and graphically depicted in Figure 17.2, namely: (1) the spatial normalization of a population of individuals in terms of their anatomy to a common coordinate system; (2) the mapping of dose distributions according to the anatomical transformation obtained; (3) a reliable methodology to perform statistical analysis of the local dose-effect relationship. Each step is crucial for achieving accurate comparisons.

Spatial Normalization for Dose Mapping

Spatial normalization is the process of obtaining a transformation between the native coordinate system and a single common coordinate system leading to meaningful correspondences across the population. This is a key step in pixel/voxel wise analysis since dose comparison results rely on anatomical alignment accuracy. In the case of DSMs, the mapping is generated by the direct relationship between a 3D coordinate system and the 2D map. As mentioned before, there are several ways for defining such a relationship. DSMs were obtained in different works (Sanchez-Nieto et al. 2001; Munbodh et al. 2008; Tucker et al. 2006; Hoogeman et al. 2004). The general idea is depicted in Figure 17.3 with a rectal DSM, where a direct relationship exists between the cylindrical coordinates and the 2D space in the O (Θ,L) space. Thus, the mapping between the 3D organ and the 2D map is performed through the link between coordinates O (R,Θ,L) of the organ wall and the cylindrical space counterpart O (R,Θ,L). After the 3D–2D relationships are obtained, the dose is propagated and interpolated, yielding a 2D image of dose on the unfolded organ.

The crucial aspects in this construction are the definition of the origin (i.e., 0,0) and the resolution and size of the 2D images. If the rectum was the organ to be studied as in Calyn et al. (2017) or Buettner et al. (2009), the cylindrical coordinate system for building the DSMs was used. In Buettner et al. (2009) the contour was thus cut at the posterior-most

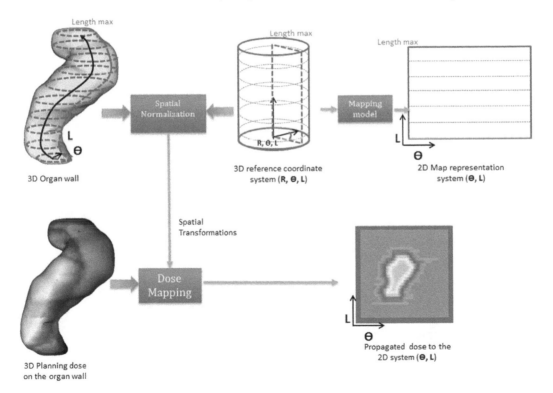

FIGURE 17.3 Construction of a dose-surface map (DSM) from the 3-dimensional representation of the organ wall in a slice wise manner. The correspondences are established to the 2-dimensional map via a parameterized representation of a geometric system. The dose is propagated via these correspondences. Thus, the DSM appears as a virtually unfolded organ.

position on each CT-slice and unwrapped to a map of 21x21 pixels. With k slices interpolated to 21 positions with the rectum typically outlined over a length of 20–22 slices of 0.5 cm. In the bladder a similar methodology was applied. Palorini et al. (2016b) and Yahya et al. (2017) generated 1 mm-resolution DSMs (cranial-caudal direction), by virtually cutting bladder contours at the points intersecting the sagittal plane passing through its center-of-mass. One of the issues is that the definition of the planes following the central path, as it can be done in a slice wise manner or can be orthogonal to the central line. Dose differences can be seen between both schemes in some tortuous configurations.

The geometric correspondences may be extended to 3D by simply spanning the third axis (i.e., R in cylindrical or spherical coordinates) to build a 3D dose map. This was done in Heemsbergen et al. (2010) and Witte et al. (2010) where the dose mapping relies on a parametric representation of the anatomy in a spherical coordinate system and mapped back again to a single anatomy to perform voxel-wise analysis.

Spatial normalization may also be performed by NRR between the population data and individual template. In that case several questions arise, such as the selection of the most representative template and the most reliable inter-individual registration method. This appears as particularly difficult given the high inter-individual anatomical variability (organ volume, artefacts, presence of gas, air, etc.) and the low contrast of soft tissues if CT scans are used for registration. The generic workflow for NRR is depicted in Figure 17.4, based upon the method proposed by Drean et al. (2016a) for the rectum case. This method

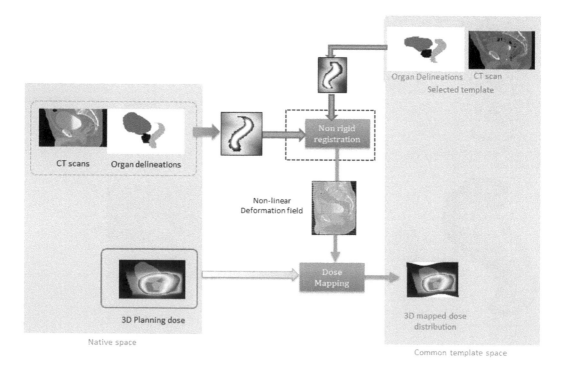

FIGURE 17.4   3-dimensional dose mapping via non-rigid registration as proposed in Drean et al. (2016a). The organs delineations are exploited to generate a structural description of rectum helping to steer deformation between the native space and template space.

was also implemented to register a whole population to a single template (Acosta et al. 2013; Coloigner et al. 2015; Chen et al. 2011; Fargeas et al. 2013, 2015; Ospina et al. 2013; Liu et al. 2015). This approach exploits a structural description of the organ from delineations performed within the clinical protocol, instead of using CTs only. To cope with some of the inter-individual variability issues, the spatial normalization and voxel-wise process was repeated on 117 different templates in Drean et al. (2016b). The use of several templates spanning different anatomies allowed the production of generic sub-regions in a patient-specific framework. In the other locations where voxel-based methods were applied (Ziad et al. 2012; Rao et al. 2012; McWilliam et al. 2016; Palma et al. 2016; Monti et al. 2017; McWilliam et al. 2017), registration was performed on grey intensity images.

INTER-INDIVIDUAL REGISTRATION AND VALIDATION

In the inter-individual setting, conventional registration methods based solely on the gray-scale of CT images might not be accurate enough for reliable mapping in the pelvis (Acosta et al. 2010). Nevertheless, individuals' CT scans have been registered using grey levels in the head and neck location (Monti et al. 2017) and in the lung (Rao et al. 2012; Palma et al. 2016; Avanzo et al. 2017) where the structures of interest offer a better contrast. In Palma et al. (2016) for instance, the log-diffeomorphic demons approach (Vercauteren et al. 2009) was used for the thoracic CT which has been shown to provide higher accuracy for landmark matching of masked lungs in serially acquired chest CT scans (Murphy et al. 2011) compared with B-splines (Rueckert et al. 1999), affine, or rigid registration. A comparison of those intensity-based methods with regards to this problem was undertaken in Monti et al. (2018).

Combining structural information from organ-like anatomical features to guide the image matching process may improve mapping accuracy for applications to population analysis. In Drean et al. (2016a) an anatomical mapping method based on a 3D structural model of the considered organ (rectum) was proposed. This work exploited the delineations that were made on treatment planning CT scans during routine clinical practice. This model makes use of the pseudo-cylindrical nature of the organ at risk of interest. It is founded on distance maps and on the solution of Laplace's equation (Jones et al. 2000) as estimated between the centerline and surface of the rectum.

Evaluating the accuracy of inter-individual mapping is particularly complex because of the nonexistence of ground truth. In the literature, different measures have been proposed to estimate intra-individual anatomical mapping accuracy and transformation validity (Castadot et al. 2008; Rigaud et al. 2015; Salguero et al. 2010). Dice similarity coefficient and Hausdorff distance are among the most commonly used evaluation metrics. However, they only reflect overall geometric overlap between transformed structures and do not show local mapping errors within the structure. It has been demonstrated that they are not very reliable for locally assessing mapping accuracy, given that a high overlap score does not necessarily imply good point-to-point mapping (Rohlfing 2012). If landmarks of the structures have been identified as in Murphy et al. (2011), the average distance leads to a more reliable quantitative score reflecting point to point distance as long as the meaningful landmarks are found.

In the setting of population analysis, a mapping error may lead to very divergent results in statistical analysis. Differences in homogeneous dose regions are of no consequence for statistical analysis. Conversely, in the presence of a high dose gradient, which is the case of most organs at risk, small shifts in registration may result in large differences in dose on the reference template. In Drean et al. (2016a), this issue is largely discussed with synthetic experiments in presence of high gradient dose. In that work the mapping accuracy was assessed not only from an anatomical view point but also from a dosimetric one, including two different metrics that assess rectal centerline dispersion (mean centerline dispersion) for the rectum during the registration process and organ overlap relative to dose distribution (dose volume overlap). This metric measures the ratio between dose distributions on the intersection and union of the considered region. The dose volume overlap is a value between 0, when structures have no voxels in common, and 1, when the dose to the structures is identical (Drean et al. 2016a). In practical terms, the score penalizes anatomical difference by taking into account the dose that would be mapped onto the reference structure. Dice scores, Hausdorff distances, and dose volume overlap scores have also been used to assess spatial normalization in recent voxel-based toxicity studies (Monti et al. 2017; Palma et al. 2016). A recent study by Monti et al. (2018) evaluated two registration methods, demons and free form deformation in well contrasted regions with respect to the capacity to highlight sub-regions with dose differences associated with pulmonary toxicity yielding similar results.

Template Selection

Selecting a single representative template as a single coordinate system from a population is an issue due to large inter-individual variability. Depending on the template the registration might not be accurate enough to ensure a reliable dose mapping at the expense of the reproducibility. Figure 17.5 illustrates this issue by depicting several candidate templates. The large interindividual variability can be seen in both the CT scans and the extracted structures. In the example, a manually delineated rectum is depicted following the clinical protocol and represented here as a surface mesh. This variability across the population can be measured via the grey level comparison of CT images or using distances between the organ shapes. Figure 17.6 shows an example of inter-individual distance using the sum of squared differences as a metric. The distance of an individual to the whole population is represented as a grey level map, indicating similarities between certain individuals and suggesting the idea of using different cluster exemplars as templates.

This template selection problem was tackled in Dréan et al. (2013), where the question of the transfer of these regions to a patient-specific anatomy and its use within planning in a personalized treatment perspective arises. In this work a clustering strategy was followed using exemplar selection (Frey et al. 2007), but it could have been done by any other clustering method (i.e., k-nearest neighbours or k-means). As the issue was to select templates in order to study rectal toxicity, the interindividual distance was computed using geometric features of the organ shapes, as shown in Figure 17.7, leading to the creation of different clusters, depicted in Figure 17.8. In this case, similar geometric characteristics are observed

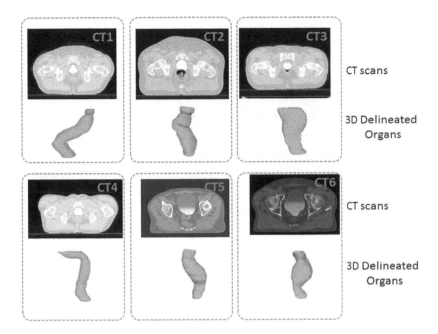

FIGURE 17.5 Inter-individual variability is an issue for the selection of the adequate reference template. The variability can be assessed via the computed tomography (CT) images or using distances between organ shapes.

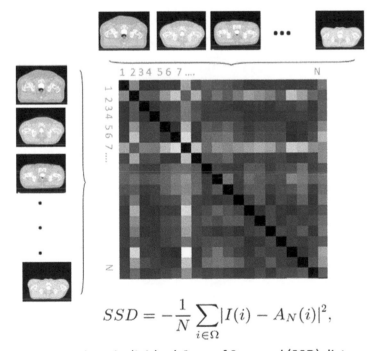

$$SSD = -\frac{1}{N}\sum_{i\in\Omega}|I(i) - A_N(i)|^2,$$

Interindividual Sum of Squared (SSD) distance

FIGURE 17.6 Example of inter-individual distance using the Sum of Squared Differences (SSD) as a metric. The black values represent distance = 0.

within the clusters and a high distance across the groups. The exemplar template can be selected by competition as the most representative shape within each cluster.

To cope with the problem of reproducibility and inter-individual variability, several templates spanning different anatomies may be used. In Drean et al. (2016b) a multi-atlas and leave-one-out cross-validation scheme was used in order to identify rectal sub-regions on 117 patients whereby each patient was iteratively chosen as the reference of a different voxel-wise analysis. Patient-specific sub-regions were thus obtained by transferring the probability map to a generic global geometric coordinate system. A reproducible rectal sub-region at risk of rectal bleeding was, therefore, identified in a single "average" anatomical template. Eventually, the knowledge acquired from a population is generalized to help identifying organ sub-regions to be spared during the planning. Based upon this work, the further study undertaken by Lafond et al. (2017a) provides a particular way forward to

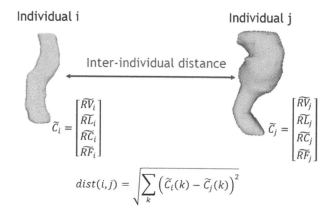

FIGURE 17.7 Geometric interindividual similarity. In Dréan et al. (2013), it was defined as a normalized distance of shape characteristics allowing to find exemplars (Frey et al. 2007) to be used as templates.

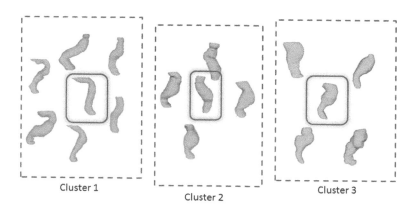

FIGURE 17.8 Example of several clusters obtained by affinity propagation (Frey et al. 2007). Similar geometric characteristics are observed within the clusters. The exemplar template can be selected as the most representative shape within each cluster.

the use of 3D constraints obtained from the before mentioned methodology into the TPS aimed at reducing the risk of toxicity for a single patient.

PIXEL/VOXEL-WISE ANALYSIS IN A COMMON COORDINATE SYSTEM

After 3D doses are spatially normalized, several comparisons can be performed in the common coordinate system. This aspect belongs rather to the statistical field and is not addressed in depth in this chapter. Extracted features from the maps or the identification of sub-regions in the common space have led to improved prediction capabilities. Very often the incorporation of spatial descriptors improves discrimination and log-likelihood of multi-variable models including dosimetric and clinical parameters (Improta et al. 2016; Buettner et al. 2011; Drean et al. 2016a; Fargeas et al. 2015).

The descriptors may be computed after extracting isodose curves, see Buettner et al. (2009) for instance. Several studies found that toxicity is related to the shape of isodoses as well as dose coverage (Calyn et al. 2017). Comparisons of dose average at the intragroup level (with and without toxicity) are frequently investigated with additional tests seeking for statistically significant differences. Either parametric or non-parametric voxel-wise hypothesis tests can be performed, depending on the data. This is the most frequently found case, i.e., where pixel/voxel two-sided t-tests and the resulting p-values map ($p < threshold$ $value$) were used for delimiting the regions better discriminating between groups (Fiorino et al. 2015; Palorini et al. 2016b; Acosta et al. 2013; Drean et al. 2016b).

Caution must be taken however because of the multiple comparison problem, arising when performing thousands of simultaneous tests that may be correlated which is the case in voxel-wise methods. Those issues have been largely treated in voxel-based morphometry studies (Winkler et al. 2014; Holmes et al. 1996; Scarpazza et al. 2015; Ashburner et al. 2000). Several correction techniques exist such as Bonferroni, false discovery rate (Ashburner et al. 2000; Chumbley et al. 2009), threshold-free cluster enhancement (TFCE) methods (Smith et al. 2009), or permutation tests (Chen et al. 2013), which have been implemented in several of the aforementioned toxicity studies (McWilliam et al. 2016; Palma et al. 2016; Yahya et al. 2017). Permutation tests allow inferences while taking into account the multiplicity of tests as described in Holmes et al. (1996). The TFCE offers an interesting spatial characteristic as it takes advantage of neighbourhood information to increase the belief in contiguous areas of the considered signal introducing spatial coherence to the findings. As reported in Palma et al. (2016) permutation testing can also be coupled to TFCE. In Yahya et al. (2017) for example, results with and without permutation-based multiple-comparison adjustments and with inclusion of clinical factors are reported. Several symptoms were evaluated (dysuria, hematuria, urinary incontinence, and an International Prostate Symptom Score increase ≥ 10). The suggestion concerning the reporting of results with and without adjustments for multiple comparisons as proposed by Yahya et al. (2017) appears in line with several recommendations aimed at adding interpretability and reproducibility to the voxel-wise studies (Ridgway et al. 2008). For an overview of simple rules in voxel-based morphometry studies that may be applied to the toxicity models, the readers may refer to the comment written in Ridgway et al. (2008).

The previously described approaches allow independent voxel by voxel analysis. Another idea that has already implemented is to relax this spatial resolution constraint by merging the dose distributions of several voxels in order to jointly use their ability to differentiate several populations. This can be achieved by projecting these voxels onto a vector subspace of smaller dimension (two, three, etc.), where both populations with and without toxicity are more separable. Following this idea, a first analysis allowed to show that voxel PCA could be used to obtain the orthogonal basis of an appropriate subspace to differentiate patients (Chen et al. 2011; Fargeas et al. 2013). Other techniques that relax the orthogonality constraint might improve the representation of the population dose by a small number of basic vectors. ICA is one of the most popular (Albera 2010). ICA assumes a complete statistical independence of the components of each individual's dose in the new representation space while PCA requires only decorrelation. ICA can be computed by achieving the canonical polyadic decomposition of a higher order cumulant tensor estimated from the data. Since the multi-voxel dose distributions show a natural multidimensionality (three space dimensions plus the individual dimension), it is also possible to canonically decompose the deterministic data tensor itself, without any statistical transformation, leading to a purely deterministic multi-way analysis (Fargeas et al. 2015). These approaches have already demonstrated performance in toxicity studies with different constraints (Fargeas et al. 2015; Coloigner et al. 2015; Liu et al. 2015). In Ospina et al. (2013) a tensor-based and a single value decomposition approach was proposed, following the matrix decomposition idea, applied to a cohort of 63 patients which was able to build up a spatial dose pattern characterizing the difference between patients presenting rectal bleeding or not after prostate cancer radiotherapy.

TOWARDS A PATIENT-SPECIFIC PLANNING

Nowadays, in the clinical routine the DVH constraints are applied as suggested by international recommendations. Some of the commercial TPS are also able to take into account predictive model constraints. Based on DVH studies, for instance, the French group, Groupe d'Études des Tumeurs Uro-Génitales (GETUG), defined a set of recommendations for the rectum and bladder dose-volume values, nevertheless corresponding to a small number of parameters (only 2 DVH threshold values for the rectum).

As seen at the bottom in Figure 17.1, there is a well-identified trend towards the definition of patient-specific constraints based upon the identification of organ sub-regions that should be spared during the treatment. This has been suggested by researchers who have demonstrated an anatomical dependence of specific GI toxicities by dividing the organs in standardized and reproducible regions (Stenmark et al. 2014; Ebert et al. 2015; Peeters et al. 2006; Heemsbergen et al. 2005). Likewise in Ghadjar et al. (2014), it is suggested that the reduction of dose to the trigone may reduce urinary toxicity in patients receiving high-dose Intesity Modulated Radiation Therapy.

What insights can be brought by pixel/voxel-wise analysis to the planning? As no prior assumptions are made regarding the location of the regions where statistically significant differences exist across the groups, the voxel-wise analysis may help the identification of new constraints based on the spatial findings. This idea is commonly found in the literature.

For example, in the studies undertaken with DSM it is suggested that the regions identified on the organ wall should be spared during the planning (Buettner et al. 2009, 2011, 2012; Calyn et al. 2017; Palorini et al. 2016b; Improta et al. 2016; Yahya et al. 2017; Wortel et al. 2015), but no further studies have been performed using these constraints.

The question arising here is then how to map the voxel-wise generated sub-regions/ structures to a patient-individual anatomy to be used during the planning. It has to be taken into account that DSMs remain limited to the organ wall description whereas voxel-wise analysis of DVM may yield volumetric regions, spanning over several sub-regions in the organ (Drean et al. 2016b; Palma et al. 2016; McWilliam et al. 2017; Monti et al. 2017). These sub-regions can be further used in a planning workflow for addition of dosimetric constraints to the organs at risk, subject to Planning Target Volume (PTV) coverage and recommendations (Monti et al. 2017; Drean et al. 2016b; Lafond et al. 2017b). In Monti et al. (2017) the sub-regions for a specific patient were transferred via non-rigid registration from the common template to the TPS application. Nevertheless, as discussed before, the use of a single template raises the problem of reproducibility with respect to the inter-individual variability. The repeated voxel-wise analysis in a leave-one-out cross-validation scheme performed by Drean et al. (2016b) allowed the identification of the inferior–anterior hemi-anorectum as highly predictive of rectal bleeding in the case of prostate cancer. The authors went further proposing a way to introduce these "generic" and reproducible regions within a TPS. Figures 17.9 and 17.10 depict this idea on a prostate cancer radiotherapy planning. First the PTV and the structure at risk (the rectum) were delineated and appear overlaid on the CT (Figure 17.9). Two different plans are then computed with and without taking into account the sub-region, but at the same time preserving PTV coverage. As shown, a plan computed with sub-region constraints decreases the dose to the rectum.

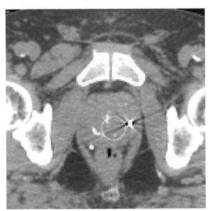 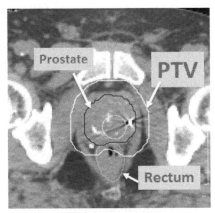

Axial view of the CT Scan used for planning   Prostate, PTV and Rectum delineations

FIGURE 17.9 Axial view of a computed tomography (CT) scan used for the planning where the prostate, the planning target volume (PTV) and the structure at risk (the rectum) have been delineated.

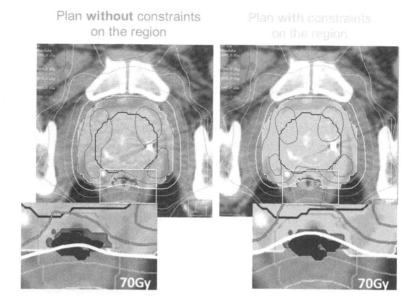

FIGURE 17.10 Two planning configurations. A rectal sub-region was found using finding by Drean et al.'s (2016a). The region was introduced within the treatment planning system (TPS). *Left*: Planning without adding 3-dimensional constraints on sub-regions. *Right*: Adding constraints to the identified rectal sub-regions. It can be observed that the 70 Gy isodose curve is displaced, sparing the region.

Sub-Region(s)

In the case of a sub-region or multiple regions found by the aforementioned methods, another question that arises is how to reach feasible constraints, sparing the computed regions while keeping a full PTV coverage. These spatial constraints may sometimes be in conflict with already segmented organs at risk. In Monti et al. (2017) different plans introducing the computed sub-regions were implemented using Volumetric Modulated Arc Therapy (VMAT) for a single patient in order to show feasibility on the head and neck location. In this study, planning was applied to an oropharyngeal cancer patient with and without the constraints to the computed voxel-based region mean dose. An overall reduction of dose was achieved by adding the *Dmean_sub-region < 30.5 Gy* constraint to the standard organs at risk optimisation. Following that direction, Lafond et al. (2017a) undertook a further study with 60 patients by introducing 3D personalized constraints for the rectum within a TPS, as proposed before in Drean et al. (2016b). Thus, the goal of this dosimetric study was to evaluate the feasibility of decreasing the dose in this rectal sub-region while keeping a high PTV coverage and measuring the potential impact on toxicity. To compute the optimal plan a dosimetric model was implemented as proposed by Moore et al. (2011). This model integrates the overlap volume between the organ at risk and the PTV. Thus, after identification of the rectal sub-regions, four strategies were compared during inverse planning, namely using or not using the rectal sub-regions and using or not using Moore's constraints. By using the rectal sub-regions in addition with Moore's constraints the dose to the rectum was decreased by 8 Gy in average and the risk of rectal bleeding decreased relatively by 23%. This work provides a way forward for the generation of personalized plans from voxel-wise models helping to achieve the reduced toxicity goals.

DISCUSSION AND CONCLUSION

Accurate models for the prediction of toxicity allow for understanding of the complexity of dose-effect relationship which, within the context of personalized medicine, will help to devise tailored radiotherapy treatments. Current models exploit a small portion of the information available at the planning stage, but as depicted in Figure 17.1, the potentiality of models is steadily increasing as they start exploiting more and more multidimensional data. Traditional models use DVHs, which are organ-wise reduced representations of the dose- distribution. Models based on DVHs present several drawbacks limiting the prediction capabilities, as previously discussed. In clinical practice the DVH constraints are applied as suggested by international recommendations but there is a trend to describe the dose-effect relationships at lower spatial scales and therefore to introduce spatial information into the planning. To enlarge the understanding of dose-effect relationships, the 3D dose distribution, together with the individual's anatomy, could be better exploited providing insights at several scales going from the voxel level to the organ and whole region levels. Oncoming models have the possibility to fully exploit the huge amount of individual data gathered not only at the planning stage but also during the whole therapy.

Emerging 2D/3D pixel and voxel models allow for the accurate description of relationships between local dose and side effects, by performing population analysis at a fine scale. As no prior assumptions on spatial information are given, these methods are unravelling the complexity of toxicity and dose-volume relationships by identifying not only sub-regions within the organs at risk but also the simultaneous involvement of different radiosensitive structures. By exploiting information already available at the planning step the prediction capabilities are being improved and would potentially help radiation oncologists to devise personalized therapies taking into account spatial considerations. To be accurate enough, population 3D dose distributions should be spatially normalized to a common coordinate system taking into account inter-individual anatomies. Normalization can be done either via a geometric mapping (to a 2D image or a 3D volume) or through non-rigid registration. The 2D mapping yields a DSM for each individual, whereby the organ anatomy is represented as a virtually unfolded surface. Conversely, non-rigid registration produces a 3D DVM in a common coordinate system (i.e., an individual average template) where anatomy is thoroughly described. Thus, voxel-wise analysis and dose-effect relationships will be directly related to the anatomically described structures.

DSM and 3D DVM have been applied in several locations and exploited in different ways (Table 17.1) providing insights into spatial dose-effect relationships for instance by extracting isodose features (Buettner et al. 2009), by pixel-wise (Improta et al. 2016; Palorini et al. 2016b; Yahya et al. 2017) or voxel-wise comparisons of dose (Drean et al. 2016b; Palma et al. 2016; Monti et al. 2017; McWilliam et al. 2017) across the population. 3D DVM has also been jointly analyzed via dimensionality reduction methods and polyadic decomposition which allow for the achievement of better classification performances (Fargeas et al. 2015; Coloigner et al. 2015) and may lead to the identification of 3D dose patterns at risk (Ospina et al. 2013).

Pixel/voxel-based methods face several challenges within a perspective of patient-specific planning and require different pre-processing steps and validations to be anatomically meaningful:

- The mapping of a population of individuals in terms of their anatomy to a common coordinate system. This step also implies the choice of a reliable template and mapping method and a thorough validation.

- propagation of dose distributions according to the anatomical transformation.

- A reliable methodology to perform local statistical analysis of dose-effect relationship.

Spatial normalization via registration is not an easy task, as shown in Dréan et al. (2013), because of the large inter-individual variability. To cope with that issue, several templates can be chosen followed by a clustering to provide a consensus region (Drean et al. 2016b). One of the major advantages of the voxel-wise analysis is that no prior assumptions are made regarding the location of regions correlating with toxicity as the whole distribution of planned doses may be compared. These regions may be then taken into account when optimizing the planning. The questions arising concern the overlap between concurrent structures and the potential inter-individual difference between the common template and a single patient. Transferring a sub-region towards a single patient can be done by registration as in Monti et al. (2017). Otherwise, a generic sub-region can be found as proposed in Drean et al. (2016b), where multi-atlas and leave-one-out cross-validation was used to identify sub-rectal regions on 117 patients whereby each patient was iteratively chosen as the template reference. Patient-specific sub-regions are thus obtained by transferring the probability map to a generic global geometric coordinate system. Based upon this work, the further study undertaken by Lafond et al. (2017a) provides a particular way forward to the use of 3D constraints obtained from the before mentioned methodology into the TPS, leading to a reduced toxicity planning.

Introducing constraints to the TPS must take into account the potentially conflicting objective functions within the optimisation, since an evident overlap may exist across the sub-regions found with these methods and the requested PTV coverage. Lafond et al. (2017a) proposed a methodology based upon experience as in Moore et al. (2011). Including knowledge can be also performed as proposed by Wu et al. (2009), where distance and constraint concepts are introduced through the "overlap volume histogram" idea to describe the spatial configuration of an organ at risk with respect to a target to evaluate the feasibility of a plan.

Knowledge-based planning trends may follow the workflow depicted in Figure 17.11. Firstly, given a single individual to be treated for whom a plan is to be devised, the 3D pixel/voxel model (iterative templates, leave-one-out, libraries, etc.) provide spatial constraints and statistical information on quality and feasibility. Secondly, a similarity is

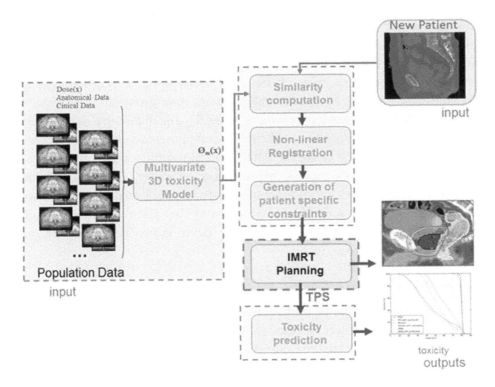

FIGURE 17.11 Workflow for knowledge-based planning, whereby similar cases previously computed are used to guide planning on a new patient. The planning considers not only the sub-regions, organs, or structures to be spared but also the feasibility from an optimisation point of view.

computed between the learnt model (Atlas, libraries, etc.) and the individual to be treated. Then, the plan is computed according to the knowledge transferred in the form of spatial constraints or plan parameters. This specific-patient plan may also include an optimisation based on other parameters such as recurrence and/or toxicity models. Thus, knowledge-based patient-specific planning will help to devise new tailored and integrative radiation treatments.

Some of the limits of voxel-based studies involve not only the validation of the registration but also the reproducibility on larger and external cohorts. Further validation for application in different locations is required.

The nature of deformable organs whose planning dose is not representative of the real delivered dose may appear as a further issue. Since organs are moving and deforming between fractions, new models in some highly deformable organs should include this information either by measuring deformations as in Shelley et al. (2017), where delivered dose to the rectum has been calculated and accumulated throughout the course of prostate radiotherapy using megavoltage computed tomography or by computing statistical models as in Rios et al. (2017) whereby a statistical model will anticipate deformations leading to simulated probability atlases.

As any relationship between local dose and toxicity does not necessarily means causality, further investigation is needed to understand the involved physio-pathological processes. The predictive approaches do not include non-dosimetric parameters, such as patient history or individual radiosensitivity, which may also impact on the risk of toxicity and should be included in oncoming models as depicted in Figure 17.1.

The endpoints to be tested, which depend on the symptoms and grading, should be carefully considered as they can considerably change inter-group distribution across the population and therefore the outcome accuracy. Besides, the differences in inter-individual radiosensitivity recall the importance of bringing complimentary data to the models (multimodal imaging, clinical and biological data, or *in silico* simulations).

Finally, as pixel/voxel-based models are providing new evidences of dose-effect relationships for toxicity, the link with tumour control may be established in a similar way. This can be done by introducing not only dose information but also multimodal imaging in a radiomics voxel-wise framework, which will help devising new therapies with increased control and yet decreased toxicity.

ACKNOWLEDGEMENTS

The works mentioned in this chapter have been partially supported by "Region Bretagne" and the French National Cancer Institute (INCA) IGRT-P. They have also received French government funding, granted to the CominLabs excellence laboratory and managed by the National Research Agency as part of the Investing for the Future program, reference number ANR-10- LABX-07-01, and were partially funded by the French Institute for Public Health Research (IreSP), as part of the 2009–2013 Cancer Plan Call (No. C13005N5).

REFERENCES

Acosta, O., Dowling, J., Cazoulat, G., et al. 2010. Atlas based segmentation and mapping of organs at risk from planning CT for the development of voxel-wise predictive models of toxicity in prostate radiotherapy. In *Prostate Cancer Imaging. Computer-Aided Diagnosis, Prognosis, and Intervention: International Workshop, Held in Conjunction with MICCAI 2010, Beijing, China, September 24, 2010. Proceedings*, edited by Anant Madabhushi, Jason Dowling, Pingkun Yan, Aaron Fenster, Purang Abolmaesumi, and Nobuhiko Hata, 42–51. Berlin, Heidelberg: Springer Berlin Heidelberg.

Acosta, O., Drean, G., Ospina, J.D., et al. 2013. Voxel-based population analysis for correlating local dose and rectal toxicity in prostate cancer radiotherapy. *Phys Med Biol* 58(8):2581–95.

Albera, L., Comon, P., Parra, L.C., Karfoul, A., Kachenoura, A. and Senhadji L. 2010. ICA and biomedical applications. In *Handbook of Blind Source Separation*, edited by P. Comon and C. Jutten, 737–77. London, UK: Academic Press.

Ashburner, J. and Friston, K. J. 2000. Voxel-based morphometry—The methods. *NeuroImage* 11(6):805–21.

Avanzo, M., Barbiero, S., Trovo M., et al. 2017. Voxel-by-voxel correlation between radiologically radiation induced lung injury and dose after image-guided, intensity modulated radiotherapy for lung tumors. *Phys Med* 42:150–6.

Benadjaoud, M.A., Blanchard, P., Schwartz, B., et al. 2014. Functional data analysis in NTCP modeling: A new method to explore the radiation dose-volume effects. *Int J Radiat Oncol Biol Phys* 90(3):654–63.

Boulé, T.P., Gallardo Fuentes M.I., Roselló, J.V., Arrans Lara R., Torrecilla, J.L. and Plaza A.L. 2009. Clinical comparative study of dose–volume and equivalent uniform dose based predictions in post radiotherapy acute complications. *Acta Oncol* 48(7):1044–53.

Buettner, F., Gulliford, S.L., Webb, S. and Partridge, M. 2009. Using dose-surface maps to predict radiation-induced rectal bleeding: A neural network approach. *Phys Med Biol* 54(17):5139–53.

Buettner, F., Gulliford, S.L., Webb, S. and Partridge, M. 2011. Modeling late rectal toxicities based on a parameterized representation of the 3D dose distribution. *Phys Med Biol* 56(7):2103–18.

Buettner, F., Gulliford, S.L., Webb, S., Sydes, M.R., Dearnaley, D.P. and Partridge, M. 2012. The dose-response of the anal sphincter region – An analysis of data from the MRC RT01 trial. *Radiother Oncol* 103(3):347–52.

Castadot, P., Lee, J.A., Parraga, A., Geets, X., Macq, B. and Grégoire, V. 2008. Comparison of 12 deformable registration strategies in adaptive radiation therapy for the treatment of head and neck tumors. *Radiother Oncol* 89(1):1–12.

Cazoulat, G., Simon, A., Dumenil, A., et al. 2014. Surface-constrained nonrigid registration for dose monitoring in prostate cancer radiotherapy. *IEEE Trans Med Imaging* 33(7):1464–74.

Chen, B., Acosta, O., Kachenoura, A., et al. 2011. Spatial Characterization and Classification of Rectal Bleeding in Prostate Cancer Radiotherapy with a Voxel-Based Principal Components Analysis Model for 3D Dose Distribution, *14th Prostate Cancer Imaging. Image Analysis and Image-Guided Interventions - International Workshop, Held in Conjunction with Medical Image Computing and Computer-Assisted Intervention - MICCAI 2011 - 14th International Conference, Toronto, Canada, September 18-22, 2011, Lecture Notes in Computer Science, vol 6963. Springer, Berlin, Heidelberg.*

Chen, C., Witte, M., Heemsbergen, W. and van Herk, Marcel. 2013. Multiple comparisons permutation test for image based data mining in radiotherapy. *Radiat Oncol* 8:293.

Chumbley, J.R. and Friston, K.J.. 2009. False discovery rate revisited: FDR and topological inference using Gaussian random fields. *NeuroImage* 44(1):62–70.

Coloigner, J., Fargeas, A., Kachenoura, A. et al. 2015. A novel classification method for prediction of rectal bleeding in prostate cancer radiotherapy based on a semi-nonnegative ICA of 3D planned dose distributions. *IEEE Journal of Biomedical and Health Informatics* 19(3):1168–77.

Dean, J.A., Wong, Kee H., Gay H. et al. 2016. Functional data analysis applied to modeling of severe acute mucositis and dysphagia resulting from head and neck radiation therapy. *Int J Radiat Oncol, Biol, Phys* 96(4):820–31.

Dréan, G., Acosta O., Ospina, J. D., et al. 2013. How to identify rectal sub-regions likely involved in rectal bleeding in prostate cancer radiotherapy. *Paper read at IX International Seminar on Medical Information Processing and Analysis, 19 November* 2013, at Mexico DF.

Drean, G., Acosta, O., Lafond, C., Simon, A., de Crevoisier R. and Haigron, P. 2016a. Interindividual registration and dose mapping for voxelwise population analysis of rectal toxicity in prostate cancer radiotherapy. *Med Phys* 43(6):2721–30.

Drean, G., Acosta, O., Ospina J.D., et al.. 2016b. Identification of a rectal subregion highly predictive of rectal bleeding in prostate cancer IMRT. *Radiother Oncol* 119(3):388–97.

Ebert, M.A., Foo, K., Haworth, A., et al. 2015. Gastrointestinal dose-histogram effects in the context of dose-volume constrained prostate radiation therapy: Analysis of data from the RADAR prostate radiation therapy trial. *Int J Radiat Oncol Biol Phys* 91(3):595–603.

Fargeas, A., Kachenoura, A., Acosta, O., Albera, L., Drean, G. and De Crevoisier, R.. 2013. Feature extraction and classification for rectal bleeding in prostate cancer radiotherapy: A PCA based method. *IRBM* 34(4):296–9.

Fargeas, A., Albera, L., Kachenoura, A., et al. 2015. On feature extraction and classification in prostate cancer radiotherapy using tensor decompositions. *Med Eng Phys* 37(1):126–31.

Fargeas, A., Acosta, O., Ospina Arrango, J.D., et al. 2018. Independent component analysis for rectal bleeding prediction following prostate cancer radiotherapy. *Radiother Oncol* 126(2):263–9.

Fiorino, C. and Palorini F. 2015. SP-0215: Advanced methods for 2D/3D dose map correlation in modelling toxicity. *Radiother Oncol* 115(S1):S108–9.

Fiorino, C., Valdagni, R., Rancati, T. and Sanguineti G. 2009. Dose-volume effects for normal tissues in external radiotherapy: Pelvis. *Radiother Oncol* 93(2):153–67.

Frey, B.J. and Dueck, D. 2007. Clustering by passing messages between data points. *Science* 315(5814):972–6.

Ghadjar, P., Zelefsky, M.J., Spratt D.E., et al. 2014. Impact of dose to the bladder trigone on long-term urinary function after high-dose intensity modulated radiation therapy for localized prostate cancer. *Int J Radiat Oncol Biol Phys* 88(2):339–44. doi: 10.1016/j.ijrobp.2013.10.042.

Gulliford, S.L., Webb, S., Rowbottom C.G., Corne D. W. and Dearnaley D.P. 2004. Use of artificial neural networks to predict biological outcomes for patients receiving radical radiotherapy of the prostate. *Radiother Oncol* 71(1):3–12. doi: 10.1016/j.radonc.2003.03.001.

Heemsbergen, W.D., Hoogeman M.S., Hart, G.A.M., Lebesque J.V. and Koper, P.C.M. 2005. Gastrointestinal toxicity and its relation to dose distributions in the anorectal region of prostate cancer patients treated with radiotherapy. *Int J Radiat Oncol Biol Phys* 61(4):1011–8.

Heemsbergen, W.D., Al-Mamgani, A., . Witte M.G, et al. 2010. Urinary obstruction in prostate cancer patients from the dutch trial(68 Gy vs. 78 Gy): Relationships with local dose, acute effects, and baseline characteristics. *Int J Radiat Oncol Biol Phys* 78(1):19–25.

Holmes, A.P., Blair, R.C., Watson, J. D. G. and Ford, I.. 1996. Nonparametric analysis of statistic images from functional mapping experiments. *J Cereb Blood Flow Metab* 16(1):7–22.

Hoogeman, M.S., van Herk, M., de Bois, J., Muller-Timmermans, P., Koper, P.C.M. and Lebesque J.V. 2004. Quantification of local rectal wall displacements by virtual rectum unfolding. *Radiother Oncol* 70(1):21–30.

Improta, I., Palorini, F., Cozzarini, C., et al. 2016. Bladder spatial-dose descriptors correlate with acute urinary toxicity after radiation therapy for prostate cancer. *Phys Med* 32(12):1681–9.

Jones, S.E., Buchbinder, B.R. and Aharon, I. 2000. Three-dimensional mapping of cortical thickness using Laplace's Equation. *Hum Brain Mapp* 11(1):12–32.

Lafond, C., N'Guessan, J., Dréan, G., et al. 2017a. PO-0841: Feasibility of dose decrease in a rectal sub-region predictive of bleeding in prostate radiotherapy. *Radiother Oncol* 123(S1):S454–5.

Lafond, C., N'Guessan, J., Dréan, G., et al. 2017b. Faisabilité d'une diminution de dose dans une sous-région rectale hautement prédictive de saignement en radiothérapie prostatique. *Cancer/Radiothérapie* 21(6):707.

Liu, L.H., Kachenoura, A., Fargeas, A., et al. 2015. Discriminant nonnegative matrix factorization for classification of rectal bleeding in prostate cancer radiotherapy. *IRBM* 36(6):355–60.

Lyman, J.T. 1985. Complication probability as assessed from dose-volume histograms. *Radiat Res Suppl.* 8:S13–9.

McWilliam, A., Faivre-Finn, C., Kennedy, J., Kershaw, L. and van Herk, M.B. 2016. Data mining identifies the base of the heart as a dose-sensitive region affecting survival in lung cancer patients. *Int J Radiat Oncol Biol Phys* 96(2):S48–9.

McWilliam, A., Kennedy J., Hodgson C., Osorio, E. V., Faivre-Fin C. and van Herk, M. 2017. Radiation dose to heart base linked with poorer survival in lung cancer patients. *European Journal of Cancer* no. 85:106–113. doi: 10.1016/j.ejca.2017.07.053.

Monti, S., Palma G., D'Avino, V.et al. 2017. Voxel-based analysis unveils regional dose differences associated with radiation-induced morbidity in head and neck cancer patients. *Sci Reps* 7(1):7220.

Monti, S., Pacelli, R., Cella, L. and Palma G.. 2018. Inter-patient image registration algorithms to disentangle regional dose bioeffects. *Sci Rep* 8(1):4915.

Moore, K.L., Brame, R.S., Low D.A. and Mutic, S. 2011. Experience-based quality control of clinical intensity-modulated radiotherapy planning. *Int J Radiat Oncol Biol Phys* 81(2):545–51.

Moulton, C.R., House M.J., Lye V., et al. 2017. Spatial features of dose–surface maps from deformably-registered plans correlate with late gastrointestinal complications. *Phys Med Biol* 62(10):4118–39.

Munbodh, R., Jackson A., Bauer J., Schmidtlein C.R. and Zelefsky, M.J. 2008. Dosimetric and anatomic indicators of late rectal toxicity after high-dose intensity modulated radiation therapy for prostate cancer. *Med Phys* 35(5):2137–50.

Murphy, K., van Ginneken, B., Reinhardt, J.M., et al. 2011. Evaluation of registration methods on thoracic CT: The EMPIRE10 challenge. *IEEE Trans Med Imaging* 30(11):1901–20.

Ospina, J.D., Commandeur, F., Ríos, R., et al. 2013. A tensor-based population value decomposition to explain rectal toxicity after prostate cancer radiotherapy. In *Medical Image Computing and Computer-Assisted Intervention – MICCAI 2013: 16th International Conference, Nagoya, Japan, September 22-26, 2013, Proceedings, Part II*, edited by Kensaku Mori, Ichiro Sakuma, Yoshinobu Sato, Christian Barillot, and Nassir Navab, 387–94. Berlin, Heidelberg: Springer Berlin Heidelberg.

Ospina, J.D., Zhu, J., Chira, C., et al. 2014. Random forests to predict rectal toxicity following prostate cancer radiation therapy. *Int J Radiat Oncol Biol Phys* 89(5):1024–31.

Palma, G., Monti, S., D'Avino, V., et al. 2016. A voxel-based approach to explore local dose differences associated with radiation-induced lung damage. *Int J Radiat Oncol Biol Phys* 96(1):127–33.

Palorini, F., Cozzarini, C., Gianolini, S., et al. 2014. Bladder dose-surface maps show evidence of spatial effects for the risk of acute urinary toxicity after moderate hypofractionated radiation for prostate cancer. *Int J Radiat Oncol Biol Phys* 90(1):S42–3.

Palorini, F., Botti, A., Carillo, V., et al. 2016a. Bladder dose–surface maps and urinary toxicity: Robustness with respect to motion in assessing local dose effects. *Phys Med* 32(3):506–11.

Palorini, F., Cozzarini C., Gianolini, S., et al. 2016b. First application of a pixel-wise analysis on bladder dose surface maps in prostate cancer radiotherapy. *Radiother Oncol* 119(1):123–8.

Peeters, S.T.H., Lebesque, J.V., Heemsbergen, W.D., et al. 2006. Localized volume effects for late rectal and anal toxicity after radiotherapy for prostate cancer *Int J Radiat Oncol Biol Phys* 64(4):1151–61.

Pella, A., Cambria, R., Riboldi, M., et al.. 2011. Use of machine learning methods for prediction of acute toxicity in organs at risk following prostate radiotherapy. *Med Phys* 38(6): 2859–67.

Rao, S.S., Saleh Z., Tam M., et al. 2012. A novel voxel-based analysis of the development of trismus following chemoradiation for Oropharyngeal(OPC) cancer. *Int J Radiat Oncol Biol Phys* 84 (S3):S30–1.

Ridgway, G.R., Henley, S.M., Rohrer, J.D., Scahill, R.I., Warren, J.D. and Fox, N.C. 2008. Ten simple rules for reporting voxel-based morphometry studies. *NeuroImage* 40(4):1429–35.

Rigaud, B., Simon A., Castelli, J., et al. 2015. Evaluation of deformable image registration methods for dose monitoring in head and neck radiotherapy. *BioMed Res Int* 2015:726268.

Rios, R., De Crevoisier, R., Ospina, J.D., et al. 2017. Population model of bladder motion and deformation based on dominant eigenmodes and mixed-effects models in prostate cancer radiotherapy. *Med Image Anal* 38:133–49.

Rohlfing, T. 2012. Image similarity and tissue overlaps as surrogates for image registration accuracy: Widely used but unreliable. *IEEE Trans Med Imaging* 31(2):153–63.

Rueckert, D., Sonoda, L.I., Hayes, C., Hill, D.L., Leach, M.O. and Hawkes, D. J.. 1999. Nonrigid registration using free-form deformations: Application to breast MR images. *IEEE Trans Med Imaging* 18(8):712–21.61. Saleh, Z., Apte, A., Sharp, G., Rao, N., Lee, N., and Deasy J. 2012. Exploring the correlation between 3D spatial dose distribution and toxicity in normal tissue. *Med Phys* 39(6,part 2):3601.

Salguero, F.J., Saleh-Sayah, N.K., Yan, C. and Siebers, J. V.. 2010. A method to estimate three dimensional intrinsic dosimetric uncertainty resulting from using deformable image registration for dose mapping. *Int J Radiat Oncol Biol Phys* 78 (S3):S736.

Sanchez-Nieto, B., Fenwick J.F., Nahum, A.E. and Dearnaley, D.P.. 2001. Biological dose surface maps: Evaluation of 3D dose data for tubular organs. *Radiother Oncol* 61(S1):S52.

Scarpazza, C., Tognin, S., Frisciata, S., Sartori, G., and Mechelli, A.. 2015. False positive rates in Voxel-based Morphometry studies of the human brain: Should we be worried? *Neurosci Biobehav Rev* 52(SC):49–55.

Shelley, L.E.A., Scaife, J.E., Romanchikova, M., et al. 2017. Delivered dose can be a better predictor of rectal toxicity than planned dose in prostate radiotherapy. *Radiother Oncol* 123(3):466–71.

Simon, A., Nassef, M., Rigaud, B., et al. 2015. Roles of deformable image registration in adaptive RT: From contour propagation to dose monitoring. *Conf Proc IEEE Eng Med Biol Soc* 2015:5215–8.

Smith, S.M. and Nichols, T.E. 2009. Threshold-free cluster enhancement: Addressing problems of smoothing, threshold dependence and localisation in cluster inference. *NeuroImage* 44(1):83–98.

Söhn, M., Alber, M., and Yan, D. 2007. Principal component analysis-based pattern analysis of dose-volume histograms and influence on rectal toxicity. *Int J Radiat Oncol Biol Phys* 69(1):230–9.

Stenmark, M.H.,. Conlon, A.S.C., Johnson S., et al .2014. Dose to the inferior rectum is strongly associated with patient reported bowel quality of life after radiation therapy for prostate cancer. *Radiother Oncol* 110(2):291–7.

Tomatis, S., Rancati ,T., Fiorino, C., et al. 2012. Late rectal bleeding after 3D-CRT for prostate cancer: Development of a neural-network-based predictive model. *Phys Med Biol* 57(5):1399.

Tucker, S.L., Zhang, M., Dong, L., Mohan, R, Kuban, D. and Thames, H.D. 2006. Cluster model analysis of late rectal bleeding after IMRT of prostate cancer: A case-control study. *Int J Radiat Oncol Biol Phys* 64(4):1255–64.

Vercauteren, T., Pennec, X., Perchant, A. and Ayache, N. 2009. Diffeomorphic demons: Efficient non-parametric image registration. *NeuroImage* 45(S1):S61–72.

Winkler, A.M., Ridgway, G.R., Webster, M.A., Smith S.M. and Nichols, T.E. 2014. Permutation inference for the general linear model. *NeuroImage* 92(SC):381–97.

Witte, M.G., Heemsbergen, W.D., Bohoslavsky, R., et al. 2010. Relating dose outside the prostate with freedom from failure in the dutch trial 68 Gy vs. 78 Gy. *Int J Radiat Oncol Biol Phys* 77(1):131–8.

Wortel, R.C., Witte M.G., van der Heide ,U.A., et al. 2015. Dose-surface maps identifying local dose-effects for acute gastrointestinal toxicity after radiotherapy for prostate cancer. *Radiother Oncol* 117(3):515–20.

Wu, B., Ricchetti, F., Sanguineti, G., et al. 2009. Patient geometry-driven information retrieval for IMRT treatment plan quality control. *Med Phys* 36(12):5497–505.

Xin, Z., Jiawei, C., Zichun, Z., et al. 2017. Deep convolutional neural network with transfer learning for rectum toxicity prediction in cervical cancer radiotherapy: A feasibility study. *Phys Med Biol* 62(21):8246–63.

Yahya, N.,. Ebert, M.A, Bulsara, M., et al. 2016. Statistical-learning strategies generate only modestly performing predictive models for urinary symptoms following external beam radiotherapy of the prostate: A comparison of conventional and machine-learning methods. *Med Phys* 43(5):2040–52.

Yahya, N., Ebert, M.A., House, Mi.J., et al. 2017. Modeling urinary dysfunction after external beam radiation therapy of the prostate using bladder dose-surface maps: Evidence of spatially variable response of the bladder surface. *Int J Radiat Oncol Biol Phys* 97(2):420–6.

CHAPTER 18

Radiobiological Models in (Automated) Treatment Planning

Ben Heijmen and Marco Schwarz

CONTENTS

Challenges in the Generation of Truly Optimal Treatment Plans 441
   Systems for Automated Planning 443
      'Knowledge-Based' (KB) Automated Planning 444
      'Wish List'-Based Lexicographic (Prioritized) Optimisation 444
      Pinnacle's 'Auto-Planning' (Philips) 447
(Automated) Planning and Prediction Models 447
   Why Is the Use of 'Biologically Related Models' in Planning Optimisation Still Not
   the Norm? 448
References 451

CHALLENGES IN THE GENERATION OF TRULY OPTIMAL TREATMENT PLANS

In treatment planning we are faced with hard planning constraints (e.g., any plan with a Planning Target Volume [PTV] coverage lower than a prescribed percentage, or exceeding the maximum spinal cord dose, is unacceptable), and clinical objectives (e.g., attain a high as possible PTV coverage, or reduce the mean lung dose to the largest possible extent). Planning basically aims at designing a dose distribution for the individual patient that is within target and organs at risk (OARs) hard constraints, while maximally attaining all objectives for these structures. The following list describes challenges in the generation of patient-specific, 'optimal' treatment plans. Some of these may be softened or even fully resolved with systems for automated treatment plan generation.

1. The computational problem is large with thousands of variables and criteria.

2. In the case of a mathematically non-convex optimisation problem, suboptimal solutions may be generated.

3. Imposed hard constraints may for individual patients result in an infeasible problem.

4. Apart from objectives and constraints for the target and the OARs, there are often objectives and (potentially less well-defined) criteria for the general shape of the dose distribution, e.g., regarding conformity, or avoidance of narrow 'spikes' of high dose going from the target far into areas of normal tissues.

5. Accurate dose-response models describing in detail the (patient-specific) impact of 3D dose distributions on the tumor and OARs are largely lacking.

6. Planning requires solving a multi-objective problem with generally conflicting objectives. The dosimetric trade-offs between objectives (e.g., what is the impact of reducing the dose in one OAR on the dose in the other OARs) can be highly patient-specific and are not a priori known. They are generally explored in the interactive trial-and-error planning process.

7. All information about the dose effect relations and the (un)acceptable trade-offs in the final dose distribution have to be expressed in a form that is manageable by a computer, i.e., a cost function. The cost function should be on the one hand specific enough, as there is no such thing as 'implicit knowledge' in computer-based optimisation, and on the other hand general enough, as it should allow to use at best the degrees of freedom available with modulated techniques.

8. The multi-objective optimisation problem has an, in principle, infinite number of solutions, including a frontier of Pareto-optimal solutions. Preferentially, the generated plan should be Pareto-optimal.

9. Apart from being Pareto-optimal, the plan needs to have clinically favourable balances between all objectives, e.g., considering the importance of one treatment complication vs. another, or a reduced target coverage vs. a high OAR dose. However, (detailed) quantitative data or models for these balances are often unknown, and choices may depend on clinicians' or patients' preferences.

Currently, Intensity Modulated Radiation Therapy (IMRT) and Volumetric Modulated Arc Therapy (VMAT) treatment plans are usually generated by a planner in an interactive trial-and-error process, aiming to steer the treatment planning system (TPS) towards a clinically acceptable plan by tweaking (dose-volume histogram [DVH] points-based) cost functions or their weights ('manual, interactive inverse planning'). Due to the challenges described in the previous list, plan quality is then dependent on the skills and experience of the planner (operator dependence) and on allotted planning time. Even within the same clinical planning protocol, a large variation in planning solutions may be proposed by different planners (Nelms et al. 2012; Batumalai et al. 2013; Berry et al. 2016; Berry et al. 2016), partly also depending on subjective preferences and priorities. Due to the manual interaction, treatment planning also creates a high workload for radiotherapy departments.

Already in 1998, in the early days of clinically applied IMRT, Reinstein et al. (1998) investigated opportunities for automation of inverse planning. For six prostate cancer patients they could generate high quality IMRT plans using a fixed set of planning parameters (template). As demonstrated in Figure 18.1, the number of publications on automated treatment plan generation started to grow dramatically around 2012. Automation in planning may relieve or resolve some of the issues mentioned in the previous list, resulting in more consistent and higher plan quality with reduced operator dependence, and substantially reduced manual labour. The goal of automated planning may be fully automatic generation of the final treatment plan of the patient, without any user interference (pushbutton system). However, often, automated planning is used to produce a good starting point for further manual fine-tuning (automated pre-planning). The choice between fully automated planning and automated pre-planning may depend on the success of configuring the available system for full automation (for a particular treatment site). Clearly, full automation results in a larger workload reduction than automated pre-planning with manual fine-tuning, but also the latter can substantially reduce the overall workload. Opportunities for plan quality enhancement with automated planning may also depend on the available software.

Systems for Automated Planning

Proposed systems for automated planning can be largely divided into three main groups, as described in the following section. Apart from these solutions, there are many reports on automated planning in the literature, often focused on a single patient group or treatment technique.

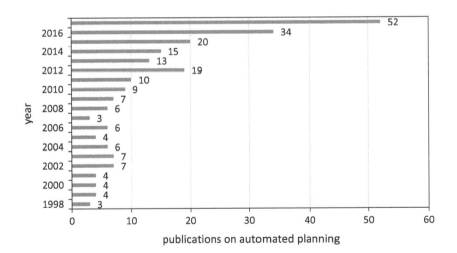

FIGURE 18.1 Yearly numbers of publications on automated treatment plan generation according to Pubmed, when searching in titles and abstracts. Applied query: (IMRT[TIAB] OR VMAT[TIAB] OR modulated[TIAB] OR Rapidplan[TIAB]) AND ((automated[TIAB] NOT (segmentation[TIAB] OR match[TIAB] OR registration[TIAB])) OR (automatic[TIAB] NOT (segmentation[TIAB] OR match[TIAB] OR registration[TIAB])) OR (auto[TIAB] NOT (segmentation[TIAB] OR match[TIAB] OR registration[TIAB])) OR knowledge-based[TIAB]) AND (planning[TIAB] OR optimisation[TIAB] OR generation[TIAB]) AND (dose[TIAB] OR radiotherapy[TIAB] OR radiation[TIAB])).

'Knowledge-Based' (KB) Automated Planning

Many systems for KB planning have been proposed. The common characteristic is that for each new patient, achievable dosimetric plan parameters are predicted using a training database of delivered (high quality) plans. These predictions are then used to automatically generate a plan. The basic assumption is that patients with similar anatomies should have similar, achievable doses. A well-known KB system is Rapidplan DVH prediction followed by automated planning in Eclipse (Varian) (Yuan et al. 2012; Appenzoller et al. 2012; Fogliata 2014; Krayenbuehl et al. 2015; Tol et al. 2015; Fogliata et al. 2015; Hussein et al. 2016). In this system, Principal Component Analysis (PCA) is used for each OAR to derive the principal components (PC) that describe the variations in DVHs in the training database. PCs of so-called Distance-To-Target (DTH) histograms (similar to Overlap Volume Histograms [OVH] applied in other systems) are used to quantify variations in distances of an OAR from the target. Regression analysis is applied for each OAR to correlate the PCs of the DVH variations in the training database with corresponding anatomical variations described by the DTH PCs and other anatomical parameters. For each new patient, the achievable DVH for each OAR is predicted by first assessing the relevant anatomical parameters (including PCs of the OAR DTH), followed by prediction of the contribution of each of the DVH PCs to the expected patient DVH, using the regression model derived from the training database. For automated planning in Eclipse, predicted DVHs are then converted into optimisation objectives with objective priorities. Both the establishment of the objectives and their priorities are configurable. The user can also add objectives. See (Hussein et al. 2016) for a detailed description of the iterative process of establishing objectives and priorities. Both the use of this system for fully automated generation of final treatment plans, and pre-optimisation, followed by manual fine-tuning have been described in the literature. Generally, average quality improvements of automatically generated plans compared to manually generated plans are modest, although negative outliers can be avoided. This seems inherent to KB systems that fully rely on the quality of the plans in the training database. There is also no guarantee that automatically generated plans are Pareto-optimal. Attempts have been made to use the Rapidplan system to homogenize plan quality in a department or between different departments (Schubert et al. 2017).

'Wish List'-Based Lexicographic (Prioritized) Optimisation

Examples of systems that follow this approach are:

1. Erasmus-iCycle (Erasmus MC, Rotterdam) + Monaco (Elekta)

2. iCycle integrated in Monaco (Elekta, work in progress)

3. Raystation Plan Explorer

With these systems, plan optimisation is based on a list of planning constraints and prioritized objectives (wish list). For each new patient, the involved objective functions are minimized sequentially following their assigned priorities. Erasmus-iCycle/Monaco is currently the most described system in literature (Breedveld et al. 2012, 2007; Voet et al.

2014, 2013; Della Gala et al. 2017; Sharfo et al. 2016; Buergy et al. 2017; Buschmann et al. 2018), with technical details mostly described in (Breedveld et al. 2012). Plan generation is multi-criterial and generated plans are Pareto-optimal. Per tumor type/treatment protocol, a fixed wish list is used for all patients. Figure 18.2 shows as an example a wish list for prostate plan optimisation with this system. Plan optimisation starts with priority 1, i.e., establishment of sufficient PTV coverage, going down to priority 10. After each objective minimization, a constraint is added to the optimisation problem to guarantee that subsequent minimizations of lower priority objectives will not result in deterioration of the higher priorities. Plan optimisation has two consecutive phases of sequential minimizations of all objectives. In the first phase, the constraint value to be added to the optimisation problem is never lower than the defined goal value (Figure 18.2), even if minimization of the cost function to a lower value was possible. In the second phase of optimisations, all added constraints are equal to the minimum achievable objective values. The purpose of the use of goal values in the first phase is to avoid that the full room for sparing of OARs is used for the top priorities, leaving no space for reducing dose in lower priority structures. Essential for the generation of high quality plans is tuning of the wish list, i.e., establishment of the constraints and the objectives with their priorities and goal values. As demonstrated in Figure 18.3, wish list configuration for a treatment site has two consecutive steps of iterative wish list improvements. The goal of the first step (I) is to develop a wish list resulting in plans that are of similar quality as the clinical plans. The first action in step II is a discussion in the treating team on most desired further improvements of achieved dose distributions. The iterative wish list adjustment in step II then aims for configuration of a

Constraints

Volume	Type	Limit
PTV	max	104% of prescribed dose
PTV shell 50 mm	max	60% of prescribed dose
Unspecified tissue	max	104% of prescribed dose
Right + Left hip	max	40 Gy

Objectives

Priority	Volume	Type	Goal
1	PTV	↓LTCP	0.5
2	Rectum	↓gEUD (parameter 12)	40% of prescribed dose
3	Rectum	↓gEUD (parameter 8)	25% of prescribed dose
4	Rectum	↓mean	33% of prescribed dose
5	External ring	↓max	40% of prescribed dose
6	PTV shell 5 mm	↓max	93% of prescribed dose
7	Anus	↓mean	10% of prescribed dose
8	PTV shell 15 mm	↓max	70% of prescribed dose
9	PTV shell 25 mm	↓max	50% of prescribed dose
10	Bladder	↓mean	60% of prescribed dose
11	Right + Left Hip	↓mean	25% of prescribed dose
12	Unspecified tissue	↓mean	10 Gy

FIGURE 18.2 An example wish list for prostate plan optimisation with Erasmus-iCycle. LTCP = Logarithmic Tumor Control Probability. See text for explanation. (From Hussein M, Heijmen BJM, Verellen D, Nisbet A. Automation in intensity modulated radiotherapy treatment planning—a review of recent innovations. *Br J Radiol* 2018; 91: 20180270).

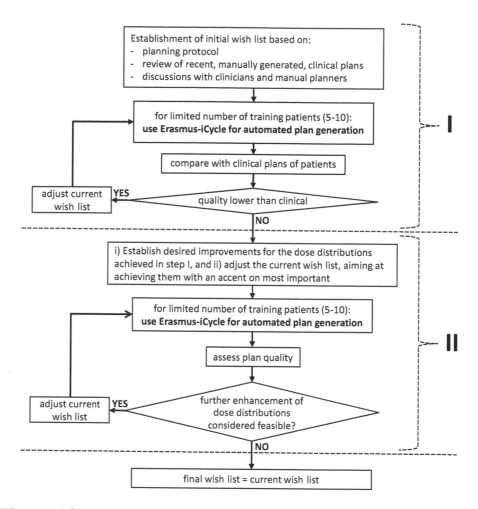

FIGURE 18.3 Schematic presentation of the procedure for configuration of a wish list (see Figure 18.2 for example) for automated planning with Erasmus-iCycle. The aim of step I is to develop a wish list that results in automatically generated plans that are of similar quality as the clinical input plans. The subsequent step II aims at a quality improvement with respect to the clinically applied plans.

final wish list that indeed results in the desired improvements in dose. An important feature of wish list configuration is that automated planning is the basis of all evaluations of intermediate wish lists, ensuring consistent evaluations without workload. So far, the system has only been described for fully automated plan generation, i.e., no manual fine tuning after automated plan generation. For various treatment sites, Erasmus-iCycle/Monaco automated planning has been compared to manual planning. In the majority of studies, automated planning resulted in substantially improved plan quality (Voet et al. 2013; Della Gala et al. 2017; Sharfo et al. 2016; Buergy et al. 2017; Buschmann et al. 2018). As a minimum, the quality was similar (Voet et al. 2014). As explained above and in Figure 18.3, the procedure for wish list tuning explicitly focusses on surpassing the clinical quality. Apart from optimisation for IMRT or VMAT, Erasmus-iCycle also features beam angle optimisation (Voet et al. 2012; Sharfo et al. 2015, 2017; Rossi et al. 2012). A specific application for this feature is automated plan generation for the Cyberknife robotic treatment unit (Rossi et al. 2012). A dedicated version for automated Intensity Modulated Proton Therapy (IMPT) planning has also been developed (van de Water et al. 2013). Wish lists developed for a certain treatment technique (e.g., VMAT) can also be used for to obtain optimal plan quality for another modality (e.g., non-coplanar IMRT with optimized beam-angles). This feature has been used for unbiased treatment technique comparisons, i.e., for both techniques, plans were generated fully automatically with exactly the same optimisation protocol defined by the common wish list (Voet et al. 2012; Sharfo et al. 2015, 2017; Rossi et al. 2012).

Pinnacle's 'Auto-Planning' (Philips)
In the Auto-Planning module, plan generation is based on a so-called 'Treatment Technique', which is basically a prescription of the dose distribution that should be generated with automated planning in terms of desired DVH parameters, mean doses, parameters describing conformality, etc. In contrast to wish list-based plan generation, a single weighted sum objective function is used for optimisation of the dose distribution. The automated plan generation is iterative, evaluating after each iteration the defined criteria and automatically adding regions of interest and associated cost functions to iteratively improve the plan (Cotrutz and Xing 2003; Xhaferllari et al. 2013; Gintz et al. 2016). For each treatment site/treatment approach a separate Treatment Technique needs to be configured. Both fully automated plan generation and pre-optimisation, followed by manual fine-tuning have been described. Large reductions in planning workload have been reported, and also considerable mean plan quality enhancements relative to manual planning have been observed (Hansen et al. 2016, 2017; Hazell et al. 2016; Speer et al. 2017; Kusters et al. 2017). The system does not have a mechanism to ensure that generated plans are Pareto-optimal.

(AUTOMATED) PLANNING AND PREDICTION MODELS

Automated planning can be used for the generation of plans with consistent ('mathematical') high quality. However, the metrics currently used for plan quality optimisation and evaluation may have large uncertainties, so possibly, the goals that we are aiming for are ('clinically') suboptimal. Sophisticated, high quality dose response models are needed to fully exploit automated planning to the benefit of radiotherapy patients.

The cost functions most commonly applied in both manual and automated planning are based on a series of DVH points. This approach, however popular, has a number of issues that becomes more evident, the more complex the cost function is.

Why is a DVH-based cost function suboptimal? In essence, because

a. DVH points are a surrogate for what we are really interested in (direct quantification of probability of local control or side effects).

b. The way DVH-based weighted optimisation works does not guarantee that we'll achieve the results we desire in terms of overall dose distribution within a volume of interest.

The dose optimisation for OARs with a large volume effect (e.g., lung, parotid or liver) is probably the best example in this respect. If one wants to control the average dose of an OAR, expressing this aim with weighted DVH points leads to these issues:

1. It is not straightforward to set a series of DVH points, each with the right volume, dose and weight, which will lead to the desired mean dose for any given patient. In practice, one will need several optimisation runs to adjust the DVH parameters and obtain the desired mean dose.

2. The weights assigned to the DVH points are working in the context of the whole cost function, and not for the OAR of interest only. When we define multiple DVH points for an OAR we (implicitly) have a volume effect in mind, but there is no guarantee that the balance between high doses to small volumes and small doses to high volume will be as desired, because there isn't anything in the cost function that will enforce it.

3. A set of DVH points actually poses a requirement on the shape of the DVH curve, which becomes an additional (and perhaps unnecessary) constraint on the optimisation process.

In addition, DVH points combined with quadratic cost functions favour a binary approach in generating and evaluating dose distributions ('good dose distribution' if the DVH point is met, 'bad dose distribution' if it isn't), while actual dose effect relations are fairly continuous in nature.

Some of the issues presented above can be tackled with automated planning, which for instance may ease the laborious task of tuning the dose volume parameters. Still, even automated planning cannot escape the fundamental problem: If the cost function is expressed with DVH points while we are actually interested in probability of cure (or risk of toxicity), no optimisation, however thorough, will compensate for that. We may even reach a Pareto optimal plan, but it will be optimal with respect to a spurious metric.

The appeal for expressing a cost function in terms of Tumor Control Probability (TCP) and Normal Tissue Complication Probability (NTCP) should therefore be quite obvious, as we could summarize the 3-dimensional (3D) dose distribution in each volume of interest with a single metric, which is also a direct estimation of the likely clinical outcome.

Why Is the Use of 'Biologically Related Models' in Planning Optimisation Still Not the Norm?

This is probably due to the combination of two factors (Allen Li et al. 2012):

1. An overestimation of the differences between 'biological' and 'physical' optimisation.

   Every radiotherapy plan optimisation is de facto 'biological.' For instance, setting a DVH point requiring that the volume of rectal wall receiving 70 Gy or more should be limited to 20% or less, is a way to incorporate in the cost function the knowledge on dose-volume thresholds for serious late effects (Michalski et al. 2010). This is per se neither more nor less 'biological' than setting an objective on generalized Equivalent Uniform Dose (gEUD) or NTCP.

A similar consideration applies to the issue of uncertainties, i.e., there is a widespread opinion that gEUD or NTCP parameters should not be used, as they are associated with larger uncertainties than DVH thresholds. It is however difficult to support this opinion with hard data. The fact that DVH thresholds have been used for much longer than NTCP parameters does not necessarily mean that they are less uncertain, and for at least a few organs at risk the number of patients included in dose effect studies deriving NTCP parameters is in the range of thousands (see Michalski et al. 2010; Marks et al. 2010; Deasy et al. 2010 and elsewhere in this book).

2. The concern that TCP/NTCP-based optimisation will enable dose distributions that are very different from usual, and therefore potentially risky.

This concern has both a general and a specific answer. The general answer is that the parameters to generate dose distributions and the parameters to evaluate it may be different, i.e., one could optimize the dose with biological indices and then verify whether these plans fulfill the usual DVH thresholds. In such a way TCP and NTCP would be used as a mathematical tool that allows the user to efficiently explore different dose distributions with very few changes in the cost function.

In addition, there is always the possibility to run in-silico studies to assess the robustness of the final dose distributions as a function of the variability of the parameters in the radiobiological model (see Schwarz et al. 2004).

For a more detailed answer we should address two situations:

2a) Protocols where the target dose (including its level of heterogencity) is predefined, and expressed in terms of physical dose, e.g., when requirements are set on the volume receiving the prescribed dose (coverage), near-minimum (e.g., $D_{98\%}$), and near maximum (e.g., $D_{2\%}$) dose.

This situation encompasses the vast majority of radiotherapy treatments. What are the benefits of TCP and NTCP optimisation in this scenario?

Concerning the target dose, the differences between a TCP-based objective and a DVH-based cost function is likely limited, in particular when a homogeneous dose escalation is sought. In case of highly heterogeneous target dose distributions, like in Sterectactic Body Radiation Therapy (SBRT), a TCP model may be useful to evaluate the actual benefit of "hotspots" in the target, e.g., to estimate whether they are an acceptable price to pay to get steep penumbra at the target edge, or a feature that improves local control.

Concerning the OAR dose, the general considerations in favour of expressing a cost function in gEUD or NTCP apply. Since the gEUD or NTCP objective ensures that the dose within the OAR is optimized in such a way to comply with the volume effect of that specific OAR, given the available degrees of freedom, one may actually argue that this approach is actually safer than a DVH-based cost function, which will act only on a very few points of the DVH. Then, if we consider all OARs individually, optimizing the gEUD or the NTCP is not conceptually different, given the monotonical relation between the gEUD and the NTCP described according to the Lyman Kutcher Burman model. In other words, by minimizing the gEUD we are also minimizing the NTCP. The potential advantage of NTCP over gEUD optimisation emerges when we also want to optimize the dose between the OARs, given the non-linear relation between gEUD and NTCP.

2b) Protocols where the target dose is variable for different patients, in terms of either dose level and/or heterogeneity.

This situation is typically encountered in clinical studies, such as Phase I trials, and represents a minority of radiotherapy treatments.

By nature, these protocols are associated with a higher level of uncertainty than normal clinical practice. Let's take for instance the case of an isotoxic dose escalation trial, where a homogeneous dose escalation is attempted (see Belderbos et al. 2006; Kong et al. 2005). This case is important because it raises a question which is both crucial and general, i.e., to what extent a knowledge on dose-effect relation obtained in a specific domain (i.e., with a given technique and a given target dose prescription) can be applied to a different domain.

Within the context of this chapter, the question is "Are there fundamental reasons why extrapolating DVH-based dose effect models for OAR is safer than extrapolating NTCP models?"

Reviewing the current knowledge on NTCP models is done elsewhere in this book. However, a few comments can be made within the context of treatment planning optimisation.

By definition, radiobiological models are always at least slightly outdated with respect to the latest radiotherapy techniques available, and models are developed because they should allow some level of careful extrapolation. Otherwise, they wouldn't be models but data fits. In this respect, models that do provide a continuous relation between dose and effect, such as NTCP models, are probably more appropriate than binary DVH thresholds that assume an 'all or nothing' effect.

The likelihood of obtaining unusual dose distributions, and therefore unexpected clinical results, is larger for complications with large volume effect than for small volume effects. This happens because for a large volume effect (say driven by the mean dose), one can think, at least in theory, of obtaining the same mean dose in quite some different ways. For instance, one could use protons instead of photons, thus being able to significantly reduce the volume of OAR receiving medium to low doses, and then significantly increase the volume receiving high doses. In this case, would an NTCP model based on mean dose maintain its validity? A general answer is likely not to exist. On the other hand, in the high dose region, which is the most important for small volume effects, the numbers of geometrical, dosimetrical and physics constraints are such that the available degrees of freedom needed to shape the dose distribution in significantly different ways are typically very limited, if they exist at all.

Last but not least, we should realize that even the best possible TCP and NTCP models may not suffice for full definition of the mathematical problem for optimal plan generation. This is related to challenges four and nine defined in the Section 'Challenges in generation of truly optimal treatment plans.' Currently, criteria are used for dose outside the OARs and the PTV, often related to dose conformity. Apparently, the healthy tissues outside explicitly defined OARs are by themselves to some extent also considered to be at risk. It is well known that high demands on dose conformity may result in enhanced OAR doses. More research is needed on the true impact of dose delivery in healthy tissues that are currently not explicitly defined as OAR. As mentioned in challenge nine, balancing risks for various treatment complications is in no way trivial. The weighting of the various complications may also depend on the individual patient's conditions and preferences. A full model considering all these balances (e.g., determined from Quality of Life studies) could possibly be used in automated plan generation for a kind of average patient. A complementary approach could be to automatically generate for each patient a set of plans with various balances between the complication risks, and choose the final plan in a shared decision making process together with the patient.

REFERENCES

Allen Li X, Alber M, Deasy OJ, Jackson A, Jee KWK, et al. The use and QA of biologically related models for treatment planning: Short report of the TG-166 of the therapy physics committee of the AAPM. *Med Phys* 2012;39(3):1386–409.

Appenzoller LM, Michalski JM, Thorstad WL, Mutic S, Moore KL. Predicting dose-volume histograms for organs-at-risk in IMRT planning. *Med Phys* 2012;39(12):7446–61.

Batumalai V, Jameson MG, Forstner DF, Vial P, Holloway LC. How important is dosimetrist experience for intensity modulated radiation therapy? A comparative analysis of a head and neck case. *Pract Radiat Oncol* 2013;3(3):e99–106.

Belderbos JS, Heemsbergen WD, De Jaeger K, Baas P, Lebesque JV. Final results of a Phase I/II dose escalation trial in non-small-cell lung cancer using three-dimensional conformal radiotherapy. *Int J Radiat Oncol Biol Phys* 2006;66(1):126–34.

Berry SL, Boczkowski A, Ma R, Mechalakos J, Hunt M. Interobserver variability in radiation therapy plan output: Results of a single-institution study. *Pract Radiat Oncol* 2016;6(6):442–9.

Berry SL, Ma R, Boczkowski A, Jackson A, Zhang P, Hunt M. Evaluating inter-campus plan consistency using a knowledge based planning model. *Radiother Oncol* 2016;120(2):349–55.

Breedveld S, Storchi PRM, Voet PWJ, Heijmen BJM. iCycle: Integrated, multicriterial beam angle, and profile optimisation for generation of coplanar and noncoplanar IMRT plans. *Med Phys* 2012;39(2):951–63.

Breedveld S, Storchi PRM, Keijzer M, Heemink AW, Heijmen BJM. A novel approach to multi-criteria inverse planning for IMRT. *Phys Med Biol* 2007;52(20):6339–53.

Buergy D, Sharfo AWM, Heijmen BJM, Voet PWJ, Breedveld S, Wenz F, et al. Fully automated treatment planning of spinal metastases – A comparison to manual planning of Volumetric Modulated Arc Therapy for conventionally fractionated irradiation. *Radiat Oncol* 2017 12(1):33.

Buschmann M, Sharfo AWM, Penninkhof J, Seppenwoolde Y, Goldner G, Georg D, Breedveld S, Heijmen BJM. Automated volumetric modulated arc therapy planning for whole pelvic prostate radiotherapy. *Strahlenther Onkol* 2018 Apr;194(4):333–342

Cotrutz C, Xing L. IMRT dose shaping with regionally variable penalty scheme. *Med Phys* 2003;30(4):544–51.

Deasy JO, Moiseenko V, Marks L, Chao KS, Nam J, et al. Radiotherapy dose-volume effects on salivary gland function. *Int J Rad Onc Biol Phys* 2010;76(3):S58–63.

Della Gala G, Dirkx MLP, Hoekstra N, Fransen D, Lanconelli N, van de Pol M, et al. Fully automated VMAT treatment planning for advanced-stage NSCLC patients. *Strahlentherapie Und Onkol* 2017;193(5):402–9.

Fogliata A, Belosi F, Clivio A, Navarria P, Nicolini G, Scorsetti M, et al. On the pre-clinical validation of a commercial model-based optimisation engine: Application to volumetric modulated arc therapy for patients with lung or prostate cancer. *Radiother Oncol* 2014;113(3):385–91.

Fogliata A, Nicolini G, Clivio A, Vanetti E, Laksar S, Tozzi A, et al. A broad scope knowledge based model for optimisation of VMAT in esophageal cancer: Validation and assessment of plan quality among different treatment centers. *Radiat Oncol* 2015;10.

Gintz D, Latifi K, Caudell J, Nelms B, Zhang G, Moros E, et al. Initial evaluation of automated treatment planning software. *J Appl Med Phys* 2016;17(3):331–46.

Hansen CR, Bertelsen A, Hazell I, Zukauskaite R, Gyldenkerne N, Johansen J, et al. Automatic treatment planning improves the clinical quality of head and neck cancer treatment plans. *Clin Transl Radiat Oncol* 2016. 1:2–8

Hansen CR, Nielsen M, Bertelsen AS, Hazell I, Holtved E, Zukauskaite R, et al. Automatic treatment planning facilitates fast generation of high-quality treatment plans for esophageal cancer. *Acta Oncol* 2017;56(11):1495–500.

Hazell I, Bzdusek K, Kumar P, Hansen CR, Bertelsen A, Eriksen JG, et al. Automatic planning of head and neck treatment plans. *J Appl Clin Med Phys* 2016;17(1):272–82.

Hussein M, Heijmen BJM, Verellen D, Nisbet A. Automation in intensity modulated radiotherapy treatment planning—a review of recent innovations. *Br J Radiol* 2018;91(1092):20180270.

Hussein M, South CP, Barry MA, Adams EJ, Jordan TJ, Stewart AJ, et al. Clinical validation and benchmarking of knowledge-based IMRT and VMAT treatment planning in pelvic anatomy. *Radiother Oncol* 2016;120(3):473–9.

Kong FM, Ten Haken RK, Schipper MJ, Sullivan MA, Chen M, et al. High-dose radiation improved local tumor control and overall survival in patients with inoperable/unresectable non-small-cell lung cancer: Long-term results of a radiation dose escalation study. *Int J Radiat Oncol Biol Phys* 2005;63(2):324–33.

Krayenbuehl J, Norton I, Studer G, Guckenberger M. Evaluation of an automated knowledge based treatment planning system for head and neck. *Radiat Oncol* 2015;10:226.

Kusters JMAM, Bzdusek K, Kumar P, van Kollenburg PGM, Kunze-Busch MC, Wendling M, et al. Automated IMRT planning in Pinnacle. *Strahlentherapie Und Onkol* 2017;193(12):1031–8

Marks LB, Bentzen SM, Deasy JO, Kong FM, Bradley JD, et al. Radiation dose-volume effects in the lung. *Int J Rad Onc Biol Phys* 2010;76(3):S70–6.

Michalski JM, Gay H, Jackson A, Tucker S, Deasy JO. Radiation dose-volume effects in radiation-induced rectal injury. *Int J Rad Onc Biol Phys* 2010;76(3):S123–9.

Nelms BE, Robinson G, Markham J, et al. Variation in external beam treatment plan quality: An inter-institutional study of planners and planning systems. *Pract Radiat Oncol* 2012;2(4):296–305.

Reinstein LE, Wang XH, Burman CM, et al. A feasibility study of automated inverse treatment planning for cancer of the prostate. *Int J Radiat Oncol Biol Phys* 1998;40(1):207–14.

Rossi L, Breedveld S, Heijmen BJM, Voet PW, Lanconelli N, Aluwini S. On the beam direction search space in computerized non-coplanar beam angle optimisation for IMRT-prostate SBRT. *Phys Med Biol* 2012;57(17):5441–58.

Schubert C, Waletzko O, Weiss C, Voelzke D, Toperim S, Roeser A, et al. Intercenter validation of a knowledge based model for automated planning of volumetric modulated arc therapy for prostate cancer. The experience of the German RapidPlan Consortium. *PLoS One* 2017; 12(5):e0178034

Schwarz M, Lebesque JV, Mijnheer BJ, Damen EM. Sensitivity of treatment plan optimisation for prostate cancer using the equivalent uniform dose (EUD) with respect to the rectal wall volume parameter. *Radiother Oncol* 2004;73(2):209–18.

Sharfo AWM, Breedveld S, Voet PWJ, Heijkoop ST, Mens JWM, Hoogeman MS, et al. Validation of fully automated VMAT plan generation for library-based plan-of-the-day cervical cancer radiotherapy. *PLoS One* 2016;11 (12):e0169202.

Sharfo AWM, Voet PWJ, Breedveld S, Mens JWM, Hoogeman MS, Heijmen BJM. Comparison of VMAT and IMRT strategies for cervical cancer patients using automated planning. *Radiother Oncol* 2015;114(3):395–401.

Sharfo AW, Dirkx ML, Breedveld S, Romero AM, Heijmen BJM. VMAT plus a few computer-optimized non-coplanar IMRT beams (VMAT+) tested for liver SBRT. *Radiother Oncol* 2017;123(1):49–56.

Speer S, Klein A, Kober L, Weiss A, Yohannes I, Bert C. Automation of radiation treatment planning: Evaluation of head and neck cancer patient plans created by the Pinnacle3 scripting and Auto-Planning functions. *Strahlentherapie Und Onkol* 2017;193(8):656–65.

Tol JP, Delaney AR, Dahele M, Slotman BJ, Verbakel WFAR. Evaluation of a knowledge-based planning solution for head and neck cancer. *Int J Radiat Oncol Biol Phys* 2015;91:612–20.

van de Water S, Kraan AC, Breedveld S, Schillemans W, Teguh DN, Kooy HM, Madden TM, Heijmen BJM, Hoogeman MS. Improved efficiency of multi-criteria IMPT treatment planning using iterative resampling of randomly placed pencil beams. *Phys Med Biol* 2013;58(19):6969–83.

Voet PW, Dirkx ML, Breedveld S, Al-Mamgani A, Incrocci L, Heijmen BJM. Fully automated volumetric modulated arc therapy plan generation for prostate cancer patients. *Int J Radiat Oncol Biol Phys Biol Phys* 2014;88(5):1175–9.

Voet PWJ, Dirkx MLP, Breedveld S, Fransen D, Levendag PC, Heijmen BJM. Toward fully automated multicriterial plan generation: A prospective clinical study. *Int J Radiat Oncol Biol Phys* 2013;85(3):866–72.

Voet PWJ, Breedveld S, Dirkx MLP, Levendag PC, Heijmen BJM. Integrated multicriterial optimisation of beam angles and intensity profiles for coplanar and noncoplanar head and neck IMRT and implications for VMAT. *Med Phys* 2012;39(8):4858–65.

Xhaferllari I, Wong E, Bzdusek K, Lock M, Chen J. Automated IMRT planning with regional optimisation using planning scripts. *J Appl Clin Med Phys* 2013;14(1):4052.

Yuan L, Ge Y, Lee WR, Yin FF, Kirkpatrick JP, Wu QJ. Quantitative analysis of the factors which affect the interpatient organ-at-risk dose sparing variation in IMRT plans. *Med Phys* 2012;39(11):6868–78

Including Genetic Variables in NTCP Models

Where Are We? Where Are We Going?

Sarah L. Kerns, Suhong Yu, and Catharine M. L. West

CONTENTS

The Importance of Biology in Predicting Radiotherapy Side Effects 455
Types of Genetic Variation 456
Approaches for Identifying Genetic Variation 457
    Candidate Gene Studies 457
    Genome-Wide Association Studies (GWAS) 459
    Gene-by-Environment Interaction Studies 459
Inclusion in Multi-Variable Models 460
    Modelling Normal Tissue Complication Probability (NTCP) 460
Current Status and Where We Are Heading 463
References 464

THE IMPORTANCE OF BIOLOGY IN PREDICTING RADIOTHERAPY SIDE EFFECTS

Radiotherapy schedules maximize tumour control probability (TCP) and minimize normal tissue control probability (NTCP), but are population-based. Radiation doses are limited so that less than ~5% of patients suffer with late side effects. Inclusion of biology in NTCP models would improve the therapeutic ratio and allow for improved biological precision to match developments in the physical precision of radiation delivery. An individual patient's risk of developing side effects following radiotherapy depends on multiple factors. Radiation dose and the volume irradiated are the most important. It is also known that additional treatments (e.g., surgery, chemotherapy), patient factors (e.g., smoking, older age) and comorbidities (e.g., collagen vascular disease) can increase risk of side effects. Genetic variation is also important.

Variation in radiosensitivity is well established from the extreme sensitivity seen in individuals with genetic syndromes such as ataxia telangiectasia (Taylor et al. 1975) to a spread in cancer patients undergoing radiotherapy (Barnett et al. 2009). The ability to include radiosensitivity data in NTCP models requires a robust approach for their measurement, and many assays have been explored. Some studies showed that measurements of normal tissue radiosensitivity predict a patient's risk of late effects following radiotherapy (West et al. 2001) but, in general, findings have been equivocal. There is sufficient evidence that radiosensitivity is important and can be measured, but available assays have lacked the sensitivity and reproducibility for clinical application. Identifying the genetic variants that determine differences in radiosensitivity is arguably a better approach because clinical tests of germline DNA are now used routinely in clinical decision making for some specialties. These DNA tests are highly reproducible (Yang et al. 2017). The challenge for the radiotherapy community is to identify sufficient variants to have a clinically useful test.

TYPES OF GENETIC VARIATION

Variation in the human genome can take several different forms including single nucleotide polymorphisms (SNPs), rare variants, small insertions and deletions (indels), copy number variants (CNVs) and epigenetic modifications. Each class of variants can affect disease risk and response to environmental exposures such as therapeutic radiation. Our knowledge of the number and type of genetic variations across the human genome has increased dramatically since the completion of the Human Genome Project in 2003 (Lander 2001), followed by genome-wide SNP genotyping and deep sequencing initiatives including the International HapMap Project in 2005 (International HapMap 2005) and the 1000 Genomes Project in 2015 (Genomes Project et al. 2015). These initiatives have greatly increased our knowledge of the architecture of the human genome and enabled rapid advances in our understanding of the genetic basis for complex diseases and traits. They have also led to technological advances in software for processing large genomic datasets and extracting meaningful information from an otherwise overwhelming set of data. In the context of radiogenomics research, much of the focus to date has been on SNPs and rare variants, though there is growing interest in CNVs and epigenetic changes.

SNPs are single base pair variable sites that are relatively common – being generally present in at least 1% of the population. The average human genome contains approximately 10 million SNPs with a minor allele frequency of 1% or greater, thus there is great potential for such variation to impact disease risk. A minority of SNPs lie in exonic regions, and even fewer are non-synonymous SNPs that alter a protein's amino acid sequence. Most SNP occur in non-coding introns and intergenic regions that may include regulatory elements such as splice sites, transcription factor binding sites, long range enhancers, or regions of epigenetic modification (Edwards et al. 2013). Over the past decade, results of large-scale genetic association studies have shown that most disease-associated SNPs are located in noncoding or regulatory regions (Maurano et al. 2012) rather than altering the amino acid sequence and function of a protein. Such disease-associated SNPs can alter the expression levels of genes nearby or far way, so-called expression quantitative trait loci (eQTL) (Nicolae et al. 2010). Perhaps not surprisingly, common SNPs tend to have only modest effects on disease susceptibility. Most individual SNPs only increase risk for disease by a few percent or explain a small fraction of the total variability in a quantitative trait.

In contrast to SNPs, rare variants are present in less than 1% of the population. They also tend to have larger effects on disease susceptibility. Mendelian disorders represent the most extreme examples of rare disease-associated variants where single mutations have nearly 100% penetrance regardless of the broader genomic or environmental context. For example, rare homozygous or compound heterozygous mutations in *ATM* almost universally result in the development of ataxia telangiectasia, characterized by cerebellar ataxia, telangiectases, immune defects, cancer susceptibility and severe radiosensitivity (Taylor et al. 1975). Other rare variants lie on a spectrum of penetrance values where risk for disease is modified by the presence or absence of other genetic variants, including SNPs, and by environmental exposure(s). In the context of radiotherapy, radiation is considered to be an environmental exposure.

Though less frequently studied in radiogenomics, CNVs and epigenetic variants likely play a role in modifying response to radiotherapy. CNVs are genomic regions of either repeated copies or loss of a given DNA sequence. While less frequent in number than SNPs, CNVs can encompass large chunks of the genome, spanning thousands to millions of base pairs. Thus, CNVs have the potential to greatly impact gene expression and downstream biology. Epigenetic variations encompass several different types of modification to either DNA itself (methylation) or to DNA-associated histone proteins (methylation, acetylation) (Taudt et al. 2016).

Because of the different types of genetic variation and different effects each have on disease susceptibility, risk prediction or stratification models will require multiple SNPs and/or rare variants to achieve clinically actionable sensitivity and specificity. Data simulation studies suggest that a high area under the receiver-operating characteristic (ROC) curve (AUC) can be achieved for a binary disease or outcome when tens to hundreds of SNPs are combined in a predictive model (Janssens et al. 2006). As expected, the number of SNPs needed depends on both the allele frequencies and effect sizes of the SNPs. While rare variants tend to increase risk for disease more so than common SNPs, they are, by definition, rare, and so few individuals in a population will be carriers. In contrast, many individuals will be carriers of multiple common risk SNPs, but each of these SNPs likely only increases disease susceptibility slightly. Thus, it is the combination of multiple genetic variants that is needed to most accurately predict risk or stratify individuals into risk groups. Polygenic scores have been developed for numerous complex diseases and phenotypes (see [Seibert 2018] for examples), and have shown promise for stratification of individuals into high and low risk groups.

APPROACHES FOR IDENTIFYING GENETIC VARIATION
Candidate Gene Studies
Early studies aiming to identify risk variants associated with radiotherapy outcomes focused on a relatively small number of genes known from radiation biologic studies to be involved in the cellular or tissue response to radiation exposure. Pathways of interest included those involved in DNA double-strand break repair, fibrosis, inflammation and apoptosis, among others. Typically, one gene or a few genes from a given pathway is selected, and then genetic variants, usually SNPs, within the gene(s) are chosen on

the basis of having a known or predicted effect on the protein. For example, variants may be selected on the basis of altering protein coding sequence or having a predicted deleterious effect on splicing or transcription factor binding. These so-called candidate gene studies had the advantage of being relatively low cost and simple in design. For example, in a case control study, the SNP(s) of interest are genotyped using germline DNA (typically from a peripheral blood sample obtained at any point in time before or after radiotherapy) collected from individuals who received prior radiotherapy and who either went on to develop toxicity ('cases') or were followed up for a comparable period of time and did not develop toxicity ('controls'). The proportion of individuals carrying each genotype for the SNP(s) of interest are then compared to determine whether there is a statistically significant difference between cases and controls, ideally after controlling for potential confounders such as ancestry or effect modifiers such as radiation dose or co-morbidities.

A major disadvantage of the candidate gene approach, however, is that it requires a priori knowledge about the biologic pathways involved in radiation response. This limits the scope of studies to genes already known to be radio-responsive and precludes discovery of novel radiosensitivity genes. Another limitation of the early candidate gene approach is that such studies generally only focused on known variants, which were assayed via targeted genotyping, rather than performing sequencing of a full gene or genomic region in order to assay all genetic variation present. The majority of selected variants were non-synonymous coding SNPs or SNPs in known regulatory sites. This is despite the observation that the vast majority of SNPs across the genome lie in regions of unknown or incompletely characterized function. With the recent development of comprehensive initiatives like the ENCODE (Consortium EP 2004) and 1000 Genomes (Genomes Project et al. 2012) projects mentioned above, critical information is emerging regarding the architecture of the human genome. These initiatives have helped the field to understand how narrowly focused the early candidate gene studies were and have enabled the shift to application of genome-wide study designs discussed below.

Despite these inherent limitations, the candidate gene approach has been successful in identifying radiosensitivity SNPs, including those in *ATM*, *TNF*, *XRCC1*, *HSPB1*, *TGFB1* and *TXNRD2*, among others (see [Rosenstein 2011] for review). Some of these SNPs have been associated with multiple late toxicity endpoints, and may represent general radiosensitivity variants. For example, rs18001516 in *ATM* was associated with an overall toxicity score that captured information on multiple endpoints in separate breast and prostate radiotherapy cohorts (Andreassen et al. 2016). Others have shown an association with a specific toxicity endpoint but not others. For example, rs1800469 in *TGFB1* was associated with esophagitis following radiotherapy for lung cancer (Zhang et al. 2010), but this SNP did not show any association with radiotherapy-induced fibrosis in breast cancer survivors despite evaluation in a large meta-analysis of individual patient data (N = 2782, from 11 independent studies) (Barnett et al. 2012). When considering incorporation of radiosensitivity SNPs into risk prediction models, it is likely that separate panels of SNPs will be needed for each disease site, and probably even for different late toxicities within a single disease site.

Genome-Wide Association Studies (GWAS)

Radiogenomics research has shifted towards the use of genome-wide study designs as knowledge of the genetic architecture of the human genome has increased, together with technical advances and decreasing cost in array-based genotyping methods and next generation sequencing methods. Early GWAS typically used array-based genotyping methods to survey a large number of tag SNPs that are selected to capture the majority of common variation across the genome. SNP arrays are designed to take advantage of the fact that chromosomes are inherited in blocks of linkage disequilibrium such that by directly genotyping a panel of several hundred thousand tag SNPs, imputation methods can be used accurately and reliably to fill in genotypes at untyped markers based on known haplotype blocks determined by genomic sequencing of a reference panel (Howie et al. 2009). Several radiogenomics GWAS have been published to date (Barnett et al. 2014) that have been successful in identifying risk loci for late toxicity. Relative to other GWAS of complex diseases, these studies have been modest in size and are only powered to detect SNP-toxicity associations with modest to large effect sizes. There is also evidence that many risk loci remain to be discovered given larger sample sizes (Barnett et al. 2014), but these associations have so far fallen short of the stringent threshold for genome-wide statistical significance due to limited statistical power.

Array-based studies are still probably the most common method for identifying disease-associated variants due to the relatively low cost and now standardized methods, but an increasing number of sequencing based studies are emerging as costs come down and analytic methods improve. Gene-based methods can test the combined effect of multiple rare variants within a gene or genomic region, each with differing effect size and directionality of association (Lee et al. 2012), and newer approaches with improved statistical power will enable their application to radiogenomics in the near future.

Gene-by-Environment Interaction Studies

Normal tissue toxicities occur specifically in response to radiation exposure, and so a logical approach to identifying genetic risk variants is to hypothesize that the impact of a variant on toxicity risk might be modified by radiation dose or volume of normal tissue exposed. Such a hypothesis can be tested in a gene-by-environment (GxE) interaction study, which posits that the combined effect of these mutations and radiation exposure exceeds that of the individual effects. This can be done by, for example, including an interaction term in a statistical model that also includes the genetic variants and exposure separately, or by stratifying a patient population according to radiation exposure and calculating separate risk ratios for the different exposure groups. One example of the use of the GxE approach is seen in a study of rare mutations in *ATM* and development of contralateral breast cancer following exposure to radiation (Bernstein et al. 2010), a very late form of radiation toxicity. This study found that carriers of deleterious mutations in *ATM* who were exposed to radiation for treatment of initial breast cancer are at increased risk of developing contralateral breast cancer compared with carriers of deleterious mutations in *ATM* who did not receive radiation for treatment of their initial breast cancer, and the risk ratio increased with increasing radiation dose.

Though genome-wide studies of GxE in late radiotherapy toxicity are only in recent years becoming feasible, initial research findings are encouraging. For example, another study investigating second malignancy after radiotherapy, but using a genome-wide approach, found one common SNP and two rare variants that modified the effect of radiation exposure on breast cancer risk after childhood cancer (Morton et al. 2017). Similarly, another GWAS identified genetic variants near *PRDM1* that were associated with increased risk of developing a second malignancy in survivors of pediatric Hodgkin's lymphoma who were treated with radiotherapy as children, and this effect was not seen in those treated with radiotherapy as adults, suggesting interaction between *PRDM1* and exposure to ionizing radiation that could also depend on the developmental stage (Best et al. 2011). In the GWAS of prostate cancer patients treated with pelvic radiotherapy described above (Fachal et al. 2014), a statistical interaction was detected for SNPs in the *TANC1* locus and total radiation dose.

Consideration of GxE interaction in radiogenomics has potential clinical relevance. Inclusion of interaction terms could improve the performance (for example, sensitivity and specificity) of predictive models for identifying those at risk for development of toxicity following radiotherapy. A more complete understanding of how genetic variants might modify the effects of radiation dose could also be used to inform modifications to dosimetric parameters in a way that minimizes toxicity risk for a given individual's genetic profile. There are challenges, however, to identifying such interactions in the context of radiotherapy toxicity. High quality and comprehensive data on radiation exposure is needed. While measures of dose and volume are routinely captured in treatment planning systems, there is inevitably some amount of error associated with these measurements due to variation in patient position over the course of multi-fraction radiotherapy, patient movement during treatment and more subtle movement of the organ at risk during treatment. The challenge of statistical power is also important to recognize. Very large sample sizes, even larger than required for simple SNP-toxicity association studies, are needed to detect GxE interaction. Newer analytic methods are being developed to increase statistical power and make such studies feasible given existing datasets (see [McAllister et al. 2017] for review), and these will become applicable to radiogenomics studies as cohort sizes increase.

INCLUSION IN MULTI-VARIABLE MODELS
Modelling Normal Tissue Complication Probability (NTCP)

NTCP models aim to describe the probability of complication in normal tissue by modelling dose-response relationships. It has been found that the response of normal tissue depends on the amount of irradiated tissue volume and the extent of such volume effects depends on the architecture of the irradiated tissue (Kallman et al. 1992). The concept of functional subunits (FSUs) was introduced to describe the structurally discrete tissue elements. Parallel organs have FSUs that function relatively independently. Therefore, whole organs can still function properly following damage to a sufficiently small region, but a volume threshold may exist. Examples of parallel organs are the lungs, kidneys and liver. By contrast, in serial organs, impairment of a single FSU leads to complications in the whole organ. Modelling normal tissue dose-volume responses is needed to establish correlations between dose-volume parameters and toxicity, and to determine safe dose distributions in normal tissues and make predications for risk of side effects. This modelling allows for potential dose escalation.

Various NTCP models have been developed based on different statistical distributions used for describing the shapes of the dose-response curve, e.g., binominal, Poisson, normal, logit and Weibull distributions. NTCP models can also be categorized into theoretically based (mechanistic) and empirically based (phenomenological) types based on how the formulation was derived. Theoretical models are built based on the existing knowledge of radiobiology incorporating statistical dose-response distributions. The relative seriality model or S model developed by Kallman et al. is one example of a mechanistic model based on Poisson distribution of the dose-response curve (Kallman et al. 1992). This model describes the response of an organ with a mixture of serial and parallel-arranged FSU. The relative contribution of the two types of tissue architecture is denoted by parameters which equals one for a fully serial organ and zero for a fully parallel organ. The NTCP equation is

$$NTCP = \left\{ 1 - \prod_i \left[1 - P(D_i)^s \right]^{v_i} \right\}^{1/s}$$

$$P(D_i) = 2^{-\exp\left[e\gamma\left(1 - \frac{D_i}{D_{50}} \right) \right]}$$

where:

v_i is the fractional organ volume that received dose D_i
$P(D_i)$ is the resulting complication probability
D_{50} is the dose which results 50% of complication probability
γ is a slope parameter which reflects the steepness of the sigmoidal dose-response curve.

In contrast, phenomenological models assume no knowledge of radiobiology, but are derived from fitting curves based on actual clinical data. The most widely used NTCP model – Layman-Kutcher-Burman (LKB) – is an example of a phenomenological model and assumes a normal distribution for the dose-response relationship. This model was initially proposed by Lyman (1985) and was designed for complications after uniform irradiation of whole or partial organs. Volume effects were taken into account by adopting the "n" factor in the equation. Practically, normal tissues are rarely irradiated uniformly, Kutcher and Burman later extended this model to incorporate heterogeneous dose distribution (Kutcher and Burman 1989). The combined formalism is the so-called LKB model.

$$NTCP = \frac{1}{\sqrt{2\pi}} \int_{-\infty}^{t} e^{\frac{x^2}{2}} dx$$

$$t = \frac{D_{eff} - D_{50}}{mD_{50}}$$

$$D_{eff} = \left(\sum_i v_i D_i^{1/n} \right)^n$$

D_{eff}, also called generalized equivalent dose (gEUD), is the dose that, if given uniformly, will result in the same NTCP as the actual non-uniformly delivered dose distribution. D_{50} is the uniform dose given to the entire organ that results in 50% complication risk. m is a measure of the slope of the NTCP curve. n is the volume effect parameter and v_i is the fractional organ volume which received a dose of D_i. Further details on NTCP models are presented in Chapter 2 of this book.

NTCP models have been shown to be clinical useful for providing guidelines for treatment planning (Marks et al. 2010). However, the probability of outcome is typically complex, involving many unknown functions of dosimetric and clinical factors. It is widely acknowledged that NTCP models relying on dosimetric factors alone are oversimplified, and benefit from the inclusion of other factors known to affect the probability of complication (e.g., genetics, lifestyle factors and information on other treatment received). An extension of the Lyman model with inclusion of dose-modifying factors was proposed (Peeters et al. 2006) and many examples of these models, including patient-related features, can be found in the chapters of this book related to organ-specific toxicity endpoint modelling. Multivariable models incorporating both dosimetric and clinical prognostic factors are more robust and can potentially improve the accuracy of the model predication (El Naqa et al. 2006).

There is both empirical and theoretical evidence supporting the notion that incorporation of genetic or other biologic data into NTCP models can improve their sensitivity and specificity for prediction of toxicity (Figure 19.1). Genetic data simulated under a variety of assumptions about minor allele frequency, effect size and toxicity prevalence showed that the addition of increasing numbers of SNPs results in increasing improvement in AUC (Kerns et al. 2015). As expected, when individual SNPs each have very small effect sizes,

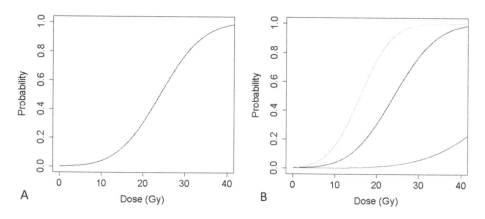

FIGURE 19.1 Normal Tissue Complication Probability (NTCP) model based on dose parameters alone (A) and with for hypothetical individuals with high, moderate, and low genetic risks of toxicity (B).

many are needed to significantly improve model performance. For example, 118 SNPs with per-allele odds ratios mostly less than 1.2 are needed to improve the AUC of a dose-only NTCP model of radiation toxicity with a 5% incidence rate from 0.7 to 0.85. The number of SNPs needed drops to just 68 when a proportion of SNPs with per-allele odds ratios of 1.2 to 2.0 are included. Inclusion of just a few higher penetrance SNPs would have a similar effect of improving model performance as many low penetrance SNPs. Tucker et al. showed that addition of 5 SNPs in *TGFB*, *VEGF*, *TNF*, *XRCC1* and *APEX1* significantly improved the performance of an LKB NTCP model of pneumonitis following radiotherapy for lung cancer (Tucker et al. 1996). Machine learning methods have also been applied to genetic datasets to generate risk prediction models for radiotherapy toxicity with some success. For example, Oh et al. (2017) developed several models of late rectal bleeding and erectile dysfunction in prostate cancer radiotherapy patients using different machine learning approaches and found that the best models included several hundred SNPs out of a genome-wide set (Oh et al. 2017). A similar approach was used to develop a model for urinary toxicity (In Press; Lee et al. *IJROBP*). There is also evidence that other biomarkers, such as cytokines (El Naqa et al. 2018) and copy number variants (Coates et al. 2015), can be incorporated into NTCP models to improve performance.

CURRENT STATUS AND WHERE WE ARE HEADING

The inclusion of genetic variables in NTCP models requires identifying sufficient variants to have a clinically useful test. It is the common variants that are of interest given rare variants such as homozygous mutations in the *ATM* gene can be associated with syndromes that are phenotypically obvious. Information on multiple SNPs can be combined to generate a polygenic risk score (Figure 19.2) to include in NTCP models. Work outside the radiotherapy field shows the potential. For example, analysis of ~89,000 prostate cancer cases and controls has identified >70 SNPs that account for ~30% of the familial risk for the disease. The variants can be combined in a polygenic risk score, which shows a normal distribution. Men with a high-risk score at the top 1% of the risk distribution had a 4.7-fold increased risk for developing prostate cancer compared with the average of the population being profiled (Eeles 2013). Individuals with the highest burden of risk variants had a 30-fold increased risk of prostate cancer compared with those with the lowest polygenic risk scores.

The radiotherapy community is some way behind efforts to develop polygenic risk tests compared with those in development for screening for cancer pre-disposition, as our cohort sizes are only now reaching a few thousand versus the tens of thousands potentially needed. However, progress has been made. The establishment of the international Radiogenomics Consortium (West et al. 2010) has facilitated collaborative efforts to increase cohort sizes, improve statistical power and identify SNPs. A challenge for the radiotherapy community is how to meta-analyze multiple cohorts where treatments and collection of side effect data are variable. Progress has also been made in showing we can combine heterogeneous cohorts to identify SNPs (Kerns et al. 2016). The number of variants identified now numbers around 10 – far fewer than the >100 possibly needed to develop polygenic risk scores that will be useful clinically. Another challenge, therefore, is in maintaining interest in

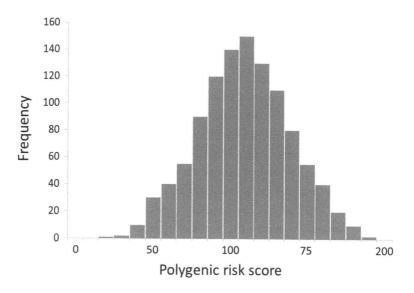

FIGURE 19.2 Distribution of hypothetical polygenic risk scores based on analysis of 100 Single Nucleotide Polymorphisms (SNPs) for 1,085 radiotherapy patients. Polygenic risk scores can be calculated as a simple sum of risk-associated alleles. As there are two alleles for each gene individuals can have two, one, or no copies of the less common SNP. A person with two copies of the less common SNP for all 100 SNPs would have a polygenic risk score of 200. The histogram shows the distribution for unweighted scores. In clinical practice weighted scores would be used with the effect sizes such as odds ratios calculated using some form of regression analysis.

continuing efforts and securing funding to carry out increasingly larger studies. Given the importance of genetic variation in determining radiosensitivity and the potential benefit to patients, continued efforts are worthwhile.

REFERENCES

Andreassen CN, Rosenstein BS, Kerns SL, Ostrer H, De Ruysscher D, Cesaretti JA, Barnett GC et al., Individual patient data meta-analysis shows a significant association between the ATM rs1801516 SNP and toxicity after radiotherapy in 5456 breast and prostate cancer patients. *Radiother Oncol.* 2016;121(3):431–9.

Barnett GC, West CM, Dunning AM, Elliott RM, Coles CE, Pharoah PD, Burnet NG. Normal tissue reactions to radiotherapy: towards tailoring treatment dose by genotype. *Nat Rev Cancer.* 2009;9(2):134–42.

Barnett GC, Elliott RM, Alsner J, Andreassen CN, Abdelhay O, Burnet NG, Chang-Claude J et al., Individual patient data meta-analysis shows no association between the SNP rs1800469 in TGFB and late radiotherapy toxicity. *Radiother Oncol.* 2012;105(3):289–95.

Barnett GC, Thompson D, Fachal L, Kerns S, Talbot C, Elliott RM, Dorling L et al., A genome wide association study (GWAS) providing evidence of an association between common genetic variants and late radiotherapy toxicity. *Radiother Oncol.* 2014;111(2):178–85.

Bernstein JL, Haile RW, Stovall M, Boice JD, Jr., Shore RE, Langholz B, Thomas DC et al., Radiation exposure, the ATM Gene, and contralateral breast cancer in the women's environmental cancer and radiation epidemiology study. *J Natl Cancer Inst.* 2010;102(7):475–83.

Best T, Li D, Skol AD, Kirchhoff T, Jackson SA, Yasui Y, Bhatia S et al., Variants at 6q21 implicate PRDM1 in the etiology of therapy-induced second malignancies after Hodgkin's lymphoma. *Nat Med.*, 2011;17(8):941–3.

Coates J, Jeyaseelan AK, Ybarra N, David M, Faria S, Souhami L, Cury F, Duclos M, El Naqa I. Contrasting analytical and data-driven frameworks for radiogenomic modeling of normal tissue toxicities in prostate cancer. *Radiother Oncol.* 2015;115(1):107–13.

Consortium EP. An integrated encyclopedia of DNA elements in the human genome. *Nature.* 2012;489(7414):57–74.

Consortium EP. The ENCODE (ENCyclopedia Of DNA Elements) Project. *Science.* 2004;306(5696):636–40.

Edwards SL, Beesley J, French JD, Dunning AM. Beyond GWASs: illuminating the dark road from association to function. *Am J Hum Genet.* 2013;93(5):779–97.

Eeles RA, Olama AA, Benlloch S, Saunders EJ, Leongamornlert DA, Tymrakiewicz M, Ghoussaini M et al., Identification of 23 new prostate cancer susceptibility loci using the iCOGS custom genotyping array. *Nat Genet.* 2013;45(4):385–91, 91e1–2.

El Naqa I, Bradley J, Blanco AI, Lindsay PE, Vicic M, Hope A, Deasy JO. Multivariable modeling of radiotherapy outcomes, including dose-volume and clinical factors. *Int J Radiat Oncol Biol Phys.* 2006;64(4):1275–86.

El Naqa I, Johansson A, Owen D, Cuneo K, Cao Y, Matuszak M, Bazzi L, Lawrence TS, Ten Haken RK. Modeling of normal tissue complications using imaging and biomarkers after radiation therapy for hepatocellular carcinoma. *Int J Radiat Oncol Biol Phys.* 2018;100(2):335–43.

Fachal L, Gomez-Caamano A, Barnett GC, Peleteiro P, Carballo AM, Calvo-Crespo P, Kerns SL et al., A three-stage genome-wide association study identifies a susceptibility locus for late radiotherapy toxicity at 2q24.1. *Nat Genet.* 2014;46(8):891–4.

Genomes Project C, Abecasis GR, Auton A, Brooks LD, DePristo MA, Durbin RM, Handsaker RE, Kang HM, Marth GT, McVean GA. An integrated map of genetic variation from 1,092 human genomes. *Nature.* 2012;491(7422):56–65.

Genomes Project C, Auton A, Brooks LD, Durbin RM, Garrison EP, Kang HM, Korbel JO et al. A global reference for human genetic variation. *Nature.* 2015;526(7571):68–74.

Howie BN, Donnelly P, Marchini J. A flexible and accurate genotype imputation method for the next generation of genome-wide association studies. *PLoS Genet.* 2009;5(6):e1000529.

International HapMap C. A haplotype map of the human genome. *Nature.* 2005;437(7063):1299–320.

Janssens AC, Aulchenko YS, Elefante S, Borsboom GJ, Steyerberg EW, van Duijn CM. Predictive testing for complex diseases using multiple genes: fact or fiction? *Genet Med.* 2006;8(7):395–400.

John S, Sabo PJ, Thurman RE, Sung MH, Biddie SC, Johnson TA, Hager GL, Stamatoyannopoulos JA. Chromatin accessibility pre-determines glucocorticoid receptor binding patterns. *Nat Genet.* 2011;43(3):264–8.

Kallman P, Agren A, Brahme A. Tumour and normal tissue responses to fractionated non-uniform dose delivery. *Int J Radiat Biol.* 1992;62(2):249–62.

Kerns SL, Ostrer H, Stock R, Li W, Moore J, Pearlman A, Campbell C et al., Genome-wide association study to identify single nucleotide polymorphisms (SNPs) associated with the development of erectile dysfunction in African-American men after radiotherapy for prostate cancer. *Int J Radiat Oncol Biol Phys.* 2010;78(5):1292–300.

Kerns SL, West CM, Andreassen CN, Barnett GC, Bentzen SM, Burnet NG, Dekker A et al., Radiogenomics: the search for genetic predictors of radiotherapy response. *Future Oncol.* 2014;10(15):2391–406.

Kerns SL, Kundu S, Oh JH, Singhal SK, Janelsins M, Travis LB, Deasy JO et al., The prediction of radiotherapy toxicity using single nucleotide polymorphism-based models: a step toward prevention. *Semin Radiat Oncol.* 2015;25(4):281–91.

Kerns SL, Dorling L, Fachal L, Bentzen S, Pharoah PD, Barnes DR, Gomez-Caamano A et al., Meta-analysis of genome wide association studies identifies genetic markers of late toxicity following radiotherapy for prostate cancer. *EBio Medicine.* 2016;10:150–63.

Kutcher GJ, Burman C. Calculation of complication probability factors for non-uniform normal tissue irradiation: the effective volume method. *Int J Radiat Oncol Biol Phys.* 1989;16(6):1623–30.

Lander ES, Linton LM, Birren B, Nusbaum C, Zody MC, Baldwin J, Devon K et al., International Human Genome Sequencing C. Initial sequencing and analysis of the human genome. *Nature.* 2001;409(6822):860–921.

Lee S, Emond MJ, Bamshad MJ, Barnes KC, Rieder MJ, Nickerson DA, Team NGESP-ELP, Christiani DC, Wurfel MM, Lin X. Optimal unified approach for rare-variant association testing with application to small-sample case-control whole-exome sequencing studies. *Am J Hum Genet.* 2012;91(2):224–37.

Lee S, Kerns S, Ostrer H, Rosenstein B, Deasy JO, Oh JH. Machine Learning on a Genome-wide Association Study to Predict Late Genitourinary Toxicity After Prostate Radiation Therapy. *Int J Radiat Oncol Biol Phys.* 2018 May 1;101(1):128–135.

Li Y, Willer CJ, Ding J, Scheet P, Abecasis GR. MaCH: using sequence and genotype data to estimate haplotypes and unobserved genotypes. *Genet Epidemiol.* 2010;34(8):816–34.

Lyman JT. Complication probability as assessed from dose-volume histograms. *Radiat Res Suppl.* 1985;8:S13–9.

Marks LB, Yorke ED, Jackson A, Ten Haken RK, Constine LS, Eisbruch A, Bentzen SM, Nam J, Deasy JO. Use of normal tissue complication probability models in the clinic. *Int J Radiat Oncol Biol Phys.* 2010;76(3 Suppl):S10–9.

Maurano MT, Humbert R, Rynes E, Thurman RE, Haugen E, Wang H, Reynolds AP et al., Systematic localization of common disease-associated variation in regulatory DNA. *Science.* 2012;337(6099):1190–5.

Mavaddat N, Pharoah PD, Michailidou K, Tyrer J, Brook MN, Bolla MK, Wang Q et al., Prediction of breast cancer risk based on profiling with common genetic variants. *J Natl Cancer Inst.* 2015;107(5), djv036

McAllister K, Mechanic LE, Amos C, Aschard H, Blair IA, Chatterjee N, Conti D et al., and new opportunities for gene-environment interaction studies of complex diseases. *Am J Epidemiol.* 2017;186(7):753–61.

Morton LM, Sampson JN, Armstrong GT, Chen TH, Hudson MM, Karlins E, Dagnall CL et al., Genome-wide association study to identify susceptibility loci that modify radiation-related risk for breast cancer after childhood cancer. *J Natl Cancer Inst.* 2017;109(11).

Nicolae DL, Gamazon E, Zhang W, Duan S, Dolan ME, Cox NJ. Trait-associated SNPs are more likely to be eQTLs: annotation to enhance discovery from GWAS. *PLoS Genet.* 2010;6(4):e1000888.

Oh JH, Kerns S, Ostrer H, Powell SN, Rosenstein B, Deasy JO. Computational methods using genome-wide association studies to predict radiotherapy complications and to identify correlative molecular processes. *Sci Rep.* 2017;7:43381.

Peeters ST, Hoogeman MS, Heemsbergen WD, Hart AA, Koper PC, Lebesque JV. Rectal bleeding, fecal incontinence, and high stool frequency after conformal radiotherapy for prostate cancer: normal tissue complication probability modeling. *Int J Radiat Oncol Biol Phys.* 2006 Sep 1;66(1):11–9.

Rosenstein BS. Identification of SNPs associated with susceptibility for development of adverse reactions to radiotherapy. *Pharmacogenomics.* 2011;12(2):267–75.

Schultheiss TE, Orton CG, Peck RA. Models in radiotherapy: volume effects. *Med Phys.* 1983;10(4):410–5.

Seibert TM, Fan CC, Wang Y, Zuber V, Karunamuni R, Parsons JK, Eeles RA et al., Polygenic hazard score to guide screening for aggressive prostate cancer: development and validation in large scale cohorts. *BMJ.* 2018;360:j5757.

Taudt A, Colome-Tatche M, Johannes F. Genetic sources of population epigenomic variation. *Nat Rev Genet.* 2016;17(6):319–32.

Taylor AM, Harnden DG, Arlett CF, Harcourt SA, Lehmann AR, Stevens S, Bridges BA. Ataxia telangiectasia: a human mutation with abnormal radiation sensitivity. *Nature.* 1975;258(5534):427–9.

Tucker SL, Geara FB, Peters LJ, Brock WA. How much could the radiotherapy dose be altered for individual patients based on a predictive assay of normal-tissue radiosensitivity? *Radiother Oncol.* 1996;38(2):103–13.

West C, Rosenstein BS, Alsner J, Azria D, Barnett G, Begg A, Bentzen S et al., Establishment of a radiogenomics consortium. *Int J Radiat Oncol Biol Phys.* 2010;76(5):1295–6.

West CM, Davidson SE, Elyan SA, Valentine H, Roberts SA, Swindell R, Hunter RD. Lymphocyte radiosensitivity is a significant prognostic factor for morbidity in carcinoma of the cervix. *Int J Radiat Oncol Biol Phys.* 2001;51(1):10–5.

Yang S, Cline M, Zhang C, Paten B, Lincoln SE. Data sharing and reproducible clinical genetic testing: successes and challenges. *Pac Symp Biocomput.* 2017;22:166–76.

Zhang L, Yang M, Bi N, Ji W, Wu C, Tan W, Zhao L et al., Association of TGF-beta1 and XPD polymorphisms with severe acute radiation-induced esophageal toxicity in locally advanced lung cancer patients treated with radiotherapy. *Radiother Oncol.* 2010;97(1):19–25.

Index

2 Gy Equivalent Dose (EQD2Gy), 27, 92, 117, 200, 201, 295, 297

A

accumulated dose, 101
accuracy, 12, 16,
accuracy (prediction), 31, 39, 40, 42, 48, 56, 59, 390, 462
acute effects / acute toxicity, 6, 25, 77, 81, 82, 88–94, 100, 101, 114–117, 124, 125, 138, 140, 142, 143, 145, 172–174, 176, 180, 218, 221, 222, 227, 231, 245, 248–251, 255, 256, 259, 270, 290–293, 295, 297, 299, 301–303, 305, 315–317, 322–326, 328, 330, 369, 376, 405, 408, 409
adaptive planning, 88, 162, 163
Akaike information criterion (AIC), 41, 59, 389
alpha/beta ratio, 18, 81, 160, 175, 177, 179, 189, 295, 296
anal canal, 79, 80, 82, 86
analytical models, 25, 29, 44, 48
ancestors, 35
ancestry, 458
animal models, 127, 316
annotated acyclic graph, 35
anorectal / anorectum, 76, 79, 80, 82, 86, 99, 101, 350
apoptosis, 316, 320, 385, 457
architecture (parallel / serial), 28
architecture (data sharing), 382, 383, 387
area under the curve (AUC), 41
array-based genotyping, 459
artificial intelligence, 48, 337, 393
artificial neural network (ANN), 32–34, 44, 46, 61, 172, 179, 254, 389, 393, 416, 418, 437, 440
aspiration, 217
atlas, 14, 70, 157, 219, 328, 330, 435
atlas-based treatment planning, 328, 331
auditory canal, 183, 187
automated planning, 441, 443, 444, 446–448, 451

B

backward (selection / elimination), 37, 60
bayesian information criterion (BIC), 41, 59
bayesian network (BN), 34, 44, 46, 61, 389
bias-variance, 389
binary (variable / outcome), 12, 29, 31, 39, 40, 57, 389, 390, 448, 450, 457
biologic (data / pathways), 457, 458, 462
biological effective dose (BED), 26, 27, 79, 145, 153, 175, 256–258, 278, 415
biology, 455, 457
biomarker, 46, 48, 156, 281, 387, 391, 463
bleeding (oesophagus), 247, 248
bleeding (gastric), 142
bleeding (rectal), 6, 44, 76, 77, 80, 81, 85–87, 99, 101, 393, 406, 407, 428, 430, 431, 433, 463
bleeding (upper gastro-intestinal), 140, 141
bone, 80, 183, 232, 259, 310–313, 320, 327,
bone marrow, 310–313, 316–323, 326–328, 330, 331
Bonferroni corrected p-value, 429
bootstrap / bootstrapping, 42, 43, 46, 62, 63
bottom-up approach, 54, 71
bowel, 76, 77, 79–82, 87–93, 100, 101, 138, 145, 406
brachial plexus, 259, 301
brain, 172–176, 178, 179, 182, 303, 338, 402
brainstem, 172, 173, 175–177, 179, 180, 182, 197, 230, 402, 403
breast, 249, 251, 252, 270, 273, 275–277, 279, 281, 282, 291, 293, 295–297, 299, 300, 302–304, 319, 373, 392, 458–460, 464
bronchi, 244, 247, 259, 260
bronchial tree, 244, 260, 261
B-splines, 425

C

calibration, 59, 63, 66, 67, 390
cardiac, 269, 270, 273–277, 279–282, 404
cardiovascular, 269, 274, 275, 279, 280, 404
carotid, 219, 280

case-control study, 5

cataract, 190, 310

categorical (variable / outcome), 12, 57, 67, 367, 389, 390

causality, 3–5, 20, 46, 68, 69, 435

censoring, 5

central nervous system (CNS), 172–177, 179–182, 219, 338, 402, 404

cerebellar, 173, 457

cerebellum, 182

cerebral, 174–176, 178, 402, 403

cerebral cortex, 402, 403

cerebrovascular, 316

cervical cord / spine, 175, 180, 313

cervical oesophagus, 231

Charlson index, 373

chest pain, 245

chest wall, 262, 282, 301

clonogen, 300

clonogenic, 24, 290, 291, 300

clustering, 70, 426, 434

Cobalt Gray Equivalent (CGE), 86

cochlea, 183–189, 203

cochlear, 183–188

cohort study, 3–5, 8, 10, 19, 20

cognitive impairment, 179

collinearity, 56, 60, 64, 220

colon, 407

Common Terminology Criteria for Adverse Events (CTCAE), 6, 7, 25, 47, 77, 82, 85–87, 89, 90, 114, 139–142, 144–146, 174, 188–190, 217, 222, 233, 245, 275, 292, 318, 377

computer-assisted theragnostics (CAT), 384

confidence intervals, 59, 61, 66

conformity index, 12

confusion matrix, 39, 41

copy number variations (CNVs), 29, 44, 456, 457

correlation coefficient, 36

coronary artery, 269, 282

coronary event, 279

coronary heart disease, 5, 274

coronary stenosis, 280

cosmetic, 277, 291

cost function, 442, 445, 447–450

cough, 232, 245, 247

cramps (abdominal), 6, 7, 85

cranial nerves, 173–175

cricoid, 232, 244

cricopharyngeal, 219, 226

criteria (inclusion / exclusion), 3, 4

cross validation, 31, 32, 42, 44, 61–63, 380, 388, 390, 399, 428, 431, 434

cross-sectional study, 10, 11, 231

cystitis, 114, 115, 407

c-statistic, 59, 273

D

data-driven models, 24, 29, 35, 44, 45, 48, 54, 64, 366, 375, 389

data-sharing, 379, 380, 382, 387, 394

decision-making, 366, 376, 391–394

deformable registration, 17, 101, 143, 163

dermatitis, 290, 292, 295, 297, 299–302

dermis, 290, 291, 300

desquamation, 290, 291, 293, 295, 297, 300, 302, 303

diarrhoea, 25, 77, 81, 85, 88–92, 406, 407

Dice coefficient, 328, 425, 426, 462

dose-modifying factor (DMF), 28, 29, 44, 117,

dose-surface histogram (DSH), 80, 127, 129, 295, 296, 300

dose-surface-map (DSM), 80, 86, 101, 157, 417–419, 422–424, 431, 433

dose-volume histogram (DVH), 19, 27, 28, 80–82, 85, 89, 90, 99, 125–127, 129, 138, 139, 141, 143, 145, 146, 156, 163, 180, 181, 185, 188, 199, 207, 208, 210–212, 230, 231, 234, 251, 259, 279, 282, 323, 330, 415–418, 422, 430, 433, 442, 444, 447–450

dose-volume map (DVM), 418, 419, 422, 431, 433

dose-wall histogram (DWH), 80

duodenal, 138, 141–146

duodenum, 82, 138–146, 156, 157, 162, 163, 406

dysphagia, 216–222, 226–231, 245, 248

dysphonia, 231, 233, 234

dyspnoea, 245, 248, 254, 392, 405, 410

dysuria, 113, 407, 429

E

early effects / early toxicity, 25, 114, 172, 173, 199, 201, 211, 212, 218, 281, 407

ears, 189

effective volume, 27, 152

endocrine, 172

enteral tube feeding, 220, 226

enteritis, 406

enterocolitis, 7

epidermal, 290, 316

epidermis, 290, 291, 299, 300

epiglottis, 232, 233

epilation, 291

epithelium, 244, 316

equivalent uniform dose (EUD/gEUD), 28, 85, 87, 126, 141, 144, 163, 177, 185, 249, 254–258, 297, 299, 415, 449, 450, 462

erectile dysfunction, 44, 45, 407, 463
erythema, 25, 290, 291, 293, 295, 297, 300
European Organization for Research and Treatment of Cancer (EORTC), 6, 7, 77, 90, 217, 226, 227, 291, 377
Expanded Prostate Cancer Index Composite (EPIC), 78, 93
eye, 189, 190, 198, 201, 326

F

faecal incontinence, 7, 77, 80, 85–87
false discovery rate, 429
false negative, 39, 66
false positive, 39, 66
feature selection, 35–38, 46–48, 212, 416
femoral, 321
femur, 311
fever, 245, 317
fibrosis, 25, 54, 76, 114, 172–174, 218, 219, 245, 249, 291, 296, 297, 299–301, 304, 404–406, 408, 457, 458
fibrotic, 221, 301
fistula, 77, 81, 114, 245, 247, 248, 406, 407
fit (regression), 38
forward variable selection, 37, 44, 60
fractionation effect, 17, 18, 26, 27, 81, 92, 93, 124, 125, 127, 144, 146, 147, 153, 156, 173, 176, 180, 181, 184, 189, 196, 201, 221, 251, 256–259, 261, 270, 277, 278, 282, 295, 297, 299, 303–305, 326, 385, 406
fractures, 310
frequency (stool), 7, 76, 80, 85, 86, 407
frequency (urinary), 113, 114, 116, 407
Functional Assessment of Cancer Therapy-Head and Neck Scale (FACT-HN), 232, 233
Functional Assessment of Cancer Therapy-Prostate (FACT-P), 78
functional sub-units (FSU), 25, 28, 116, 146, 247, 300, 460, 461

G

gastric, 76, 141, 142, 244
gastritis, 406, 407
gastroduodenal, 141, 144, 145
gastroduodenum, 141, 144, 161
gastrointestinal, 7, 76, 77, 91, 140, 142–145, 219, 222, 316, 324, 325, 392, 407, 418, 440
gastrostomy, 220, 228
gene, 44, 101, 253, 373, 456–458, 463
gene interaction study, 459, 460

genetic, 13, 24, 31, 44, 45, 70, 101, 221, 253, 385, 387, 391, 392, 455–460, 462–464
genitourinary, 94, 392, 407, 418
genome, 456–459
genome-wide, 101, 391, 456, 458–460, 463
genome-wide association study (GWAS), 391, 459, 460
genomic, 24, 70, 101, 253, 456–459
genotype, 29, 458, 458
glottic, 219, 233
glottis, 232, 234
goodness of fit, 41, 58, 59, 61, 63

H

haematuria, 113–115, 124, 125, 407, 429
haemorrhage, 6, 7, 85, 141, 145, 146, 173, 198, 199, 317
hazard ratio (HR), 88, 273, 276, 338, 390
hazards regression, 85, 88, 141, 389
hearing, 173, 183–189
heart disease / morbidity, 11, 25, 270, 273, 274, 275, 279, 280, 283, 404
heart failure, 4
heart-lung interaction, 54, 280
hematologic, 140, 310, 316, 318, 321–328, 331, 338, 363
haematopoiesis, 310, 313, 319, 322
haemoptysis, 259
hematopoietic, 310, 311, 313, 315, 316–320, 326
hepatic, 143, 146, 147, 152, 406
hepatitis, 139
hepatobiliary tract, 138, 153
hepatomegaly, 139
hepatotoxicity, 147, 164
hip, 311
hippocampal, 174, 179
hippocampus, 14, 175, 179, 180, 182
Head and Neck Quality of Life instrument (HNQOL-C), 232
hoarseness, 232
hyperplasia, 291
hypofractionated / hypofractionation, 81, 89, 92–94, 99, 100, 115, 124, 125, 130, 138, 142, 144, 145, 176, 177, 180–184, 186, 188–191, 196–198, 200, 201, 244, 258, 259, 262, 277, 278, 338
hypopharynx, 219, 230

I

ileum, 138, 406, 407
iliac, 280, 322
iliococcygeal, 86, 87
imputation, 56, 387, 389, 459

Incontinence Modular Questionnaire Short Form (ICIQ-SF), 115
Independent Component Analysis (ICA), 416, 430
International Consortium for Health Outcomes Measurement (ICHOM), 102, 115, 372
induration, 291
infarction, 273
infections, 259, 317
inflammation, 5, 54, 114, 218, 290, 315, 369, 407, 457
International Prostate Symptoms Score (IPSS), 114, 125
interpretability (model), 31, 34, 48, 429
intestinal, 76
ischaemia, 114, 174, 276

J

Jack-knife, 42
jaw, 310
jejunum, 138, 406

K

kidney, 5, 25, 311, 326, 338, 406, 460
knowledge-based automated planning, 444
Kutcher–Burman dose-volume histogram reduction, 27, 461
K-fold cross validation, 42, 390

L

laryngeal, 216, 226, 227, 231–234, 404
larynx, 219, 220, 222, 226–233, 404
lasso, 31, 37, 61
late effects / toxicity, 2, 6, 25, 44, 77, 80–82, 85, 86, 88–94, 99, 102, 114–117, 124–128, 138, 140, 142, 145, 172, 174, 176, 177, 183, 189, 198, 212, 217, 218, 222, 226, 227, 232, 245, 262, 270, 274, 276, 277, 280, 290, 291, 293, 296, 297, 299, 300, 302, 304, 305, 315–317, 319, 323, 330, 376, 389, 402, 404, 448, 455, 456, 458–460, 463
Layman-Kutcher-Burman model (LKB), 27, 28, 44, 81, 86, 87, 99, 141, 143–145, 152, 153, 156, 176, 249, 250, 254, 255, 259, 260–261, 295, 386, 415, 450, 461, 462, 463
leakage (faecal), 86
learning method, 24, 29–31, 33–38, 44, 47, 48, 54, 57, 68–71, 211, 280, 337, 371, 383–385, 393, 394, 416–418, 463
least-square approach, 30
leave-one-out cross validation (LOOCV), 42, 43, 61, 428, 431, 434

late effects normal tissue task force - subjective, objective, management, and analytic (LENT-SOMA), 6, 25, 47, 77, 86, 94, 217, 245
leukocytes, 313, 319
leukopenia, 322, 323
likelihood, 28, 30, 41, 59, 67, 85, 389, 429
linear-quadratic model, 26, 81, 124, 141, 152, 145, 153, 175, 187, 201, 257, 260, 293, 277
liver, 25, 138, 139, 141, 143, 144, 146, 147, 152, 153, 156, 157, 162–164, 301, 326, 406, 448, 460
logistic regression, 30–32, 37, 44, 47, 57, 70, 86, 87, 146, 208, 209, 221, 222, 226, 256, 274, 275, 295, 377, 384, 389, 393
long-term effects / toxicity, 4, 139, 173, 177, 180, 212, 216, 220, 227, 228, 270, 273, 276, 282, 291, 311, 317, 322, 372, 376, 377
lung, 5, 14, 25, 44, 54, 153, 244, 245, 247–260, 270, 274–280, 282, 301, 310, 319, 323, 324, 326, 373, 377, 380, 390, 392, 393, 405, 406, 409, 410, 419, 425, 441, 448, 458, 460, 463
lung-heart interaction, 54, 280
lymphocytes, 316, 319
lymphopenia, 323

M

machine-learning, 254, 384, 440
Markov chain Monte Carlo (MCMC), 35, 46
masticatory, 404, 409
medulla, 173, 175
medullary canal, 320
mucosa, 8, 218, 219, 221, 222, 227, 230
mucositis, 7, 25, 76, 218, 406
muscle, 80, 86, 87, 101, 114, 116, 127, 216, 218, 219, 222, 226–231, 320, 404, 409
myelitis, 181
myeloablative, 326
myelopathy, 174, 177, 181, 182
myocardial, 269, 273

N

nausea, 173, 221
necrosis, 173, 177, 178, 253, 291, 296, 300, 301, 310, 369, 406
necrotic, 178, 385, 403
nerve, 15, 173–175, 183, 184, 186–191, 196–201, 219
neuritis, 201
neurocognitive, 172, 174, 176, 178, 182
neurological, 173, 177, 179, 182, 197
neuropathy, 172, 177, 189, 196–199, 201, 301

neurotoxicity, 172, 176, 177
neuro-optic, 196
neutropenia, 322–324, 328
nocturia, 115, 393
Normal Tissue Complication Probability (NCTP),
        2, 4–7, 11, 13–18, 20, 24, 25, 27–30, 34,
        35, 39, 44, 48, 56, 81, 86, 87, 99, 101, 138,
        141, 144, 145, 152, 153, 156, 157, 163, 176,
        181, 183–185, 187, 188, 199–201, 230, 234,
        249–251, 254–256, 258, 262, 270, 273, 274,
        279, 281, 282, 293, 295–297, 303, 323, 331,
        386, 391, 394, 415, 417, 448–451, 455, 456,
        460–463

O

obstruction, 3, 4, 77, 81, 92, 113–115, 145, 406, 407
obstructive, 245
occipital cortex, 201
ocular, 189–191, 196, 199, 201
odds ratio (OR), 89, 126, 226, 252, 275, 323, 338,
        404, 463
odynophagia, 245, 248
oedema, 114, 173, 196, 198, 199, 218, 227, 232, 233,
        291, 404
oesophageal, 217, 219, 220, 222, 227, 230, 231, 247,
        249, 251, 252, 260–262, 324, 404
oesophagitis, 25, 245, 248–251, 253, 255, 256, 367,
        379, 404, 405, 409, 458
oesophagus, 5, 219, 230, 231, 244, 247–251, 259–261,
        323, 379, 404, 405
olfactory bulbs, 182
ontology, 378, 380–382, 384
ophthalmologic, 190, 198, 201
ophthalmology, 190
optic, 15, 189–191, 196–201
oral cavity, 210, 219, 227, 229, 230
overfit / overfitting, 31, 34, 36, 37, 39, 43, 46, 58,
        57–60, 63, 389

P

pain, 7, 85, 245, 262, 291, 301, 407
paraesthesia, 173
paraplegia, 181
Pareto, 255, 256, 442, 444, 445, 447, 448
parotid, 54, 172, 207–212, 226, 228, 230, 403,
        409, 448
penetration, 216, 217, 226, 227
penile bulb, 407
percutaneous endoscopic gastrostomy (PEG), 220,
        222, 226, 227
perforation, 81, 145, 244, 245, 247, 261, 406

pericardial, 269, 270, 276, 277, 404
pericarditis, 270, 276, 404
pericardium, 270, 276–278, 404
perineum, 408
peritoneal cavity, 79, 81, 90, 91
pharyngeal, 217–220, 222, 226–229, 231, 404
pharynx, 220, 244, 404
pigmentation, 291
pixel-wise, 418, 422, 433
platelet, 152, 313, 317
pneumonia, 216, 228, 245
pneumonitis, 44, 46, 245, 247, 249–256, 258–260,
        377, 393, 402, 405, 409, 463
polymorphism, 29, 44, 101, 221, 253, 391, 456
principal component analysis (PCA), 30, 38, 416,
        430, 444
pyloric, 138

Q

quadriplegia, 181
Quality of Life (QoL), 7, 8, 25, 71, 76, 78, 85,
        113–115, 177, 183, 216, 218, 227, 231, 232,
        372, 391, 451

R

radiation-induced liver disease (RILD), 139, 143,
        156, 245, 247–249, 252, 253, 361
radiation-induced optic neuropathy (RION),
        189–191, 196–198, 200, 201
radiomics, 24, 70, 208, 391, 402, 409, 410, 436
radionecrosis, 173, 402, 403
radiosurgery, 24, 175, 184, 197
real-world data, 365, 370–372, 375, 378, 379, 384,
        385, 387, 391
receiver-operator-characteristics curve (ROC
        curve), 40, 41, 44–46, 87, 140, 142, 145,
        208, 222, 226, 233, 253, 296, 390, 403, 457
rectal, 6–8, 44, 76, 77, 79–82, 85–87, 90–92, 94, 99,
        138, 323, 393, 407, 423, 426, 428, 430–433,
        448, 463
rectum, 7, 14–16, 76, 79, 80, 82, 86, 87, 90, 92, 94, 99,
        101, 407, 418, 419, 423–426, 430–433, 435
regression, 30–32, 37–39, 44, 47, 57, 61, 70, 85–87,
        141, 185, 191, 221, 226, 256, 273, 275, 323,
        384, 389, 393, 417, 444
regression coefficient, 256, 383, 384, 386, 389
regularization, 31, 32, 37, 61–63, 390
relative biological effectiveness (RBE), 178–180, 191,
        196, 198–200, 293, 294, 297, 300
relative seriality, 185, 461
respiratory, 25, 163, 175, 259

reticular, 313
retinopathy, 190, 198, 200
ridge regression, 31, 37, 61

S

sacral, 81, 175
sacrum, 326
saliva, 5, 212, 403
salivary, 17, 207–210, 229, 403
sensorineural, 183, 184
single-fraction, 111, 176, 190, 191, 197
single-nucleotide polymorphism (SNP), 29, 44, 101,
        253, 456–460, 462, 463
sinuses, 172, 182, 313
skin, 25, 245, 259, 289–293, 295–297, 299–305,
        316, 338
small-bowel, 91
speech, 20, 183, 217, 232, 234, 367
sphincter, 86, 87, 101, 116, 138, 219, 227, 244, 407
spinal cord, 14, 25, 172–175, 177, 180–182, 230, 259,
        261, 321, 441
spine, 144, 181, 312, 313, 323
splenic, 153
stenosis, 116, 144, 173, 275, 280
Stereotactic Body Radiotherapy (SBRT), 82,
        125–127, 130, 138, 139, 144–147, 152, 153,
        156, 157, 163, 168, 174, 197, 252, 258–260,
        277–279, 301, 303–305, 449, 453
sternum, 311, 313
sticky, 403
stomach, 82, 138–144, 156, 157, 162, 244, 406
stool, 7, 80, 85, 86, 90, 99, 407
stricture, 113, 114, 145, 146, 220, 228, 248, 404, 407
study design, 2–4, 11, 12
subcutaneous, 290–292, 299, 304, 320
submandibular, 210, 211, 219, 403
support vector machine (SVM), 31, 32, 37
supraglottic, 219, 226, 227, 230, 232, 341
suprahyoid, 232
suprapubic, 114
survival fraction, 26
swallowing, 216–222, 226, 228–231
systolic, 269, 275, 281, 404

T

telangiectasia, 173, 290–292, 296, 297, 299, 301, 304,
        455, 457
tenesmus, 77, 86
thoracic, 175, 244, 245, 251, 259, 261, 269, 270, 273,
        275, 277, 279, 281, 282, 301, 304, 312, 323,
        373, 404, 409, 425
thoracic cavity, 244

thorax, 244, 245, 254, 338
thrombocytopenia, 322
thrombosis, 173
tongue, 219, 221, 228
total body irradiation (TBI), 311, 317, 322, 326, 331
trachea, 244, 247, 259, 261
trigeminal, 180
trigone, 116, 125, 127, 418, 430
trismus, 5, 240, 404, 409

U

ulceration, 144, 245, 247, 248, 291, 296, 300, 406
ulcers, 145
underfitting, 59
unsupervised, 30
urethra, 116, 127, 418
urethral, 114, 116
urgency, 77, 85, 86, 99, 101, 114, 116, 407
urgent, 116, 130, 246
urinary incontinence, 113–116, 124, 125, 429
urinary symptoms / toxicity, 3, 4, 113–117, 123–130,
        407, 416, 429, 430, 463
urothelium, 114

V

vaginal, 407, 408
validation, 20, 46, 48, 58, 61–68, 126, 188, 209, 212,
        228, 229, 253, 258, 262, 297, 370, 385, 388,
        391, 403, 425, 434, 435
valve, 269, 274, 275, 404
valvular, 270, 275, 404
variance, 31, 36, 38, 39, 61, 64, 126, 210, 212, 221,
        389, 409
variants, 24, 101, 102, 253, 456–460, 463
vascular, 76, 114, 117, 172, 173, 253, 280, 455
vein, 153, 290, 291, 313, 391, 406
ventilation, 252
ventricle, 270, 273, 275, 404
ventricular, 281, 404
vertebrae, 175, 244, 311, 323
vertebral, 181, 323
vessels, 15, 259, 280, 281, 291, 313
vocal cords, 230–234
voice, 231–234, 237
vomiting, 173, 221, 311
voxel-based, 70, 163, 231, 259, 281, 282, 418, 419,
        422, 425, 426, 429, 432, 434–436

X

xerostomia, 5, 207–210, 212, 216, 221, 230, 403,
        404, 409